The Practical Use of Fracture Mechanics

The Practical Use of Fracture Mechanics

by

DAVID BROEK

FractuREsearch Inc., Galena, OH, USA

Kluwer Academic Publishers

Dordrecht / Boston / London

Library of Congress Cataloging in Publication Data

Broek, David.
The practical use of fracture mechanics.

Bibliography: p.
1. Fracture mechanics. I. Title.
TA409.B773 1988 620.1'126 88-9336
ISBN 90-247-3707-9

Published by Kluwer Academic Publishers,
P.O. Box 17, 3300 AA Dordrecht, The Netherlands

Kluwer Academic Publishers incorporates
the publishing programmes of
D. Reidel, Martinus Nijhoff, Dr W. Junk and MTP Press

Sold and distributed in the U.S.A. and Canada
by Kluwer Academic Publishers,
101 Philip Drive, Norwell, MA 02061, U.S.A.

In all other countries, sold and distributed
by Kluwer Academic Publishers Group,
P.O. Box 322, 3300 AH Dordrecht, The Netherlands.

Printed in the Netherlands

Preface

This book is about the use of fracture mechanics for the solution of practical problems; academic rigor is not at issue and dealt with only in as far as it improves insight and understanding; it often concerns secondary errors in engineering. Knowledge of (ignorance of) such basic input as loads and stresses in practical cases may cause errors far overshadowing those introduced by shortcomings of fracture mechanics and necessary approximations; this is amply demonstrated in the text.

I have presented more than three dozen 40-hour courses on fracture mechanics and damage tolerance analysis, so that I have probably more experience in teaching the subject than anyone else. I learned more than the students, and became cognizant of difficulties and of the real concerns in applications. In particular I found, how a subject should be explained to appeal to the practicing engineer to demonstrate that his practical problem can indeed be solved with engineering methods. This experience is reflected in the presentations in this book. Sufficient background is provided for an understanding of the issues, but pragamatism prevails. Mathematics cannot be avoided, but they are presented in a way that appeals to insight and intuition, in lieu of formal derivations which would show but the mathematical skill of the writer. A practicing engineer does care little about how a crack tip stress field is derived; he accepts that it can be done, as long as he can understand that the result must be of the form it is. His real concern is what it means for the solutions to practical problems.

Mathematical background is of use to future scientists, but few engineering students taking fracture mechanics courses will become researchers in fracture mechanics. My advice is that indeed very few should. Fracture mechanics has matured to a useful engineering tool as has e.g. buckling analysis. Certainly, it is not perfect, but no engineering analysis is. Not much buckling research is practiced today; the present number of researchers in fracture mechanics is far out of proportion to the remaining engineering problems.

Despite the acclaimed solid education of engineers, it is my experience in teaching fracture mechanics to literally hundreds of practicing engineers, that most have only a vague idea of such subjects as plastic deformation and design; to many Mohr's circle is an enigma; at most one in a class knows the stress concentration factor of a circular hole; fewer even have ever heard of yield criteria and their significance. For this reason Chapter 2 of the text discusses the effect of notches and local yielding and provides a simplified look at yield criteria. The treatment is necessarily compromising rigidity, but it serves the purpose of providing the insight without which fracture mechanics cannot be understood.

My research work covered fundamental fracture and fatigue mechanisms, experimental evaluation of criteria for fatigue, fracture, and combined mode loading, the development of engineering procedures for arrest analysis in stiffened panels, collapse conditions, and damage tolerance analysis in general. My engineering background however, has always prevailed and forced me to consider the practicality of procedures. This book reflects a lifetime of experience in research and practical applications. No subject is discussed on the basis of hearsay. Instead the basis is "hands-on" experience with virtually every issue from the fundamental to the practical.

I am aware of my shortcomings, prejudices and opinionations, but believe to be entitled to these on the basis of my engineering experience. This text reflects them, and I do not apologize. Too many "refinements" in engineering solutions pertain to secondary errors; they increase the complexity, but do not improve the solution. One does not improve the strength of a chain by improving the strong links. The weak links in the fracture mechanics analysis are the unknowns, not the procedures. This book is for engineering students and for engineers, who must solve urgent problems yesterday. Engineering solutions are always approximative, no matter what the subject is. Such is the nature of engineering. Necessary assumptions are far more influential than those due to limitations of fracture mechanics.

The text is intended for the education of 'engineers'. At the same time it serves as a reference. For this reason there is some duplication and extensive cross-references are provided. This may be objectionable to the reader going through the text from A to Z, but it will be of help to those who read sections here and there. It is not perfect as no human effort ever is, and I shall welcome constructive criticism with regard to the engineering applications. My haste in accomplishing things (enforced by the unfortunate situation that I have to make a living, while writing a book is an extraneous effort which is not very profitable) may be reflected in the text. Again, I am not apologizing, just explaining.

I am grateful to my wife, Betty, for putting up with my preoccupations and moods while writing this text, and for submitting all writing to a word-processor. I am also thankful to my son Titus, who spent numerous hours in producing solutions to exercises and in drawing figures.

I dedicate this book to the memory of my father, Harm Broek. Many sons see their father as the ultimate example. So do I. His unfailing support has always been a driver of my ambitions.

Galena, Ohio, February 1988

Notice

Extensive computer software for fracture mechanics analysis was developed by the author of this book. This software is capable of

- performing residual strength analysis in accordance with Chapters 3, 4 and 10 both for LEFM and EPFM.
- performing fatigue crack growth analysis for constant amplitude, random loading and semi random loading in accordance with Chapters 5 and 10, with or without retardation. There are options for various retardation models, rate equations and tabular rate data (Chapter 7).
- automatically generating semi-random stress histories on the basis of exceedance diagrams (Chapter 6) and performing clipping and truncation upon command.
- determining inspection intervals and cumulative probability of detection in accordance with the procedure discussed in Chapter 11, using the calculated crack growth curves, and accounting for specificity and accessibility.
- providing professional plots.

An extensive library of materials data is included, as well as an extensive library of geometry factors. Besides a pre-processor can generate geometry factors, using most of the procedures discussed in Chapter 8.

The above software is available for personal computers as well as for VAX computers. Because of the large size of the software, it is split up in seven modules, each of which fits in a personal computer. The modules communicate through disket files that are generated automatically. The software can be obtained from FractuREsearch Inc, 9049 Cupstone Drive, Galena, OH 43021, USA.

A much simplified version of the same software (also by the author) is available from the American Society of Metals (ASM), Metals Park, OH 44073, USA. This simplified version has no data library, no preprocessor for geometry factor, cannot do retardation, and does not generate semi-random stress histories.

Contents

CHAPTER 1

Introduction

1.1. Fracture control

Fracture control of structures is the concerted effort by designers, metallurgists, production and maintenance engineers, and inspectors to ensure safe operations without catastrophic fracture failures. Of the various structural failure modes (buckling, fracture, excessive plastic deformation) fracture is only one. Very seldom does a fracture occur due to an unforeseen overload on the undamaged structure. Usually, it is caused by a structural flaw or a crack: due to repeated or sustained "normal" service loads a crack may develop (starting from a flaw or stress concentration) and grow slowly in size, due to the service loading. Cracks and defects impair the strength. Thus, during the continuing development of the crack, the structural strength decreases until it becomes so low that the service loads cannot be carried any more, and fracture ensues. Fracture control is intended to prevent fracture due to defects and cracks at the (maximum) loads experienced during operational service.

If fracture is to be prevented, the strength should not drop below a certain safe value. This means that cracks must be prevented from growing to a size at which the strength would drop below the acceptable limit. In order to determine which size of crack is admissible, one must be able to calculate how the structural strength is affected by cracks (as a function of their size); and in order to determine the safe operational life, one must be able to calculate the time in which a crack grows to the permissible size. For this, one must first identify the locations where cracks could develop. Analysis then must provide information on crack growth times and on structural strength as a function of crack size. This type of analysis is called damage tolerance analysis.

Damage tolerance is the property of a structure to sustain defects or cracks safely, until such time that action is (or can be) taken to eliminate the cracks. Elimination can be affected by repair or by replacing the cracked structure or component. In the design stage one still has the options to select a more crack resistant material or improve the structural design, to ensure that cracks will not become dangerous during the projected economic serivce life. Alternatively, periodic inspections may be scheduled, so that cracks can be repaired or components replaced when cracks are detected. Either the time to retirement (replacement), or the inspection interval and type of insepction, must follow from the crack growth time calculated in the damage tolerance analysis.

Inspections can be performed by means of any of a number of non-destructive inspection techniques, provided the structure is inspectable and accessible; but destructive techniques such as proof-testing are essentially also inspections. If a burst occurs during hydrostatic testing of e.g. a pipe line, then there was apparently a crack of sufficient size to cause the burst. Although this may be troublesome, the proof test is intended to eliminate defects under controlled circumstances (e.g. with water pressure) to prevent catastrophic failure during operation when the line is filled with oil or gas. After the burst and repair, the line can continue service. If no burst occurs during the proof test, then apparently any cracks were smaller than the critical size in the proof test. a certain period of safe operation is then possible before such cracks would grow to the permissible size.

Fracture control is a combination of measures such as described above (including analysis), to prevent fracture due to cracks during operation. It may include all or some of these measures, namely damage tolerance analysis, material selection, design improvement, possibly structural testing, and maintenance/inspection/replacement schedules. The extent of the fracture control measures depends upon the criticality of the component, upon the economic consequences of the structure being out of service, and last but not least, the consequential damage caused by a potential fracture failure (including loss of lives). Fracture control of e.g. a hammer may be as simple as selecting a material with sufficient fracture resistance. Fracture control of an airplane, includes damage tolerance analysis, tests, and subsequent inspection and repair/replacement plans.

Damage tolerance analysis and its results form the basis for fracture control plans. Inspections of whichever nature, repairs and replacements, must be scheduled rationally using the information from the damage tolerance analysis. This book deals with practical damage tolerance analysis. Fracture control measures are discussed in general, but the execution, use, and interpretation of the damage tolerance analysis for scheduling fracture control measures are discussed in detail.

The mathematical tool employed in damage tolerance analysis is called fracture mechanics; it provides the concepts and equations used to determine how cracks grow and how cracks affect the strength of a structure. During the last 25 years fracture mechanics has evolved into a practical engineering tool. It is not perfect, but no engineering analysis is. The equation for bending stress ($\sigma = Mh/I$) is rather in error when used to calculate structural strength, because it ignores plastic deformation. Nevertheless it has been used successfully for many years in design. Similarly, fracture mechanics can be used successfully. Acclaimed inaccuracies are due to inaccurate inputs much more than due to inadequacy of the concepts, which will become abundantly clear in the course of this book. Naturally, the results of damage tolerance analysis must be used judiciously, but this can be said of any other engineering analysis as well. Although further improvements of fracture mechanics concepts may well be desirable from a fundamental point of view, it is unlikely that damage tolerance analysis can be much improved, as its accuracy is determined mostly by the accuracy of material data and predicted loads and stresses.

Fracture mechanics can give useful answers to questions that hitherto could not be answered at all. The answers may not be perfect, but a reasonable anwser is better than none. Unfortunately, it is rather easy to obtain a wrong answer: pitfalls are numerous. This book is intended to explain the engineering usage of fracture mechanics, and to point out the pitfalls in detail. Some of the pitfalls are so treacherous that one sees things more often done wrong than right. These and other things have led to some myths about fracture mechanics and its uses, which are hard to eliminate. They will be addressed in this book.

Although the basis of fracture mechanics concepts will be discussed, this book focuses on how the analysis should be performed, on how to solve practical problems and on how to avoid errors. Attainable accuracy, and the factors affecting accuracy, are discussed in detail. A brief recapitulation is given of those concepts of fracture mechanics that are actually used in practical damage tolerance analysis. For more in-depth treatment of those subjects, the reader is referred to other text-books on the matter [1, 2, 3]. Yet, the present text provides sufficient background for a proper understanding of the practical methods discussed.

As this book deals with the practical use of established fracture mechanics concepts, references have been kept to a minimum. They can be found in more extensive texts. It is not necessary to reference established concepts; e.g. references to Hooke's law are superfluous, and so are references to accepted and established fracture mechanics concepts. Hence, references are provided only in those cases where relevance or extent do not warrant complete treatment in this text, so that use of the original publication might be desirable.

1.2. The two objectives of damage tolerance analysis

Establishment of a fracture control plan requires knowledge of the structural strength as it is affected by cracks, and of the time involved for cracks to grow to a dangerous size. Thus, damage tolerance analysis has two objectives, namely to determine

1. the effect of cracks on strength (margin against fracture)

2. the crack growth as a function of time.

These two objectives are discussed below.

Figure 1.1 shows diagrammatically the effect of crack size on strength. In fracture mechanics crack size is generally denoted as a. In Figure 1.1 the strength is expressed in terms of the load, P, the structure can carry before fracture occurs (fracture load). Supposing for the moment that a new structure has no significant defects ($a = 0$), then the strength of the new structure is P_u, the (ultimate) design strength (load). It should be emphasised that the strength of the new, crack-free structure is finite. Fracture will, and must occur when it is subjected to a load P_u, otherwise the structure was over-designed.

In every design a safety factor (ignorance factor) is used. This factor may be applied in different ways, but the result is always the same. In some areas of technology the safety factor is applied to load. For example if the maximum anticipated service load is P_s, the structure is actually designed to sustain $jP_s = P_u$, where j is the safety factor. The designer sizes the structure in such a manner that the stress is equal to or slightly less than the ultimate tensile strength when the load is P_u (checks against plastic deformation are usually necessary as well). Alternatively, the safety factor is applied to the allowable stress: if the actual material strength (ultimate tensile strength) is F_{tu}, the structure is sized in such a way that the stress at the highest service load, P_s, is less than or equal to F_{tu}/j

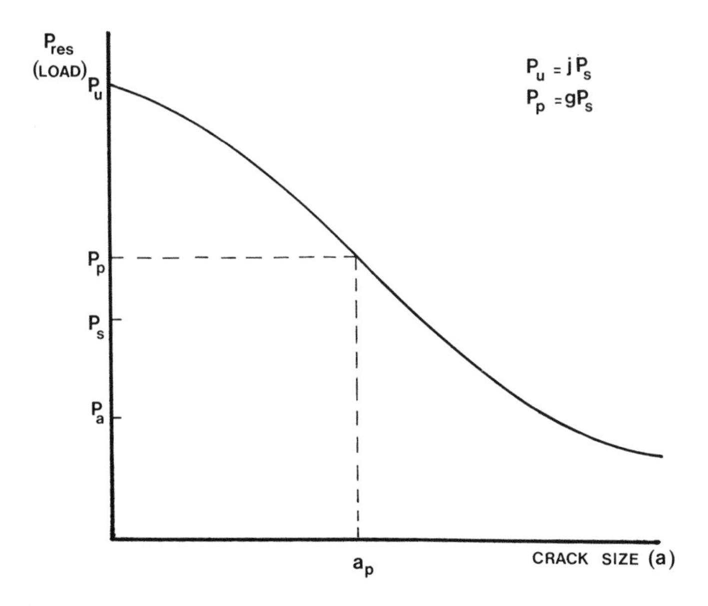

Figure 1.1. Residual strength in the presence of cracks; strength of new structure ($a = 0$) is $P_u = jP_s$.

where j is again the safety factor. Hence, since load and stress are usually proportional, the structure is actually capable of carrying $jP_s = P_u$. Plasticity may well prohibit proportionality, but since plasticity is generally limited to small areas at notches and stress concentrations, the above is approximately correct. But, even if it is not correct in actual numbers, it is true in spirit: the structure is designed to carry a load higher than the highest anticipated service load by a factor j, and the structural strength is $P_u = jP_s$. The value of j is between 3 (many civil engineering structures) and 1.5 (airplanes).

It is emphasized that P_s is the highest service load. If the service load varies, the load may well be much less than P_s during most of the time. For example the loads on cranes, bridges, off-shore structures, ships and airplanes are usually much less than P_s. Only in exceptional circumstances (e.g. storms) does the load reach P_s. At other times the load may be only a fraction of P_s, so that the margin against fracture is much larger than j, except in extreme situations. The loads on some structures, e.g. pipelines, pressure vessels, rotating machinery are reaching more nearly always the same level (P_s), as shown in Figure 1.2.

The new structure has a strength P_u with safety factor j. Its strength is finite, so that the probability of fracture is not entirely zero. If the load should reach P_u (e.g. in a storm) the structure fails. The probability of this occurring is non-zero, but experience has shown that it is acceptably low. If cracks are present the strength is less than P_u. This remaining strength under the presence of cracks is generally referred to as the 'residual strength', P_{res}; the diagram in Figure 1.1 is called the residual strength diagram. With a residual strength $P_{res} < P_u$ the safety factory has decreased: $j = P_{res}/P_s$ which is less than $j = P_u/P_s$. In concert, the probability of fracture failure has become higher.

Fracture is the catastrophic break-up of the structure into two or more pieces. With a crack of size a, the residual strength is P_{res}. Should a load $P = P_{res}$ occur then fracture takes place. The fracture process may be slow and stable initially, the crack extending (by fracture), but the structure still hanging together. Eventually, the fracture becomes unstable and the structure breaks into two or more pieces. The whole process of stable–unstable fracture may take place in a fraction of a second. If the load $P = P_{res}$ does not occur, service loading continuing at loads at or below P_{res}, the crack will continue to grow, not by fracture but by cracking mechanisms such as fatigue, stress-corrosion or creep.

Due to continual growth the crack becomes longer, the residual strength less, the safety factor lower, and the probability of fracture higher. If nothing is done and the structure remains in service, the residual strength evenutally will become equal to P_s (or even equal to the average service load P_a in Figure 1.2). Then the safety factor is reduced to 1 and fracture occurs already at P_s, i.e. at the (highest) service load, or even at P_a. This is what must be prevented: the crack should not be allowed from becoming so large that fracture occurs at the service loads.

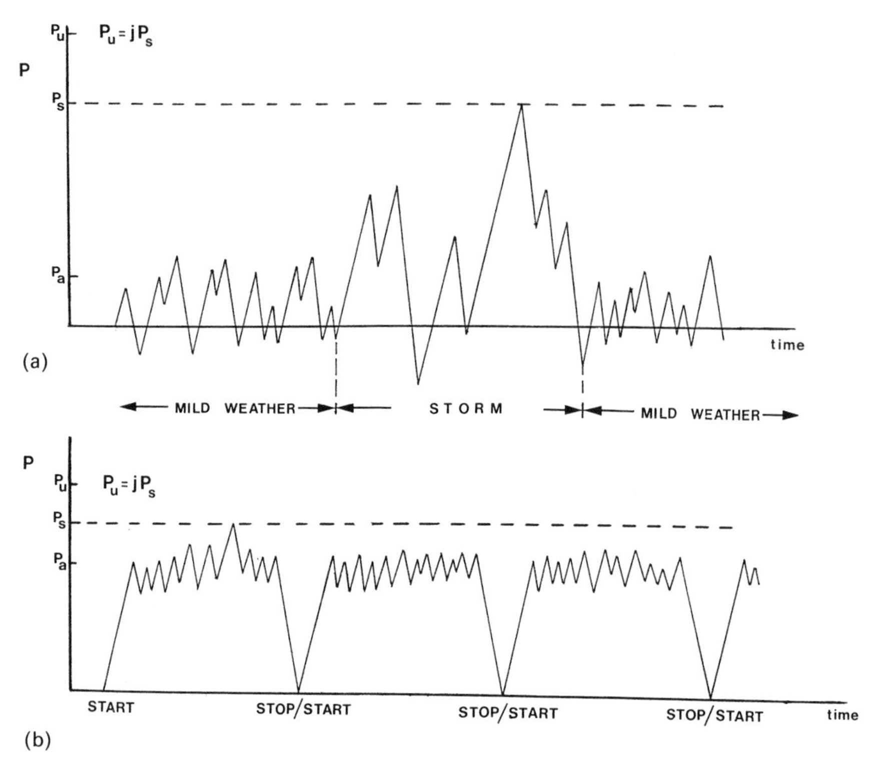

Figure 1.2. Schematic examples of load histories. (a) Typical loading of offshore structures, ships, airplanes; (b) typical loading of rotating machinery.

Hence, the structure or component must be replaced before the crack becomes dangerous, or the crack must be detected and repaired before such time.

The above implies that the limit should be set somewhat above P_s. For example, one may require that the residual strength never be less than $P_p = gP_s$, where g is the remaining safety factor, and P_p the minimum permissible residual strength. The design engineer or user does not decide what should be the initial safety factor j. This factor is prescribed by rules and regulations issued by engineering societies (e.g. ASME) or Government authorities. Similarly, these rules or requirements should prescribe g. This has not been done for all types of structures yet, while e.g. the ASME rules approach the problem somewhat differently (rules and regulations are discussed in Chapter 12). However, some rule or goal must be estabished, some decision made, to set the minimum permissible residual strength, so that the maximum permissible crack size, a_p, can be determined from the residual strength diagram.

Provided the shape of the residual strength diagram is known, and P_p

prescribed, the maximum permissible crack size follows from the diagram. In order for damage tolerance analysis to determine the largest allowable crack, the first objective must be the calculation of the residual strength diagram of Figure 1.1. If a_p can be calculated directly from P_p it may not be necessary to calculate the entire residual strength diagram, but only the point (a_p, P_p). However, this is seldom possible and rarely time saving. In general, the calculation of the entire diagram is far preferable. The maximum permissible crack size follows from the calculated residual strength diagram and from the prescribed minimum permissible residual strength, P_p. The residual strength diagram will be different for different components of a structure and for different crack locations; permissible crack sizes will be different as well.

The permissible crack size is sometimes called the critical crack size. However, the objective of fracture control is to prevent 'critical' cracks. A critical crack is one that would cause fracture in service. Cracks are not allowed to grow that long. Instead, they are permitted to grow only to the permissible size a_p. They would be critical only in the event that a load as high as P_p would occur.

Knowing that the crack may not exceed a_p is of little help, unless it is known when the crack might reach a_p. The second objective of the damage tolerance analysis is then the calculation of the crack growth curve, shown diagrammatically in Figure 1.3. Under the action of normal service loading the cracks grow by fatigue, stress corrosion or creep, at an ever faster rate leading to the convex curve shown in Figure 1.3.

Starting at some crack size a_0 the crack grows in size during time. The permissible crack a_p following from Figure 1.1 can be plotted on the curve in Figure 1.3. Provided one can calculate the curve in Figure 1.3 one obtains the

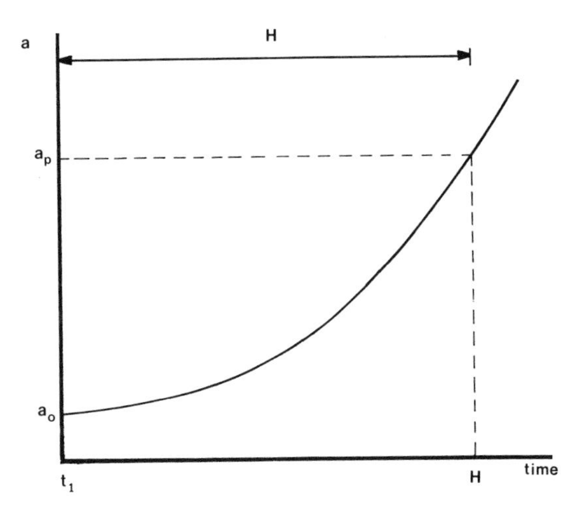

Figure 1.3. Crack growth curve (schematically).

time H of safe operation (until a_p is reached). If a_0 is for example an (assumed or real) initial defect, then the component or structure must be replaced after a time H. Alternatively, a_0 may be the limit of crack detection by inspection. This crack a_0 will grow to a_p within a time H. Since crack growth is not allowed beyond a_p, the crack must be detected and repaired or otherwise eliminated before the time H has expired. Therefore, the time between inspections must be less than H. At an inspection at time t_1, the crack will be missed, because a_0 is the detection limit. If the next inspection were to take place H hours later, the crack would have reached a_p already, which is not permitted; i.e. the inspection interval must be less than H; it is often taken as $H/2$. In any case, the time of safe operation by whatever means of fracture control follows from H. In turn, H emerges from the damage tolerance analysis, provided both the residual strength diagram (a_p) and the crack growth curve can be calculated to obtain H.

1.3. Crack growth and fracture

The residual strength and crack growth diagrams are essentially different, not only in shape but also in significance. Fracture is the final event, often taking place very rapidly, and resulting in a breaking-in-two. Crack growth on the other hand occurs slowly during normal service loading. Also the mechanisms of crack growth and fracture are different.

Crack growth takes place by one of five mechanisms:

a. Fatigue due to cyclic loading.
b. Stress corrosion due to sustained loading.
c. Creep.
d. Hydrogen induced cracking.
e. Liquid metal induced cracking.

Of these, the first two, and combinations thereof are the most prevalent, while the last is hardly of interest for load-bearing structures. Crack growth is sometimes referred to as 'sub-critical crack growth', a pleonasm. Fracture is critical, and fracture is not the same as crack growth.

A crack may cause a fracture. There are only two mechanisms by which fracture can occur, namely:

a. Cleavage.
b. Rupture.

A third 'mechanism', namely intergranular fracture, requires operation of some form of either cleavage or rupture.

A mechanism for fatigue crack growth is shown in Figure 1.4. Other mechanisms are possible but not essentially different [4, 5, 6]. Even at very low loads there is still plastic deformation at the crack tip because of the high stress concentration. Plastic deformation is slip (due to shear stresses; see Chapter 2) of

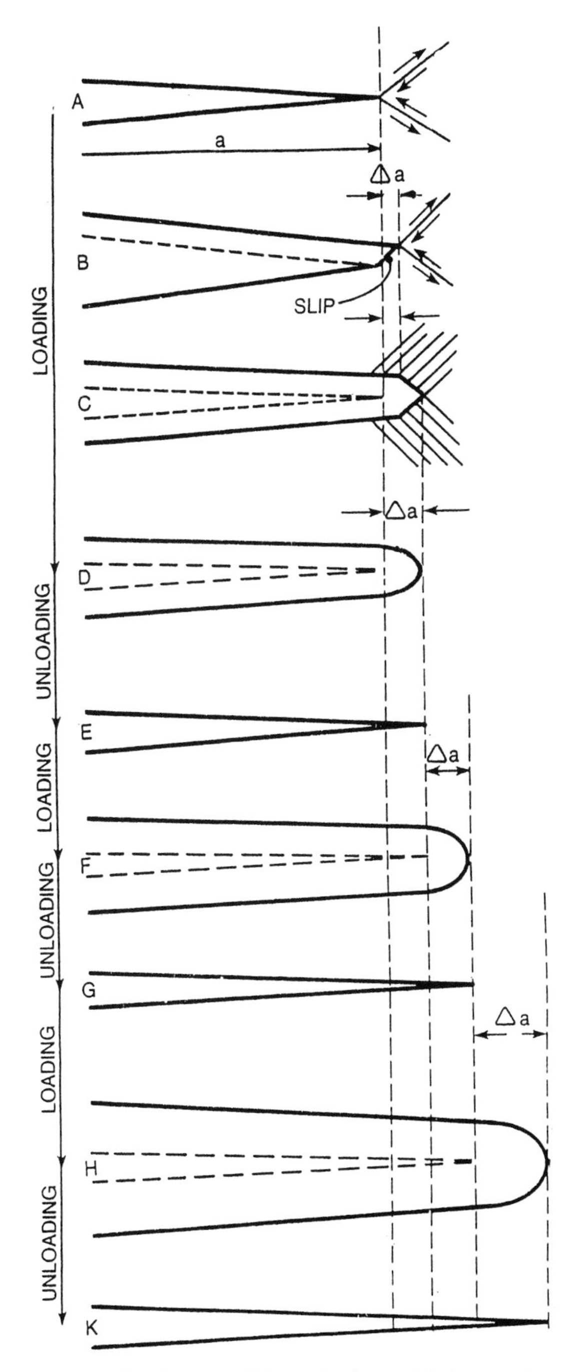

Figure 1.4. One of various possible mechanisms of fatigue crack growth.

atomic planes, depicted in Figure 1.4, stage B. Continued slip on complementary planes results in a blunted crack tip (Figure 1.4, stages B–D). The very first slip step in stage 2 has already caused a very small crack extension Δa. Upon unloading (or if necessary compressive loading) the crack tip again becomes sharp. Mechanistically, the whole process of slip could be reversed so that the end result after unloading would again be as stage A. However, because of oxidation of the freshly exposed material along the slip steps, and the general disorder due to the slip, the process is irreversible in practice; the crack extension Δa remains. In the next load cycle the process is repeated: the crack grows again by Δa.

Growth per cycle, Δa, is extremely small as can be judged immediately from the mechanism in Figure 1.4. Typically, the growth is on the order of 10^{-8}–10^{-4} inches (10^{-7}–10^{-3} mm); however if the load is cycled for 10^4–10^8 cycles, the crack will have grown by an inch. The repeated blunting and sharpening gives rise to marks on the crack surface (often erroneously referred to as fracture surface), which can be made visible at high magnification in an electron microscope, as shown in Figure 1.5. The marks, called fatigue striations, represent the successive positions of the crack front, i.e. the blunting/sharpening steps. If the crack grows by 10^{-5} inch, then the striation spacing is 10^{-5} inch (i.e. 0.05 inch at a magnification of 5000 $\times$). Not all materials exhibit striations as regular as in Figure 1.5. [4, 5, 6] (Chapter 13).

Crack growth by stress corrosion is a slow process as well. The crack extends due to corrosive action (often along the grain boundaries) facilitated by the high stretch and consequent atomic disarray at the crack tip. A common mechanism of creep cracking is the diffusion of vacancies (open atomic places), a conglomerate of vacancies forming a hole, which subsequently joins up with the crack tip.

A crack by itself is only a partial failure, but it can induce a total failure by fracture. Fracture occurs by either of two mechanisms, cleavage or ductile rupture. Cleavage is the splitting apart of atomic planes. From grain to grain the preferred splitting plane is differently oriented causing a faceted fracture. (Figure 1.6). The facets by themselves are flat and, therefore, good reflectors of incident light. This causes the cleavage fracture to sparkle when fresh, but the glitter may soon fade due to oxidation.

The alternative fracture mechanism of ductile rupture is shown in Figure 1.7. All structural materials contain particles and inclusions. These particles are generally complex compounds of the alloying elements. Some alloying elements are used to improve castability and machineability; others are specifically included to improve the alloy's strength. First the large particles let loose or break, forming widely spaced holes close to the crack tip. In the final phase, holes are formed at myriads of smaller particles; these holes or voids join up to complete the fracture. Because of its irregularity the fracture surface diffuses the

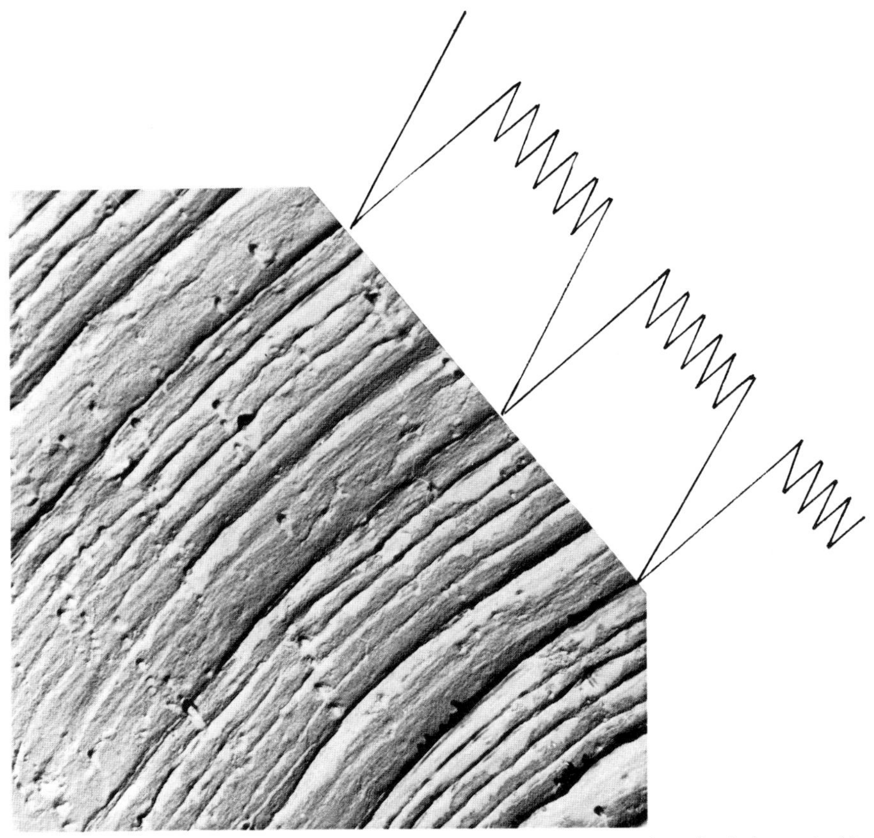

Figure 1.5. Fatigue striations on crack surface of aluminium alloy. Spacing of striations coincides with cyclic loading (insert). Magnification 12 000 ×.

light and looks dull grey. The holes of the large particles and those of the small particles (Figure 1.7) are visible at high magnification in the electron microscope, as shown in figure 1.8. As the fracture surface shows the halves of the holes, the fracture is referred to as dimple rupture or just rupture. Elimination of the large particles (by selective alloying and heat treatment) can improve the fracture resistance of an alloy [4]. On the other hand, alloying is necessary to provide strength in the first place, so that not all particles can be avoided. That they play a role in the final fracture process becomes rather of secondary importance; without the particles, the strength would be less.

Both cleavage fracture and rupture are fast processes. A cleavage fracture may run as fast as 1 mile/sec (1600 m/s), a dimple rupture as fast as 1500 ft/sec (500 m/s), although it may be slower. Fracturing is sometimes stable. The crack

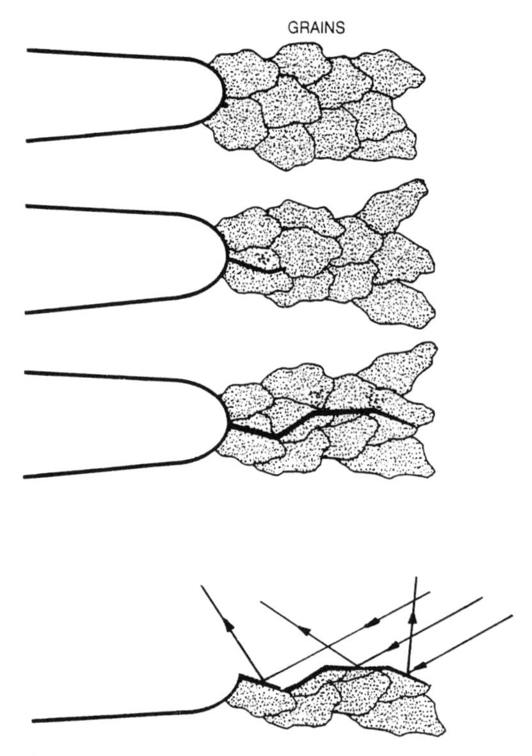

Figure 1.6. Cleavage fracture starting at (blunted) crack tip. Bottom: flat facets glitter due to reflection of incident light.

(fracture) then extends by one of the fracturing processes (cleavage or rupture) instead of by one of the cracking mechanisms. Usually stable fracture is immediately followed by the final unstable event.

The infamous brittle fracture of steel below the transition temperature occurs by cleavage. Cleavage is often referred to as brittle fracture, while dimple rupture is described as ductile fracture. This may be adequate for metallurgists, but in general one should be very careful using these terms. By far the majority of service fractures occur by dimple rupture, but most of these exhibit very little overall plastic deformation, so that they are brittle from an engineering point of view (see Chapter 2). Reference to cleavage and rupture avoids confusion. A brittle fracture is one with little (overall) plastic deformation, whether cleavage or rupture.

Similarly, fractographers sometimes refer to rupture as the 'overload fracture' to distinguish it from the fatigue or stress corrosion crack surface (Chapter 13). This can also cause confusion. Distinction between 'fracture surface' and 'crack surface' is more appropriate. Practically all service failures are due to cracks or

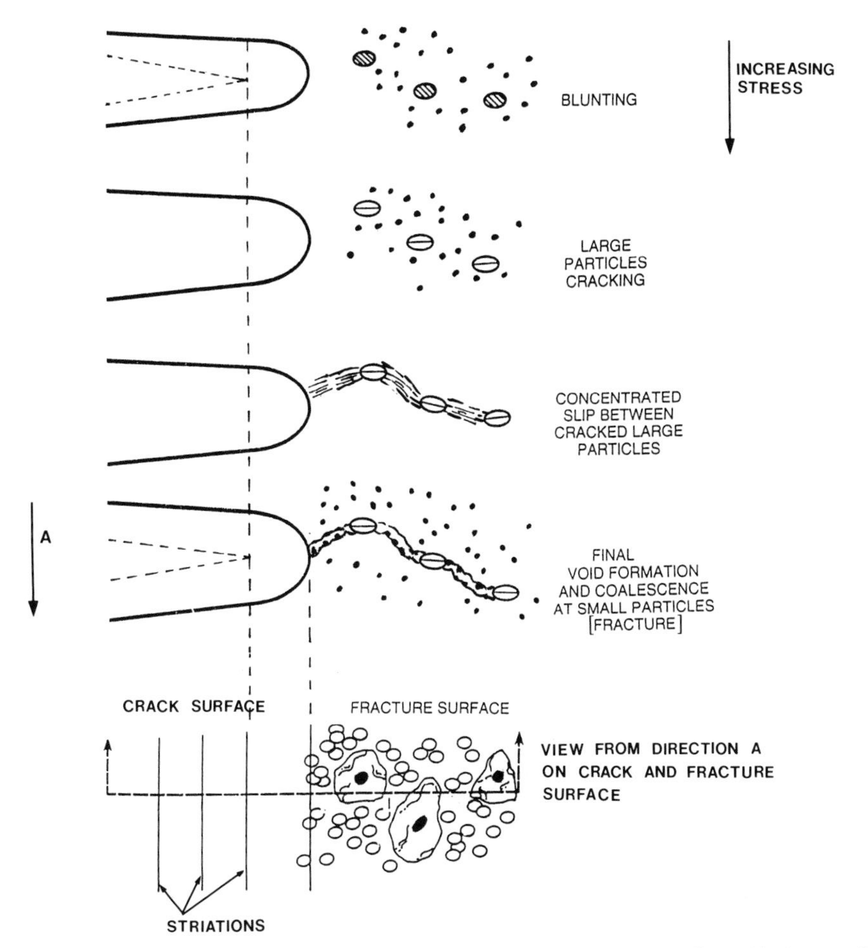

Figure 1.7. Four stages of ductile rupture. Bottom: fracture surface shows holes formed by large and small particles (dimples).

defects, a true overload fracture being very rare indeed (Chapter 13). It is equally wrong (though not as confusing) to speak of fatigue fracture or stress corrosion fracture. The fracture may be a consequence of a fatigue crack or a stress corrosion crack, but it occurs by cleavage or rupture.

There is a third fracture type, namely intergranular fracture. The distinction here is on the basis of fracture path (along the grain boundaries). Nevertheless, intergranular fracture is not an altogether different type. Whether intergranular or transgranular, the mechanism of separation either resembles cleavage or rupture, be it that cleavage at a grain boundary is not as well defined as transgranular cleavage.

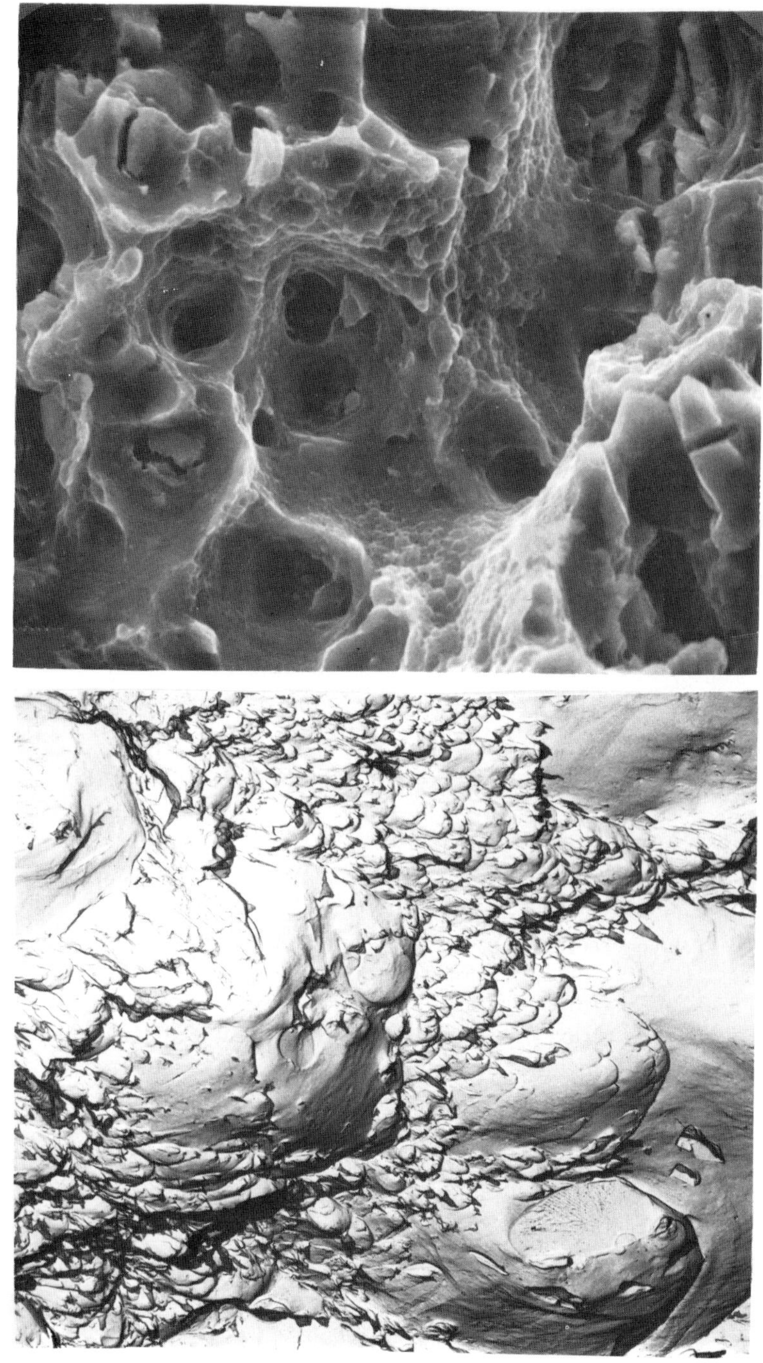

Figure 1.8. Fracture surface of ductile rupture showing large holes with cracked large particles and interconnecting dimples; Aluminium alloy; Top: Scanning electron microscope; 1400 × . Bottom: Transmission microscope (replica): 5000 × .

1.4. Damage tolerance and fracture mechanics

Damage tolerance analysis must provide a capability for the calculation of both the residual strength diagram (fracture due to cracks) and of the crack growth curve. Fracture mechanics methods have been developed to analyze fracture to obtain the fracture stress (residual strength), and to analyze fatigue crack growth and stress corrosion crack growth. The treatment of creep crack growth by fracture mechanics is still largely in the research phase. As this book is dealing with practical and applicable methods and not with potential methods, creep crack growth will not be discussed.

Although it is possible to deal with stress corrosion cracking in principle, the crack growth times are usually so short (from 1 to 1000 hrs) that crack growth is rather uninteresting from a practical point in view. Fracture control in the case of stress corrosion is often aimed at crack growth prevention. Fracture mechanics methods to accomplish this will be discussed.

In the treatment of crack growth emphasis will be on fatigue cracking. In many structures fatigue cannot be prevented altogether; cracks then will occur. These must be dealt with by means of fracture control, i.e. they must be eliminated before they can cause a fracture. The time to the initiation of a fatigue crack will not be considered for two reasons; first because fracture mechanics deals with existing cracks, and second because it is not necessary to know exactly when cracks appear to exercise control of cracks (Chapter 11). Besides, in cases of fracture control by means of replacement and retirement, an initial crack (hypothetical or real) is usually assumed present in the new structure (Chapter 12). In some cases rules and regulations actually specify that an initial crack be assumed.

Fracture mechanics (as all engineering mechanics) uses stresses, rather than loads. Thus, the residual strength diagram of Figure 1.1 is normally based upon σ_{res}, the stress the structure can sustain (instead of load), before fracture occurs. The residual strength diagram based upon stress is shown in Figure 1.9. For this purpose the engineering stress is used, as in the case of the uncracked structure, the residual strength being given as σ_{res} (Figure 1.9). Note that σ_{res} is a strength (like the tensile strength or the yield strength) and not a stress. Fracture occurs when the stress equals σ_{res}: fracture if $\sigma = \sigma_{res}$). Residual strength should not be confused with residual stress. (Residual stress is a stress residing in a structure while there are no loads applied).

Stress can be used as a basis for the analysis if there is a relationship between the applied stress and the processes taking place at the crack tip. As the crack tip events are governed by the local stresses at the crack tip, it is required that the local crack tip stress be described as a function of the applied stress. Such relationships can be derived, provided the problem is defined clearly. For this purpose one must distinguish the 3 modes of loading shown in Figure 1.10.

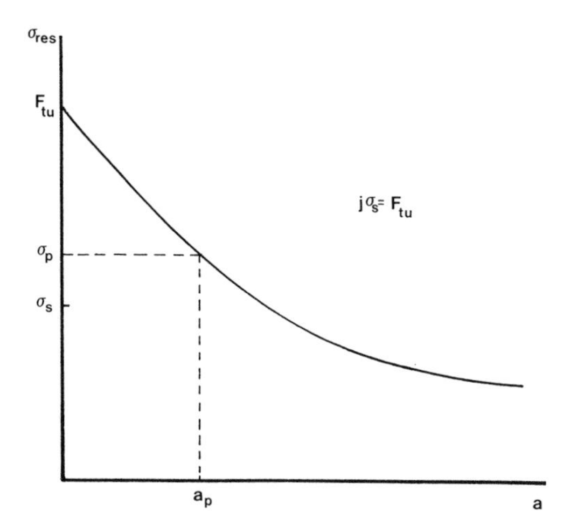

Figure 1.9. Residual strength diagram on the basis of nominal engineering stress.

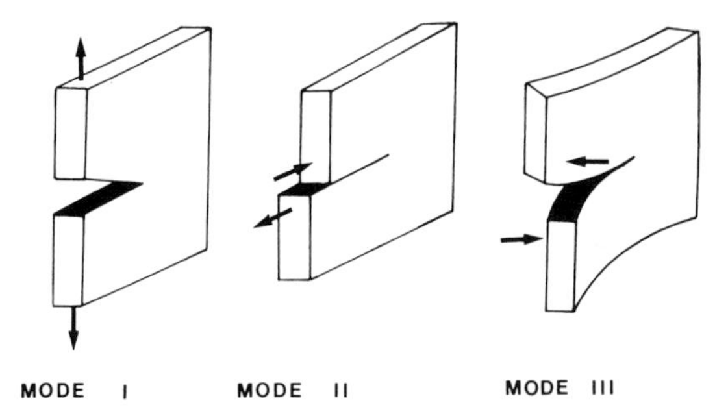

Figure 1.10. The modes of loading. Mode I opening mode. Mode II shear mode. Mode III tearing mode.

These modes of loading (not modes of cracking), are usually referred to simply by Roman numerals I, II and III. Other descriptions used are opening mode or tension mode for mode I, (in-plane) shear mode for mode II, and (out-of-plane) shear mode or tearing mode for mode III.

It turns out that the crack-tip stress equations are very similar for each of the modes. As a matter of fact the format of the equations is exactly the same. Consequently, the fracture and crack growth analysis procedures for each of the modes individually, turn out to be identical, be it that different numbers apply. If one knows how to deal with mode I, one knows essentially how to analyse modes II and mode III individually. This is one reason why this book deals primarily with mode I. However, there are other more compelling reasons.

In practice, by far the majority of cracks result from mode I loading. the other two modes do not occur individually, but they may occur in combination with mode I, i.e. I–II, I–III or I–II–III. However, if the loading of these modes is in phase, cracks will rapidly choose a direction of growth in which they are subjected to mode I only (Chapter 9). Thus, the majority of apparent combined mode cases are reduced to mode I by nature itself. There are few cases left then which cannot be treated as pure mode I. These occur when for example in a mode I–II combination (where the crack would normally select a mode I path), there is a direction in the material where the resistance to cracking and fracture is substantially lower. A case in point is a circumferential weld in a torque tube (Chapter 9). Normally, a crack in a torque tube will develop and grow under 45 degrees (tension only) and thus be in mode I. But if a circumferential weld confines the crack to the circumferential direction (shear plane) mode II must be considered. Another case occurs when mode I and e.g. mode II loading are out of phase (bending and torsion cycles with different frequencies; see Chapter 9).

Although individual loading modes are easy to deal with, a combined mode loading case is rather more difficult. Several combined mode (or mixed mode) analysis concepts have been proposed. A practical approach will be presented in Chapter 9 for the few instances of 'true' mixed mode loading described above.

1.5. The need for analysis: purpose of this book

During the first half of the industrial era structural failures were numerous; railroad accidents, elevator accidents, boiler explosions, etc., were a common occurrence; accidents due to structural fracture were reported weekly if not daily. The first powered flight was postponed because of a broken propellor shaft. Orville Wright returned from Kitty Hawk to Dayton to machine a new shaft, which caused a 2-week delay of the first flight. Due to improved materials, and refinements of design procedures, and not in the least, due to the enforcement of design safety factors and quality control measures, the number of strucutral failures has abated, but not been reduced to zero.

More than two dozen major bridges have collapsed during this century. Since World-War II in excess of 200 civil airplanes had fatal accidents due to fatigue cracks; the number of cracks discovered in commercially operated jets is estimated at well over 25 000. Many cracks have been reported in nuclear power structures. Although less frequent, catastrophic structural failures have far more serious consequences now than in the past. At the same time society has become less tolerant and abundantly more litigious. Thus, no manufacturer, nor operator of larger structures, can afford to ignore the issue of fracture control. Rational fracture control being based on damage tolerance analysis, the use of fracture mechanics and damage tolerance analysis is well-nigh imperative.

Rules and requirements have been issued for Military airplanes as well as for commercial airplanes. Damage tolerance analysis of an airplane subject to these rules, typically requires 20 000–60 000 man hours. (An equal or larger effort is in the associated testing for material's data and analysis substantiation). The ASME boiler and pressure vessel code requires damage tolerance analysis, at least for nuclear power structures. Before long, requirements are likely to be implemented for railroads, ships, bridges, pipelines and other structures, but many manufacturers are already using some form of damage tolerance analysis.

Fracture mechanics concepts and damage tolerance analysis may not be ideal (they never will be), but they provide answers where none were available before. Judicious use of these can certainly reduce the risk of fracture. (No risk in life is ever reduced to zero). The purpose of this text is to present the procedures of practical damage tolerance analysis as they are (or can be) applied to almost any structure, including ships, pipelines, aircraft, pressure vessels, cranes, bridges and rotating machinery. Each of these structures presents its own specific problems, but there is enough common ground for a universal procedure, which can be modified, simplified or extended to serve a certain purpose. The various components of the universal procedure will be discussed. Specific use and interpretation for various types of structures will be reviewed; examples of applications given.

A theoretical text this is not. The emphasis is on how damage tolerance analysis is and should be performed, on how the necessary information is obtained and used, on how the results of the analysis are employed to exercise fracture control. The reliability of the analysis and the major sources of error are discussed. Engineering approaches and approximations are given ample space: in view of the general inaccuracies caused by material data and projected loads and stresses, approximative solutions are quite adequate as the errors so introduced are secondary. In many instances, approximative solutions are the only feasible ones from an economic point of view. It may well be reasonable to spend large sums on the analysis of one crack in a nuclear pressure vessel, the cost of a power plant being extreme and the consequential cost of a failure beyond imagination. However, the analysis of a crack in a hand-tool must be cheap and the best engineering methods are indicated. For an aircraft structure with hundreds of potential crack locations in very complex details, so many cases must be considered that approximations and engineering judgement must be used. The same holds for ships, offshore structures, bridges, etc. it will be shown in this text why and where approximations are useful and suitable. The order of magnitude of the error in the final solution will be the guideline.

The 'how' cannot be understood without some knowledge of the 'why'. It would be a disservice to the reader to refer her or him to other texts for the 'why', in particular where these texts are often intended for theoreticians rather than practitioners. Thus the present text provides the concepts and theoretical background in an abridged and simplified form, adequate for the understanding of applications and use. The background is contained in Chapters 2–5 which present the concepts of fracture mechanics for the analysis of fracture (elastic as well as elastic-plastic) and for the analysis of crack growth. These chapters, although providing sufficient basis for what follows, can serve as an introduction to more comprehensive texts [e.g. 1, 2, 3] for readers desiring more theoretical information.

Chapter 6 is devoted to the analysis and interpretation of load and stress histories (the sequence of cyclic stresses causing fatigue crack growth). The stress history is of eminent importance in damage tolerance analysis, and yet, this subject is rather ignored in theoretical texts.

Mathematics have been kept to the barest minimum. Lengthy derivations do not appear if they are irrelevant to the understanding of the 'why'. Instead, dimensional arguments are used to show why an equation must be of the form in which it appears. Nevertheless, as in any analysis, a certain amount of mathematics is unavoidable and necessary.

Following the background are chapters dealing with obtaining the ingredients for the analysis, and the performance of the analysis. Chapters 7, 8 and 9 respectively provide information on the interpretation and use of the materials data needed for input, simple methods to obtain geometry factors for complex structures and cracks, and the effects of special conditions such as mixed mode loading, cold work of holes, interference fits, residual stresses, etc. Approximations and the assessment of their errors are discussed in extenso. The analysis procedure is illustrated on the basis of examples of applications to different types of structures, cracks and load spectra (Chapters 10 and 14). Chapters 11, 12 and 13 show how damage tolerance analysis is used for the implementation of fracture control measures, while addressing sources of error, pitfalls, common misconceptions, rules and regulations, and the establishment of fracture control plans.

The reader may find that certain subjects and problems are touched upon at several places in this book. Cross references are then made to the pertinent chapter where the subject is discussed in detail. This was done on purpose, at the risk of offending those readers who go through the entire text in sequence. It is anticipated that many readers will use this as a reference text; cross references to related subjects and problems are included for their convenience. For those readers and students desiring to sharpen their skill, exercise problems are provided at the end of each chapter; solutions are given in Chapter 15.

A final remark about the use of units is appropriate. Although metric units become more and more accepted, English units and kg-force are still used at many places. Since the unit system used is of no consequence whatsoever for the essence of the engineering procedures, various unit systems are being used in this text. No attempt is made to provide other units between parentheses, as this requires a lot of space and subtracts more than it adds. The only units of interest are those of length, stress, force, and stress intensity. Should the reader want to do so, she or he can easily make the conversions.

Material properties such as tensile strength, yield strength and collapse strength are consistently referred to as strength and denoted as F_{tu}, F_{tv} and F_{col} respectively. Although they are expressed as stress, denoted as σ, they are critical values (material properties).

1.6. Exercises

1. Design a tension member with circular cross section to carry a load of 1 000 000 lbs (1000 kips) with a safety factor of 2 against yield, and a safety factor of 3 against static fracture. Consider two materials: Material A with $F_{ty} = 50$ ksi and $F_{tu} = 60$ ksi; Material B with $F_{ty} = 50$ ksi and $F_{tu} = 80$ ksi. What is the ultimate design load in each case.

2. Design the same member as in problem 1 with a safety factor on load of 2.5 for static fracture, while no yielding is allowed at 1.3 times the maximum service load.

3. Assume that for the members designed in problems 1 and 2 a damage tolerance rule applies requiring that the residual strength may never drop below 1.2 times the maximum service stress. For all cases calculate σ_p and P_p. How large is the remaining safety factor in each case?

4. Define the difference between cracking and fracture. By which mechanisms do cracks develop and grow?

5. Assume Material A and the design of problem 1, and assume the damage tolerance requirement of problem 3 is in effect. Further assume that the maximum permissible crack size following from the damage tolerance requirement is 2 inch. Sketch the residual strength diagram in terms of stress and load, and identify all known points (nomenclature and numbers) in the diagram.

6. Using the result of problem 5, and assuming that cracking occurs by fatigue, estimate the fatigue crack size that would cause a fracture if a load equal to 1.5 times the maximum anticipated service load would occur (is this likely to happen?). Also estimate the crack size which would cause a fracture at the maximum anticipated service stress (under which circumstances might such a fracture occur?).

References

[1] D. Broek, *Elementary engineering fracture mechanics*, 4th Edition, Nijhoff (1985).

[2] M.F. Kanninen and C.H. Popelar, *Advanced fracture mechanics*, Oxford University Press (1985).
[3] J.F. Knott, *Fundamentals of fracture mechanics*, Butterworths (1973).
[4] D. Broek, Some contributions of electron fractography to the theory of fracture, *Int. Met. Reviews, Review 185*, **9** (1974), pp. 135–181.
[5] C.Q. Bowles and D. Broek, On the formation of fatigue striations, *Eng. Fract. Mech.*, **8** (1972), pp. 75–85.
[6] V. Kerlins, Modes of fracture, *Metals Handbook*, **12** (1987), pp. 12–71.

CHAPTER 2

Effects of cracks and notches: collapse

2.1. Scope

In Linear Elastic Fracture Mechanics (LEFM; Chapter 3) as well as in Elastic-Plastic Fracture Mechanics (EPFM; Chapter 4) the analysis of fracture is based on a parameter representing the crack tip stress field; while the quantity used in EPFM is actually the strain energy release rate, this can be shown to be equivalent to a stress field parameter. In neither LEFM nor EPFM the possibility of so-called plastic collapse is implicitly evaluated. Fracture mechanics analysis may provide a fracture stress (residual strength) higher than the stress for plastic collapse; since the maximum load carrying capacity is reached at the time of collapse, the fracture stress calculated with fracture mechanics may be in error (too high). Similarly, in such a situation, the calculated critical crack size would be too large. Fracture parameters measured in a test where fracture occurs due to collapse would be too low.

No fracture analysis is complete, without the evaluation of collapse conditions. Collapse and fracture are competing conditions and the one satisfied first will prevail. Before the fracture mechanics analysis procedures are discussed in Chapters 3, 4 and 5, the conditions for collapse will be reviewed in this chapter.

2.2. An interrupted load path

Discontinuities, fillets and notches, and cracks in particular, give rise to a stress concentration, i.e. a local region where the stresses are higher than the nominal or average stress. The following is a brief and limited discussion of stress concentrations due to geometrical discontinuities in the structure. Stress concentrations play a decisive role in virtually all structural cracking problems.

Consider two parallel bars of the same size and of the same material, as in Figure 2.1. Each carries half of the total load; the strain in the bars is equal, causing an elongation ΔL. If the left bar is cut in two, the right bar will have to

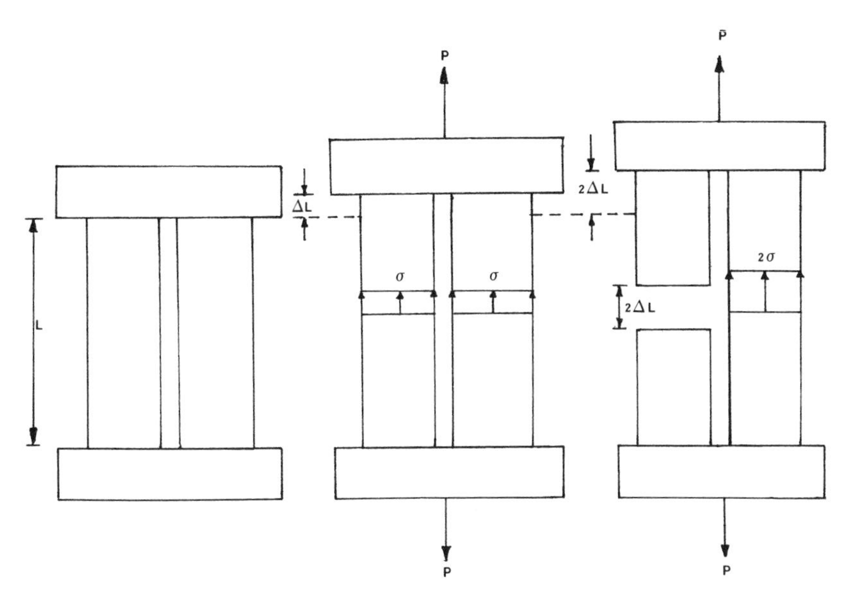

Figure 2.1. Effect of cutting one of two parallel bars.

carry all the load: the stress, the strain and the elongation will be twice as high as before. Consequently, the gap between the two halves of the left bar will be $2\Delta L$. The left bar carries no stress at all (Figure 2.1.c).

Next consider the case where the left and right bar are attached (e.g. welded) as depicted in Figure 2.2. If the bars are intact, the situation is identical to the previous. However, if the left bar is cut a different situation develops. If the top half of the right bar is strained, the top half of the left bar must necessarily undergo approximately the same strain. Both bars being of the same material (same modulus of elasticity, E), equal strain in the bars dictates equal stress ($\sigma = \varepsilon E$). Thus, each bar still carries the same stress and therefore each bar carries half the load. However, since the left bar is cut, the right bar alone must carry all load across the cut. Below the cut the two bars are again attached and must strain equally. Consequently the bottom halves of the bar again share the load equally.

The attachment sets the condition for approximately equal strain and equal stress in both bars, almost all the way to the cut. Close to the slit, the load of the left bar must be transferred to the right bar which, over a short distance, will then carry the total load. The nominal stress in the right bar in the section of the slit is then P/A. However, the load from the left must be transferred to the right and back over such a small distance that the additional load cannot be distributed evenly. Instead, most of the extra load will be carried by a small portion of the right half cross section, so that higher stresses occur close to the cut (Figure 2.2c): there is a stress concentration at the cut. It does not make a

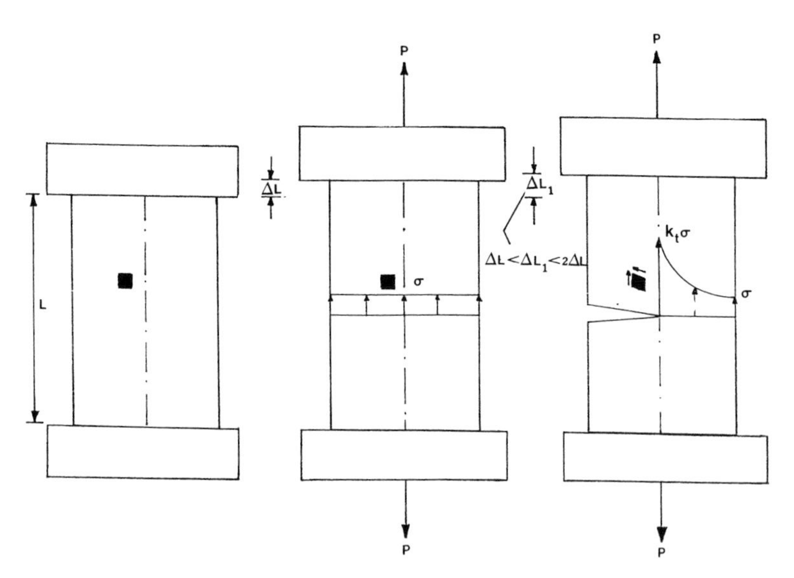

Figure 2.2. Effect of cutting one of two welded bars or half of one (wider) integral bar.

difference whether the bars are welded together or whether they are one piece, cut until midway. Transfer of the load from the cut half to the right half takes place by shear. This can be appreciated by imagining how a material element of the left deforms in the area of the slit as depicted in figure 2.2.

As long as the bar is intact, the strain is uniform and the total elongation is ΔL. If the bar is half-way cut the stress and strain will still be uniform in the extreme upper and lower portions; they will be the same as in the bar without the cut. Only in the area of the slit, where the total load must be carried by half of the bar, the stresses and strains are higher. Hence, the total elongation will be somewhat larger than ΔL, but much less than $2\delta L$ (compare Figures 2.1 and 2.2).

It is helpful to consider load-path (load-flow) lines: imaginary lines indicating e.g. how one unit of load is transferred from one loading point to the other (Figure 2.3). For uniform load the flow lines are straight and equally spaced, indicating that the load is evenly distributed (uniform stress). If the load path is interrupted by a cut, the flow lines must go around this slit within a short distance as shown in Figure 2.3. Were the load lines rubber hoses and the cut a wedge, a similar pattern would develop.

At the tip of the cut the flow lines are closely spaced, indicating that more load is flowing through a smaller area, which means higher stress. Load flow lines are also useful for obtaining a rough indication of the direction of stress. In by-passing the cut the flow lines are bent, i.e. the load changes direction. The direction of the load flow line is an indication of the direction of the local tensile

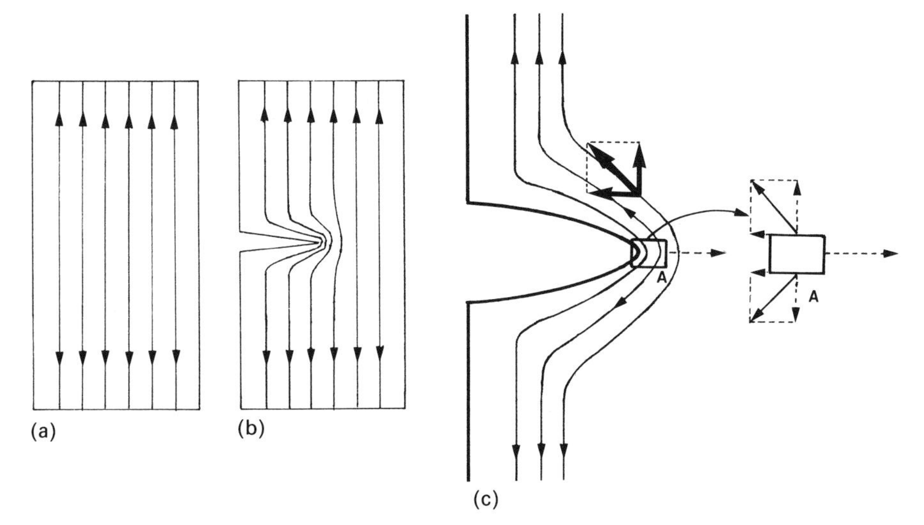

Figure 2.3. 'Load-flow' lines.

stress, as indicated in Figure 2.3.c. It appears that the direction of load around the notch is not the same as in the uniformly loaded part: the local stress has a vertical as well as a horizontal component, and it must be concluded that in the vicinity of the cut the stress field is biaxial (σ_y and σ_x) while the applied load is uniaxial. In the absence of the cut the stress field is uniform, the state of stress uniaxial throughout. Due to the slit a biaxial stress field develops locally. Not only does the slit cause a stress concentration, it also gives rise to a transverse stress.

2.3. Stress concentration factor

In the case of a notch (instead of a sharp cut) the situation is very similar. Every discontinuity forms an interruption of the load path, will therefore deviate the load-flow lines and, hence, cause a stress concentration (Figure 2.4). If the notch is blunt, e.g. a round hold, its dimension in the direction of load is larger, causing and earlier deviation of the load-flow lines. Redistribution of this load can then take place over a larger distance. As a consequence, the area of stress concentration is more extended than in the case of a cut, and the highest stresses are less.

As a general rule, blunt notches produce lower local stresses, sharp notches cause higher local stresses. The highest local stress σ_l is a number of times higher than the nominal stress σ_{nom}. The ratio between local stress and nominal stress is called the theoretical stress concentration factor, denoted as k_t (elastic stresses). The local stress is

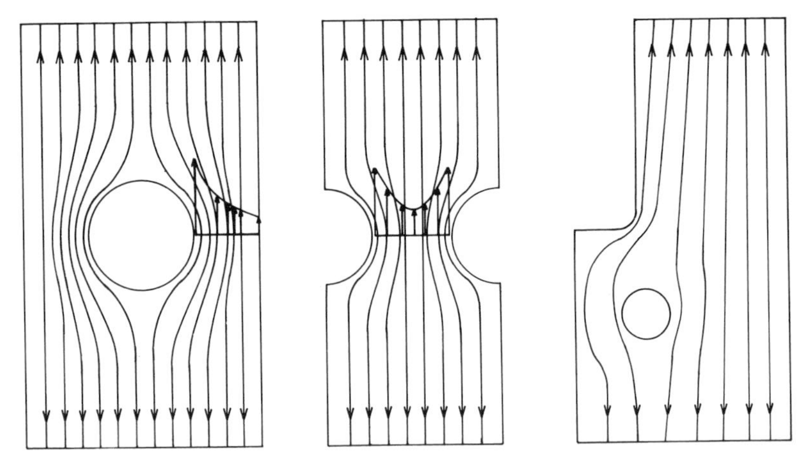

Figure 2.4. 'Load-flow' lines around notches and consequent stress concentrations.

$$\sigma_l = k_t \sigma_{\text{nom}}. \tag{2.1}$$

The nominal stress is not always defined in the same manner. Sometimes one uses the (uniform) stress in the full section away from the notch, sometimes the average stress in the section through the notch. As the local stress is the same in either case, the use of a different definition for the nominal stress leads to different values of the stress concentration factor. This presents no problem as long as the proper combinations are used to obtain the local stress.

For an elliptical notch (Figure 2.5), the stress concentration factor is:

$$k_t = 1 + 2\frac{b}{a} \tag{2.2}$$

where a and b are defined as in Figure 2.5. The radius of curvature, ϱ, of the ellipse at the end of the transverse axis is $\varrho = a^2/b$, so that Equation (2.2) can also be written as:

$$k_t = 1 + 2\sqrt{\frac{b}{\varrho}} = 1 + \alpha\sqrt{\frac{b}{\varrho}}. \tag{2.3}$$

In the case of a circle, $b = a$, (and $b = \varrho = R$; R being the radius of the hole) so that the stress concentration factor of a circular hole is equal to $k_t = 3$.

Stress concentration factors for many notch shapes have been determined. They can be found in handbooks, the best known being the one by Peterson [1]. Equation (2.3) is a more or less general form for the stress concentration factor if b is interpreted as a relevant geometrical dimension (α depends upon notch geometry). The equation shows the large effect of the notch-root radius ϱ; the

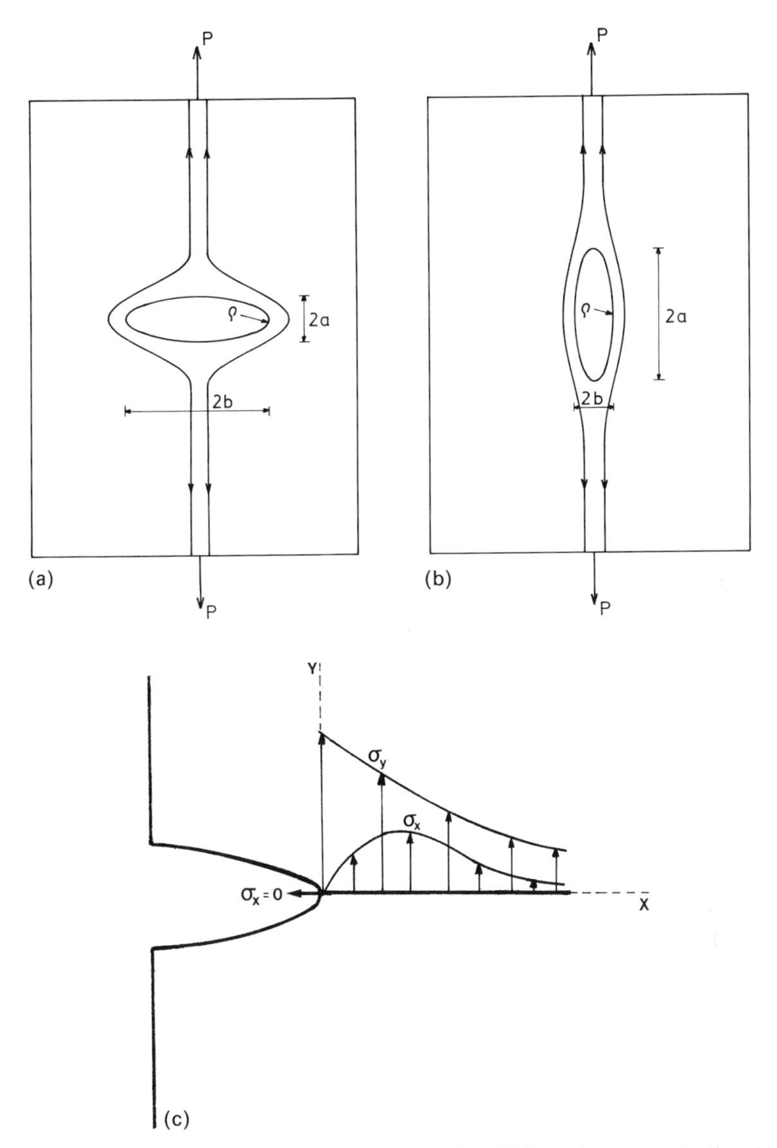

Figure 2.5. Elliptical notches. (a) High stress concentration; (b) Low stress concentration; (c) Stress free notch faces.

sharper the notch (smaller ϱ), the larger is k_t. For an ellipse with $b/a = 3$ ($\varrho = b/9$) the stress concentration factor is $k_t = 7$, while for an ellipse with $a/b = 3$, the stress concentration is only $k_t = 1.67$.

2.4. State of stress at a stress concentration

As was shown in Section 2.2, a stress concentration also causes a change in the state of stress; even if the stress is uniaxial throughout the remainder of the body, the state of stress in the area of the notch will be at least biaxial (Figure 2.3). At the free surface where no external loads are acting the state of stress will be plane stress (there are no stresses on a free surface). Since the free surface carries no shear either, it is a principal plane with a principal stress equal to zero. A state of stress in which one of the principal stresses is zero is a state of plane stress.

The face of the notch (Figure 2.5c) is a free surface: it carries no stress (plane stress). The root of the notch (if there is a radius) is also a free surface. This means that at the free surface of the notch root σ_x must be zero, because there is no stress on (perpendicular to) that free surface. Slightly inwards from the notch root, however, σ_x will be non-zero. The faces of the plate are still stress free, so that $\sigma_z = 0$. The state of stress is biaxial (plane stress). At some small distance from the notch root, the state of stress may be triaxial as explained below.

Because the surface (face of plate) is a principal plane, and because the three principal planes are mutually perpendicular, it follows that the stresses σ_x and σ_y at the notch root in the plane of the notch are the principal stresses σ_1 and σ_2. Due to the stress concentration the local values of σ_1 and σ_2 are very high and so are the strains ε_1 and ε_2. According to Hooke's law, this will lead to a strain in Z-direction, given by:

$$\varepsilon_z = -\nu \frac{\sigma_x}{E} - \nu \frac{\sigma_y}{E} \tag{2.4}$$

assuming that the stress $\sigma_z = 0$. This negative strain, ε_z, indicates a thinning of the plate.

The stresses are high only in the vicinity of the notch root. Further away σ_x vanishes and σ_y is much lower. Hence further away ε_z would become very small. The faces of the notch are stress free, so that $\varepsilon_z = 0$ along the faces. It appears that there will only be a small amount of material at the notch root for which ε_z should be very large, while around it ε_z is either zero (notched faces), or small (further inwards). Assuming that the material wanting to undergo large ε_z is approximately a cylinder (roll), the situation is as shown in Figure 2.6.

If the roll is very long (large thickness) and thus relatively thin, such a large ε_z cannot take place. The surrounding material, being attached to this roll, will not permit the contraction to occur, apart from a little bit at the face of the plate. Imagine that the roll of material is a steel bar cast in concrete, where the concrete represents the surrounding material. If the steel is cooled, it wants to contract. However, contraction will be prevented by the concrete, which would lead to

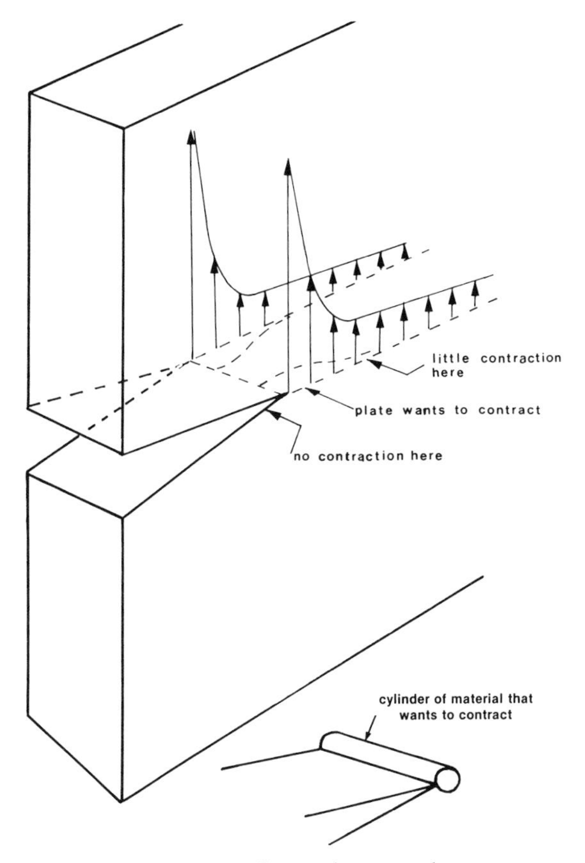

Figure 2.6. Contraction at notch.

(thermal) tension stresses in the bar. The same happens to the roll of material in Figure 2.6: contraction is prevented by the surrounding material and a tensile stress develops in the roll.

Apparently, constraint of the contraction causes a tension σ_z in the Z-direction. Should the contraction be completely constrained then the strain ε_z is zero. Writing the complete equation for ε_z provides:

$$\varepsilon_z = \frac{\sigma_z}{E} - \nu \frac{\sigma_x}{E} - \nu \frac{\sigma_y}{E} = 0 \tag{2.5}$$

and then:

$$\sigma_z = \nu (\sigma_x + \sigma_y) \approx 0.3 (\sigma_x + \sigma_y) \tag{2.6}$$

Equation (2.6) shows that a high tension develops in thickness direction when no contraction is permitted. This stress is exerted on the roll by surrounding material acting to constrain the contraction. At the free surface this stress cannot exist, but it builds up rapidly, going inward (Figure 2.7). Due to the absence of a σ_z at the surface, an ε_z occurs there, so that a small dimple develops at the surface.

It follows then that at the surface, with stresses only in X and Y-direction, there is a biaxial state of stress (plane stress). Further inward there is a triaxial state of stress. Should there be complete constraint then this triaxial state of stress is plane strain, because $\varepsilon_z = 0$ for complete constraint. (A state of stress where one of the principal strains is zero is called plane strain).

Now consider a thin plate with a notch as in figure 2.7b. In this case the roll

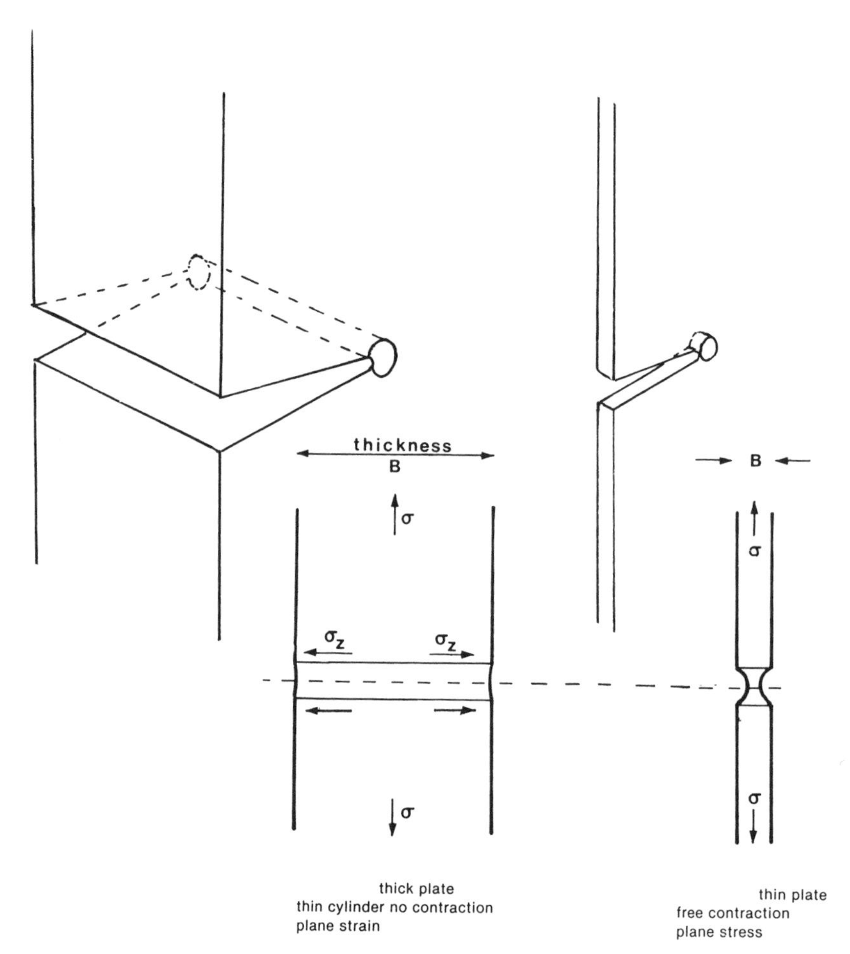

Figure 2.7. Contraint of long, thin cylinder (large thickness) and free contraction of short, thick cylinder (small thickness).

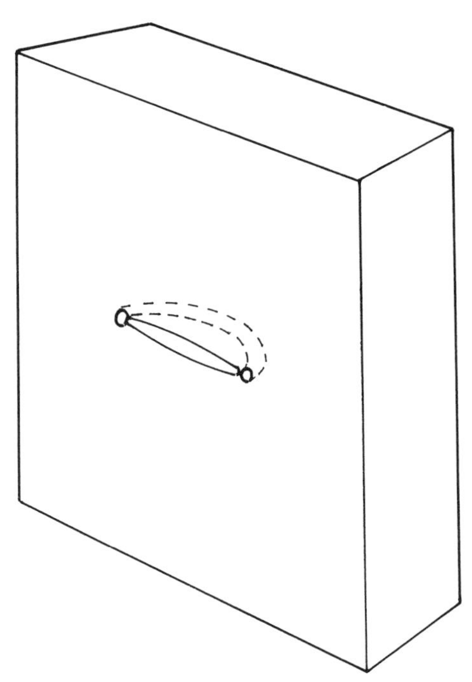

Figure 2.8. Length of contracting cylinder bears no relation to thickness in case of surface notch.

of material wanting to undergo contraction is short and relatively thick. contraction can occur freely and will be in accordance with Equation (2.4). The stress in thickness direction will be zero ($\sigma_z = 0$; plane stress).

In cases between those of Figure 2.7a and b, there will be some, but not complete constraint. These transitional cases have a triaxial state of stress, but not one of plane strain.

Apparently the state of stress at the notch root depends upon the length of the roll. As the length of the roll is equal to the thickness, one can argue that the state of stress depends upon thickness. However, this dependence upon thickness is a coincidence. If the notch is of the type of Figure 2.8, the length of the roll has no relation to the thickness; in such a case the thickness is irrelevant for the state of stress. Although this may seem trivial, its importance is emphasized; in the use of fracture mechanics serious errors are possible if it is assumed that the state of stress is always dictated by thickness.

2.5. Yielding at a notch

The stress required for plastic deformation depends strongly upon the state of stress. In plane stress yielding occurs when the highest principal stress is approximately equal to the yield strength, but much higher stresses are required in the case of a triaxial state of stress.

For readers not familiar (any more) with yield criteria, the following brief summary may be of help. Yielding is plastic deformation which takes place by slip; it is therefore caused by shear stresses. Plastic deformation will not take place unless the shear stress is sufficient to cause slip. In a tensile bar (uniaxial tension) the state of stress is as shown in Figure 2.9a. Yielding occurs when $\sigma = F_{ty}$ where F_{ty} is the uniaxial yield strength as measured in a tensile test. Since plastic deformation takes place in the case of Figure 2.9a, the shear stress

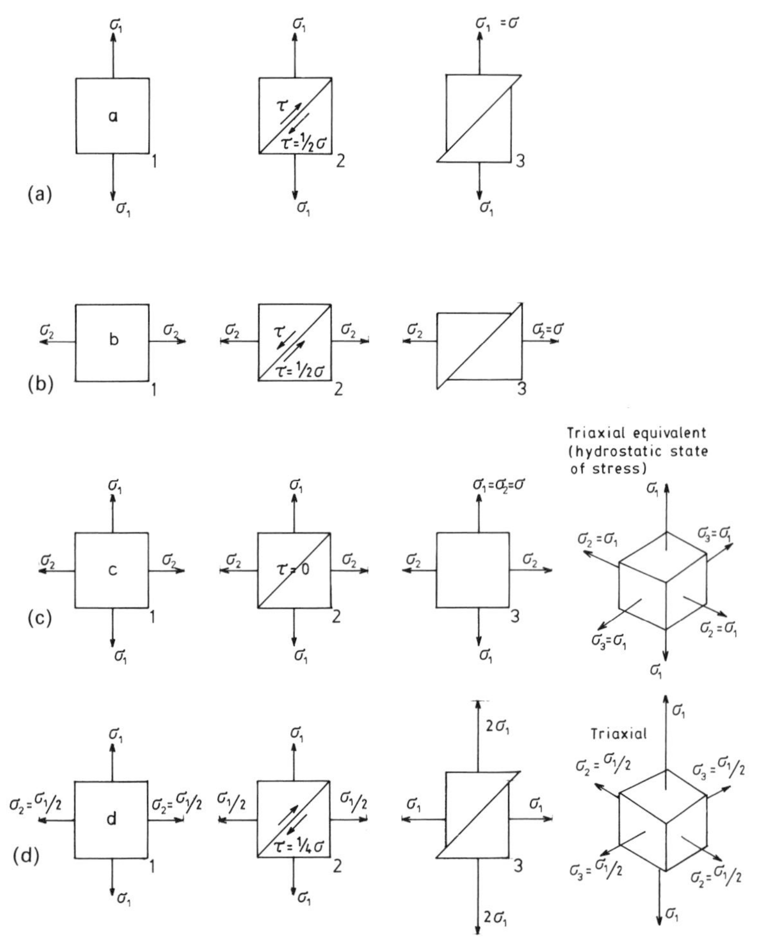

Figure 2.9. Postponement of plastic deformation due to combined stresses. (a) Uniaxial (longitudinal); shear followed by slip (plastic deformation); (b) Uniaxial (transverse); shear followed by slip; (c) Biaxial; no shear, no slip, no plastic deformation; (d) Biaxial ($\sigma_2 = \sigma_1/2$; plastic deformation, but not until stress is twice as high as in case a.

required for slip must have been exceeded. The maximum shear stress acting on a plane at 45 degrees is $\tau = \sigma/2$. Apparently then the shear stress required for slip (yield) equals $\tau_{ty} = F_{ty}/2$.

Figure 2.9b shows a similar case of uniaxial loading; again yielding occurs when the shear stress equals $F_{ty}/2$. Next consider Figure 2.9c assuming for the moment that the material is two-dimensional (existing in the plane of the figure only). Since there is equal tension in both directions the two shear stresses of Figures 2.9a and b cancel, so that there is no shear stress in the case of Figure 2.9c. Without shear stress there can be no slip and no plastic deformation. In order for slip to occur one of the stresses must be larger than the other by F_{ty}, in which case the same shear stress will be present as in Figure 2.9a. For example, if the horizontal stress is F_{ty}, the vertical stress must be $2F_{ty}$ for yielding to occur, so that the highest stress must be twice the yield strength before plastic deformation begins. If the horizontal stress were $3F_{ty}$, the vertical stress would have to be $4F_{ty}$ before the shear of Figure 2.9a would be restored. In both of these examples the difference between the two principal stresses is F_{ty}.

Naturally, a real material is three dimensional as in Figure 2.9d, and out of plane shear stresses are possible. But it is easy to see now that there still will be no shear when all three principal stresses are equal (Mohr's circle becoming a point), and hence, there will be no plastic deformation. Apparently, yielding requires that the difference between the largest and smallest principal stress is equal to F_{ty}, because only in that case will there be a shear stress $\tau = F_{ty}/2$ as required for yield. Thus, yielding (slip) will take place when ($\tau = F_{ty}/2$):

$$(\sigma_1 - \sigma_3) = F_{ty} \quad \text{or} \quad (\sigma_1 - \sigma_2) = F_{ty} \tag{2.7}$$

depending upon which is lower σ_2 or σ_3. Were for example $\sigma_2 = \sigma_3 = 0.8\,\sigma_1$, then it would follow from Equation (2.7) that the stress required for yielding is:

$$\sigma_1 - 0.8\,\sigma_1 = F_{ty} \quad \text{or} \quad \sigma_1 = 5F_{ty}$$

i.e. the stress would have to be 5 times the yield strength for plastic deformation to occur. If $\sigma_1 = \sigma_2 = \sigma_3$, the shear stress is zero, and yielding will never occur, regardless how high the stress. The above is the Tresca yield criterion. Instead, the Von Mises yield criterion is more generally used, but its results differ very little from those of Tresca. As it makes no difference for the essence of the discussion, the Tresca criterion is used here, because it is the easiest to understand.

The above has important repercussions for the yielding at a notch, since the state of stress may be triaxial. As an example, consider a case of plane strain, where the third stress σ_z is given by Equation (2.6). Note again that for a material element in the section through the notch, the principal stresses σ_1, σ_2, σ_3 are σ_x, σ_x, σ_z respectively. Consider a case in which $\sigma_x = \sigma_y$ as would occur at a very sharp notch. Given that $\nu \approx 0.33$, the lowest principal stress $\sigma_3 = \sigma_z$

would be $\sigma_z = 0.33\ (\sigma_x + \sigma_y) = 0.66\ \sigma_y$. Plastic deformation would then require a stress $\sigma_y = 3F_{ty}$ according to Equation (2.7).

Let Figure 2.10 represent the notch and the distribution of σ_y, on the section through the notch. At the notch root the local stress is σ_y. At low loads the local stress is less than $3F_{ty}$, and therefore the strains remain elastic. Further increase of the load will raise all stresses (Figure 2.10b) until $\sigma_y = 3F_{ty}$, upon which plastic deformation commences.

Next consider a notch in a thin plate with plane stress as in Figure 2.11. In this case plastic deformation occurs when $\sigma_y = F_{ty}$, because $(\sigma_y - \sigma_z) = (\sigma_y - 0) = \sigma_y$. During further increase of the load the stress distribution develops in a similar manner as in the previous case. Eventually the entire remaining section will yield, unless fractur occurs earlier.

It is important that the notch tip stresses are much higher in plane strain than in plane stress. In the latter case they are limited to F_{ty}, in the former case to $3F_{ty}$. Thus a plane strain condition is more severe and can more easily lead to fracture and cracks as will be shown in detail in later chapters.

If the notch is blunt, the stress $\sigma_x(\sigma_2)$ cannot exist at the notch root (Figure 2.5). Although there could be a σ_z (σ_3), the state of stress would still be plane stress, because of $\sigma_2 = 0$. Hence, at the root of the notch the stress σ_y would be limited to the yield strength F_{ty}. Further inwards σ_x will exist; in the thick plate with constraint, the state of stress will then be triaxial and possibly plane strain. Further inward therefore the stress might reach as high as $3F_{ty}$, as shown in Figure 2.12, before yielding occurs.

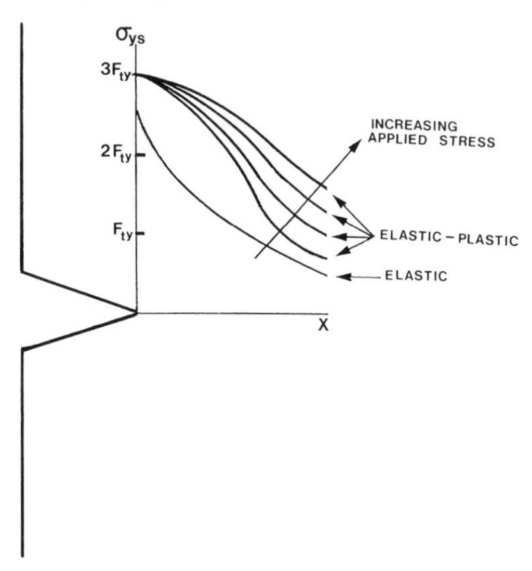

Figure 2.10. Yielding at sharp notch in plane strain.

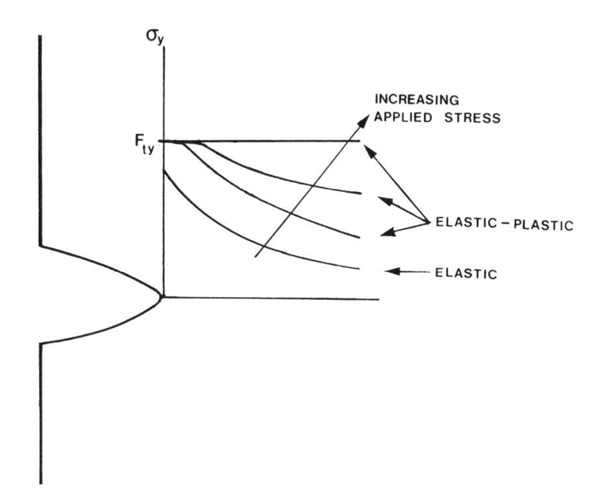

Figure 2.11. Yielding at notch in plane stress.

Assuming the material does not exhibit strain hardening (horizontal stress-strain curve beyond yield), the stress cannot increase further after yielding occurs. Away from the notch there is again a uniaxial state of stress ($\sigma_x = \sigma_z = 0$) so that the stress will again be limited to F_{ty}. Hence, by the time the entire section is yielding (fracture may occur before this can happen) the stress distribution in the section is as shown in Figure 2.12. Whether there is plane stress or plane strain at the notch, the final stress distribution will be about the same (compare Figures 2.11 and 2.12 and note that triaxial stress area is very small), apart from a small area where the stress reaches $3F_{ty}$ in the plane strain case. Note that these stress distributions may not be reached if fracture occurs before the entire ligament yields.

2.6. Plastic collapse at a notch

Not only may the high local stresses cause cracks by fatigue, stress corrosion (or creep) which may eventually lead to a fracture, they may cause fracture to proceed immediately from the notch, in particular when the notch is sharp (e.g. crack). Alternatively, failure can occur by plastic collapse which is always followed by fracture. If the stress distributions of figure 2.11 or 2.12 can be reached before fracture then plastic collapse could occur. Obviously, with the given stress–strain curve without work hardening, the cross section with the notch cannot carry any more load once the entire cross section is yielding, because the yielding will continue unihibited until fracture results. This is called plastic collapse. Thus, in plane stress where the stress in the entire cross section is equal to yield strength at the time of collapse, (Figure 2.11) the maximum load

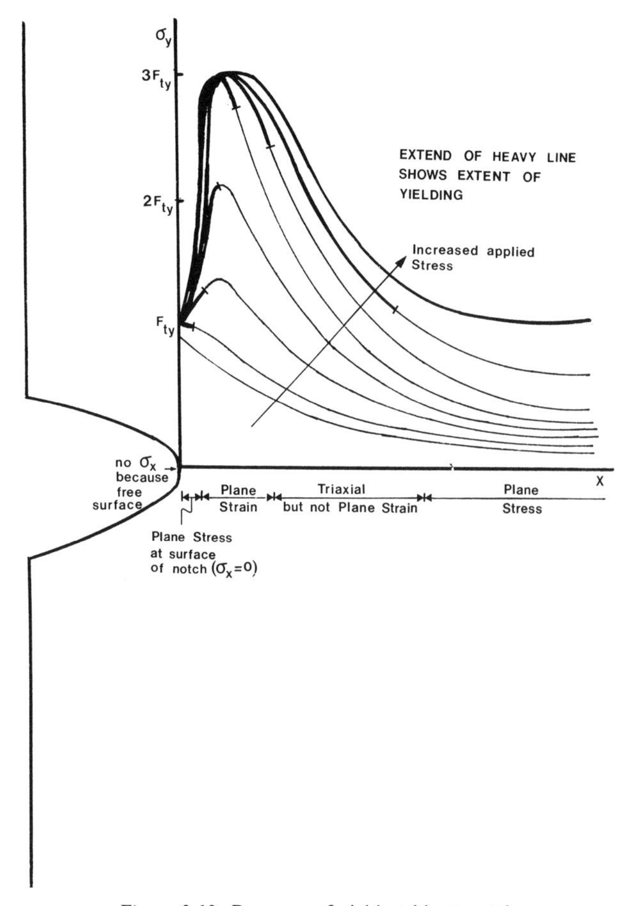

Figure 2.12. Progress of yield at blunt notch.

carrying capability is:

$$\text{Collapse: } P_{\max} = B(W - a)F_{ty} \tag{2.8}$$

if a is the notch depth, W the total width and B the thickness. This failure load is called the collapse load or limit load. The nominal stress in the full width part is $\sigma = P/BW$. Hence, the part in Figure 2.11 fails when the nominal stress is:

$$\text{Collapse: } \sigma_{fc} = P/W = \frac{W - a}{W} F_{ty}. \tag{2.9}$$

This is the equation of a straight line (as a function of the notch depth a. If $a = W$, failure will occur already when the nominal stress is $\sigma_{fc} = 0$. This failure stress is plotted as a function of notch depth in Figure 2.13. If fracture indeed

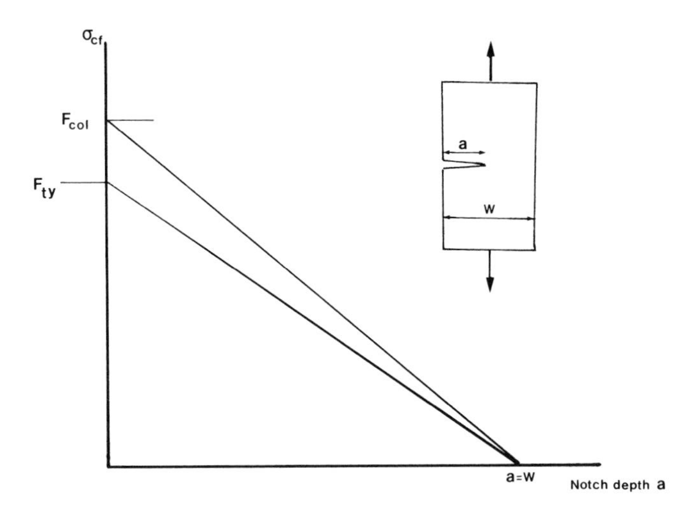

Figure 2.13. Net section collapse.

occurs as a consequence of collapse, the strength, σ_{fc}, would be the residual strength as defined in Chapter 1.

If the material work hardens, the notched cross section can carry a higher load. In general however, it cannot reach a situation where the entire cross section carries a stress equal to the tensile strength. This can be understood if it is realized that the strains in the cross section are not uniform. Yielding starts at the notch root (Figure 2.11) and proceeds through the ligament. Hence, the strains at the notch root are always much higher than elsewhere. This is depicted in Figure 2.14. Clearly, even though after yielding the stress is almost uniform, there is still a strain concentration at the notch.

As long as the stress is elastic there is a stress concentration given by $\sigma_l = k_t \sigma_{\mathrm{nom}}$. In the elastic case, $\sigma = \varepsilon E$. therefore the strain concentration will be:

$$\varepsilon_l = k_t \frac{\sigma_{\mathrm{nom}}}{E} = k_t\, \varepsilon_{\mathrm{nom}} = k_\varepsilon \varepsilon_{\mathrm{nom}} \tag{2.10}$$

where the strain concentration, k_ε, is equal to the theoretical stress concentratin, $k_\varepsilon = k_t$. When the whole ligament is yielding, the stress concentration has disappeared ($k_\sigma = 1$ in Figure 2.14b), but there is still a strain concentration. Neuber [2], writes for the stress concentration factor k_σ, and for the strain concentration factor k_ε, and postulates that:

$$k_\sigma k_\varepsilon = k_t^2. \tag{2.11}$$

In this equation k_t stands for the theoretical stress concentration factor, as discussed. In the elastic case $k_\sigma = k_\varepsilon$, so that both are then equal to k_t. The

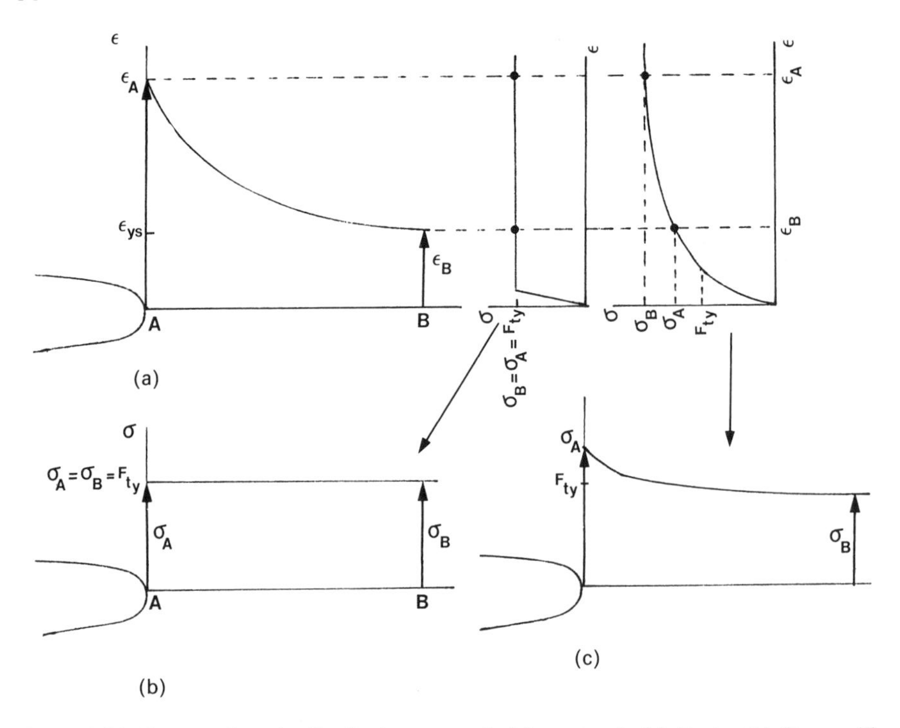

Figure 2.14.. Stress and strain distribution at notch (plane stress). (a) Strain; (b) Stress with horizontal stress-strain curve; (c) stress with rising stress-strain curve.

stress concentration factor, k_σ, is reduced when the material yields, but the strain concentration increases. Taking the case of Figure 2.14b where k_σ has become $k_\sigma = 1$, the strain concentration factor has become $k_\varepsilon = k_t^2/k_\sigma = k_t^2/1 = k_t^2$. In the case of e.g. $k_t = 3$, the stress concentration decreases from $k_\sigma = k_t = 3$ in the elastic case to $k_\sigma = 1$ in the fully plastic case; the strain concentration on the other hand increases from $k_\varepsilon = k_t = 3$ in the elastic case to $k_\varepsilon = k_t^2 = 9$ in the plastic case.

Now consider a work hardening material in which the stress in the ligament could be raised somewhat above F_{ty}. Some stress concentration will remain (Figure 2.14c) but the strain concentration will become very high. As the whole ligament is plastic, even the strain at the plate edge is above yield, but at the notch the (plastic) strain is many times higher. Eventually the high notch root strain will cause fracture long before the stress further away has reached values much above F_{ty}. Once fracturing has begun a much sharper notch has formed, so that the situation becomes worse and fracture continues.

Given the strain distribution and the stress–strain curve, the stress distribution can be sketched as shown in Figure 12.14. When tearing or plastic collapse commences at the notch tip, the stresses in most of the ligament are still close to F_{ty}, because the strain gradient is very steep. Thus, even in a strain hardening material and plane stress, the (average) stress in the cross section cannot reach F_{tu}, where F_{tu} is the tensile strength (see stress–strain curve in Figure 12.14). Apparently, collapse will occur at an average ligament stress somewhat higher than F_{ty} but less than F_{tu}. The average ligament stress at which collapse occurs is called the collapse strength F_{col}. Note that for a non-workhardening material $F_{col} = F_{ty}$, and that at best $F_{col} = F_{tu}$. It is not very well possible to determine the values of F_{ty} and F_{tu} other than by a (tensile) test; similarly, it is not very well possible to determine F_{col} other than by a test on a notched sample. However, its value depends upon the severity of the notch (k_σ and k_ε). For circular holes with $k_t = 3$, the collapse strength is often close to the tensile strength ($F_{col} \approx F_{tu}$), but for a sharp crack, the value is very close to the yield strength ($F_{col} \approx F_{ty}$).

It follows that for a work hardening material Equation (2.9) changes into:

$$\sigma_{fc} = \frac{W - a}{W} F_{col} \tag{2.12}$$

which is also a straight line as a function of notch depth (Figure 2.13).

If there is a plane strain, or in general non-plane-stress, the stress distribution after yield is not uniform (Figure 2.12). Since the stress peak is local, the average stress in the section cannot become much higher than in the case of plane stress. In most cases — depending upon notch acuity – Equation (2.12) still applies.

It should be noted that the above discussion was strictly for the case of uniform applied loading. If there is bending (or other stress gradients in the applied stresses), the conditions for collapse are slightly more complicated. The fully plastic stress distribution in the ligament for the case of bending is shown in Figure 2.15c. The maximum bending moment occurs when the net setion stress is equal to F_{col}. Taking the moment around point A one obtains:

$$M_{max} = 2F_{col} B \left(\frac{W - a}{2}\right)\left(\frac{W - a}{4}\right) = \frac{1}{4} F_{col} B(W - a)^2. \tag{2.13}$$

Using the collapse strength of e.g. 50 ksi and $W = 2$ in, $B = 0.5$ in (Figure 2.15) and $a = 0.5$ in the collapse moment is:

$$M_{max} = \tfrac{1}{4} \times 50 \times 0.5\,(2 - 0.5)^2 = 14.1 \text{ inkips} = 14\,100 \text{ inlbs.}$$

One can establish the entire collapse curve by solving Equation (2.13) for various crack sizes, as shown in Figure 2.15d.

Analysis of combined bending and tension requires determination of the point of stress reversal, D, in Figure 2.15e. Since the stresses in the cross section have to be in equilibrium with the external load, one can establish 2 equilibrium

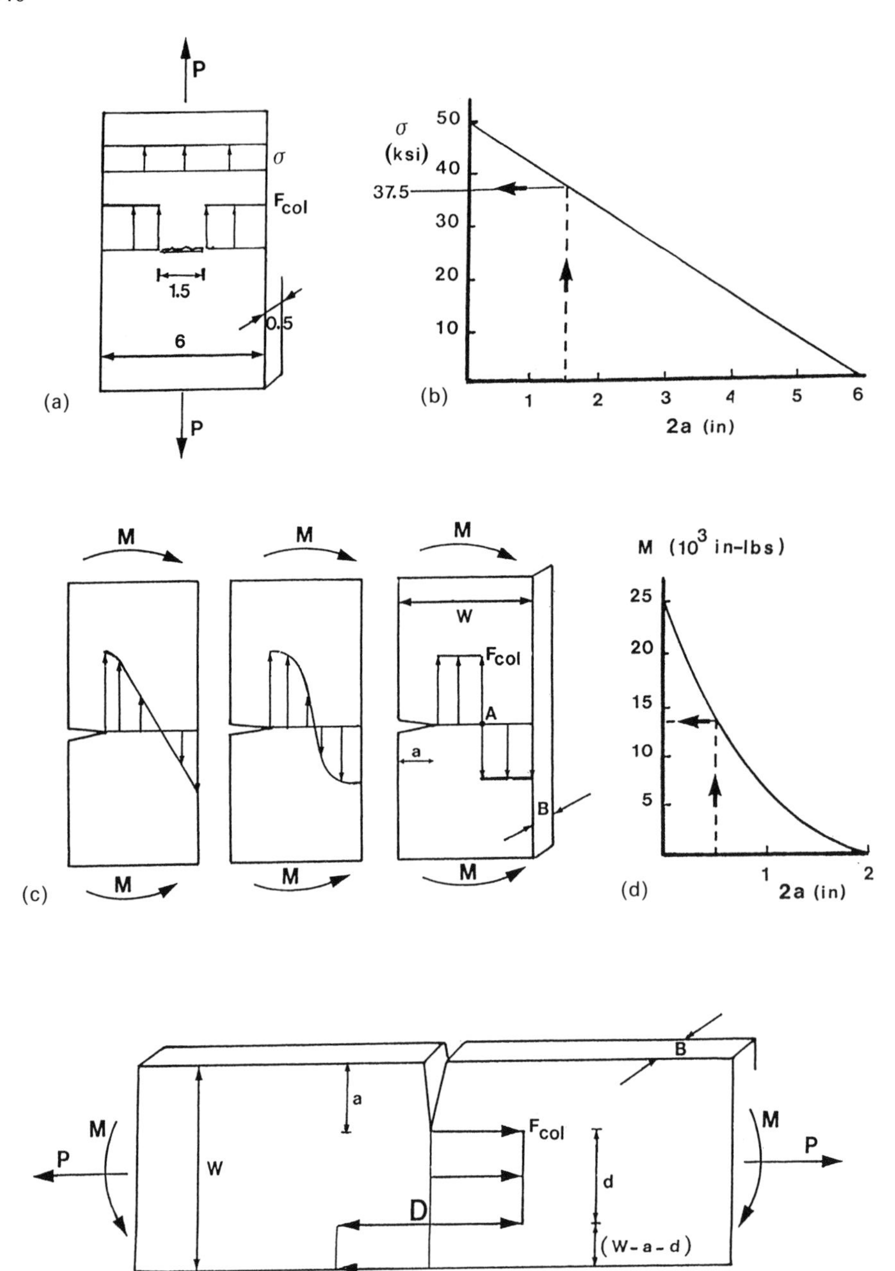

Figure 2.15. Collapse analysis. (a) Center crack; (b) Collapse stress; (c) Stress distribution for increasing M. (d) Collapse strength; (e) Combined bending and tension.

equations, one for P and one for M. For a given P one finds D and then M by solving the equilibrium equations. This is shown in Table 2.1.

Generally speaking, in engineering all solutions must be obtained in terms of the nominal applied stress, as was done going from Equation (2.8) to (2.9). For the bending case the same can be done, because the nominal bending stress at M_{max} is $\sigma = 6\ M_{max}/BW^2$. Substitution in Equation (2.13) leads to:

$$\sigma_{fc} = \frac{3}{2}\left(\frac{W-a}{W}\right)^2 F_{col}. \tag{2.14}$$

Note that in contrast to Equations (2.9) and (2.12), the Equation (2.14) is not a straight line, but a curve (Figure 2.15d). For complete derivation see Table 2.1.

2.7. Fracture at notches: brittle behavior

The collapse load or limit load is the highest load that can ever be reached, i.e. the absolute maximum load carrying capability is defined by the collapse load. Fracture follows automatically when the collapse conditions are reached. However, fracture may occur already before the collapse conditions are attained. (This is the concern of fracture mechanics analysis as discussed in Chapters 3 and 4). In cases where fracture occurs due to collapse, the residual strength (Chapter 1) is determined by Equation (2.9) or (2.14): $\sigma_{res} = \sigma_{cf}$. The residual strength can never be higher. If fracture occurs before collapse, the residual strength is lower than the nominal stress at collapse ($\sigma_{res} < \sigma_{cf}$).

Strains and, up to point, stresses are higher at the notch than anywhere else. In the case of very high k_t this may lead to local fracture long before the remainder of the section reaches collapse. As a matter of fact, with very sharp notches (cracks) the condition for fracture at the notch root may already be met while virtually the entire notched section is still elastic. Once fracturing starts, the notch has beome longer and sharper, so that the fracture proceeds throughout the ligament. Cracks due to fatigue or stress corrosion constitute very sharp notches; in some cases a cracked part will be capable of reaching the collapse load, but in many cases fracture occurs at much lower loads (stresses). This problem will be discussed in detail in Chapters 3 and 4. Nevertheless, collapse, if it occurs before the fracture condition is reached, will cause failure.

According to Equation (2.9) collapse occurs when the stress in the full section has not yet reached the yield stress. Thus all plastic deformation is confined to the section through the notch. If the collapse strength is higher than the yield strength Equation (2.12) applies. In that case the stress in the full section could reach or exceed the yield if the collapse stress is high and the notch size small. In general however, the stress at failure in the full section will still be below yield. Thus, in the vast majority of fractures occuring at notches and cracks, plastic deformation takes place in the notched section only, in particular if fracture occurs before collapse conditions can be attained (Chapters 3 and 4).

Table 2.1. Collapse conditions for cases with bending

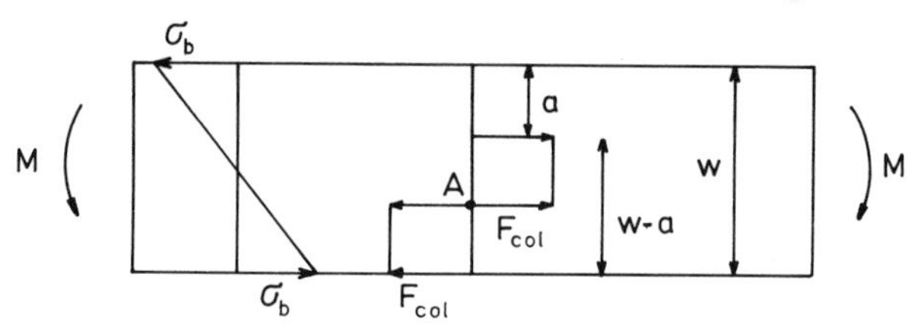

Pure bending: unit thickness:

$$\sigma_b = \frac{M}{I} = \frac{6M}{W^2} \tag{a}$$

From Moment equilibrium around A:

$$\text{Collapse: } M = 2\frac{W-a}{2} \times \tfrac{1}{4}(W-a)\,F_{col}$$

$$M = 0.25\,(W-a)^2\,F_{col}. \tag{b}$$

Substitute (a) in (b) for ($M = \sigma_b W^2/6$):

$$(\sigma_b)_{fc} = 1.5\left(\frac{W-a}{W}\right)^2 F_{col}. \tag{c}$$

Note: quadratic curve; also $\sigma_{fc} = 1.5\,F_{col}$ for $a = 0$, collapse curve starts at $\sigma = 1.5\,F_{col}$ (not at F_{col} as in tension).

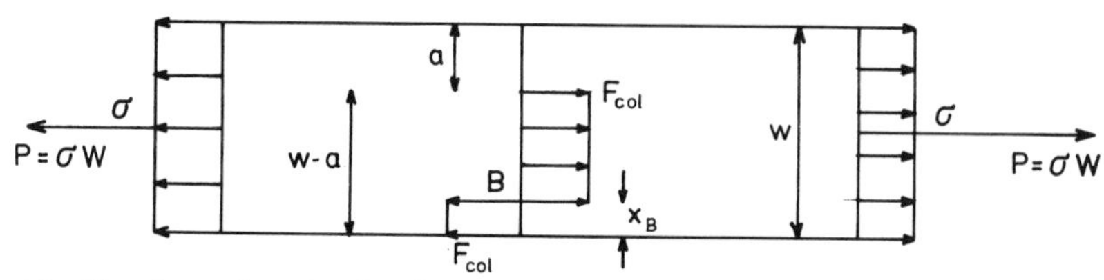

Combined bending and tension (bending due to excentricity; unit thickness, total load is $P = \sigma W$ in center; hence eccentricity with respect to cracked section.
Point B (i.e. X_B) is unknown:
Horizontal equilibrium:

$$\sigma W = (W - a - x_B)F_{col} - x_B F_{col} \tag{d}$$

Moment equilibrium point B:

$$\sigma W\left(\frac{W}{2} - x_B\right) = \frac{(W-a-x_B)^2}{2}F_{col} + \frac{x_B^2}{2}F_{col} \tag{e}$$

Solve (d) and (e) to obtain $x_B = 0.5\,W - 0.25\sqrt{4W^2 - 8a(W-a)}$, and more important: $\sigma_{fc} = (a/W + 0.5\sqrt{4 - 8a(W-a)/W^2})\,F_{col}$.

The limitation of yielding to the notched section has considerable effect on the total elongation at the time of failure. In a normal tensile test on an unnotched sample the strain at the time of necking is often on the ordr of 10% or more. Thus the bar has become 10% longer (Figure 2.16). During necking still higher strains occur. In a notched sample plastic deformation occurs only in the notched section. Even if this small section would deform 10% before fracture, the total elongation would be much smaller than of the unnotched bar (Figure 2.16). Consequently, elongation after fracture is hardly noticeable. If little overall plastic deformation occurs, the fracture is called 'brittle' from an engineering point of view. The text in this book strictly adheres to the engineering use of the word. In that sense virtually all service fractures are brittle: fractures occur almost always at notches or cracks. Thus plastic deformation is always restricted to the notched section and the fractures are brittle in the engineering sense. The fracture mechanism may be either by dimple rupture (often called ductile) or by cleavage (often called brittle), but the great majority of fractures occurs by rupture; from an engineering point of view they are brittle because of little overall deformation in the cracked section (Figures 2.16).

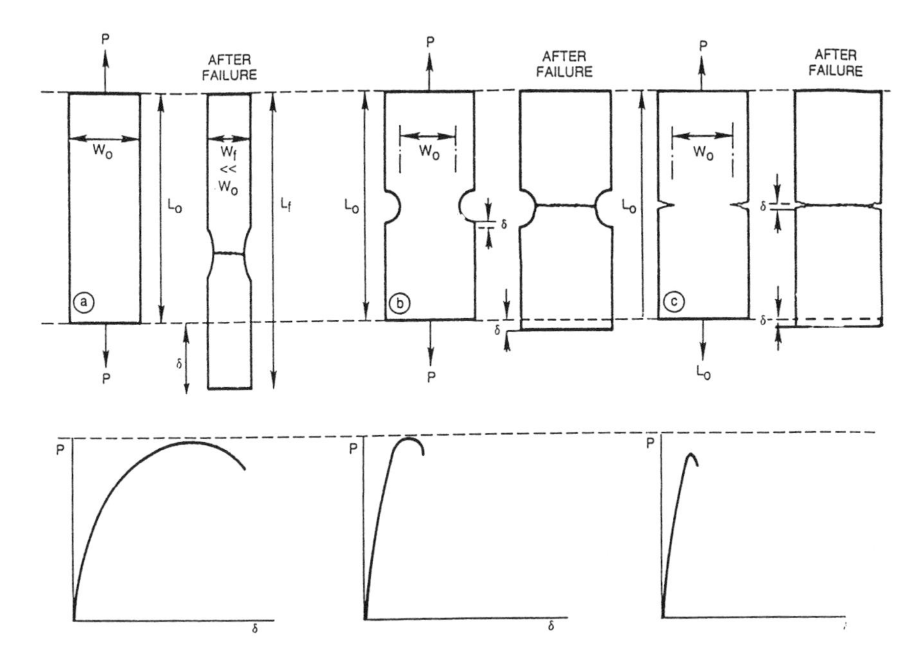

Figure 2.16. Apparent 'brittleness' due to notches and cracks as a consequence of localized plastic deformation in area of high stress.

2.8. Measurement of collapse strength

The easiest way to measure the collapse strength, F_{col}, is by means of a center cracked panel. In view of the fact that constraint will still play a role, the specimen should have a thickness close to the thickness used in the structure of interest. As this may lead to unwieldy specimens, a compact tension specimen (Chapter 3) can be used where circumstances so dictate, provided it is analyzed for combined bending and tension (Table 2.1).

It should be noted here, that in the case of center cracks, the definition of crack size is 2a instead of a. This is a convention, by which all cracks with one tip are defined as a, all cracks with two tips as 2a (see also Chapter 3). Note that with this definition Equations (2.9) and (2.12) become for the center cracked panel:

$$\sigma_{fc} = \frac{W - 2a}{W} F_{col}. \tag{2.15}$$

The specimen may be fatigue cracked, but blunting before collapse usually is so extensive that a sharp saw cut and a fatigue crack are undistinguishable at the onset of fracture. The load at fracture (maximum load) is recorded. From this the net section stress can be calculated directly as:

$$F_{col} = \frac{P_{fracture}}{W - 2a} \frac{1}{B} \tag{2.16}$$

Note that the measured F_{col} should be greater than or equal to the yield strength, otherwise fracture occurred before collapse was reached and the measured net section stress is not the collapse stress. In that event the test should be analyzed using fracture mechanics as discussed in the following chapters.

Instead of calculating the net section stresses directly, one may calculate the remote stress, $\sigma = P/A$, at the time of fracture, which is residual strength. By plotting the latter in a diagram such as in Figure 2.17, and by drawing a straight line through the data point and the point (W, 0), the collapse strength F_{col} is found at the intercept of this straight line with the vertical axis, as can be seen from Equation (2.15). The advantage of this procedure is that results obtained from specimens with different crack sizes can be plotted as well, thus providing a check of the applicability of the collapse criterion. If all data fall on the same straight line as per Equation (2.15) then collapse occurred in all tests; if they do not, the tests below the line represent cases where fracture occurred before collapse and these tests should be analyzed using fracture mechanics concepts (Chapters 3 and 4).

Figure 2.17c presents test data [3] for 304 stainless steel showing a collapse strength of 67 ksi. (The yield strength of this material is 30 ksi and the ultimate strength is 90 ksi). Indeed, the collapse strength is considerably higher than

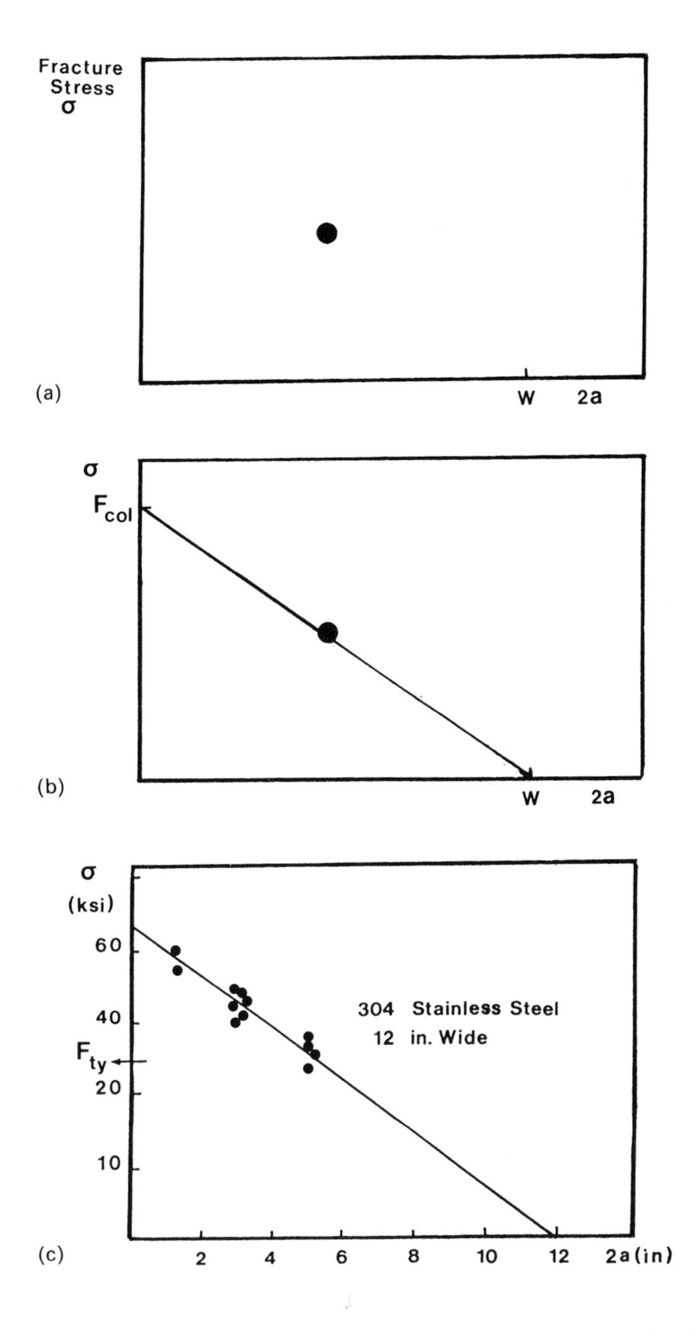

Figure 2.17. Measurement of F_{col}. (a) Plot of test data; (b) Determination of F_{col}; (c) Actual test data (Ref. 3).

yield. For all cracks smaller than 5-in the residual strength is more than 30 ksi. This means that the entire panel was above yield (σ_{cf} is the nominal remote stress) and not just the net section. Data points at various crack sizes all fall on or close to the straight line, thus confirming that collapse occurred in all tests.

2.9. Exercises

1. Calculate the theoretical stress concentration factor of an elliptical notch with a semi major axis of 2 inch perpendicular to the applied load, and a semi minor axis of 0.4 inch. What is the stress concentration factor; what is the strain concentration factor. What are these factors after the notched section has fully yielded in plane stress, assuming no strain hardening?

2. For the case of Exercise 1 calculate the nominal stress in the full section at the time of collapse if the yield strength is 50 ksi. Do the same for a material with a collapse strength of 75 ksi. Calculate the fracture load in the two cases if the notch is a center notch and $W = 12$ inch, $B = 0.5$ inch.

3. Cut a strip of paper 4 inches wide and 10 inches long. At three inches from one end cut a circular hole of 1.6 inch diameter in the center. At three inches from the other end cut a slit of 1.6 in long and two holes at both ends of the slit of 0.2 inch diameter. After cutting the holes cut the notch into an elliptical shape (semi minor axis 0.2 in). Calculate the stress concentration factors of the two notches and predict where the strip will fail. Roll the ends around 2 pencils and pull the strip to failure to prove your prediction.

4. If a crack appears in service one sometimes drills a so called stop hole at the crack tip as a temporary repair. Suppose a crack has started at the edge of a strip. Its size is a. A stop hole is drilled with diameter d, its center exactly on the crack tip. The crack tip root radius is almost zero. Assume the crack with the stop hole to be an ellipse (why is this permissible). Calculate the theoretical stress concentration factor before and after stop drilling for the general case. If the crack is 1 inch long calculate which size stop hole is needed to give a theoretical stress concentration factor of 5; and 7?

5. What is the stress at yield at a notch in plane strain if $F_{ty} = 30$ ksi? What is it at the free surface of the notch root? Sketch the stress distribution.

6. Using the solution of Exercise 5 calculate the nominal stress in the full section at which yielding begins at the free surface of the notch. Sketch the stress distribution in these two cases. Assume $k_t = 4$.

7. Consider a cylindrical bar with a circumferential groove of half circular shape. In order to reduce the stress concentration one sometimes machines relief grooves: similar grooves but of smaller depth at short distance on both

sides of the main groove. Demonstrate by means of load-flow lines that this indeed gives a lower stress concentration factor.

8. Using load-flow lines demonstrate the high stress concentration factor (as high as 10) at the fillet radius of the head of a tension bolt.

9. Calculate and construct the diagram for fracture due to collapse for a center cracked panel as well as for an edge cracked panel. $W = 600\,\mathrm{mm}$; $F_{col} = 350\,\mathrm{MPa}$. At which stress will collapse occur if the panels contain a crack of 150 mm? For the edge cracked panel assume that the side edges are constrained (kept straight, so that no bending occurs).

References

[1] R.E. Peterson, *Stress concentration design factors*, John Wiley (1953).
[2] H. Neuber, Theory of stress concentration for shear strained prismatical bodies with arbitrary non-linear stress-strain law of *J. App. Mech.* **28** (1961) pp. 544–550
[3] M.F. Kanninen, et al., Towards an elastic-plastic fracture mechanics capability for reactor piping, *Nuclear Eng. and Design* **(48)** (1978) pp. 117–134

CHAPTER 3

Linear elastic fracture mechanics

3.1. Scope

In this and the following chapter, the fracture mechanics concepts for the analysis of fracture will be discussed. With these concepts it will be possible to obtain the residual strength diagram and the maximum permissible crack size (Chapter 1). Following the arguments used in Chapter 1, the discussion will be limited to mode I loading (mixed mode loading is considered in Chapter 9).

Materials with relatively low fracture resistance fail below their collapse strength and can be analysed on the basis of elastic concepts through the use of Linear Elastic Fracture Mechanics (LEFM). Such materials are, among others, practically all high strength materials used in the aerospace industry, high-strength-low-alloy steels, cold worked stainless steels, etc. For the fracture analysis of many other materials, the use of Elastic-Plastic Fracture Mechanics (EPFM) as discussed in Chapter 4, is often indicated. Understanding of EPFM concepts requires familiarity with LEFM concepts.

3.2. Stress at a crack tip

Consider (Figure 3.1) a body of arbitrary shape with a crack of arbitrary size, subjected to arbitrary tension, bending, or both, as long as the loading is mode 1 (Chapter 1). The material will be considered elastic, following Hooke's law. For such a case the theory of elasticity can be used to calculate the stress field. The stresses, σ_x and σ_y can be obtained as well as the shear stress τ_{xy}. Details of the derivation can be found in more extensive texts [1, 2, 3]. In the following only the solution will be given, and it will be shown that the result is in accordance with what would be expected.

As was explained already in Chapter 2, the crack tip stress field is at least biaxial (load-flow lines) and it may be triaxial if contraction in thickness direction is constrained. Hence, there will be stresses in at least X and Y direction, σ_x and σ_y. From the stress field solution it appears that the stresses on a material element as shown Figure 3.1 can be described by (in the absence of constraint):

$$\left.\begin{aligned}
\sigma_x &= \frac{K}{\sqrt{2\pi r}} \cos\frac{\theta}{2}\left(1 - \sin\frac{\theta}{2}\sin\frac{3\theta}{2}\right) \\
\sigma_y &= \frac{K}{\sqrt{2\pi r}} \cos\frac{\theta}{2}\left(1 - \sin\frac{\theta}{2}\sin\frac{3\theta}{2}\right) \\
\sigma_z &= 0 \\
\tau_{xy} &= \frac{K}{\sqrt{2\pi r}} \cos\frac{\theta}{2}\sin\frac{\theta}{2}\cos\frac{3\theta}{2}
\end{aligned}\right\}. \tag{3.1}$$

Indeed, both σ_x and σ_y exist (as should be the case as discussed in Chapter 2). For the case that $\theta = 0$ (plane through the cracked section), the shear stress, τ_{xy}, is zero, as should be expected for a plane of symmetry. Although it does not make any difference for the following discussion, it is convenient to confine the considerations to the plane through the crack with $\theta = 0$; in that case the functions of θ will be either 0 or 1, so that they essentially disappear: (note also that $x = r$ for $\theta = 0$):

$$\left.\begin{aligned}
\sigma_x &= \frac{K}{\sqrt{2\pi x}} \\
\sigma_y &= \frac{K}{\sqrt{2\pi x}}
\end{aligned}\right\}. \tag{3.2}$$

It appears that, at least along the plane $Y = 0$ for which Equations (3.2) hold, the transverse stress, σ_x, is equal in magnitude to the longitudinal stress, σ_y. The stresses depend upon the distance x from the crack tip; note that at greater distances (larger x) the stresses are lower. The stresses also depend upon a parameter K, which is as yet undefined.

The expressions for the stresses appear to be remarkably simple. Even more remarkable is that a stress field solution could be obtained at all, since the body, crack, and the loading are arbitrary. Because of this arbitrariness it is not surprising that the equations contain an unknown parameter K. As Equations (3.2) are for an arbitrary case, they must describe the stresses at each and every crack in each and every elastic body in the universe. Apparently, the equations are universal and can be used for all crack problems. Therefore, also the parameter K will appear in all crack problems, reason to give it a name: K is called the stress intensity factor, not to be confused with the stress concentration factor, k_t, discussed in Chapter 2.

Similar solutions can be obtained for other modes of loading (Chapter 9), be

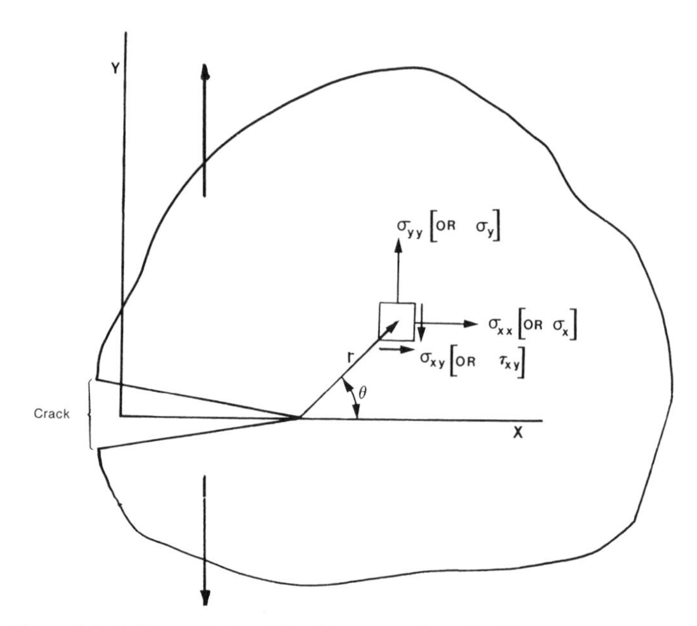

Figure 3.1. Arbitrary body with arbitrary crack and arbitrary mode I loading.

it that the goniometric functions are different for different modes. A stress intensity factor appears in all solutions, but the definition of K is slightly different for different modes. For this reason, the stress intensity factors are often labeled in accordance with the mode of loading (Chapter 1), namely K_{I}, K_{II}, and K_{III}. In the above equations K should then be labeled as K_{I}. However, when there can be no confusion about which mode is meant, the subscript is often omitted; this is what will be done in the following. Should more than one loading mode be operable, the labels must be carried to avoid confusion, as will be done in Chapter 9.

Since Equations (3.2) are for all crack problems, there is no objection against selecting a familiar geometry. Let this be a very large (infinite) panel, subjected to uniform unaxial loading with a nominal stress σ, and a central crack, as in Figure 3.2a. The size of the crack will be called $2a$. This is a convention: In fracture mechanics all cracks with 2 tips are called $2a$, all cracks with one tip a. There is no reason why it could not be done differently, but this convention is necessary to compare notes. The convention should be adhered to if one wants to use data generated elsewhere, otherwise erroneous results will be obtained.

Equations (3.2) apply to the problem of Figure 3.2a, and it is now possible to examine the significance of the stress intensity factor, K. It should be noted then that the stresses everywhere in an elastic body are proportional to the applied load. If the load is increased by a factor 2, all stresses everywhere will

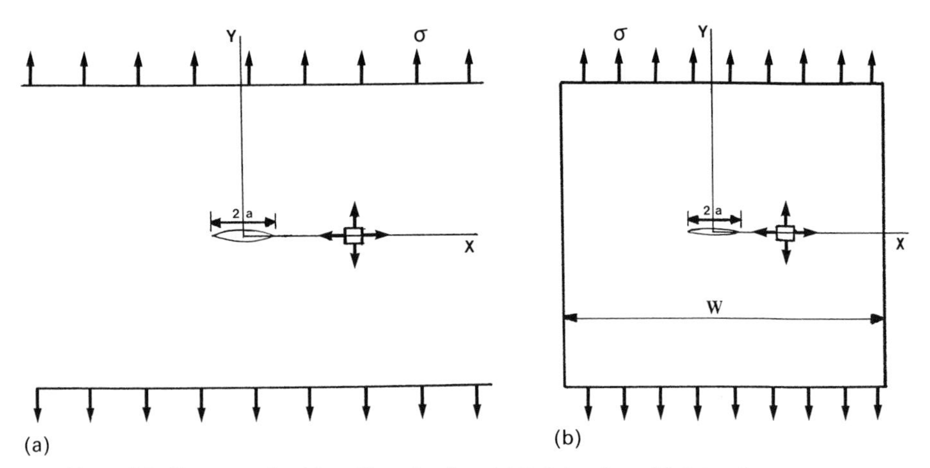

Figure 3.2. Center crack with uniform loading. (a) Infinite plate; (b) finite plate.

increase by a factor 2. Thus, it can be concluded that the crack tip stress must be proportional to the applied stress, σ, i.e. (see Equation (3.2))

$$\sigma_y \div \frac{\sigma}{\sqrt{2\pi x}} . \tag{3.3}$$

It stands to reason that the crack tip stresses will also depend upon crack size. The stresses will certainly he higher when a is larger; hence, the crack size, a, must appear in the numerator in Equation (3.3). There is only one way in which this can happen. Both sides of the equation must have the dimension of stress, but in the denominator at the right there appears a square root of a length, the distance x. In order to cancel the square root of length in the denominator, the crack size must appear in the numerator as square root of a:

$$\sigma_y \div \frac{\sigma\sqrt{a}}{\sqrt{2\pi x}} . \tag{3.4}$$

Equation (3.4) still contains a proportionality sign instead of an equal sign, because dimensional analysis does not show whether there is a dimensionless number involved. Calling this dimensionless number C, one finally arrives at an equation:

$$\sigma_y = \frac{C\sigma\sqrt{a}}{\sqrt{2\pi x}} . \tag{3.5}$$

Clearly, it is simple to find the format of Equation (3.5), but a formal solution [1, 2, 3] would be necessary to obtain the actual value of C. It turns out that $C = \sqrt{\pi}$, for the case depicted in Figure 3.2a. Hence:

$$\sigma_y = \frac{\sigma\sqrt{\pi a}}{\sqrt{2\pi x}}. \tag{3.6}$$

Comparison of Equations (3.6) and (3.2), leads to the conclusion that for the configuration of Figure 3.2a:

$$K = \sigma\sqrt{\pi a}. \tag{3.7}$$

3.3. General form of the stress intensity factor

The manner used in the previous section to demonstrate the significance of K is not limited to the case shown in Figure 3.2a. For example consider (Figure 3.2b): the plate of finite width W. From the same arguments as used above, it follows that:

$$\sigma_y = \frac{C\sigma\sqrt{a}}{\sqrt{2\pi x}}. \tag{3.8}$$

In this case it must be expected that the size of the plate also will affect the crack tip stresses. It must be anticipated that the stresses will increase when W becomes smaller. The only manner in which the effect of W can appear is in the factor C. Hence, C must be a function of (depend upon) the width W. However, C must be dimensionless, and hence, it can depend upon W only through dependence upon a dimensionless parameter such as W/a or a/W. For the configuration of Figure 3.2b the expression for C appears to be [1]:

$$C = \sqrt{\pi \sec \frac{\pi a}{W}} \tag{3.9}$$

so that:

$$\sigma_y = \frac{\sqrt{\pi \sec \frac{\pi a}{W}}\, \sigma\sqrt{a}}{\sqrt{2\pi x}} \tag{3.10}$$

and

$$K = \sqrt{\pi \sec \frac{\pi a}{W}}\, \sigma\sqrt{a}. \tag{3.11}$$

If W is very large, or a very small, the value of $\sqrt{\sec(\pi a/W)} = 1$. This is in accordance with anticipation, because for small a/W the configuration of Figure 3.2b is identical to that of Figure 3.2a. Indeed, in that case Equation (3.11) reduces to Equation (3.7).

Now it becomes obvious that for ANY configuration the crack tip stress will always be:

$$\sigma_y = \frac{C\left(\frac{a}{L}\right)\sigma\sqrt{a}}{\sqrt{2\pi x}} = \frac{K}{\sqrt{2\pi x}} \tag{3.12}$$

and the stress intensity factor is always:

$$K = C\left(\frac{a}{L}\right)\sigma\sqrt{a} \tag{3.13}$$

where L is a (unified) length dimension describing the geometry of the cracked part.

In the practical use of these equations all C's are divided by $\sqrt{\pi}$, and $\sqrt{\pi a}$ is substituted for $\sqrt{a}$ to compensate. The function $C(a/L)/\sqrt{\pi}$ is then renamed β. The geometry factor (often erroneously called correction factor):

$$\left.\begin{aligned} \sigma_y &= \frac{\beta\sigma\sqrt{\pi a}}{\sqrt{2\pi x}} = \frac{K}{\sqrt{2\pi x}} \\ K &= \beta\left(\frac{a}{L}\right)\sigma\sqrt{\pi a} \end{aligned}\right\} . \tag{3.14}$$

Note that Equations (3.14) are identical to the previous equations. The Equations (3.14) present the stress and the stress intensity factor in comparison to those for the infinite panel: $\beta = 1$ for the infinite panel, but for the finite width panel:

$$\beta = \sqrt{\sec\frac{\pi a}{W}} . \tag{3.15}$$

It must be emphasized that Equations (3.14) represent the crack tip stresses and stress intensity for all crack problems, the equations having been derived from the general solution for an arbitrary crack in an arbitrary body with arbitrary mode I loading. For any crack in any practical problem only the function β, or its functional value, need be derived. For many configurations the function, β, has been calculated already; the results can be found in handbooks [4, 5, 6]. Examples of β functions for a few common crack cases are shown in Figure 3.3. Note that in the Equations (3.1) through (3.14) the stress, σ, is the nominal stress in the uncracked section. The fact that the stresses are higher in the cracked section when e.g. W becomes smaller, is wholly accounted for by β. Although handbooks are a source for β expressions for many generic configurations, obtaining β for a practical problem may be rather involved. Fortunately there are many simple ways in which it can be obtained with good accuracy. Such simple procedures to derive β for practical cases are discussed in Chapter 8. At

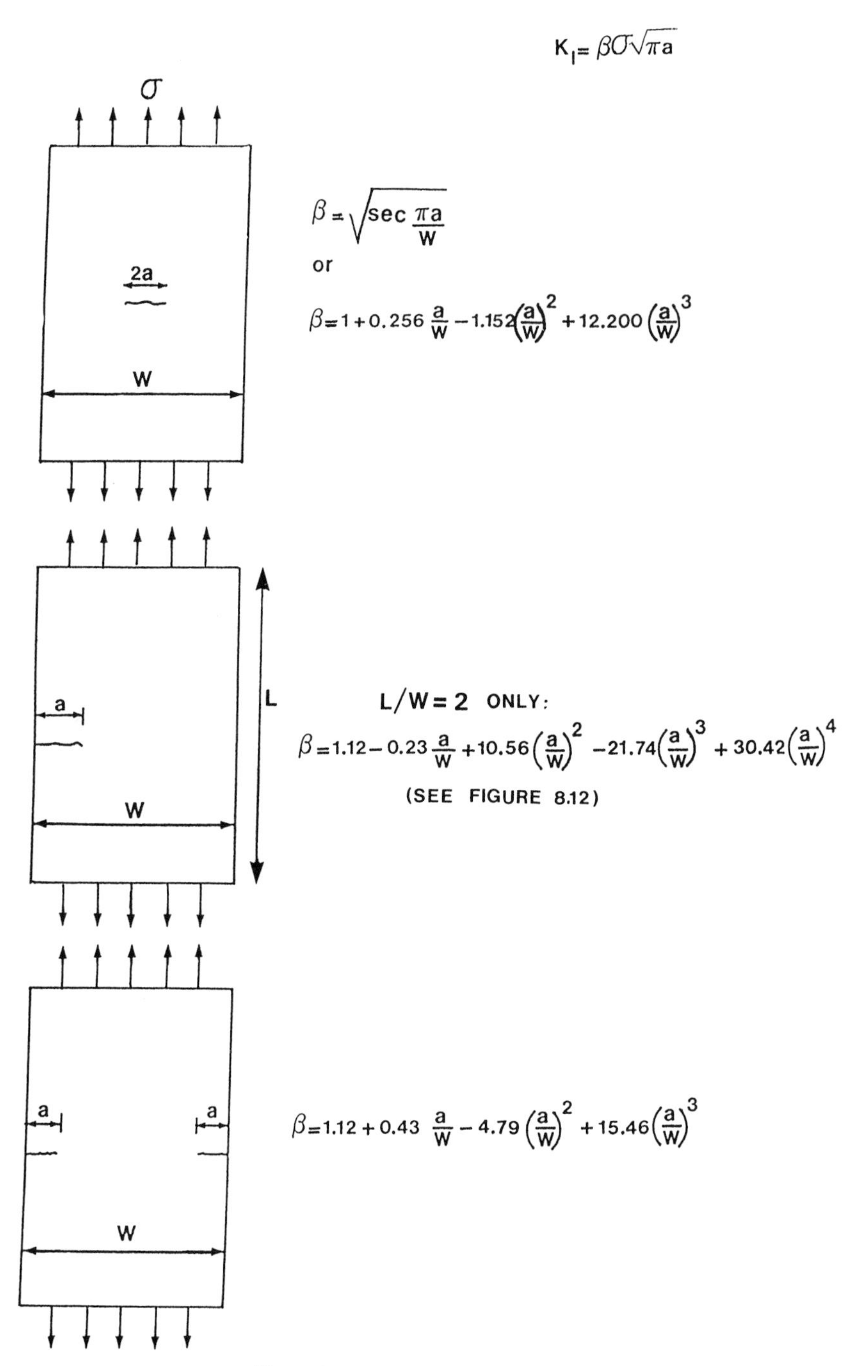

$K_I = \beta\sigma\sqrt{\pi a}$
σ
2a
W
$\beta = \sqrt{\sec \frac{\pi a}{W}}$
or
$\beta = 1 + 0.256\,\frac{a}{W} - 1.152\left(\frac{a}{W}\right)^2 + 12.200\left(\frac{a}{W}\right)^3$
a
W
L
L/W = 2 ONLY:
$\beta = 1.12 - 0.23\,\frac{a}{W} + 10.56\left(\frac{a}{W}\right)^2 - 21.74\left(\frac{a}{W}\right)^3 + 30.42\left(\frac{a}{W}\right)^4$
(SEE FIGURE 8.12)
a
a
W
$\beta = 1.12 + 0.43\,\frac{a}{W} - 4.79\left(\frac{a}{W}\right)^2 + 15.46\left(\frac{a}{W}\right)^3$

Figure 3.3. Examples of β-functions.

this point it is sufficient to note that β can always be obtained and that Equations (3.14) are universal.

3.4. Toughness

Several difficulties arise with the direct use of Equations (3.14), and these may have serious consequences, as discussed in subsequent sections. Nevertheless, although somewhat prematurely, the practical use of the equations for the analysis of fracture problems, will be considered first, be it that some of the conclusions will have to be modified later.

Fracture will occur when the stresses at the crack tip become too high for the material to bear. As the stress intensity factor determines the entire crack tip stress field, the above statement is equivalent to: fracture will occur when K becomes too high for the material. How high the stress intensity can be depends upon the material; it must be determined from a test.

For example let a test be performed on a 30 inches (762 mm) wide plate of a certain steel X, with a central crack $2a = 4$ inches (101.6 mm). Let the plate have a thickness of 0.2 inch (5.08 mm). The plate is pulled to fracture in a tensile test machine. Suppose that fracture occurs at a load of 180 kips or 180 000 lbs (800 000 N). The nominal stress in the uncracked section at the time of fracture was then $180/(0.2 \times 30) = 30$ ksi $(800\,000/(5.08 \times 762) = 207\ \text{N/mm}^2 = 207$ MPa).

This information permits calculation of the value of the stress intensity at fracture. Note from equation (3.15) that $\beta \approx 1$ for the small a/W used, and that $a = 2$ in (0.0508 m). Hence:

$$K = 1 \times 30 \times \sqrt{\pi \times 2} = 75\ \text{ksi}\sqrt{\text{in.}}$$

or

$$K = 1 \times 207 \times \sqrt{\pi \times 0.0508} = 83\ \text{MPa}\sqrt{\text{m.}}$$

or

$$K = 1 \times 207 \times \sqrt{\pi \times 51} = 2620\ \text{N/mm}^{3/2}.$$

Apparently, when K reached the value of 75 ksi $\sqrt{\text{in}}$ (83 MPa $\sqrt{\text{m}}$), the stresses at the crack tip were too high for the material and fracture ensued. This value of K is called the 'toughness' of the material. Hence, the toughness of a material is the highest stress intensity that can be supported by a cracked component made of that material.

The unit of toughness is ksi $\sqrt{\text{in}}$ or MPa $\sqrt{\text{m}}$; it follows directly from the dimension of K, which is stress $\times \sqrt{\text{crack length}}$. Conversion of units from English to metric provides: 1 ksi $\sqrt{\text{in}} = 6.89$ MPa $\sqrt{0.02540\ \text{m}} = 1.09$ MPa $\sqrt{\text{m}}$, or 1 MPa $\sqrt{\text{m}} = 0.92$ ksi $\sqrt{\text{in}}$.

Once the toughness of a material has been measured in a test, its value can be recorded in a data sheet or data handbook as in the case of e.g. F_{ty}. The latter is the value of the stress at which yielding occurs. Toughness is the value of the stress intensity at which fracture occurs in the case of cracks.

Now the universality of Equations (3.14) becomes important: the equations hold for ALL cracks. At a crack in a structure made of steel X, the stresses at the crack tip will be equal to those in the test specimen at the time of fracture when K is equal to the tougness. Since the material of the test specimen could not sustain this stress field, it follows that the structure of the same material also will fracture when K is equal to the toughness, because then the crack tip stress field in the structure will be identical to that in the specimen; if the specimen failed as a consequence of this stress field, the structure will fail when this same stress field occurs.

Consider a simple structure in uniform tension with an edge crack of $a = 3$ inch (76.2 mm), a width of 40 inches (1016 mm), and a thickness of 0.2 inch (5.1 mm), made of the same material as the above specimen. For this configuration and small a/W, it is found from Figure 3.3 that $\beta \approx 1.12$. The fracture condition is:

$$\text{Fracture if:} \quad K = \text{Toughness} \tag{3.16}$$

with $K = \beta\sigma\sqrt{\pi a}$, it follows that fracture occurs when:

$$\text{Fracture if:} \quad \beta\sigma\sqrt{\pi a} = \text{Toughness.} \tag{3.17}$$

The nominal stress at which fracture takes place will be denoted as σ_{fr}. It follows from Equation (3.17) that:

$$\sigma_{fr} = \frac{\text{Toughness}}{\beta\sqrt{\pi a}}. \tag{3.18}$$

Substitution of $\beta \approx 1.12$, $a = 3$ inch (0.0762 m) and toughness $= 75\,\text{ksi}\,\sqrt{\text{in}}$ (83 MPa $\sqrt{\text{m}}$), gives:

$$\sigma_{fr} = \frac{\text{Toughness}}{\beta\sqrt{\pi a}} = \frac{75}{1.12\sqrt{\pi \times 3}} = 22\,\text{ksi}$$

or

$$\sigma_{fr} = \frac{\text{Toughness}}{\beta\sqrt{\pi a}} = \frac{83}{1.12\sqrt{\pi \times 0.076}} = 151\,\text{MPa}.$$

The fracture stress, σ_{fr}, is the residual strength (remaining strength under the presence of a crack; Chapter 1). Using the equation for β in Figure 3.3, one can calculate the residual strength from Equation (3.18) for a number of crack sizes. A plot of the results provides the residual strength diagram shown in Figure 3.4.

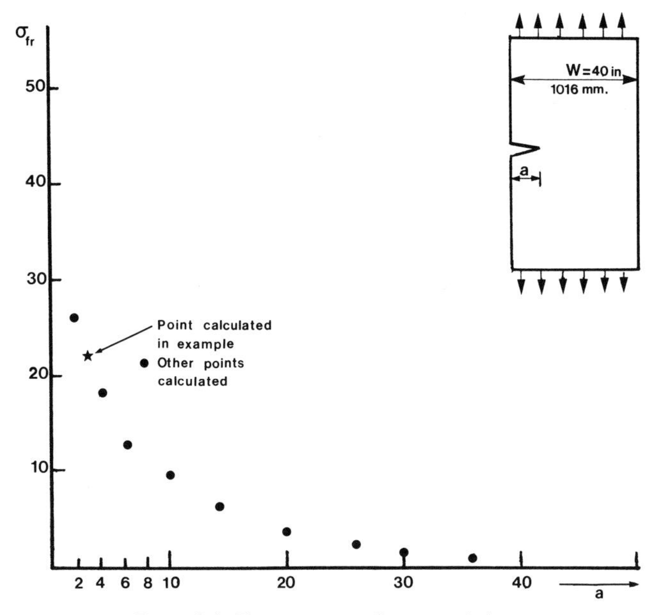

Figure 3.4. Fracture stress for example in text.

(The fracture stress, σ_{fr}, would indeed be the residual strength defined in Chapter 1 unless failure by collapse as discussed in Chapter 2 prevails. As this remains to be established, σ_{fr} is not yet called σ_{res}).

3.5. Plastic zone and stresses in plane stress and plane strain

Equations (3.2) and (3.14) predict that far from the crack tip where x is large, the stresses σ_x and σ_y become zero. That the transverse stresses, σ_x, becomes zero is not surprizing but σ_y becoming zero is incorrect because, e.g. in Figure 3.2 the stress σ_y far away from the crack is equal to the applied stress σ. The reason is that Equations (3.2) do not provide the complete solution of the stress field. The complete solution is a series of terms:

$$\sigma_y = \frac{K}{\sqrt{2\pi x}} + Cx^0 + Dx^{1/2} + Ex^1 + \dots \quad . \tag{3.19}$$

In particular the second term will ensure that $\sigma_y = \sigma$ for large x.

Approaching the crack tip, x becomes very small, so that all terms except for the first become very small. Actually, for $x = 0$ all terms become zero, except for the first term, which then becomes infinite. Hence, for small x, close to the crack tip, all terms can be neglected with respect to the first term, and for $x = 0$

Equations (3.1) are the exact solution. Fracture begins at the crack tip and not somewhere else, so that the use of only the first term is justified.

For $x = 0$ the stresses become infinite. This is correct from the point of view of the theory of elasticity, because Hooke's law ($\sigma = \varepsilon E$) puts no limitations on stress nor strain. The crack (tip) is a sharp discontinuity, which causes very high stress (according to Chapter 2, a notch with a tip radius of zero causes an infinite stress concentration). Note however, that the infinite stress occurs at only one single point (i.e. in an infinitely small area) and not over a certain distance. Yet, in a real material plastic deformation will occur so that stresses cannot increase much further after yielding begins.

As explained in Chapter 2, the stress field at notches and cracks and the state of stress depend upon thickness. Contraction in thickness direction of the highly stressed roll at the crack tip can take place freely when the plate is thin, leading to plane stress. In thick plates however, contraction is constrained which leads to plane strain (see Chapter 2, Figures 2.6 and 2.7). According to Equations (3.2) the transverse stress, σ_x, equals the longitudinal stress, σ_y, at the crack tip. Plane strain (Chapter 2) then leads to:

$$\varepsilon_z = \frac{\sigma_z}{E} - \nu\frac{\sigma_x}{E} - \nu\frac{\sigma_y}{E} = 0 \tag{3.20}$$

which provides (with $\nu \approx 0.33$):

$$\sigma_z = \nu(\sigma_x + \sigma_y) \approx 0.66\sigma_y. \tag{3.21}$$

According to the Tresca yield criterion (Chapter 2) yielding begins when the difference between maximum and minimum principal stress is equal to the yield strength, F_{ty}. This leads to the following condition for plastic deformation in plane strain:

$$\sigma_y - 0.66\sigma_y = F_{ty} \quad \text{or} \quad \sigma_y = 3F_{ty}. \tag{3.22}$$

Hence, the stress σ_y must rise to three times the yield strength before plastic deformation will commence. In plane stress on the other hand, where $\sigma_z = 0$, yielding will being when $\sigma_y = F_{ty}$, (Chapter 2).

Assuming that the stress does not increase much after yielding (again, see Chapter 2), the stress distributions will be as shown in Figure 3.5. The curved part of the stress distribution is still described by Equations (3.2), and it is the same in plane stress and plane strain, because further away from the crack tip constraint is not a problem (Chapter 2). Apparently, within some distance r_p, from the crack tip the yield conditions are met, so that there is always a small area at the crack tip where plastic deformation takes place (Note that σ_y is infinite according to Equation 3.14). This area is called the plastic zone. It is possible to estimate the distance from the crack tip over which plastic deforma-

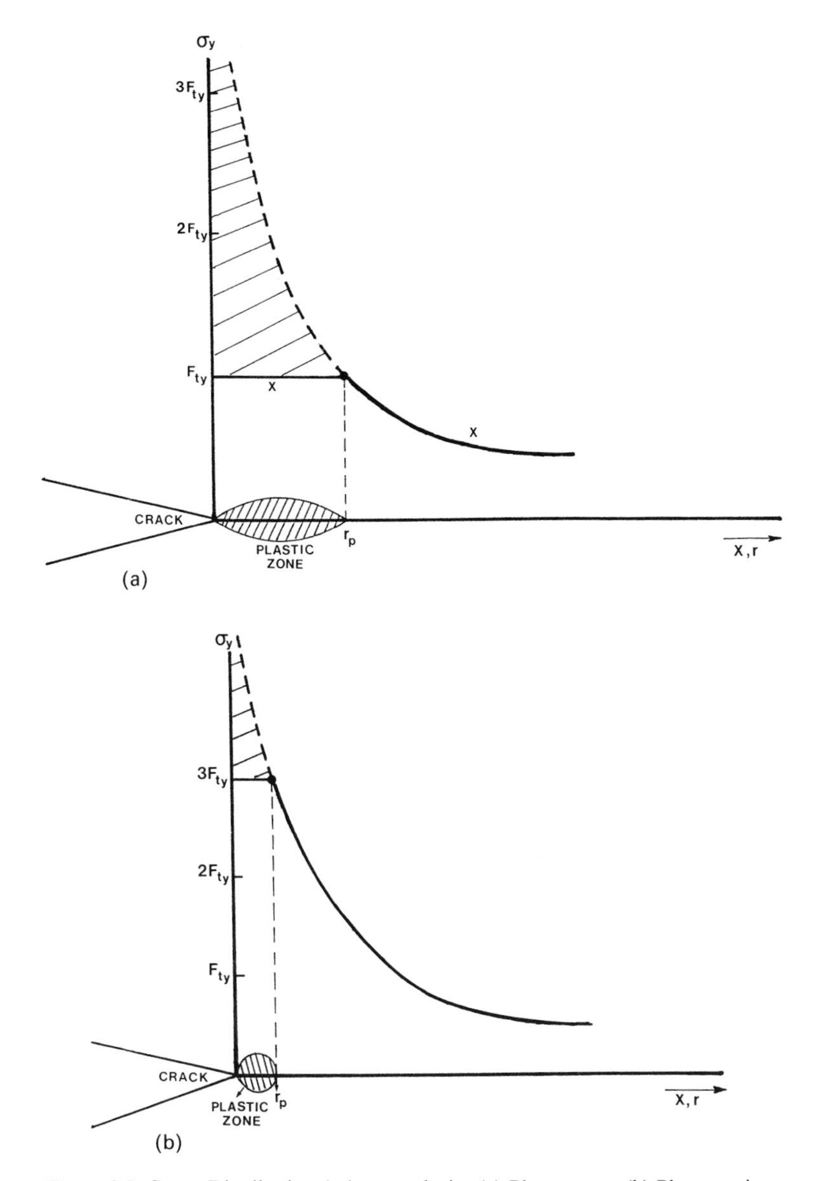

Figure 3.5. Stress Distribution (σ_y) at crack tip. (a) Plane stress; (b) Plane strain.

tion occurs, i.e. the size of the plastic zone. For plane stress this follows from the Tresca yield criterion (Chapter 2) and the condition $\sigma_y = F_{ty}$, and for plane strain from the condition $\sigma_y = 3F_{ty}$. Using Equation (3.2) for σ_y one obtains:

$$\left.\begin{aligned} &\text{for plane stress: } \frac{K}{\sqrt{2\pi r_p}} = F_{ty} \quad \text{or} \quad r_p = \frac{K^2}{2\pi F_{ty}^2} \\ &\text{for plane strain: } \frac{K}{\sqrt{2\pi r_p}} = 3F_{ty} \quad \text{or} \quad r_p = \frac{K^2}{18\pi F_{ty}^2} \end{aligned}\right\}. \tag{3.23}$$

In reality, the plastic zones are about twice as large as in Equations (3.23). This is due to the fact that actually a much higher stress should have been carried by the plastic zone. This extra stress must be bypassed around the plastic zone thus increasing the stress there and causing more material to yield [1]. By limiting the crack tip stresses because of plastic deformation, the dashed part of the stress distribution in Figure 3.5 was essentially eliminated. However, the total stress distribution, including the dashed part, was calculated from the theory of elasticity on the basis of equilibrium. Cutting off the dashed part is violating equilibrium. This must be mended by restoring equilibrium. The dashed areas in Figure 3.5 represent a load which cannot be carried by the material in the plastic zone because the load carrying capacity of that material is limited by yielding. However, the cut load must still be carried through (equilibrium) and it must bypass the crack, or it must bypass r_p, as the plastic zone cannot carry it. This implies that yielding actually extends beyond r_p. (Note that the high stresses were due in the first place due to a load path interruption by the crack, so that the load had to bypass the crack as shown in Chapter 2. Now it appears that the load represented by the shaded areas in Figure 3.5 must bypass further away, because once it is loaded to yield, the area r_p is no longer a viable load path either). The consequence is that the plastic zones are actually larger than those represented by Equations (3.23). Restoring equilibrium will lead to plastic zones exactly twice as large [1], as those in Equations (3.23). For other reasons [1], the plastic zones are usually taken as:

$$r_p = \frac{K^2}{\alpha\pi F_{ty}^2} \tag{3.24}$$

with $\alpha = 2$ for plane stress and $\alpha = 6$ for plane strain.

It appears from Figure 3.5 and Equations (3.23) and (3.24) that the plastic zone in plane strain is much smaller than in plane stress. Also, the crack tip stresses are higher in plane strain than in plane stress. The longitudinal stress, σ_y, differs by a factor of 3, while $\sigma_z = 0.66\sigma_y$ in plane strain and $\sigma_z = 0$ in plane stress (see also Chapter 2).

In the example of the hypothetical test given in the previous section, fracture took place at a value of K of 75 ksi $\sqrt{\text{in}}$. Supposing that the material had a yield strength of $F_{ty} = 100$ ksi, the size of the plastic zone at the time of fracture would have been according to Equation (3.24):

$$\text{plane stress:} \quad r_p = \frac{75^2}{2\pi \times 100^2} = 0.09 \text{ inch}$$

$$\text{plane strain:} \quad r_p = \frac{75^2}{6\pi \times 100^2} = 0.03 \text{ inch.}$$

These are very small plastic zones indeed. At this point it cannot be decided whether there was plane stress or plane strain in the example in Section 3.4. This question is discussed in the following section.

3.6. Thickness dependence of toughness

The true size of the plastic zone is not important for the following discussion, but the format of Equation (3.24) is relevant; it shows that the size of the plastic zone depends upon K only. Thus for all cracks in any configuration, and at the same K, the stress distributions at the crack tip will still be completely dictated by K and by K only. It follows that the fracture criterion then still holds regardless of the plasticity; if fracture occurs in a test specimen at a certain value of K (the toughness) then it will do so in any other configuration at the same K, because in that other configuration the stress distribution as well as the size of the plastic zone will still be the same as in the test specimen, provided there is equal constraint.

One problem has arisen however. The stress distributions are not the same in plane stress and plane strain; as shown in Figure 3.5. Also σ_z differs in the two cases. As a consequence, the above fracture criterion cannot be used indiscriminately. If fracture in plane strain occurs at a certain value of K, then it will not in plane stress at that same value of K, because then the stress distributions will be different (different σ_z and different σ_y at the crack tip due to different plasticity as shown in Figure 3.5). It follows then that the toughness will depend upon the state of stress. As the latter depends upon thickness, the toughness must be expected to depend upon thickness.

It should be noted here that Equation (3.24) was derived using only the first term of the stress distribution in Equation (3.19). Should the plastic zone be so large that the other terms come into play then its size will depend not only upon K, but also upon C, D etc. In such a situation equal K is not enough to guarantee equal stress distribution and fracture is not likely to occur at always the same value of K. Apparently, for the fracture criterion to be useful the plastic zone must be very small. Since r_p depends upon K, this implies that the stress intensity at fracture, i.e. the toughness, must be low. As r_p also depends upon the yield strength, the fracture criterion will be better applicable to materials of high yield strength and low toughness. Although the criterion does not work when the yield strength is very low and the toughness very high, satisfactory results can be obtained for many such cases, provided one realizes that plastic collapse

(Chapter 2) may prevail. This will be discussed in detail in a later section. In the case of extremely high toughness and/or low yield strength, EPFM may have to be used (Chapter 4).

Consider two plates, a thin one with plane stress, and a thick one with plane strain. Both have cracks and are loaded to have equal stress intensity factors. In the case that the prevailing stress intensity is equal to the toughness of the thick plate fracture ensues in the thick plate. The stress distributions in the thin plate will be much more benign (see Figure 3.5 and note the difference in σ_z) at this same K, so that it must be anticipated that the stress intensity in the thin plate can be further increased; i.e. the expectation is that the toughness is higher in plane stress than in plane strain (Figure 3.6a).

In plates of intermediate thickness the state of stress gradually changes from plane stress, via more and more triaxility, to plane strain in thick plates. Therefore, the toughness will not suddenly, at one particular thickness, drop from its high plane stress value to its low(er) plane strain value, but the change will be gradual; i.e. the toughness decreases with thickness as depicted in Figure 3.6b. Beyond a certain thickness there will be plane strain. As the state of stress cannot become worse than plane strain (complete constraint of contraction), the toughness will not decrease further.

For plane strain cases the toughness is called the plane strain fracture toughness. It is generally denoted as K_{Ic}, i.e. the critical value of the mode I stress intensity, K_{I}, at which fracture occurs. (Compare: F_{ty} is the critical value of the stress σ at which yield occurs). Similarly one speaks of the plane stress fracture toughness, and of the transitional fracture toughness for intermediate situations. Plane stress and transitional toughness are generally denoted as K_c or $K_{\mathrm{I}c}$. Unfortunately, this introduces a confusing inconsistency. In plane stress or in transitional cases, the loading is still mode I, the toughness is still the critical value of the mode I stress intensity. Thus, the general denotation of toughness should be K_{Ic}, regardless of the state of stress. One could speak of THE toughness K_{Ic}, and indicate whether the value is for plane stress, plane strain, or a transitional case. Be that as it may, the above denotations have become the 'accepted' ones. Once the user gets used to these denotations, they are, in a way, useful. If the toughness is given as K_{Ic}, it is immediately clear that this toughness is for plane strain; while a toughness K_c or $K_{\mathrm{I}c}$ is obviously for a thickness in which full constraint does not occur (plane stress or transitional).

The toughness as a function of thickness can be measured in tests of the type discussed in Section 3.4 on plates of different thickness. It can then be used to calculate the residual strength of cracked structures:

$$\left.\begin{array}{ll}\text{Fracture if:} & K = K_c \quad \text{or} \quad K = K_{Ic} \\ \text{i.e. Fracture if:} & \beta\sigma\sqrt{\pi a} = K_c \quad (\text{or } K_{Ic})\end{array}\right\} \tag{3.25a}$$

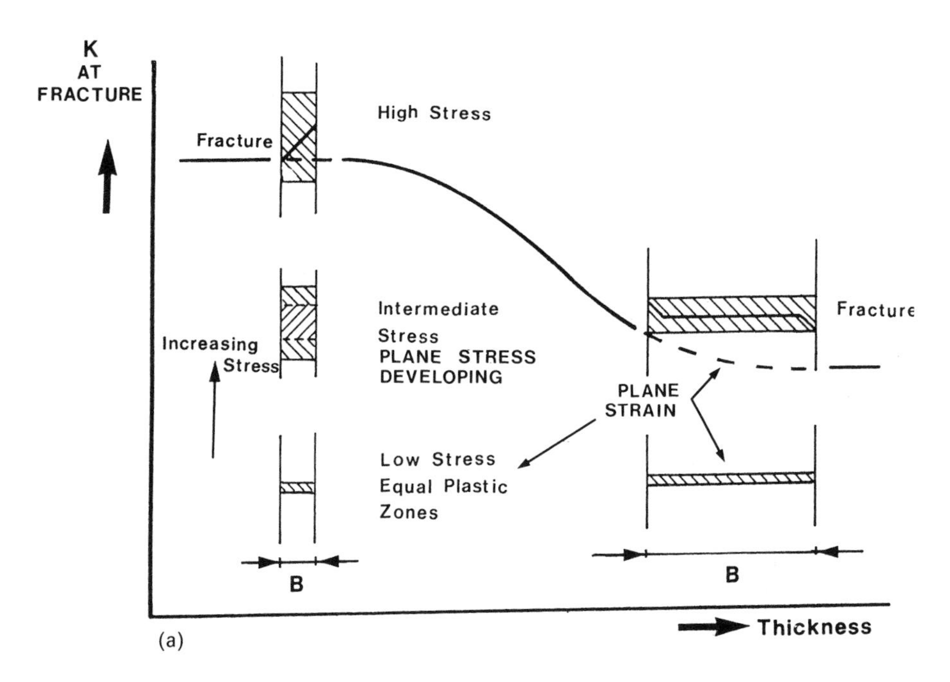

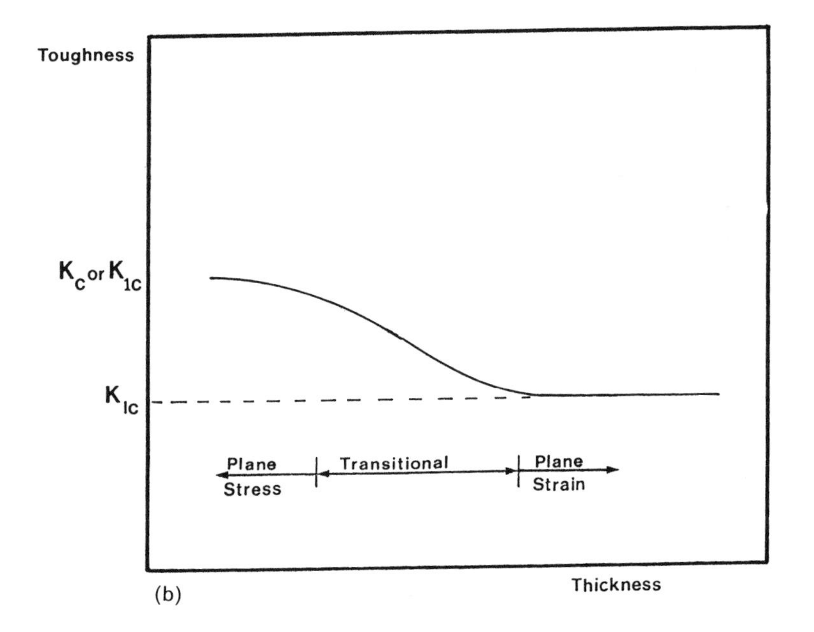

Figure 3.6. Dependence of toughness upon thickness. (a) Effect of thickness on plastic zone and state of stress; (b) Effect of thickness on toughness.

from which the fracture strength follows as

$$\sigma_{fr} = \frac{K_c \text{ (or } K_{Ic})}{\beta\sqrt{\pi a}}. \tag{3.25b}$$

For the configuration at hand β can be obtained from a handbook or otherwise (Chapter 8), and the residual strength calculated provided one uses the toughness for the thickness at hand, in the same manner as discussed in Section 3.4.

It is of interest to know at which thickness full constraint and plane strain occur. As discussed in Chapter 2, plane strain develops when the roll of material at high stress at the crack tip is long and thin (Figure 3.7). Hence, L/D, where L is the roll's length and D its thickness, must be large, e.g. larger than a certain value Q. When the crack is all the way through the thickness, the length of the roll is equal to the thickness, B, i.e. $L = B$. Taking D equal to the size of the plastic zone, $D = r_p$, from Equation (3.24) one obtains as the condition for plane strain:

$$\frac{L}{D} = \frac{B}{(K_{Ic}^2/\alpha F_{ty}^2} > Q. \tag{3.26}$$

In this equation α and Q are numbers. Rewriting $Q/\alpha = q$, the condition can be given as:

$$B > q\frac{K_{Ic}^2}{F_{ty}^2}. \tag{3.27}$$

The value of q cannot be derived mathematically. It is obtained from toughness tests on plates of different thicknesses as described, and by finding the thickness at which the curve levels off. Such experiments have shown [7] that this happens for $q = 2.5$, but the results are not very conclusive as can be seen in Figure 3.8. The scatter is considerable and the most that can be said from Figure 3.8 is that the leveling off occurs somewhere between $q = 2$ and $q = 4$. Nevertheless the number 2.5 is often considered sacrosanct. The reader is advised that it is but a rough indication.

It should be emphasized that the thickness is only of relevance for through-the-thickness cracks. Constraint is determined by the length of the roll of highly stressed material. In the case of a part through crack such as in Figure 3.9, contraction is always fully constrained and the thickness has no relevance to the problem (note that the thickness came into the problem because the length L of the roll happened to be equal to the thickness in the case of a through-the-thickness crack, and in that case only; this was discussed already in Chapter 2).

Figure 3.7. Thickness and State of Stress. (a) Attempt to contraction at crack tip; (b) Effect of thickness upon contraction. →

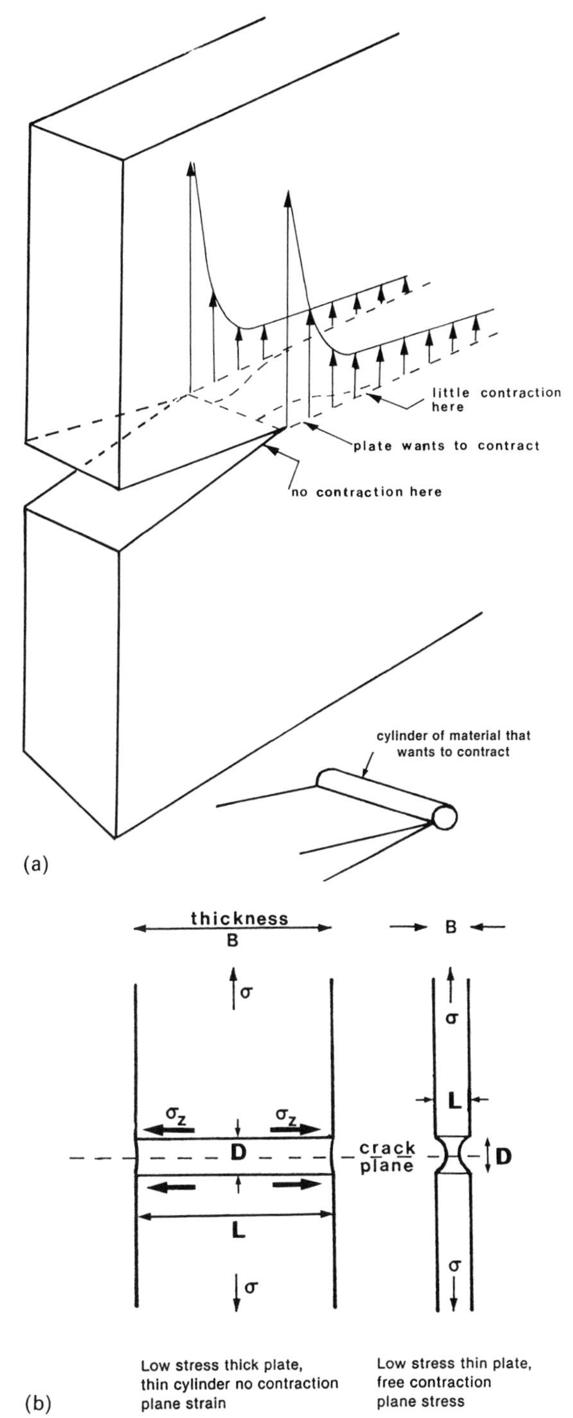

little contraction here
plate wants to contract
no contraction here
cylinder of material that wants to contract
(a)
thickness
B
σ
σ_z
σ_z
D
L
σ
crack plane
B
σ
L
D
σ
Low stress thick plate, thin cylinder no contraction plane strain
Low stress thin plate, free contraction plane stress
(b)

Figure 3.7.

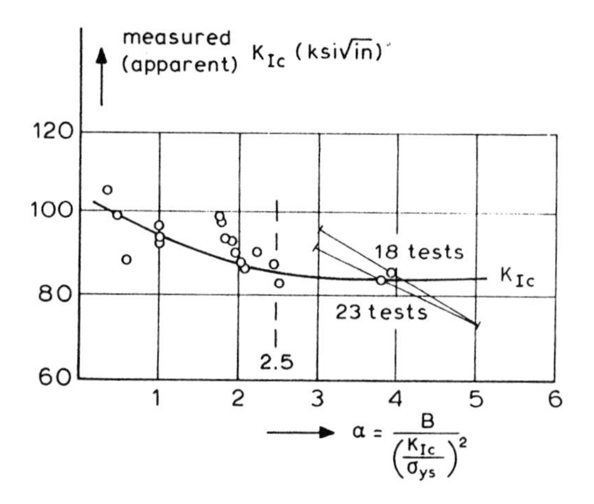

Figure 3.8. Effect of thickness on measured K_{Ic} of a Marageing Steel [7].

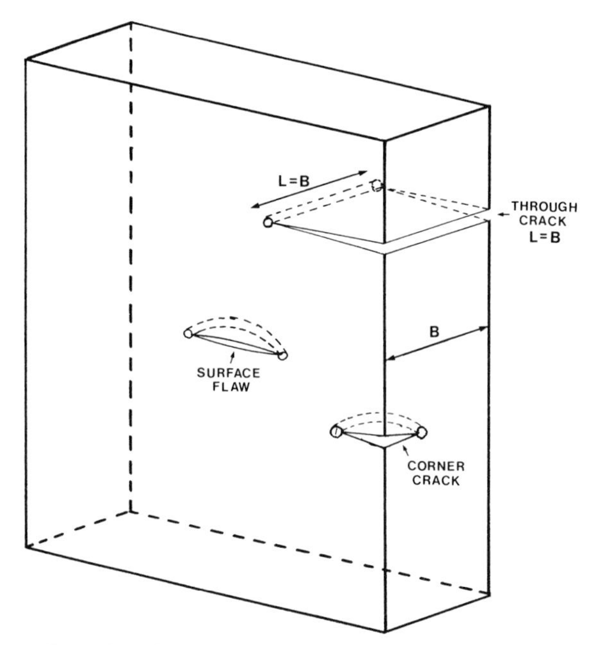

Figure 3.9. Constraint condition for part-through cracks.

Therefore part-through cracks (surface flaws and corner cracks) are in plane strain, at least in the center. For such cracks, one must use the plane strain fracture toughness, K_{Ic}, regardless of the thickness.

3.7. Measurement of toughness

In principle the toughness can be measured on any kind of cracked specimen. Fracture occurs when Equation (3.25) is satisfied. With knowledge of β for the specimen at hand, and the measured fracture stress (load) the stress intensity at fracture can be calculated; the result is the toughness. In the reverse operation, Equation (3.25b) is used to calculate the residual strength; the toughness then being known, the crack size given, and β obtained for the structure at hand, σ_{fr} follows from Equation (3.25b).

Although any kind of specimen can be used a few simple test specimens have been standardized by ASTM [7, 8] for the measurement of plane strain fracture toughness. The most universally used is the so-called compact tension (CT) specimen depicted in Figure 3.10. It contains a notch in the form of a chevron

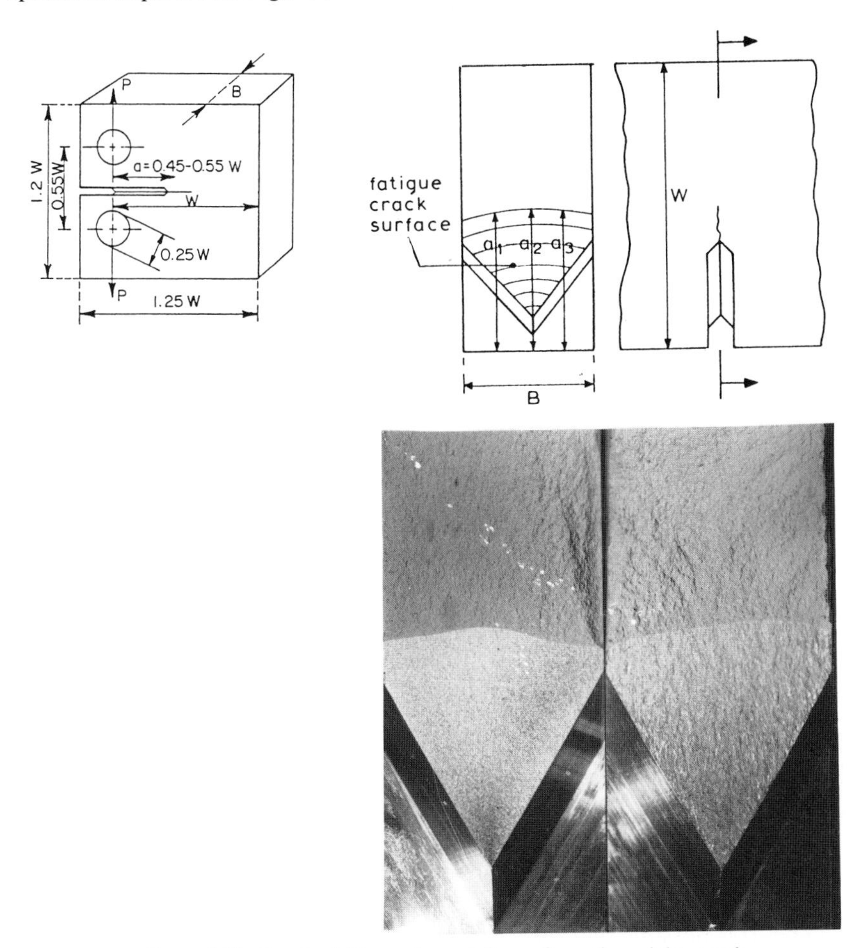

Figure 3.10. Standard K_{Ic} specimen; left: configuration; right: notch.

and is loaded through two pin's in a tensile machine. Cyclic loading is applied to introduce a fatigue crack. When the crack is at the desired length the cycling is stopped, and the load is raised until fracture occurs. The stress intensity at fracture can then be calculated; i.e. the toughness obtained.

The stress intensity factor for the compact specimen is usually written as:

$$K = \frac{P}{BW^{1/2}}\left[29.6\left(\frac{a}{W}\right)^{1/2} - 185.5\left(\frac{a}{W}\right)^{3/2} + 655.7\left(\frac{a}{W}\right)^{5/2} - 1017\left(\frac{a}{W}\right)^{7/2} + 639\left(\frac{a}{W}\right)^{9/2}\right]. \quad (3.28)$$

This expression seems to be rather different from Equation (3.14), but it is essentially the same. It is written in this manner because the nominal stress is hard to define. However, there is no objection against defining a (somewhat hypothetical) nominal stress as $\sigma = P/BW$. Then Equation (3.28) indeed reverts to:

$$K = \left[16.7 - 104.6\left(\frac{a}{W}\right) + 370\left(\frac{a}{W}\right)^{2} - 574\left(\frac{a}{W}\right)^{3} + 361\left(\frac{a}{W}\right)^{4}\right]\sigma\sqrt{\pi a} = \beta\sigma\sqrt{\pi a}. \quad (3.29)$$

During the test the so-called crack mouth opening displacement is measured by inserting a gage in the notch (Figure 3.11). Since there is hardly any strain in the material between the loading pins, the crack mouth opening displacement is essential equal to the displacement of the loading pins. Load and displacement are plotted in a load-displacement diagram (Figure 3.11). The maximum load is substituted in Equation (3.28) to find the toughness.

The test standard prescribes that the load-displacement line must be straight or nearly so. When extensive plastic deformation occurs the diagram will become non-linear. However, yielding is often confined to such a small plastic zone that non-linearity is hardly perceivable. Should the diagram be significantly non-linear then the plastic zone is large: as discussed in the previous sections the fracture criterion based on K is then no longer valid and "the" toughness cannot be defined. Hence, the standard sets a limitation on non-linearity for the measurement of 'valid' toughness numbers. Should the entire ligament yield then collapse may occur as discussed in Chapter 2.

The standard test is specifically for the measurement of the plane strain fracture toughness K_{Ic}. This means that there must be plane strain. Therefore the thickness of the specimen must be sufficient to satisfy Equation (3.27) with $q = 2.5$. The toughness is not known in advance (otherwise measurement would not be necessary), so that one cannot determine how thick the specimen should be. It is possible that the specimen was made too thin, in which case the

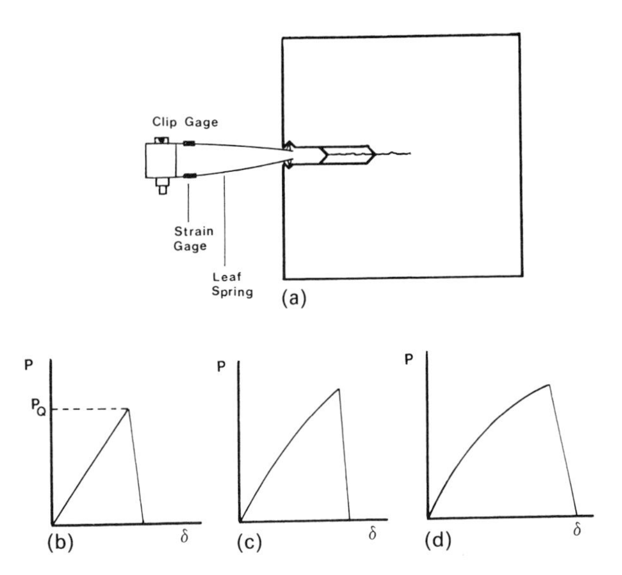

Figure 3.11. Standard K_{Ic} test. (a) Clip gage mounted in specimen; (b) Straight $P - \delta$ record; (c) Curvature due to plasticity; (d) Curvature due to plasticity and growth.

measured toughness would not be the plane strain fracture toughness. For this reason the toughness obtained from Equation (3.28) is initially called the candidate value, K_Q, of the plane strain fracture toughness, K_{Ic}. One must check whether $B > 2.5K_Q^2/F_{ty}^2$. If this is the case then indeed $K_{Ic} = K_Q$; if not then a new test must be done on a thicker specimen if one insists on knowing the plane strain toughness. It should be noted however, that $K_c = K_Q$ (Figure 3.6), is a perfectly useful toughness number for the actual thickness of the specimen. (See also Chapter 7).

As was pointed out already, the number $q = 2.5$ is rather dubious, and therefore rigorous application of Equation (3.27) somewhat ludicrous. The number 2.5 was obtained from an interpretation of a limited number of test data (Figure 3.8). It was then agreed upon by a committee that the number was reasonably representative, and as a consequence it appears in the test standard [8]. However, a committee decision, nor an ASTM standard, is a sufficient guarantee that the number is indisputable. If there is general concurrence that the number is useful then there are no objections against its use, but it still remains an arbitrary number. Also the uncommon significance attached to the standard test is largely exaggerated. Toughness can be measured on any cracked specimen, provided the plastic zone is small. If it could not, the result could not be used to calculate residual strength of any other structural crack (based upon the generality of Equations (3.2)), which is the purpose of the measurement. In that case the standard test would have no use in the first place. (See also Chapter 7).

Although the compact specimen is cheap and simple, it is totally unsuitable for the measurement of plane stress and transitional toughness. For such toughness measurements a center cracked panel is the most advisable. In order to obtain useful numbers, the plastic zone at fracture must be kept small. This often requires very large panels, as will appear in the following section. Due to the fact that stable fracture often preceeds fracture instability, a revised definition of toughness is needed as well (Section 3.12).

3.8. Competition with plastic collapse

In Chapter 2 the conditions for collapse were discussed. It turned out that collapse occurs in a center cracked panel when:

$$\sigma_{fc} = \frac{W - 2a}{W} F_{col}. \tag{3.30}$$

In this equation F_{col} is the collapse strength which may be as low as the yield strength and almost as high as the tensile strength. Recall from Chapter 2 that Equation (3.30) is a straight line as a function of crack size a.

Also discussed in Chapter 2 is the fact that Equation (3.30) represents the absolute highest load carrying capability: at collapse plastic deformation becomes unbounded and fracture follows, regardless of the toughness. Hence, (fracture) failure by collapse may occur before K reaches the toughness. Naturally, K_{Ic} is the result of a semi-elastic concept while at collapse the entire cracked section is yielding. Therefore, the two failure conditions cannot be directly compared; but they can be in an indirect way.

If the toughness is high, Equations (3.25b) will predict a very high fracture stress (residual strength). This may be as high or higher than the stress for failure by collapse. Then failure by collapse will prevail, because, of two competing failure modes, the one which first becomes possible will prevail. This may happen when the toughness is high, but under other conditions as well.

Should the fracture stress σ_{fr}, be higher than the stress causing failure by collapse, then collapse will prevail. This is the case when the result of Equation (3.25b) is more than the result of Equation (3.30), i.e. when (for a center crack):

$$\left.\begin{aligned} & \sigma_{fr} > \sigma_{fc} \\ \text{or} \quad & \frac{W - 2a}{W} F_{col} < \frac{K_c \text{ (or } K_{Ic})}{\sqrt{\sec \frac{\pi a}{W}}\sqrt{\pi a}} \end{aligned}\right\} \tag{3.31}$$

The lowest of the two is the actual residual strength discussed in Chapter 1.

From Equation (3.31) it appears that there are 3 situations in which a collapse failure could prevail, namely when

- the toughness is very high or;
- the crack is very small ($a \to 0$);
- the width W is very small.

These situations are illustrated in Figure 3.12. For the material with $K_{Ic} = 160\,\text{ksi}\sqrt{\text{in}}$ a panel, of 12 inch width always fails by collapse, but a 60 inch wide panel of the same material fails by K_{Ic}. Apparently, for a particular structural configuration failure may occur by collapse (unconditionally leading to fracture) in certain instances and by fracture in other instances. Collapse always prevails if the cracks are small, regardless of how low the toughness. At any toughness Equation (3.25b) puts the fracture stress at infinite when the crack size approaches zero. Thus the residual strength curve will always rise

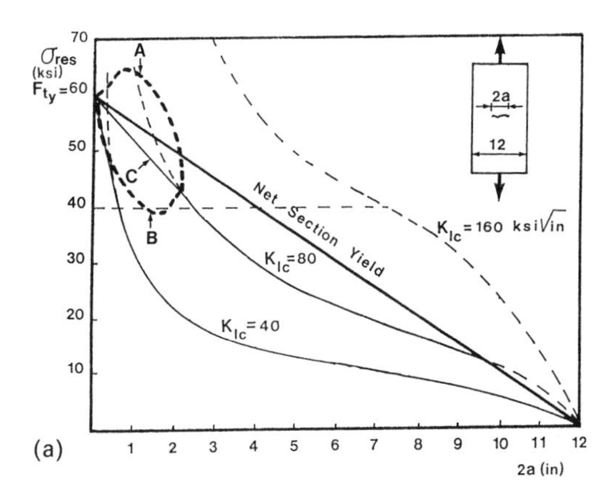

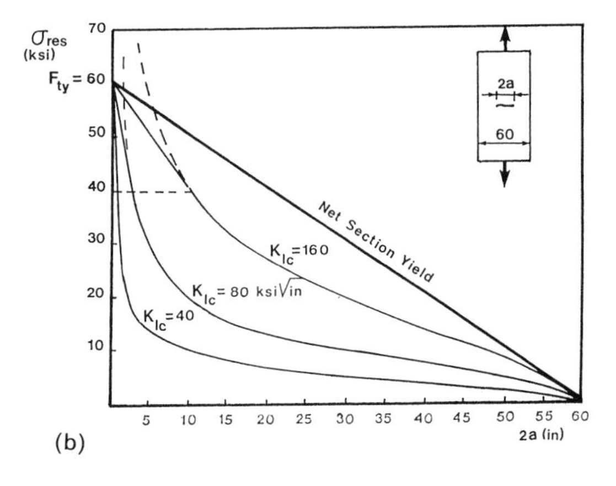

Figure 3.12. Effect of panel size on Residual Strength. (a) Residual strength of 12-in wide panels; (b) Residual strength of 60-in wide panels.

asymptotically to infinity for small a; this means that there will always be a point at a certain small a, where collapse failures will prevail, regardless how low the toughness.

For a crack size $a \to 0$, the fracture equations predict that the fracture stress will become infinite. Clearly, when $a = 0$ the fracture stress is equal to the ultimate tensile strength F_{tu}, or less. Thus, the most left-hand part of the curve represented by Equation (3.25b) will always be in error, whether the toughness is high or low. For $a = 0$, the strength is F_{tu} (or less), while for large a the Equation (3.25b) applies (which is the curve in Figure 3.12). Obviously, the behavior between $a = 0$ with $\sigma \approx F_{ty}$, and the curve for large a, cannot be as A or B in Figure 3.12a. Hence, if one assumes the 'eye-balled' curve C (tangent to the curve), the assumption cannot be far from the truth; it certainly will be adequate for engineering analysis.

It is important that of two plates of different sizes – but of the same material – one may fail by collapse, the other by fracture. This is depicted also in Figure 3.12 as well as in Figure 3.13. From this it can be concluded that a specimen for the measurement of toughness must be of sufficient size. If in the case of Figure 3.13 a plate of the smaller size W_3 would be used in a test, the failure would occur by collapse. Hence, the value of the stress intensity at the time of failure would still be lower than the toughness, because K has not yet reached the toughness.

This problem also exists for plane strain toughness testing with the CT specimen. The specimen may fail by collapse if the toughness is very high or if the specimen is too small (specimen size is rather ignored in the test standard). If one insisted to calculate K_Q from such a test by using Equation (3.28), the K_Q

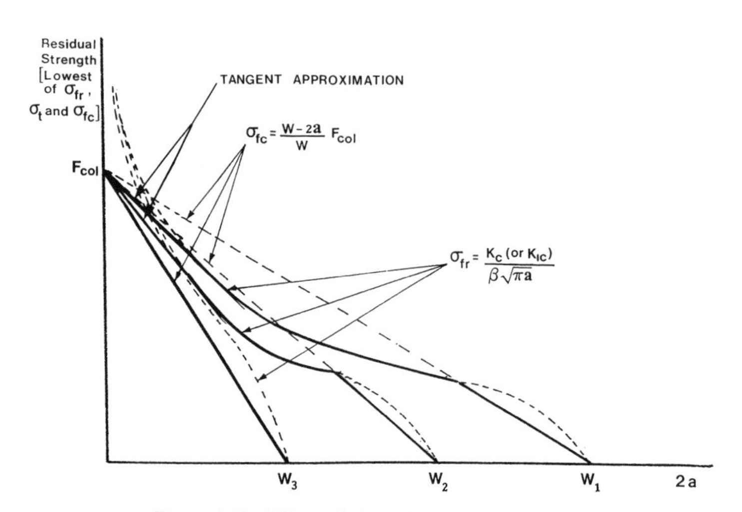

Figure 3.13. Effect of plate size on failure mode.

would be LESS than the actual K_{Ic}. Fortunately, the test would be invalid because the standard requires a 'straight' load displacement diagram. If the specimen were to fail by collapse, the load displacement diagram would not be straight because of large scale plastic deformation in the ligament.

Apparently, Equation (3.25b) is not sufficient to determine the residual strength diagram, nor is Equation (3.25a) enough to determine the toughness. It is always necessary that the stress for collapse be determined as well, because the lower of the two prevails, and the residual strength, σ_{res}, is the lower of σ_{fc} and σ_{fr}. For short cracks, the tangent approximation, σ_{ft}, may prevail if it is lower than σ_{fc} and σ_{fc}. Determination of the collapse curve is simple in the case of uniform tension loading. If the loading is non-uniform such as for the compact tension specimen, the collapse condition is somewhat more difficult to evaluate (see Table 2.1). The calculation of residual strength diagrams is discussed further in Chapter 10.

3.9. The energy criterion

This and the following sections present another look at the fracture criterion. This is of interest for understanding LEFM, but it is essential for the understanding of EPFM as discussed in Chapter 4. Finally, it will appear that the definition of toughness may have to be revised (Section 3.12).

Conservation of energy demands that the work, F, done by the load on a body, is not lost. It is conserved as strain energy, U, so that:

$$F - U = 0. \tag{3.32}$$

The work done by the load is $F = \int P \, \mathrm{d}\delta$ (where P is the load and δ the load displacement), which is the area under the load-displacement curve. As long as the material is elastic, the load-displacement diagram is a straight line, so that the work done by the load is $\frac{1}{2}P\delta$ (see Figure 3.14). It follows from Equation (3.32) that $U = \frac{1}{2}P\delta$ as well. But U can be determined in a different manner.

Consider a small material element of unit size, subjected to a uniaxial tension σ, as shown in Figure 3.14b. The stress does work to deform the material over $\mathrm{d}\varepsilon$ the total work done is $\int \sigma \, \mathrm{d}\varepsilon$, which for a linear elastic material is the area of the triangle shown in Figure 3.14b, namely $\frac{1}{2}\sigma\varepsilon$. By substituting $\varepsilon = \sigma/E$ from Hooke's law, the work becomes $\frac{1}{2}\sigma^2/E$. As this work must be equal to the strain energy, the value of $\frac{1}{2}\sigma^2/E$ also represents the strain energy in the material element.

The stress on each volume element is not always the same. Therefore the total strain energy in the entire body or structure is obtained by taking the integral over the volume of the body:

$$U = \iiint \frac{\sigma^2}{2E} \, \mathrm{d}x \, \mathrm{d}y \, \mathrm{d}z \tag{3.33}$$

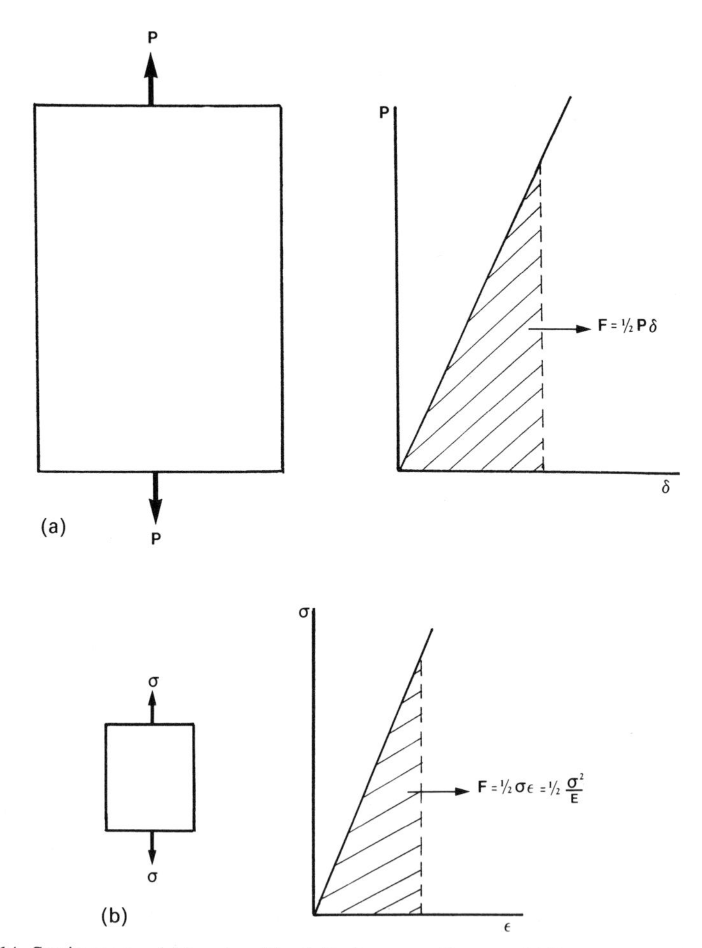

Figure 3.14. Strain energy. (a) Load and load-displacement diagram; (b) Stress on material element.

in which σ may depend upon x, y and z. In the case of a simple tensile bar the stress is the same on each and every volume element. In that case the total strain energy is simply equal to $\frac{1}{2}\sigma^2/E$ times the volume. If the cross section of the bar is A and its length L, then the strain energy is:

$$U = \frac{\sigma^2}{2E} LA. \tag{3.34}$$

With knowledge of F $(= \frac{1}{2}P\delta)$ one can substitute both F and U in Equation (3.32) to obtain:

$$\tfrac{1}{2}P\delta - \frac{\sigma^2}{2E} LA = 0. \tag{3.35}$$

By noting that, in this case, $\delta = \varepsilon L = \sigma L/E$, and $P = \sigma LA$, it can be readily seen that Equation (3.35) is true. As already stated, the criterion provides $U = \frac{1}{2}P\delta$.

Equation (3.32) also holds when the body has a crack, $2a$. In the case of limited plasticity the load-displacement diagram is still a straight line as in Figure 3.15. If a crack of somewhat larger size, $a + \mathrm{d}a$, exists then it will take less load to cause the same displacement: the stiffness is less. This, the load displacement line is lower as shown in Figure 3.15d.

Let the load increase until it is P_1, the displacement δ_1, as in Figure 3.15d. In the event that fracture takes place at this load, the crack extends over a small increment $\mathrm{d}a$ from a to $a + \mathrm{d}a$. During this process there must be energy conservation, but instead of two energy terms F and U there is now a third energy term W, where W is the work expended in fracturing material over $\mathrm{d}a$. The energy conservation equation now covers only the changes of the energy. During fracture over $\mathrm{d}a$ the load may do some work $\mathrm{d}F$, the strain energy may change somewhat, $\mathrm{d}U$, and some energy, $\mathrm{d}W$, will be required for fracture. Hence, the energy conservation criterion reads:

$$\frac{\mathrm{d}}{\mathrm{d}a}(F - U - W) = 0 \tag{3.36}$$

which can be written as:

$$\frac{\mathrm{d}}{\mathrm{d}a}(F - U) = \frac{\mathrm{d}W}{\mathrm{d}a}. \tag{3.37}$$

The equality must hold when fracture occurs. Conversely when the equality cannot hold, fracture does not yet take place and Equation (3.32) remains valid. Apparently then, Equation (3.37) is a fracture criterion. Fracture will occur when enough energy can be delivered to provide for the fracture energy $\mathrm{d}W/\mathrm{d}a$. Energy delivery must come from a surplus of $\mathrm{d}(F - U)/\mathrm{d}a$. If this surplus is sufficient to cover the energy required for fracture, $\mathrm{d}W/\mathrm{d}a$, then fracture will occur; if not Equation (3.32) continues to govern in which case $\mathrm{d}(F - U)/\mathrm{d}a$ is zero.

The fracture criterion of Equation (3.37) can be used if the energy delivery $\mathrm{d}(F - U)/\mathrm{d}a$ can be quantified. This will be done in the next section.

3.10. The energy release rate

Consider again the situation of Figure 3.15d with a load, P_1, resulting in a displacement δ_1. First examine the situation in which the displacement does not change when the fracture occurs over $\mathrm{d}a$. In that case the load decreases to P_2, because it is easier to maintain the displacement δ_1, with a longer crack. However, the load does not move and therefore it does no work: $\mathrm{d}F = 0$. The strain energy does change from $\frac{1}{2}P_1\delta_1$ (area OAD) to $\frac{1}{2}P_2\delta_1$ (area OBD) after the

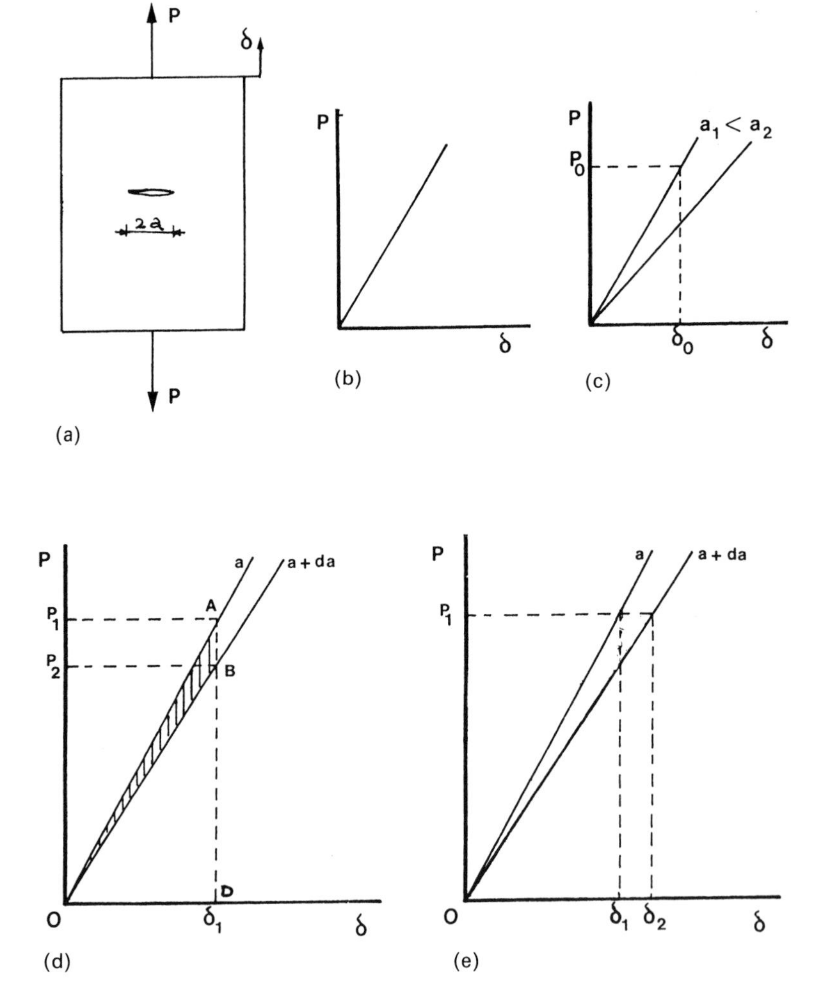

Figure 3.15. Energy release during constant displacement and constant load. (a) Plate with crack a under load P; displacement δ; (b) Load displacement record; (c) $P\delta$ record for small and larger crack; (d) Fracturing at constant displacement; (e) Fracturing at constant load.

process. The strain energy decreases, so the change is negative. Hence, it turns out that the deliverable energy is $-\mathrm{d}U/\mathrm{d}a$, and Equation (3.37) becomes:

$$-\frac{\mathrm{d}U}{\mathrm{d}a} = \frac{\mathrm{d}W}{\mathrm{d}a} \quad \text{where} \quad \frac{\mathrm{d}U}{\mathrm{d}a} < 0. \tag{3.38}$$

Next examine the case in which the load remains constant during the fracture over da. In that case the displacement increases from δ_1 to δ_2 as shown in Figure

3.15e. Now the load does work, namely an amount $P(\delta_2 - \delta_1)$; note that the load remains constant during the process. The strain energy also changes, namely from $\frac{1}{2}P_1\delta_1$ to $\frac{1}{2}P_1\delta_2$. This time the strain energy increases, so its change is positive. Equation (3.37) becomes:

$$\left.\begin{aligned} P_1(\delta_2 - \delta_1) - \tfrac{1}{2}P_1(\delta_2 - \delta_1) &= \frac{\mathrm{d}W}{\mathrm{d}a} \\ \text{or} \qquad \tfrac{1}{2}P_1(\delta_2 - \delta_1) &= \frac{\mathrm{d}W}{\mathrm{d}a} \end{aligned}\right\} . \tag{3.39}$$

In one case the deliverable energy is $-\mathrm{d}U/\mathrm{d}a$, in the other case $\frac{1}{2}P(\delta_2 - \delta_1)$. Note that the latter amount is exactly equal to $\mathrm{d}U/\mathrm{d}a$. It appears then that the deliverable energy is always the same, namely $\mathrm{d}U/\mathrm{d}a$, regardless of whether fracture occurs under constant load or constant displacement. In either case the deliverable energy is equal to the change in strain energy $\mathrm{d}U/\mathrm{d}a$. For constant displacement the deliverable energy is coming directly from a release of strain energy. In the other case the strain energy increases, but the load does twice that much in work, so that the surplus is still equal to the change in strain energy, be it with opposite sign

Clearly then, the deliverable energy is always equal to the change in strain energy regardless of its sign. Instead of $\mathrm{d}(F - U)/\mathrm{d}a$ one may then use the absolute change of the strain energy $\mathrm{d}U/\mathrm{d}a$ in Equation 3.37. Then the fracture criterion is:

$$\frac{\mathrm{d}U}{\mathrm{d}a} = \frac{\mathrm{d}W}{\mathrm{d}a} . \tag{3.40}$$

The left hand side is called the strain energy release rate, the right hand side the fracture energy or fracture resistance. Fracture will occur if, due to the extended slit by $\mathrm{d}a$, sufficient strain energy is released to cover the energy to create a fracture over $\mathrm{d}a$.

The fracture criterion is somewhat simpler now than before, but it remains necessary to find an expression for $\mathrm{d}U/\mathrm{d}a$. As the strain energy is affected by the crack one can write:

$$U = U_{\text{body with no crack}} + U_{\text{due to crack}} . \tag{3.41}$$

Consider a very large plate of length L, width W and thickness B with a small center crack of $2a$. If the loading is uniform tension then the strain energy of each element of the uncracked body is the same and equal to $\frac{1}{2}\sigma^2/E$, so that the strain energy of the uncracked body becomes:

$$U_{\text{body with no crack}} = \frac{\sigma^2}{2E} LBW. \tag{3.42}$$

The stress due to the crack will depend upon the crack tip stresses. These crack tip stresses are proportional to the applied stress σ. Thus the strain energy due to the crack must be proportional to σ^2/E. The term will also be proportional to the thickness B (twice as thick a plate has twice the energy). Hence,

$$U_{\text{due to crack}} \div \frac{\sigma^2}{E} B. \tag{3.43}$$

The contributions by Equations (3.42) and (3.43) must have the same dimension. They differ by a length squared (LW). Since the contribution due to the crack must depend upon crack size, a, it becomes inescapable that it depends upon a^2, otherwise the dimensions would not be correct. Therefore:

$$U_{\text{due to crack}} = C \frac{\sigma^2}{E} Ba^2. \tag{3.44}$$

There can be a dimensionless coefficient C in the equation, the value of which cannot be obtained from simple dimensional analysis. Formal calculation of the strain energy will show that $C = \pi$. Using this and combining Equations (3.42) and (3.44) gives the total strain energy U. As everything is evaluated for unit thickness, one must divide by B to obtain:

$$U = \frac{\sigma^2}{2E} LW + \frac{\pi\sigma^2 a^2}{E} \tag{3.45}$$

Now $\mathrm{d}U/\mathrm{d}a$ follows from differentiation as:

$$\frac{\mathrm{d}U}{\mathrm{d}a} = \frac{2\pi\sigma^2 a}{E} \tag{3.46}$$

This is for a crack with two tips. Since all consideration are for one crack tip, one must use

$$\frac{\mathrm{d}U}{\mathrm{d}a} = \frac{\pi\sigma^2 a}{E} \tag{3.47}$$

per crack tip and per unit thickness.

This permits writing the fracture criterion of Equation (3.40) as:

$$\text{Fracture if:} \quad \frac{\pi\sigma^2 a}{E} = \frac{\mathrm{d}W}{\mathrm{d}a}. \tag{3.48}$$

Usually, the fracture energy per unit crack extension $\mathrm{d}W/\mathrm{d}a$ is denoted by R (for fracture Resistance), while the energy release rate $\mathrm{d}U/\mathrm{d}a$ is denoted as G. Then Equation (3.48) in shorthand is given as:

$$\frac{\pi\sigma^2 a}{E} = R \quad \text{or} \quad G = R \tag{3.49}$$

which is the same as Equation (3.48) except for the short hand notation.

3.11. The meaning of the energy criterion

Equation (3.49) shows that fracture occurs when ($\pi\sigma^2 a$) reaches a certain value, namely ER. Now notice that $\pi\sigma^2 a$ is exactly the square of the stress intensity factor: K. Hence, the equation pronounces that fracture occurs when K^2 reaches a certain value, which is the same as K reaching a certain value:

$$\text{Fracture if:} \quad K = \sqrt{ER}. \tag{3.50}$$

It follows then that $\sqrt{ER}$ must be the toughness K_c, and apparently the fracture resistance is $R = K_c^2/E$.

It may be concluded that, fortunately, the fracture criterion derived from the energy conservation law appears to be identical to the one derived before on the basis of crack tip stress. Had this not been the case, one of the two would have to be declared wrong. Apply this criterion to the same test and example as in Section 3.4. From the test one can obtain the material's fracture energy R. Assuming a steel for which $E = 30\,000$ ksi, and knowing that $\beta \approx 1$, $a = 2$ inch, and the fracture stress 30 ksi, Equation (3.49) provides the fracture energy:

$$R = \frac{\pi \times 30^2 \times 2}{30\,000} = 0.19\,\frac{\text{ksi}^2\,\text{in}}{\text{ksi}}$$

$$= 0.19\,\frac{\text{in kip}}{\text{in}^2} = 0.19\,\text{kip/in} = 190\,\text{lbs/in}.$$

Note that indeed $R = K_c^2/E$, because with $K = 75\,\text{ksi}\sqrt{\text{in}}$ (Section 3.4) one obtains $R = 75^2/30\,000 = 0.19$ kips/in. The fracture energy is an energy (in/lbs or Nm) per unit crack extension (in or m) and per unit thickness (in or m), so that its unit is in lbs/in or $\text{Nm}^2/\text{m} = \text{N/m}$.

One can apply this result to calculate the residual strength for the same case as in Section 3.4. The fracture stress follows from Equation (3.49) as:

$$\sigma_{fr} = \sqrt{\frac{ER}{\beta^2 \pi a}}\,. \tag{3.51}$$

Substituting the value of $R = 0.19$ kips/in obtained above, $\beta = 1.12$, and $a = 3$ in, one obtains $\sigma_{fr} = 22$ ksi, which is the same result as obtained in Section 3.4. The two fracture criteria are indeed equivalent.

3.12. The rise in fracture resistance: redefinition of toughness

It would seem reasonable to assume that the fracture resistance R is constant and independent of how far fracture has progressed. This is assuming that it takes equally much energy to fracture the first da as the adjacent, next da. It would cause R to be a horizontal line as a function of Δa, as shown in Figure 3.16a.

The deliverable energy (dU/da), usually denoted as G is equal to K^2/E. For

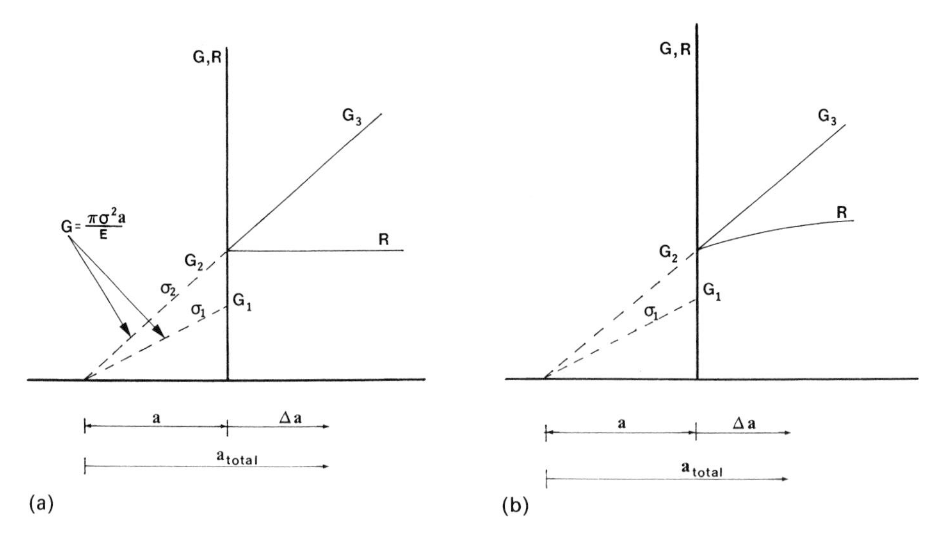

Figure 3.16. *R*-Curve. (a) Horizontal *R*-curve; (b) Rising *R*-curve.

the case that $\beta = 1$, G would be $G = \pi\sigma^2 a/E$, which is a straight line as a function of a, with a slope depending upon σ. Assume a crack of size a present, and take a second origin in the diagram by taking a to the left. One can then draw the dashed line for G, the actual value of G for crack size a being given by G_1 on the ordinate.

Clearly G_1 is less than R, and fracture will not take place at the stress σ_1. Raising the stress to σ_2 will increase the value of G to G_2. Since $G_2 = R$, fracture will indeed occur. The 'crack' extends during fracture, so that G increases from G_2 to G_3, while R remains the same. Hence, G remains larger than R and fracturing will continue uncontrollably.

In reality R is not a horizontal line. The fracture energy appears to increase somewhat during fracture [1]. If the increase is small, such as in Figure 3.16b, the fracture will still proceed as above. This is generally the case when the toughness is low or in plane strain. When the toughness is high, the R-curve rises more steeply, as shown in Figure 3.17.

At the stress σ_i, the energy release G_i, is equal to R for the first time. Fracturing will commence but not continue; it is stable. If the fracture proceeds over Δa, the value of G rises from G_i to G_1. But R rises steeper, and although G_i was equal to R, the new G_1 is less than (the new increased) R. Apparently, the fracture cannot proceed (it is stable).

Further increase of the stress to σ_2 will cause fracture to continue. At G_2 the deliverable energy is again equal to R. Finally, at the stress σ_{fr}, the fracture will be unstable, because G_3 increasing to G_4 remains larger than R. Fracture will then proceed uncontrollably (instability).

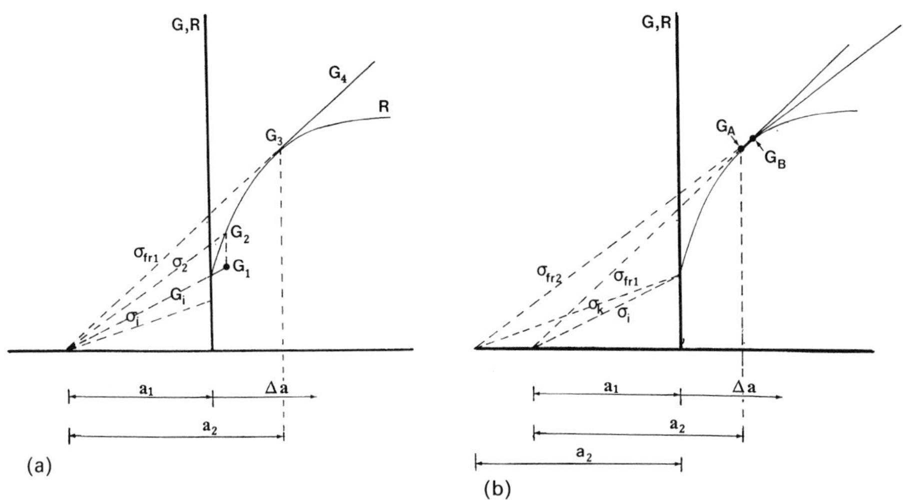

Figure 3.17. Rising *R*-Curve. (a) Stable fracture from G_i to G_3; (b) Stable fracture at cracks of different length.

During the stable phase of the fracture the crack extends from a_1 to a_2. This should not be interpreted to mean that a crack of size a_2 has a strength σ_{fr1}. Although this seems trivial, serious errors in fracture calculations often occur due to this misinterpretation as shown later in this section. Extension of a_1 to a_2 occurs by fracture and not by the process by which a_1 was formed as e.g. by fatigue or stress corrosion. A fatigue crack of length a_2 cannot sustain the same high stress. Fracturing will commence at a lower stress than σ_{fr1}, namely σ_{fr2}, (Figure 3.17b). Fracture instability occurs at σ_{fr2}, which is lower than σ_{fr1}. Note that the smaller slope of the *G*-lines indicates lower stress.

The interpretation is that a fatigue crack of size a_1 causes the strength to be σ_{fr1}. That the fracture proceeds to a_2 before it becomes unstable is rather immaterial for the end result (two half structures). Similarly, a crack of size a_2 has a lower strength σ_{fr2}. Given the presence of a certain fatigue crack say a_1, in e.g. an airplane wing, the strength of the wing is σ_{fr1}. That the fracture proceeds first to a_2 before the wing comes off is rather immaterial for the passenger (and for the engineer). It is the strength that counts. Naturally, if the stress would rise above σ_i (Figure 3.17) but not reach σ_{fr} then the wing would stay together and the crack would be somewhat longer than a_1, due to some stable fracture. However, this is trivial: if the fracture stress is not reached the structure always stays together, whether there is stable fracture or not.

Clearly, fracture instability occurs when the *G* line is tangent to the *R*-curve, i.e. when the two have equal slopes (Figure 3.17). As the slope of the line is described by the first derivative, the condition for instability is:

$$\left.\begin{aligned} G &= R \\ \text{and also} \quad & \\ \mathrm{d}G/\mathrm{d}a &= \mathrm{d}R/\mathrm{d}a\,. \end{aligned}\right\} \tag{3.52}$$

If only the first equation is satisfied fracture will occur, but in order for fracture to become unstable (uncontrollable), the second equation must be satisfied as well.

This has important consequences for the definition of toughness. In previous sections toughness was defined as the stress intensity at which fracture occurs. It now appears that (unstable) fracture occurs at different values of G for different crack sizes. Figure 3.17 shows that with crack of size a_1 fracture occurs at G_A, while with a crack size a_2, fracture occurs at G_B. Since $K^2 = \sqrt{EG}$ it follows that the values of K are different at the time of fracture in the two cases. In previous sections it was assumed that fracture always takes place at the same value of K, namely the toughness K_c. Apparently, this is not true if there is stable fracture first. This might cause serious complications for the analysis of fracture, but it will be shown below that an engineering approach can solve the problem.

It should first be mentioned that Figures 3.16 and 3.17 are for the case that $\beta = 1$, namely the infinite panel. For real structural cracks β is not equal to one, so that the G-lines are curved (note that $G = K^2/E = \beta^2\sigma^2\pi a/E$ (The curvature being larger for large cracks, the values of G, and hence of K, at fracture tend to be more or less equal to different crack sizes as illustrated in Figure 3.18a. From this one might conclude that the assumption of a constant K (namely K_c) at fracture was not so objectionable from an engineering point of view.

Next consider the residual strength diagram as shown in Figure 3.18b. In the presence of a fatigue crack of say one inch, fracture starts at a stress indicated by A, and fracture instability occurs at a stress indicated by B (compare previous figures). This means that if a crack of 1 inch is present, the strength of the structure is defined by B. It does not mean that the residual strength with a crack size of 3 inches is defined by B. The figure shows that such a crack would cause fracture instability at D, and its residual would therefore be defined by D. Although the one inch crack shows stable fracture until 3 inches, a fatigue crack of three inches would cause fracture at D, and not at B.

The two curves in Figure 3.18b can be defined by a 'critical' stress intensity. The top curve is determined by the critical G (or K) for fracture instability, the bottom curve by the value of G (or K) for the onset of fracture:

$$\left.\begin{aligned} \text{Onset of fracture:} \quad \beta\sigma_i\sqrt{\pi a_i} &= K_i \\ \text{Instability:} \quad \beta\sigma_c\sqrt{\pi a_c} &= K_c \end{aligned}\right\}. \tag{3.53}$$

From this the stresses at onset of fracture, σ_i, and the stresses at fracture instability, σ_c, would follow from:

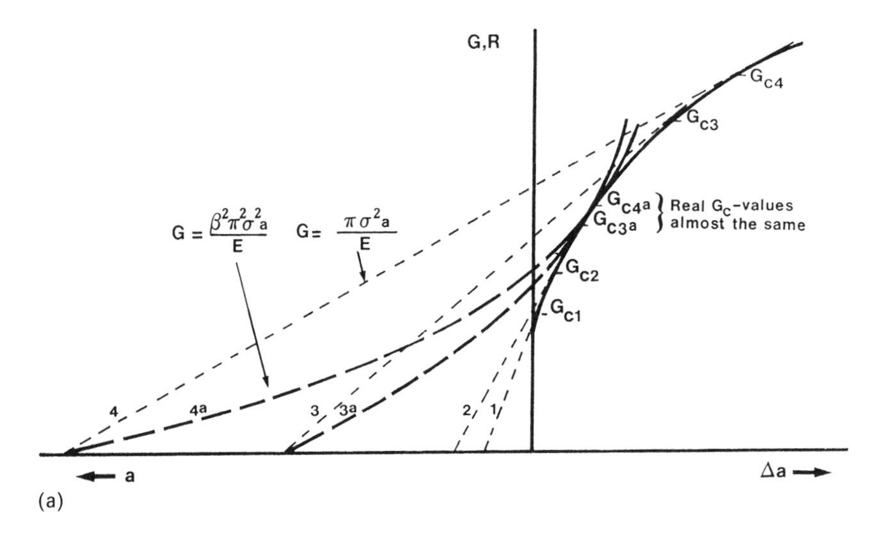

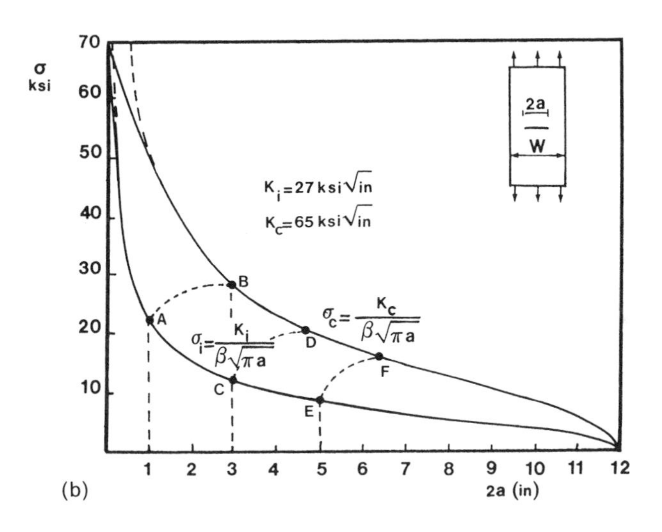

Figure 3.18. Approximate Constancy of G_i and K_c. (a) Curved as opposed to straight G-lines; (b) Residual strength curves for plane stress.

$$\left.\begin{aligned} \sigma_i &= \frac{K_i}{\beta\sqrt{\pi a_i}} \\ \sigma_c &= \frac{K_c}{\beta\sqrt{\pi a_c}} \end{aligned}\right\} . \tag{3.54}$$

The curves represented by Equations (3.54) are shown as the top and bottom curve in Figure 3.19a. Clearly, the value of K_c in the above equation is not a constant (see previous figures), but the value of K_c depends upon crack size. On the other hand, as shown in Figure 3.18a, the value of K_c will not vary a great deal in actual structures.

Neither the top curve, nor the bottom curve in Figure 3.19a provides the actual residual strength. On the basis of Figure 3.18, it was shown that the actual strength of a structure with a crack of say one inch is B. Hence, to obtain the actual residual strength of this crack one should plot the strength B at the original crack size. This has been done in Figure 3.19.

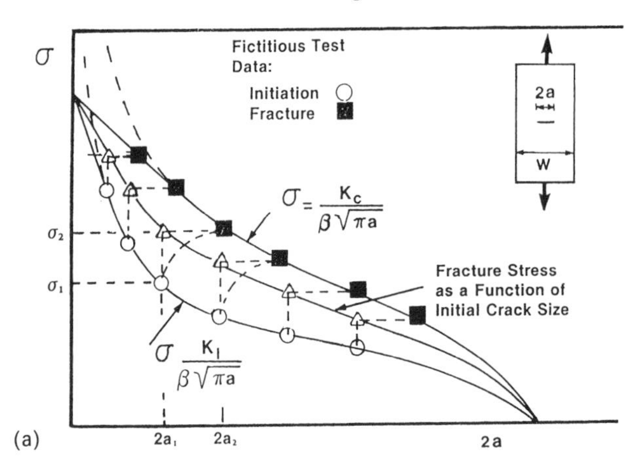

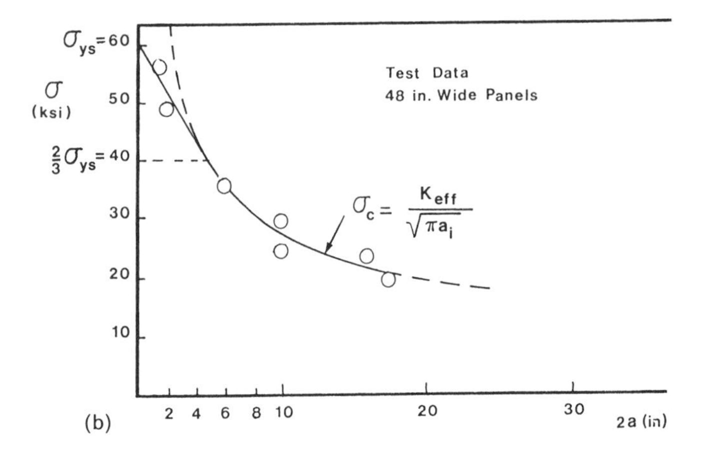

Figure 3.19. Use of K_{eff} in Fracture Analysis. (a) Residual strength and initial crack size; (b) Predicted curve and test data on basis of K_{eff}.

The residual strength in the presence of cracks is determined neither by the top curve, nor the bottom curve in Figure 3.19a, but by the middle curve. The latter gives the actual residual strength for the case that a crack of size a is present. One could assign a critical K_{eff} (effective toughness value) to this curve, and describe the curve by the following equations:
Fracture if:

$$\left.\begin{aligned} K &= K_{\text{eff}} \\ \beta\sigma_c\sqrt{\pi a_i} &= K_{\text{eff}} \\ \text{and} \qquad & \\ \sigma_{fr} &= \frac{K_{\text{eff}}}{\beta\sqrt{\pi a_i}} \end{aligned}\right\} \tag{3.55}$$

In these equations, indeed a_i is the length of the crack as caused by fatigue or stress corrosion, the fracture stress, σ_{fr}, is indeed the strength of the structure with this crack present. However, the Equations (3.55) combine a crack and stress not occurring simultaneously. As such, the equations are physically in error. However, from an engineering point of view, they are perfectly acceptable, as long as K_{eff} is more or less a constant (material property). As was shown in Figure 3.18a this is indeed the case.

The upper curve in Figure 3.19a could be defined by:

$$\beta\sigma_{fr}\sqrt{\pi a_c} = K_c. \tag{3.56}$$

This would be a good definition if K_c were approximately constant, which is the case as shown in Figure 3.18a. However, if this were used in a fracture analysis, an erroneous result would be obtained:

$$\sigma_{fr} = \frac{K_c}{\beta\sqrt{\pi a}}. \tag{3.57}$$

The fracture stress calculated by Equation (3.57) would be for e.g. a fatigue crack size of three inches in Figure 3.18b, and therefore would be B, while the REAL fracture stress is at D. Clearly, the use of K_c as defined would lead to wrong answers. The use of Equation (3.55) would lead to the correct answer if an engineering approximation is accepted.

It must be concluded that if the toughness is defined as K_{eff}, the engineering answer will be correct, while erroneous answers are obtained with K_c. Nevertheless, in many instances reviewed by the author, K_c was used instead of K_{eff}. As it is somewhat cumbersome to use K_{eff} for toughness, the discussion of residual strength analysis in Chapter 10 will make use of the denotation K_c, but it is strongly emphasized that K_{eff} is meant; i.e. the numbers used for the toughness are K_{eff}, and NOT K_c.

It seems almost superfluous to mention that the above discussion is not restricted to plane stress or transitional cases. All arguments hold for plane strain, although this is often not realized. As the R-curve in plane strain usually rises only moderately, the problem is not acute; it nevertheless exists. Hence, all of the above applies to plain strain as well as to plane stress.

Clearly, the above engineering approximations need not be made if fracture analysis is based on the R-curve. The procedure would then be to solve the problem of Figures 3.15 through 3.18 either graphically or mathematically. This can be done rather easily (Chapter 10). The question is whether the result will be any more reliable. From an academic point of view, it will be, but from a practical point of view it is not. Such a solution requires knowledge of the R-curve. Certainly, the R-curve can be measured, but it requires the measurement of the progress of fracture. Such measurements by their nature are so inaccurate that the R-curve can be determined only in an approximate way. Its subsequent use in analysis does not provide more accurate answers than the engineering solution discussed above. Academic rigor does not necessarily lead to better answers; it only complicates problems. This will become abundantly clear in the next chapter.

3.13. Exercises

1. Calculate the fracture toughness of a material for which a plate test with a central crack gives the following information: $W = 20$ in, $B = 0.75$ inch, $2a = 2$ inch, failure load $P = 300$ kip. The yield strength is $F_{ty} = 70$ ksi. Is this plane strain? Check for collapse. How large is the plastic zone at the time of fracture?

2. Using the result of problem 1, calculate the residual strength of a plate with an edge crack $W = 5$ inch, $a = 2$ inch. Do the same for $W = 6$ inch and $a = 0.5$ inch (See Figure 3.3 for β). Check for collapse.

3. Given is a toughness of $K = 70\,\text{ksi}\sqrt{\text{in}}$ and a collapse strength equal to the yield strength with $F_{ty} = 75$ ksi. Determine the residual strength of a center cracked plate of 20 inch width with a crack of $2a = 3$ inch, and of a center cracked plate of two inch width with a crack of $2a = 1$ inch; check for collapse.

4. In a toughness test on a center cracked plate one obtains the following results: $W = 6$ inch, $B = 0.2$ inch, $2a = 2$ inch, $P_{max} = 41$ kips; $F_{ty} = 50$ ksi. Calculate the toughness. How large is the plastic zone at fracture? Can this problem be solved with Equation (3.25), and why not? Is the calculated toughness indeed the true toughness; why not? What caused the fracture?

5. In a plane strain fracture toughness test on a compact tension specimen the failure load is 5 kip. $W = 2$ in, $a = 1$ in, $B = 1$ inch, $F_{ty} = 80$ ksi. Calculate the plane strain fracture toughness, using both Equations (3.28) and (3.29). Is this a valid number? What is the size of the plastic zone at fracture?

6. In a plane strain toughness test one obtains a value of $K_Q = 50\,\text{ksi}\,\sqrt{\text{in.}}$ Is this a valid number if the material's yield strength is 100 ksi and the specimen thickness is 0.5 inch? What is the maximum toughness this specimen can measure? If the toughness is not valid, then estimate the plane strain fracture toughness with an accuracy of 5%. How thick should one take the specimen in order to measure the plane strain toughness under any conditions?

7. Cut a strip of paper four inches wide. With a knife or scissors cut a center crack (slit) of length $2a = 1.5$ inches in the top half. In the bottom half cut two edge cracks with $a = 0.60$ inch. Predict where the strip will fail in tension. Confirm your answer by rolling the ends around pencils and by pulling.

8. (a) Calculate the complete residual strength diagram for a center cracked plate of a material with $F_{ty} = 70$ ksi, $K_c = 90\,\text{ksi}\,\sqrt{\text{in}}$ and $W = 24$ inch.
(b) Do the same for a panel of $W = 6$ inch.
(c) Do the same for a panel with a single edge crack assuming β of Figure 8.3 is applicable; $W = 10$ inch. Note: Think of tangent and collapse. For each case determine the permissible crack size for a stress of 58 ksi, a stress of 50 ksi, and a stress of 10 ksi, by reading from your diagrams.

References

[1] D. Broek, *Elementary engineering fracture mechanics*, 4th Edition, Nijhoff (1985).
[2] M.F. Kanninen and C.H. Popelar, *Advanced fracture mechanics*, Oxford University Press (1985).
[3] J.F. Knott, *Fundamentals of fracture mechanics*, Butterworths (1973).
[4] H. Tada, P.C. Paris and G.R. Irwin, *The stress analysis of cracks handbook*, Del Research (1973).
[5] G.C. Sih, *Handbook of stress intensity factors*, Lehigh University Press (1973).
[6] D.P. Rooke and D.J. Cartwright, *Compendium of stress intensity factors*, Her Majesty's Stationery Office (1976).
[7] W.F. Brown and J.E. Srawley, *Plane strain crack toughness testing of high strength metallic materials*, ASTM STP 410 (1966).
[8] Anon., *The standard K_{Ic} test*, ASTM Standard E-399.

CHAPTER 4

Elastic-plastic fracture mechanics

4.1. Scope

If fracture is accompanied by considerable plastic deformation, a concept known as Elastic-Plastic Fracture Mechanics (EPFM) is used. The literature on this subject is very confusing. The fracture parameter used is often referred to as the 'J-integral'. However, J is simply the strain energy release rate, and on this basis the equation for J can be readily obtained, without the lengthy derivations, from a generalization of the energy criterion discussed in the previous chapter.

In the following the reader is assumed to be familiar with the strain energy release criterion for LEFM. Instead of following the historical development of EPFM, it will be shown that the strain energy release criterion can be used directly for elastic-plastic fracture by the simple expedient of an equation for the non-linear stress-strain curve. This leads directly to an expression for J. Its application and use will be discussed. In order to avoid confusion, the argumentation will be in small steps at a time. This leads to some repetition but will facilitate understanding. It also leads to a few more equations, but these can be followed easily because no steps are omitted. In a later section the integral form of J will be considered. The use of the crack opening displacement (COD) criterion will be discussed as well.

4.2. The energy criterion for plastic fracture

Whether there is plasticity or not, the energy conservation criterion must hold. It was demonstrated in Chapter 3 that this leads to the following fracture criterion:

$$\frac{\mathrm{d}U}{\mathrm{d}a} = \frac{\mathrm{d}W}{\mathrm{d}a} \quad \text{or} \quad G = R. \tag{4.1}$$

For elastic behavior it could be demonstrated that

$$\frac{\beta^2 \pi \sigma^2 a}{E} = \frac{\mathrm{d}W}{\mathrm{d}a}. \tag{4.2}$$

With the use of Hook's law ($\sigma = \varepsilon E$), the energy release rate, $\mathrm{d}U/\mathrm{d}a$, in the above equation can be written in terms of stress and strain:

$$\beta^2 \pi \sigma \varepsilon a = R \quad \text{or} \quad G = R. \tag{4.3}$$

The strain energy release is usually denoted in shorthand by G, and the fracture energy by R. Although the expression for G was derived formally in the previous chapter, it is important to notice that Equation (4.3) must be true on the basis of dimensional arguments. The factor β is dimensionless. Since strain energy is $\int \sigma \, \mathrm{d}\varepsilon = C\sigma\varepsilon$ (Chapter 3), the equation for G, the strain energy release rate, must contain the factor $\sigma\varepsilon$. Obviously, G depends also upon the crack size a. As G is an energy per unit thickness and per unit extension, its dimension is in-lbs/in^2 = lbs/in. The dimension of σ is lbs/in^2 (ε is dimensionless). Therefore, the crack size must appear in linear form as a (not a^2 or a^3 or otherwise). In that case the dimension of G becomes indeed lbs/in^2 $\times$ in = lbs/in. Thus is can be understood without analysis that Equation (4.3) must be of the format shown.

Equation (4.3) can be changed into (4.2) because the stress-strain equation ($\varepsilon = \sigma/E$) is known. Expressing G in terms of stress only, permits the use of the criterion in engineering analysis where stresses are of interest:

$$\frac{\pi \beta^2 \sigma^2 a}{E} = R \tag{4.4}$$

so that the fracture stress can be calculated from:

$$\sigma_{fr} = \sqrt{\frac{ER}{\pi \beta^2 a}} \left(= \sqrt{\frac{K_c^2}{\pi \beta^2 a}} = \frac{K_c}{\beta\sqrt{\pi a}} \right). \tag{4.5}$$

Provided β is known for the geometry at hand, Equation (4.5) can be used for engineering fracture analysis to calculate the fracture stress of a cracked structural component.

In the case of plastic deformation the geometry factor β may change, but there will still be a dimensionless geometry factor; let this factor be denoted as H. Whether the stress-strain curve of the material is linear or not, the strain energy release rate is still given by Equation (4.3) (above dimensional arguments)

$$H \sigma \varepsilon a = R. \tag{4.6}$$

Note again that the equation must contain a factor $\sigma\varepsilon$, the dimension of which is lbs/in, so that the crack size must appear in linear form (a) for the dimension to become lbs/in (strain energy per unit thickness and per unit crack extension). Indeed, Equations (4.3) and (4.6) are identical regardless of the form of the strain-stress curve, be it that another dimensionless geometry factor may appear.

For a linear-elastic material $\mathrm{d}U/\mathrm{d}a$ is generally called G, but for a non-linear material the shorthand denotation J is used, which is a little confusing and for which there is no good reason. Since the denotation J is used throughout the literature it will be used here as well. Similarly, while the symbol R is used for the fracture energy of a linear-elastic material, the fracture energy for a non-linear material is denoted in shorthand as J_R. The fracture criterion is still $\mathrm{d}U/\mathrm{d}a = \mathrm{d}W/\mathrm{d}a$, which is written in shorthand as $G = R$ for elastic materials, but in accordance with the above new symbols, it becomes for a non-linear material:

$$J = J_R \tag{4.7}$$

or alternatively:

$$H\sigma\varepsilon a = J_R. \tag{4.8}$$

The previous Equations (4.1) and (4.6) have not changed; only the symbols have.

It was possible to use Equation (4.3) for fracture analysis because $\varepsilon = \sigma/E$ so that Equation (4.3) becomes (4.5), thus permitting calculation of the fracture stress. By the same token, Equation (4.8) can be used for fracture analysis, if ε can be expressed in σ, i.e. if an equation is available for the non-linear stress-strain curve.

Any form of stress-strain equation will be useful, provided it properly describes the material's stress-strain curve. It will have to be a curve-fitting equation (empirical). There is no objection against this; also Hooke's law is empirical as it is no more than a linear curve fit, where E follows from the slope of the line.

The most convenient curve fit for a non-linear stress-strain curve (Figure 4.1) is a power function. The resulting stress-strain equation is known as the Ramberg–Osgood equation:

$$\varepsilon = \frac{\sigma}{E} + \frac{\sigma^n}{F} \quad \text{or} \quad \varepsilon = \varepsilon_{el} + \varepsilon_{pl}\,. \tag{4.9}$$

For most materials, but not all, Equation (4.9) provides a good fit of the stress-strain curve. The equation can be used in various forms, but Equation (4.9) is the most convenient. Other forms of the equation will be discussed in Section 4.7.

4.3. The fracture criterion

The Ramberg–Osgood stress-strain equation can be used to evaluate the fracture criterion of Equation (4.8). First note that the first part of Equation

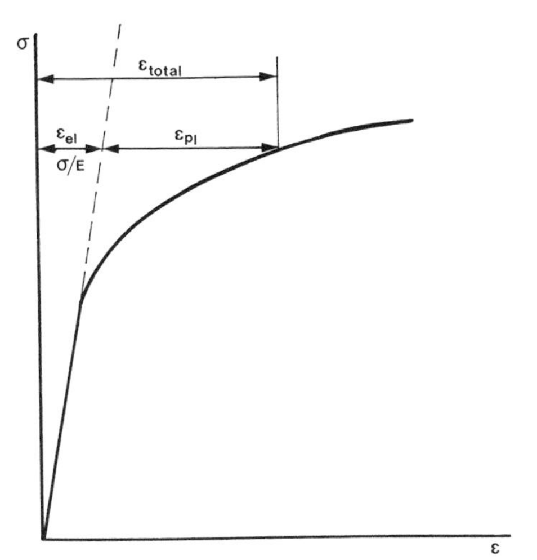

Figure 4.1. Stress-Strain diagram.

(4.9) represents the linear (Hooke's) part of the stress-strain curve. For this part it is already known that $\mathrm{d}U/\mathrm{d}a = \pi\beta^2\sigma^2 a/E$. Therefore, only the effect of the plastic part needs consideration:

$$\varepsilon_{pl} = \frac{\sigma^n}{F}. \tag{4.10}$$

Note that the equation simplifies to Hooke's law for the case that $n = 1$ $(F = E)$.

Substitution of Equation (4.10) in Equation (4.8) provides:

$$\frac{H\sigma^{n+1}a}{F} = J_R. \tag{4.11}$$

For $n = 1$ $(F = E)$ the Equation (4.11) indeed reduces to Equation (4.4), as it should; this implies that for $n = 1 : H = \pi\beta^2$.

Now $\mathrm{d}U/\mathrm{d}a$ for the elastic part, G, and $\mathrm{d}U/\mathrm{d}a$ for the non-linear part, J, are both known. The fracture condition is:

$$\frac{\pi\beta^2\sigma^2 a}{E} + \frac{H\sigma^{n+1}a}{F} = \frac{\mathrm{d}W}{\mathrm{d}a} \tag{4.12}$$

or

$$G + R = J_R.$$

It has become general practice to denote the fracture energy $\mathrm{d}W/\mathrm{d}a$ as J_R. This is again confusing because actually it is $R + J_R$, which is now shorthanded as J_R.

Because G is equivalent to J for $n = 1$, as pointed out on the basis of

Equation (4.11), it is more consistent to call it J_{el}, and to drop the shorthand denotation G altogether:

$$\text{Fracture if:} \quad J_{el} + J_{pl} = J_R \tag{4.13}$$

or

$$\text{Fracture if:} \quad \frac{\pi \beta^2 \sigma^2 a}{E} + \frac{H \sigma^{n+1} a}{F} = J_R. \tag{4.14}$$

This equation indeed provides a useful fracture criterion for engineering applications, because it is expressed in stress. Fracture will occur when the stress is sufficiently high to satisfy Equation (4.14). The fracture energy J_R in the equation represents the material's fracture resistance; it is therefore essentially the material's toughness expressed in a somewhat different form. It can be measured in a test, if the geometry factors β and H are both known for a certain specimen configuration. The modulus E, as well as n and F can be derived from a stress-strain curve measured in a regular tensile test. By performing a fracture test on a cracked specimen, the fracture stress can be measured. Values for all parameters in the left hand side of Equation (4.14) are then available, so that the 'toughness' J_R can be calculated.

For example, assume that for a certain material $E = 30\,000$ ksi, $n = 7$ and $F = 2 \times 10^{12}$. (Section 4.7 explains how n and F are obtained from a tensile test). Also assume that for a certain specimen with a crack of $a = 2$ inch and the values of β and H are 1.1 and 6 respectively. The specimen is tested and it is observed that fracture starts at a stress $\sigma = 30$ ksi. Substitution of all data in Equation (4.14) provides:

$$\frac{\pi \times 1.1^2 \times 30^2 \times 2}{30\,000} + \frac{6 \times 30^{7+1} \times 2}{2 \times 10^{12}} = J_R$$

so that

$$J_R = 0.23 + 3.94 = 4.17 \text{ in-ksi} = 4170 \text{ lbs/in.}$$

The fracture energy now being known for the material at hand, Equation (4.14) can be used to calculate the fracture stress of a cracked structure by solving for σ. Naturally, both geometry functions β and H must be known for the structure at hand. Unfortunately, Equation (4.14) cannot be solved directly for σ; an iteration procedure must be employed, but this is merely a tedium. As the solution usually is obtained by a computer, there is no serious difficulty. In the event that J_{el} is small compared to J_{pl}, the first term in Equation (4.14) can be neglected; it reduces to:

$$\frac{H \sigma^{n+1} a}{F} = J_R \tag{4.15}$$

which can be solved directly for the fracture stress:

$$\sigma_{fr} = \left(\frac{FJ_R}{Ha}\right)^{1/(n+1)}. \tag{4.16}$$

Once more, note that for $n = 1$ ($F = E$) this equation reduces to Equation (4.5).

Assume that for a certain structure with a three inch crack the geometry factors are $\beta = 1.2$ and $H = 7$. The structure is made of the material on which the above hypothetical test was done, so that E, F and n are as given above. The material's fracture resistance was found to be 4.17 kips/in = 4170 lbs/in. Indeed the elastic part of J was small, so that Equation (4.16) can be used to calculate the fracture stress of the structure as:

$$\sigma_{fr} = \left(\frac{2 \times 10^{12} \times 4.17}{7 \times 3}\right)^{1/(1+7)} = 28.2\,\text{ksi}.$$

Clearly then EPFM is no more difficult to use than LEFM; the only complication is the necessary iterative solution when Equation (4.14) must be used. (One other complication arises as discussed in Section 4.4.) Of course the 'toughness' J_R is to be measured in a test, but also the LEFM toughness must be measured. The geometry factors β and H must be obtained for the structure at hand. The elastic geometry factor, β, has been calculated for many geometries and compiled in handbooks [1, 2, 3], or can be derived using one of the simple procedures discussed in Chapter 8. Similarly, the plastic geometry factors, H, have been calculated for a number of geometries. These are also available in handbooks [4, 5]. One complication arises here because H depends upon n also: $H = H(a/L, n)$. Obtaining H is more cumbersome than obtaining β, but once calculated H can be presented in handbooks for later use.

4.4. The rising fracture energy

EPFM deals with high toughness materials. For such materials the fracture energy, J_R, tends to increase during the fracture process. This was discussed already on the basis of R in Chapter 3; its consequence is that fracture may be slow and stable initially, until at some point an instability occurs causing fracture to become fast and uncontrollable. The procedure to distinguish between stable and unstable fracture in EPFM is the same as in LEFM, and also in this respect EPFM presents no new problems. The LEFM procedure for R-curves as reviewed in Chapter 3 will be assumed known for the following discussion.

Figure 4.2 shows the rising J_R-curve. Fracture commences when $J = J_R$. For a certain existing crack, the curve for J can be drawn for a number of values of the stress, on the basis of the left part of Equation (4.14). Several such curves

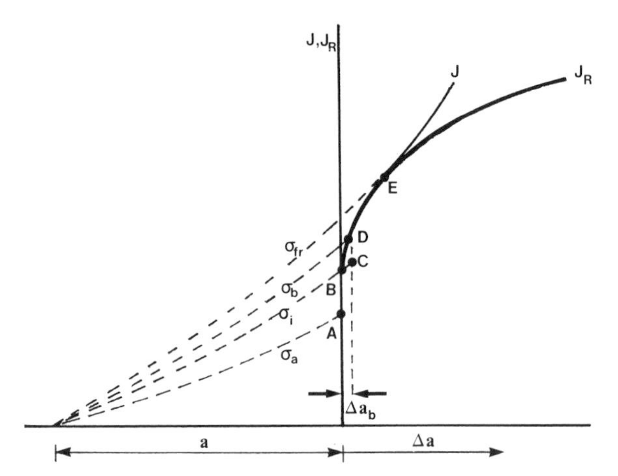

Figure 4.2. J-curves for different stresses and J_R-curve.

are shown in Figure 4.2. At the stress σ_a the value of $J(a)$ is given by point A, which is lower than B. Hence $J < J_R$, and fracture is not possible. Increase of the stress to σ_i raises $J(a)$ to point B. Fracture now takes place because $J = J_R$. But this fracture is stable; it cannot proceed if the stress remains equal to σ_i, because then J would increase to point C, while J_R would increase to point D, so that J would be less than J_R. For fracture to continue the stress must be increased to σ_b, which brings J to point D. During the increase of the stress from σ_i to σ_b the fracture proceeds in a stable manner from a to $a + \Delta a_b$. Further increase of the stress causes fracture to proceed . The fracture is controllable (stable); it can be stopped at any time simply by keeping the stress constant.

Finally, when the stress reaches σ_{fr}, the fracture become unstable (point E), because from there on J will remain larger than J_R. Fast, complete fracture will cause the structure to break in two. If at any time between B and E the structure would be unloaded, it would stay intact, although the damage would be larger by Δa.

The stable fracture is often referred to as stable crack growth. This is a misnomer, because the crack extends by fracture and not by one of the crack growth mechanisms discussed in Chapter 1. Stable fracture is fracture in progress, although it is slow and can be stopped if the stress is not further increased. After instability fracture is uncontrollable. Also the term 'sub-critical crack growth' is sometimes used for the stable fracture. The description may be clear in the context, but in its general use it refers to cracking by fatigue and stress corrosion. The crack becomes critical when fracture begins, whether stable or not. The use of more consistent language in fracture mechanics would certainly avoid confusion. One may argue about the preciseness of the terms used here, but at least different phenomena are identified consistently by different names.

It is apparent from Figure 4.2 that the condition for instability is:

$$\left.\begin{aligned} J &= J_R \\ \frac{\mathrm{d}J}{\mathrm{d}a} &= \frac{\mathrm{d}J_R}{\mathrm{d}a} \end{aligned}\right\} \tag{4.17}$$

which signifies tangency between the $J(a)$ curve and the J_R-curve. The instability point can be found graphically by plotting the J_R-curve and the $J(a)$ curves for a number of stress values as in Figure 4.2. Algebraic solutions are discussed in Chapter 10. (See also Eqs. 3.52)

The procedure is valid when there is a condition of load control; i.e. if the stress does not drop once the fracture becomes unstable. If there is displacement control, the stress may drop. For example, when the loading is due to thermal stress, the end displacement is fixed. During fracturing the stiffness decreases which may cause a very rapid decrease of the stress, because less stress is needed to maintain the fixed displacement. Since J depends upon σ, it may not be possible to maintain $J > J_R$, so that the instability is postponed. In that case the drop in stress must be calculated from the decreased structural stiffness, and the above calculation must be continued until Equation (4.17) is satisfied (Chapter 10).

Both in load control as well as in displacement control, the stress will increase to a certain maximum. Instability in load control coincides with the maximum stress. In displacement control the fracture may remain stable after the maximum is reached (Figure 4.3); instability may or may not ensue later. Up till the maximum stress there is no difference between load control and displacement control. Displacement control is governed by the stiffness of the system (loading + structure); instability is therefore system dependent and not a 'material property' (Figure 4.3).

Most structures are under load control. For example, the wave load on a ship, the gust load on an airplane or the load on a crane does not abate when the structure is fracturing. Hence, a load control analysis is usually applicable. In the event of displacement control, the only thing of interest often still is the maximum stress that can be sustained. In that case no account for displacement control is needed either, because up till the maximum stress there is no distinction anyway.

In most cases the difference between σ_i and the stress at instability is very small, and solving Equation (4.14) once for σ_i (point B in Figure 4.2) may be adequate (conservative). Examples of analysis are provided in Chapter 10.

Finally it should be mentioned that Equation (4.17) has been modified by

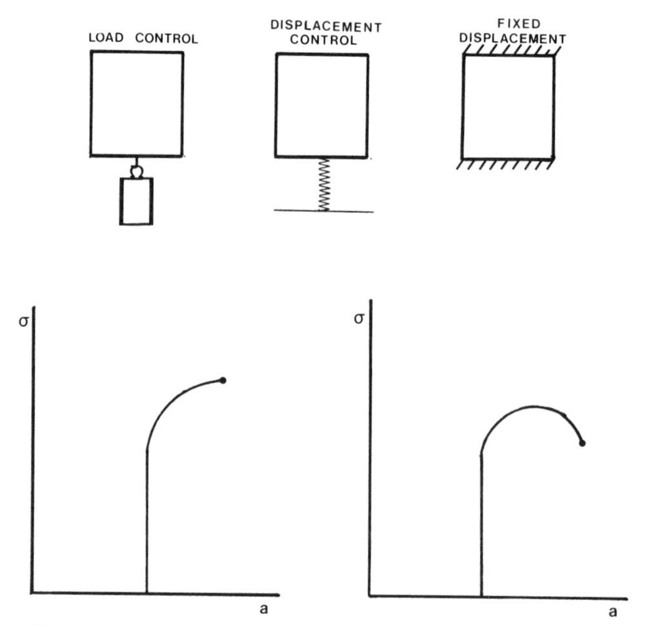

Figure 4.3. Load control (left) versus displacement control (right).

multiplying both sides by E/F_{ty}^2 leading to:

$$\frac{E}{F_{ty}^2}\frac{\mathrm{d}J}{\mathrm{d}a} = \frac{E}{F_{ty}^2}\frac{\mathrm{d}J_R}{\mathrm{d}a} \quad \text{or in shorthand} \quad T = T_{\text{mat}}. \tag{4.18}$$

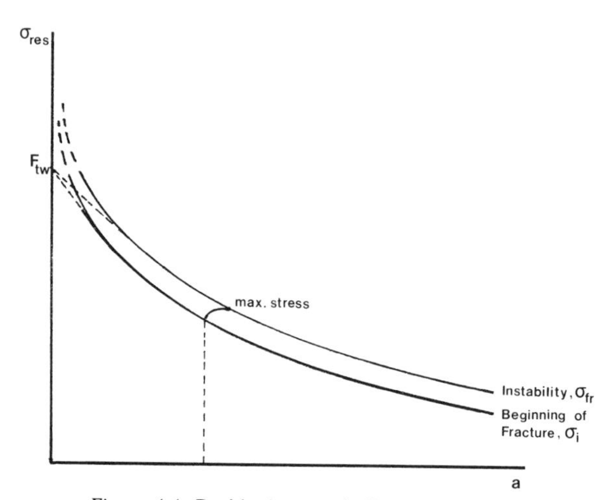

Figure 4.4. Residual strength diagram in EPFM.

The quantities at both sides are then renamed. The left side in shorthand is called the tearing modulus, T, the right side T_{mat} for lack of a descriptive word. Hence, the instability equation is written as $T = T_{mat}$, but nothing has changed. To the uninitiated only the fog intensified.

4.5. The residual strength diagram in EPFM; collapse

By solving Equation (4.14) for a number of crack sizes one obtains the residual strength diagram, as shown in Figure 4.5. Solution of Equation (4.17), either graphically or otherwise (Chapter 10), provides the stresses for instability as well. Hence, there will be a line for the onset of fracture, and one for instability (compare Figures 3.19 and 3.20). For convenience only the lower of these (Figure 4.5) is taken into account in the following discussion, because in general there is little difference between σ_i and σ_{fr}.

Now consider again Equation (4.14). For the case of $a \rightarrow 0$, the calculated residual strength will be infinite. This same problem was encountered in LEFM; it has not been solved by EPFM. An approximation is still necessary as in LEFM. For small crack sizes the calculated fracture stresses will be very high; for very small cracks it will be infinite, i.e. higher than F_{tu}, which is clearly wrong. Apparently, as in LEFM, collapse as a competing fracture condition cannot be ignored. Collapse determines the highest load carrying capability of

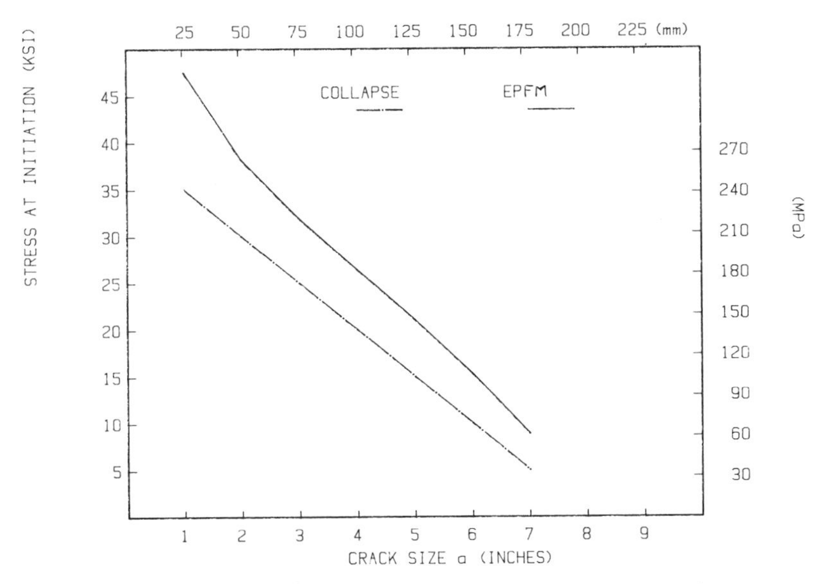

Figure 4.5. EPFM analysis for center cracked panel; $W = 16$ inch, $E = 23\,000$ ksi, $F = 1\,E\,7$, $n = 3.4$, $F_{ty} = 40$ ksi. Collapse prevails (lower curve).

the structure, regardless of any fracture criterion, whether LEFM or EPFM. Thus collapse must be analysed as a separate condition, as in the case of LEFM. This problem was discussed extensively in Chapters 2 and 3. The condition satisfied first (at the lowest stress) determines the strength of the structure.

For an ideally plastic material (horizontal stress-strain curve beyond F_{ty}), the exponent in the Ramberg–Osgood equation (Equation 4.9) is $n = \infty$. Thus the left hand side of Equation (4.14) becomes indeterminate; J is undefined and attains any value necessary (collapse) to satisfy the equation. This always will occur at plastic instability. When plastic deformation becomes uncontrollable J attains the necessary value for collapse. Attempts have been made to incorporate the collapse condition in the EPFM criterion, but this cannot be done by just considering J. It would require a separate evaluation of plastic instability criteria. Although this might be possible in principle, it is hardly feasible for engineering analysis. Concocted modifications of J are doomed to failure as are plastic zone corrections to K [6]. The practical solution is to treat collapse in the manner discussed in Chapters 2 and 3, and to assess both the collapse condition and Equation (4.14). The one leading to the lowest failure stress determines the residual strength. Then the problem is the same as in LEFM, discussed extensively in the previous chapter.

4.6. The measurement of the toughness in EPFM

Any kind of specimen can be used to measure the K_c or K_{Ic}; If this were not the case, applications to structures would not be possible either. There is no objection against a standard specimen for which β is known very accurately, but any other specimen with known β is appropriate. In the same vein, J_R can be measured with any specimen for which β and H are known. An example was given already in Section 4.3. In order to obtain the J_R-curve the stress should be measured during the stable fracture as well as the progress of fracture $a = a + \Delta a$. Equation (4.14) then provides J_R. There is definitely a specimen size requirement, because J becomes indeterminate when collapse occurs, as discussed.

It is unfortunate that a test procedure to measure the J_R-curve was standardized [7] prematurely, at a time Equation (4.14) had not evolved. Thus the test seems to bear no relation to what was discussed in the previous sections.

Recall from Figure 3.12 that the energy release rate is equal to $\mathrm{d}U/\mathrm{d}a$, the small area between two load-displacement curves for a and $a + \mathrm{d}a$, regardless of whether the load or the displacement is constant over $\mathrm{d}a$. If the displacement is constant there is a decrease of strain energy (negative), so that the available energy is $-\mathrm{d}U/\mathrm{d}a$ (see Chapter 3). In the elastic case the area between the curves is easily determined as $G = -\mathrm{d}U/\mathrm{d}a = -\frac{1}{2}P\delta$. This is still the case when the load-displacement curve is non-linear as in Figure 4.6; the area between the curves is $\mathrm{d}U/\mathrm{d}a$ (negative), so that $J = -\mathrm{d}U/\mathrm{d}a$ is:

$$J = -\int_0^\delta \left(\frac{\partial P}{\partial a}\right)_\delta \mathrm{d}\delta. \tag{4.19}$$

If an equation for the load-displacement curve is available, J can be obtained from Equation (4.19). The standard for J measurements is based upon an estimated equation for the load-displacement curve.

Consider a bend specimen in which the remaining ligament is fully plastic as in Figure 4.7. The elastic deformations can then be neglected. The specimen will deform as two rigid bars connected by a plastic hinge. The bend angle ψ depends upon the bending moment, which is equal to $PL/2$ (Figure 4.6). If the stress-strain curve is assumed horizontal beyond F_{ty} the resistance of the plastic hinge is equal to F_{ty}. The equation for ψ then must be as shown in Figure 4.7. Obviously, ψ is proportional to P and inversely proportional to F_{ty}, the resistance of the hinge. If the specimen is twice as thick, ψ will be half as large for the same load, so that ψ must be inversely proportional to B. Hence, for the equation to be dimensionless the ligament must appear as b^2 in the denominator. A dimensionless constant C will complete the equation. By noting that $\psi = \delta/L$, the equation for the dependence of δ on P is obtained as in Figure 4.7. Apparently, this is the sought equation for the load-displacement curve.

Differentiation of the load displacement equation with respect to b provides $\partial P/\partial b$, and it follows from substitution of P that $\partial P/\partial b = 2P/b$ (Figure 4.7). When the crack extends over $\mathrm{d}a$ the ligament will decrease by $\mathrm{d}b$, which means that $\mathrm{d}a = -\mathrm{d}b$, and therefore $\partial P/\partial a = -2P/b$. Substitution of this in Equation (4.19), and noting that J must be determined for a unit thickness (which requires division by B), the equation for J becomes:

$$J = \frac{2}{bB}\int P\,\mathrm{d}\delta \quad \text{or} \quad J = \frac{2A}{bB}. \tag{4.20}$$

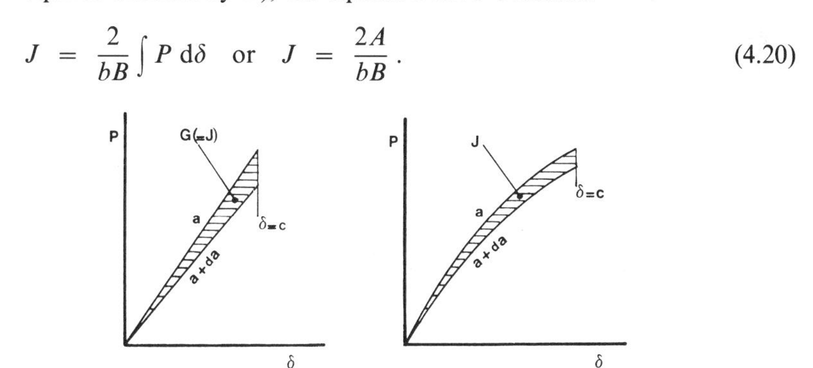

Figure 4.6. Strain energy release at constant displacement (infinitesimal da). left: linear; right: non-linear. Compare with Figure 3.15.

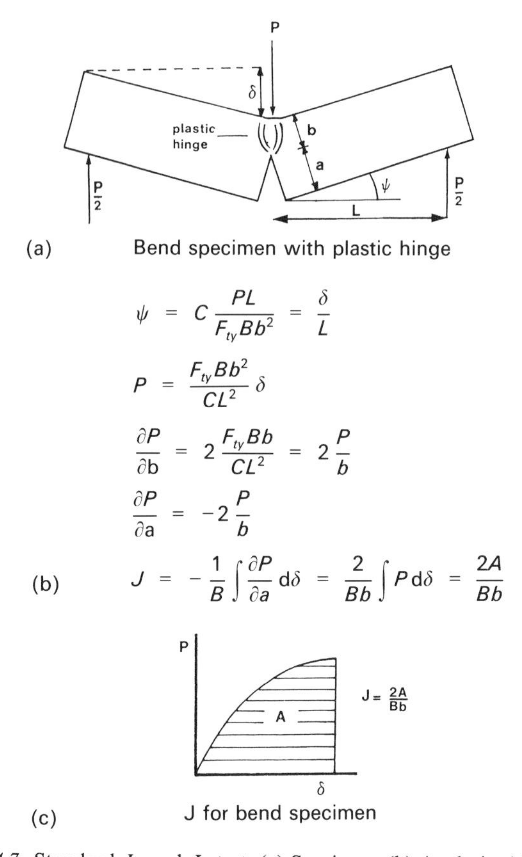

Figure 4.7. Standard J_{Ic} and J_R test. (a) Specimen; (b) Analysis; (c) Result.

The integral part of the equation is simply the area, A, under the load displacement diagram (Figure 4.7c), which can be measured – by using a planimeter or otherwise – from the test record of $P - \delta$.

Surprisingly, the small area between the curves in Figure 4.6b is found by taking twice the total area, A, under the curve and by dividing by b and B. Naturally, this is true only under the assumption made, namely that the entire ligament is yielding and that the stress-strain curve is horizontal beyond F_{ty} (note that n does not appear in the equation). These are in fact the collapse conditions under which J becomes indeterminate as argued above. The latter means that Equation (4.20) is unsuitable for the independent calculation of J, but since $J = J_R$ during fracture, the equation can be used to determine J_R. Figure 4.7 is for a ligament in bending; it applies to a compact tension specimen as well, be it that a small correction must be made [7] as there is some tension superposed on the bending.

At the time of fracture $J = J_R$. Hence the fracture resistance J_R is derived from the load-displacement diagram measured in a test on a compact specimen by applying Equation (4.20). The fracture resistance increases (Section 4.4). In a test fracture is first stable. At various stages during the test the amount of stable fracture, Δa, is measured and marked on the load-displacement diagram (Figure 4.8). By the use of Equation (4.19), J_R can be obtained as a function of Δa, yielding the sought J_R-curve as in Figure 4.8b (compare with Figures 3.18, 3.19 and 4.6).

The standard recommends use of a so-called 'blunting' line, which forces the J_R-curve to go zero for $a = 0$. Obviously, with $a = 0$ in Equation (4.14), the stress for fracture would be calculated as zero. Thus the blunting line cannot be used in a fracture analysis. It is an artificiality; blunting is due to plastic deformation (Figure 1.4), not fracture (Figure 1.7). It has no place in the J_R-curve, because it ignores the physics of the problem.

Constraint is essentially ignored in the standard. In LEFM the toughness K_c or R depends upon constraint (thickness), and the standard for K_{Ic} tests puts great emphasis on the constraint condition through the thickness requirement.

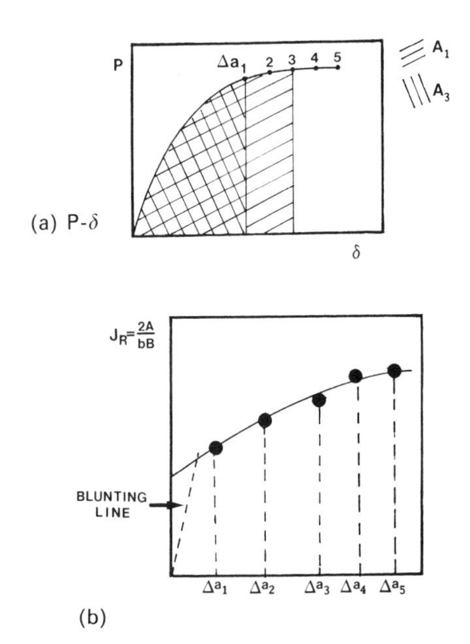

Figure 4.8. Measurement of J_R. (a) Load displacement diagram with indications of fracture extension; (b) J_R curve.

Clearly, this problem exists equally in EPFM. The fracture resistance, J_R, depends upon whether contraction in thickness direction can take place freely or whether it is constrained by the surrounding material (Chapters 2 and 3).

The standard for K_{Ic} tests requires the thickness to be larger than $B > 2.5 (K_{Ic}/F_{ty})^2$. Since $J = K^2/E$, this translates into:

$$B > 2.5\ EJ_R/F_{ty}^2. \tag{4.21}$$

There is no reason why this requirement should be less severe in elastic plastic fracture. Instead, the following condition has been proposed [8]:

$$B > 25\ J/F_{ty}. \tag{4.22a}$$

When this is applied in LEFM, it yields with $E = 30\,000$ ksi, and e.g. $F_{ty} \approx 50$ ksi and Equation (4.21):

$$B > 0.04 \left(\frac{K_{Ic}}{F_{ty}}\right)^2. \tag{4.22b}$$

Obviously, this is an inadequate condition (Figure 3.8). The condition of Equation (4.22a) puts $C \approx 1500$ rather than $C \approx 25$, in order for the requirement to be as stringent as the one for LEFM. There is no reason why it should be any less for J_R tests. Naturally, regardless of constraint, the test will provide J_R for the thickness used, like an LEFM test would provide K_c (Chapter 3).

It seems strange that ASTM embarked on standardizing a test while the subject was still in the research phase (a standard may be useful for engineering applications; it has no place in research), and at a time Equation (4.14) was not available. Knowledge of J_R is of no use but for calculation of the fracture stress of a structure. The latter became possible only due to Equation (4.14). Last but not least, Equation (4.14) has made the equation in the standard obsolete. A thorough revision of the standard for engineering rather than research seems opportune.

4.7. The parameters of the stress-strain curve

This section could be very short if the stress-strain equation used in EPFM had been given the simple form of Equation (4.9). Unfortunately, new confusion has been introduced as will appear in Section 4.8. It is for this reason that a longer discussion of the stress-strain equation is necessary here.

Hooke's law is a simple mathematical description of the experimental fact that elastic stress and strain are proportional; it is an empirical law. In the same vein one can use mathematical equations that fit the remainder of the stress-strain curve. The most useful equation is the Ramberg–Osgood equation, which covers the plastic strain as:

$$\varepsilon_{pl} = \frac{\sigma^n}{F}. \tag{4.23}$$

Its exponent n is called the strain hardening exponent. Further, F is *a* proportionality constant like E is a proportionality constant. For lack of a better word F might be called the 'plastic modulus'. For $n = 1$ and $F = E$ the equation covers elastic behavior as well.

Combination of the elastic and plastic strain provides the total strain:

$$\varepsilon_{tot} = \varepsilon_{el} + \varepsilon_{pl} = \frac{\sigma}{E} + \frac{\sigma^n}{F}. \tag{4.24}$$

The modulus E is measured as the slope of the linear part of the stress-strain curve. Values for n and F can be obtained rather easily as well. A tensile test provides the stress-strain curve in terms of strain. At any stress the plastic strain is obtained from the measured total strain as $\varepsilon_{pl} = \varepsilon_{tot} - \sigma/E$ (Figure 4.1); for an example see the solution to Exercise 1).

Taking the logarithms in Equation (4.23) yields:

$$\log \varepsilon_{pl} = n \log \sigma - \log F. \tag{4.25}$$

Hence, in a plot of log(stress) versus log(plastic strain), the data should fall on a straight line as shown in Figure 4.9a. The slope of the line is n, the intercept with the abscissa is $-\log(F)$ as shown. If the data do not fall on a straight line, the material does not obey a Ramberg–Osgood equation, which is sometimes the case. There is nothing that can be done about this.

With n and F, the total stress-strain curve can be calculated with Equation (4.24). If the data (Figure 4.9) are reasonably on a straight line Equation (4.24) will be a good representation of the measured stress-strain curve as is demonstrated in Figure 4.9b: the curve through the data points was obtained using n and F from Figure 4.9a.

The Ramberg–Osgood equation was intended to describe the true stress–true strain curve. However, up to the point of maximum load, the equation can be used just as well for the engineering stress-strain curve (Figure 4.9), be it that n and F are different even though the material is the same. The latter is demonstrated in Figure 4.10 which is for the same data as Figure 4.9. The only restriction is that the equation cannot be used beyond maximum load in the case of the engineering stress-strain curve. It can be readily demonstrated [9] that n is equal to the inverse of the true plastic strain at maximum load. Although this would be an easier way to find n, its accuracy is poor. The value of n so obtained from Figure 4.10 would be $n = 8.49$ which is different from the n derived from the slope of the line in Figure 4.9. As F would still have to be determined as in Figure 4.10, the plot is necessary anyway. Besides, this simpler method would not apply to the engineering stress-strain curve.

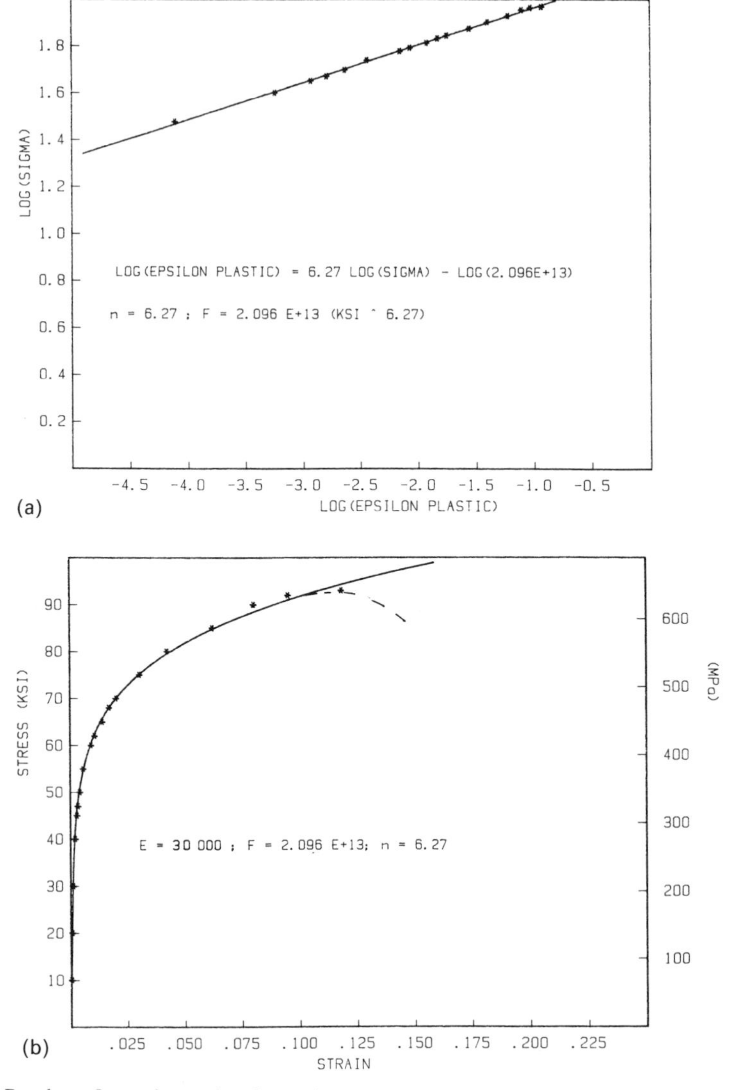

Figure 4.9. Ramberg-Osgood equation for engineering stress-strain curve. (a) Log (plastic strain) versus log (stress); (b) Engineering stress-strain curve equational fit.

The 'plastic modulus', F, tends to have rather unwieldy values, so that it may be more convenient to replace F by $(S_f)^n$:

$$\varepsilon_{pl} = \left(\frac{\sigma}{S_f}\right)^n. \tag{4.26}$$

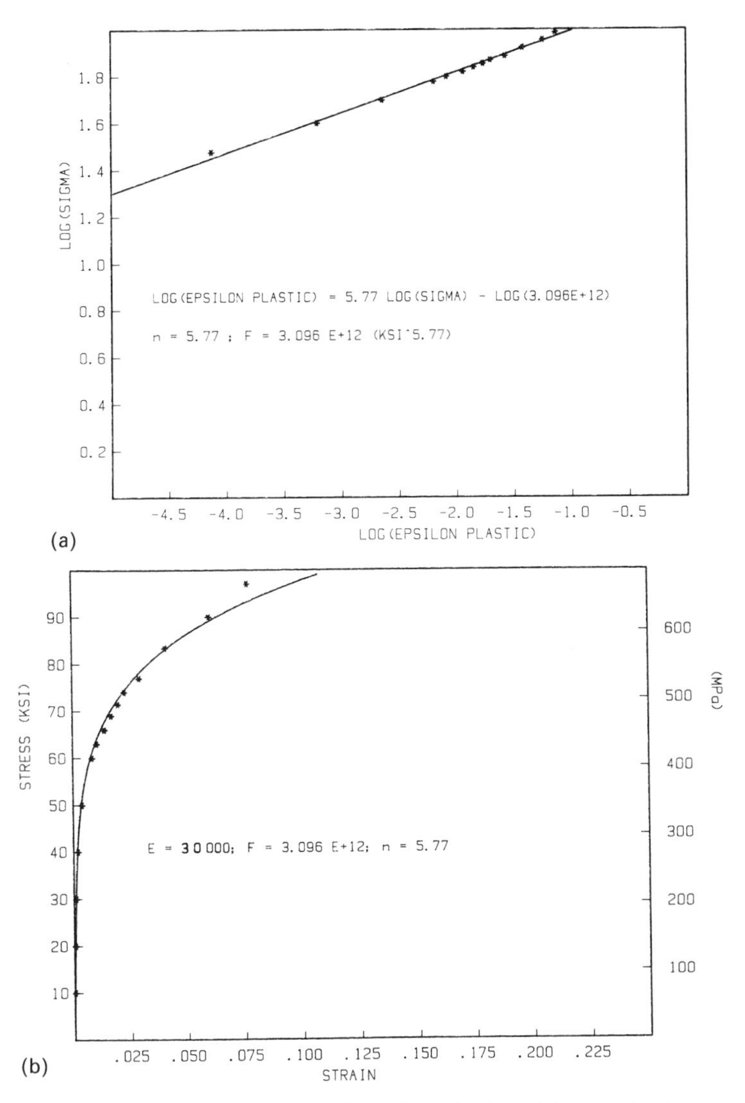

Figure 4.10. Fit of true stress-strain curve. (a) Long (true plastic strain) versus log (true stress); (b) True stress – true strain equational fit.

It appears from Equation (4.26) that for $\sigma = S_f$, the strain reaches 1 (100% strain). Sometimes S_f is called the 'flow stress', a somewhat unfortunate name, because it suggests that S_f has physical significance as a stress. As a strain of 100% is hardly ever possible, the stress will never reach the value S_f.

This section could end here if the developers of EPFM geometry factors had used the Ramberg–Osgood equation in the form discussed. Instead they employed a more complex form, the implications of which must be understood by the users of the geometry factors.

It should be realized that the Ramberg–Osgood equation has only three parameters, namely E, F and n or E, S_f, n. No further parameters are needed, or even permitted. Nevertheless, a fourth parameter was introduced as follows:

$$\varepsilon_{pl} = \alpha \left(\frac{\sigma}{\sigma_0}\right)^n. \tag{4.27}$$

There is no objection against replacing F by $(\sigma_0)^n$ as was done before in Equation (4.26). However, introduction of the fourth parameter is permitted only if α and σ_0 are dependent, because by definition: $\sigma_0^n/\alpha = F$. As a matter of fact, one may now take any value for σ_0, as long as α is adjusted accordingly to $\alpha = \sigma_0^n/F$.

For no reason at all, also the modulus was eliminated by DEFINING a strain ε_0, such that $\varepsilon_0 = \sigma_0/E$. Thus, Equation (4.27) was written as:

$$\frac{\varepsilon_{tot}}{\varepsilon_0} = \frac{\sigma}{\sigma_0} + \alpha \left(\frac{\sigma}{\sigma_0}\right)^n \qquad \varepsilon_{pl} = \alpha\varepsilon_0 \left(\frac{\sigma}{\sigma_0}\right)^n. \tag{4.28}$$

The above equation is permissible as long as one strictly adheres to the dependence of parameters, namely:

$$\varepsilon_0 = \sigma_0/E \quad \text{and} \quad \alpha = \frac{\sigma_0^n}{\varepsilon_0 F}. \tag{4.29}$$

The literature on EPFM often refers to σ_0 as the flow stress, which is a very disturbing and confusing misnomer, because σ_0 can be given any arbitrary value as long as Equations (4.29) are adhered to. It is generally taken equal to the yield strength, and the suggestion is raised that the latter is significant for the equation. Nothing could be further from the truth when any arbitrary value is appropriate.

This can be readily demonstrated by an example. From Figure 4.9 it appears that with $E = 30\,000$ ksi, $n = 6.27$ and $F = 2.1$ E13. The stress-strain equation is as shown in Table 4.1. Taking arbitrary values of 50, and 100 ksi for σ_0, and using the mandatory Equations (4.29) the stress-strain equations become as shown in the table. Any of these equations leads to the same strain for the same stress, as shown. Indeed, σ_0 can be chosen at will. It may be taken equal to the yield strength if so desired, as long as it is realized that this is an arbitrary choice.

It can be argued that the yield strength depends upon the shape of the curve (E, F, n), and vice versa, the yield strength depends upon F and n. This is true, and as such taking σ_0 equal to the yield strength is certainly defendable.

4.8. The *h*-functions

The geometry factors $H(a/L, n)$ were developed [4, 5] for a number of structural geometries and n-values, using the plastic stress-strain equation of Equation

Table 4.1. Effect of different definitions of Ramberg-Osgood equations.

$E = 30\,000\,\text{ksi}; \quad F = 2.1\ E\ 13\ \text{ksi}^{6.27}; \quad n = 6.27$

$$\varepsilon = \frac{\sigma}{30\,000} + \frac{\sigma^{6.27}}{2.1\ E\ 13} \qquad \text{As Equation (4.9)} \qquad \text{(A)}$$

With $s_f^{6.27} = F$ one obtains $s_f = 133\,\text{ksi}$

$$\varepsilon = \frac{\sigma}{30\,000} + \left(\frac{\sigma}{133}\right)^{6.27} \qquad \text{As Equation (4.26).} \qquad \text{(B)}$$

With arbitrary $\sigma_0 = 50\,\text{ksi}$; $\varepsilon_0 = 50/30\,000 = 0.00167$
$\alpha = \sigma_0^{6.27}/\varepsilon_0 F = 50^{6.27}/0.00167 \times 2.1\ E\ 13 = 1.28,$

$$\frac{\varepsilon}{0.00167} = \frac{\sigma}{50} + 1.28\left(\frac{\sigma}{50}\right)^{6.27} \qquad \text{As Equation (4.28)} \qquad \text{(C)}$$

With arbitrary $\sigma_0 = 100\,\text{ksi}$; $\varepsilon_0 = 100/30\,000 = 0.0033$
$\alpha = 100^{6.27}/0.0033 \times 2.1\ E\ 13 = 50$

$$\frac{\varepsilon}{0.0033} = \frac{\sigma}{100} + 50\left(\frac{\sigma}{100}\right)^{6.27} \qquad \text{As Equation (4.28).} \qquad \text{(D)}$$

Results

σ ksi	ε Equation (A)	ε Equation (B)	ε Equation (C)	ε Equation (D)
50	0.00381	0.00383	0.00381	0.00379
70	0.01998	0.02021	0.01996	0.01994
80	0.04342	0.04395	0.04339	0.04336

(4.28). This requires adherence to Equations (4.29) as discussed in the previous section.

The equation for J_{pl} was written as:

$$J_{pl} = \alpha\,\sigma_0\,\varepsilon_0\,c\,h_1\left(\frac{P}{P_0}\right)^{n+1}. \qquad (4.30)$$

In this equation P is the load, P_0 is the load at collapse supposing σ_0 were the collapse strength. Instead of the crack size, a, the unbroken ligament, c, is used. Finally, h_1 is the geometry factor.

Clearly, the load P is related to the stress, P_0 to σ_0 and the ligament c to the crack size a:

$$\left.\begin{aligned} P &= g\sigma \\ P_0 &= k\sigma_0 \\ c &= \left(\frac{W}{2} - 1\right) a = la \end{aligned}\right\} \tag{4.31}$$

where g, k, and l are just functions of the geometry. With this knowledge, and with Equation (4.29) the complicated Equation (4.31) readily turns into:

$$J_{pl} = \frac{\sigma_0^n}{\varepsilon_0 F} \sigma_0 \varepsilon_0 \, l a h_1 \left(\frac{g}{k}\right)^{n+1} \left(\frac{\sigma}{\sigma_0}\right)^{n+1} \tag{4.32}$$

which immediately reduces to:

$$J_{pl} = \frac{H\sigma^{n+1} a}{F} = \frac{H\sigma^{n+1} a}{S_f^n} \tag{4.33}$$

with $H = lh(g/k)^{n+1}$. Equation (4.33) is the basic form of the equation already known as Equation (4.8) from simple arguments. Equation (4.30) is just a complicated version of the same. Obviously, α, σ_0, and ε_0 can be divided out; they are superfluous. Indeed, the solution to Exercise 6 shows that the same results are obtained with Equation (4.32) regardless of the choice of σ_0 (arbitrary). Equation (4.30) suggests that J depends upon a collapse load P_0, but obviously σ_0 is divided out as well. This should be expected, because a collapse load cannot enter into J since the stress-strain equation used has no limit. The elastic energy release, G, could be expressed in the same manner by using $P = g\sigma$, $P_0 = k\sigma_0$ and $\sigma_0 = \varepsilon_0 \mathrm{E}$

$$G = \pi \beta^2 \sigma_0 \varepsilon_0 \frac{k^2}{g^2} a \left(\frac{P}{P_0}\right)^2. \tag{4.34}$$

Bringing in the collapse load or the collapse strength does not make G dependent upon same; it merely amounts to multiplying numerator and denominator by the same number, which does not change the basic equation. Collapse does not enter LEFM equations, nor does it enter EPFM equations in their present form. An artificial introduction does not change this fact. Collapse is a competing condition which must be assessed separately, at least for the time being.

Two other objections can be raised against Equation (4.30). Instead of just one geometry parameter H, it must use four geometry parameters: h, g, k, and l. Every time a calculation is performed double work is necessary: parameters must be derived and are subsequently divided out. Naturally, one could, once and for all, calculate H from $H = lh(g/k)^{n+1}$ and from then on use Equation

(4.33), but a computer does not object to unnecessary work and for it the form of Equation (4.30) need not be changed.

The second objection is that J is expressed in the load P, while in engineering one works with stress. Therefore Equation (4.33) is more useful; all other fracture mechanics equations are expressed in stress for this very reason. For complicated structures the conversion from load to stress is done in the design stage not at the time of fracture analysis.

4.9. Accuracy

Researchers have expressed great concern about the large variability of J and (consequently) J_R. The reason for the large variability is obvious. As J depends upon stress to the n-th power, according to Equation (4.11), a slight difference of 5% in stress with e.g. $n = 9$, leads to a difference of $(1.05)^{10} = 1.63$ or a difference of 63% in J_R. (Note that this occurs also in a standard tests: the load-displacement diagram becoming almost horizontal, the area under the curve, which determines J_R, changes dramatically with a slight change in δ.

This may seem bothersome but it is of little practical importance. The value of J_R is of no interest as long as the predicted fracture stress is reasonably accurate. This the case, because in a fracture analysis the situation is reversed: a difference of 63% in J with $n = 9$ will lead to only a difference of 5% in the predicted fracture stress: $(1.63)^{1/10} = 1.05$.

For a difference in J by a factor of 2, and for $n = 7$, the predicted fracture stresses would be different by $2^{1/8} = 1.09$; hence the error (difference) would be 9% only. This is clearly demonstrated in Figure 4.11, showing the results of two calculations with exactly the same input and $n = 7$. Two J_R-curves were used differing by a factor of approximately two throughout. The predicted fracture stresses differ only by a small amount (9%). In general the stresses in a structure will not be known with better accuracy, so that any of the predictions in Figure 4.11 would be satisfactory from an engineering point of view. Even the predicted amounts of stable crack growth at maximum load do not differ appreciably, as shown in Figure 4.12.

Most alloys satisfy the Ramberg–Osgood equation fairly well. However, a material of great interest to some industries, namely annealed 304-SS, exhibits a stress-strain curve that cannot be fitted with the equation. Yet it must be fitted to such an equation, otherwise an EPFM fracture analysis cannot be performed. In that case there is a choice as to whether the equation should fit the lower or the upper part of the stress-strain curve.

There is no categoric answer to this problem. Any conclusions reached are necessarily material specific. Although it may be argued that most of the crack tip material is subject to relatively small plastic strains, so that a fit of the lower part of the stress-strain curve is the most important, it cannot be denied that the

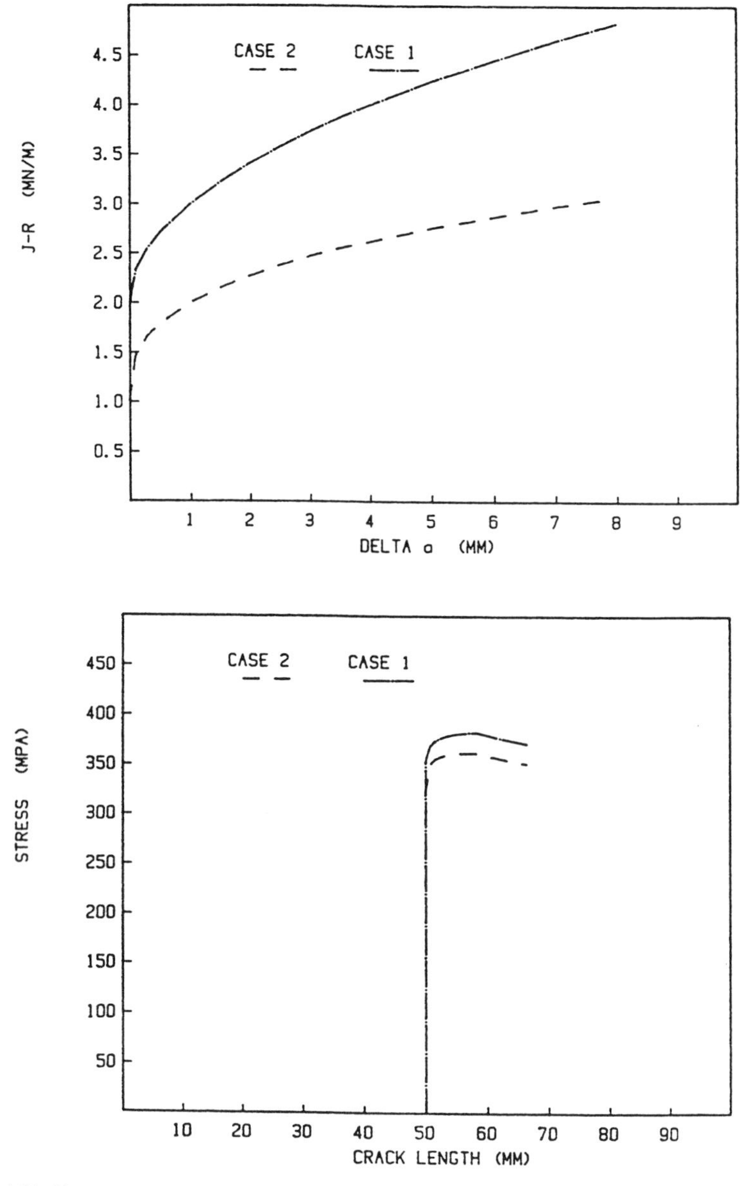

Figure 4.11. Fracture stress prediction for different R-Curves for center cracked panel [15]. Above: R-curves; Below: predicted fracture stress and fracture progress (Courtesy EMAS).

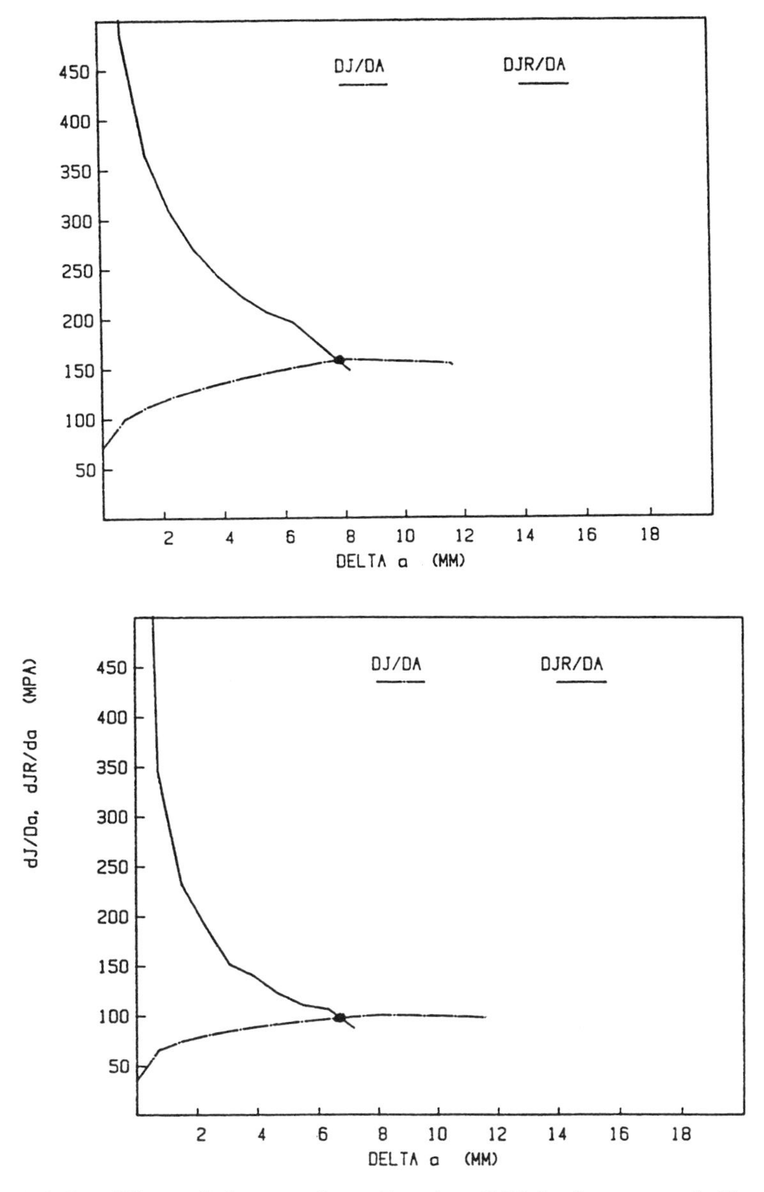

Figure 4.12. Instability prediction according to Equations (4.17) for the two cases in Figure 4.11 [15]; Above: Case 1; Case 2 of Figure 4.11 (Courtesy EMAS).

crack tip material is subject to strains in accordance with the upper part of the stress-strain curve. Common sense indicates that the decision must be made on a material-by-material basis; there is no categoric answer. Any conclusions arrived at certainly should not be generalized, and if they are, they are still restricted to materials not obeying an exponential stress-strain curve; for other materials there is no choice. When there is a choice, the criterion for the choice is whether the fracture stress is predicted correctly; the value of J is irrelevant. The problem would not exist if the measurement of J_R would use Equation (4.15) instead of (4.20), because the reverse operation of Equation (4.16) would then automatically lead to the correct answer (use of same F and n as in Equation (4.15).

4.10. Historical development of J

Eshelby [10] defined a number of contour integrals which are path independent by virtue of the energy conservation theorem. The two-dimensional form of one of these integrals can be written as:

$$\left.\begin{aligned} &\oint V\,\mathrm{d}y - T\frac{\partial u}{\partial x}\,\mathrm{d}s = 0 \\ &\text{where} \\ &V = \int_0^{\varepsilon} \sigma\,\mathrm{d}\varepsilon \end{aligned}\right\}. \tag{4.35}$$

V being strain energy per unit volume.

The integral is taken along a closed contour, S, followed counter clockwise (Figure 4.13a) in a stressed solid, T is the tension perpendicular to S, u is the displacement, and $\mathrm{d}s$ is an element of S.

Although the equation is somewhat elusive, it can be seen that the first term represents strain energy, while in the second term T is the 'force', and $\mathrm{d}u/\mathrm{d}x$ a strain, so that $(\mathrm{d}u/\mathrm{d}x)\,\mathrm{d}s$ is a displacement. As 'force' times displacement equals the work, F, done by the force, the equation essentially states that $\int V - F = 0$, which is energy conservation criterion of Equation (3.32).

Applying this integral to a cracked body [11] one can construct a closed contour ABCDEF around the crack tip, as shown in Figure 4.13b. The integral of Equation (4.35) along this contour must equal zero; it consists of the sum of four parts:

$$\int_{\Gamma_1} + \int_{CD} + \int_{\Gamma_2} + \int_{FA} = 0. \tag{4.36}$$

Since $T = 0$ and $\mathrm{d}y = 0$ along CD and FA the contribution of these parts is zero, so that for the remaining parts:

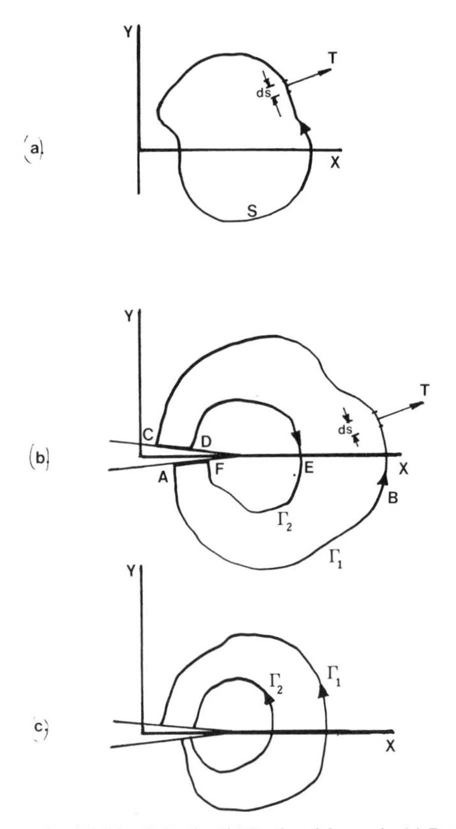

Figure 4.13. Contour integrals. (a) Elastic body; (b) Body with crack; (c) Path independent contour.

$$\int_{\Gamma_1} + \int_{\Gamma_2} = 0 \quad \text{or} \quad \int_{\Gamma_1} = -\int_{\Gamma_2}. \tag{4.37}$$

Therefore, the contribution of ABC must be equal (but opposite in sign) to the contribution of DEF. Note that one is clockwise, the other counter-clockwise. This means that the integral, if taken in the same direction along Γ_1 and Γ_2 will have the same value: $\int_{\Gamma_1} = \int_{\Gamma_2}$ in (Figure 4.13c). As Γ_1 and Γ_2 were arbitrary paths, the integral over Γ is apparently path-independent (one may take any path, beginning and ending at opposing crack faces, and the integral will always have the same value). The value of the integral was called J:

$$\int_{\Gamma} V\,\mathrm{d}y - T\frac{\partial u}{\partial x}\,\mathrm{d}s = J. \tag{4.38}$$

It should be noted that there is no proof as yet that this J is the same as the one used in previous sections; thus, for the time being it should be considered as the definition of a new quantity, defined only by the value of the above integral. The path-independence is not of apparent relevance to fracture.

If the integral is path independent any convenient contour may be taken to determine its value. The simplest contour is a circle, as in Figure 4.14, with radius r, its center at the crack tip. For this case $y = r \sin \theta$, so that $\mathrm{d}y = r \cos \theta \, \mathrm{d}\theta$, and $\mathrm{d}s = r \, \mathrm{d}\theta$. Then Equation (4.38) becomes:

$$J = \int_{-\pi}^{\pi} \left\{ \left(\int_{0}^{\varepsilon} \sigma \, \mathrm{d}\varepsilon \right) \cos \theta - T \frac{\partial y}{\partial x} \right\} r \, \mathrm{d}\theta. \tag{4.39}$$

No matter what the relationship between σ and ε, the integral $\int \sigma \, \mathrm{d}\varepsilon$ at any point always evaluates to $\alpha_1 \sigma \varepsilon$, where α_1 is dimensionless. Since T is a stress, it can always be expressed as $C_1 \sigma$, and $\mathrm{d}u/\mathrm{d}x$ being a strain can always be expressed as $C_2 \varepsilon$. Then $T(\mathrm{d}u/\mathrm{d}x)$ evaluates as $C_1 C_2 \sigma \varepsilon = \alpha_2 \sigma \varepsilon$. Both α_1 and α_2 may depend upon θ, but regardless of how complicated this dependence, the integral of Equation (4.39) will be:

$$J = \int_{-\pi}^{\pi} \{\sigma \varepsilon \alpha_1(\theta) \cos \theta - \sigma \varepsilon \alpha_2(\theta)\} \, r \, \mathrm{d}\theta. \tag{4.40}$$

No matter what the functions of θ are, the integral reduces to:

$$J = \sigma \varepsilon r \int_{-\pi}^{\pi} f(\theta) \, \mathrm{d}\theta = \sigma \varepsilon r F(\theta) \Big|_{-\pi}^{\pi} = \sigma \varepsilon r Q. \tag{4.41}$$

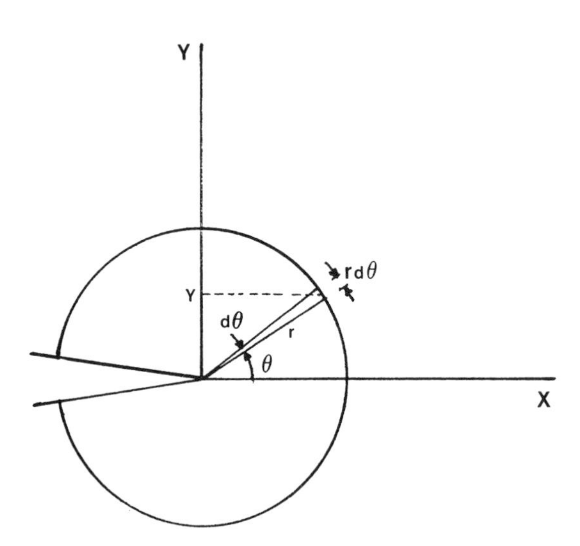

Figure 4.14. Simple circular path for contour integral.

The determined integral between $-\pi$ and π will evaluate to a dimensionless number. The solution therefore will be as shown in Equation (4.41), while σ and ε may be defined at any convenient point. Taking another point as a reference will merely change α_1 and α_2, and therefore Q, but Q is only a number anyway.

If one defines σ and ε at the location $y = 0$, $x = r$ (Figure 4.14) then:

$$J = \sigma_r \varepsilon_r r \bar{Q}. \tag{4.42}$$

By using the Ramberg–Osgood equation for the stress-strain curve, namely $\varepsilon = \sigma^n/F$, Equation (4.42) yields:

$$J = \sigma_r^{n+1} r \bar{Q}/F. \tag{4.42}$$

Then finally:

$$\sigma_r = \left(\frac{FJ}{\bar{Q}r}\right)^{1/(n+1)}. \tag{4.43}$$

If $n = 1$ $(F = E)$ this reduces to:

$$\sigma_r = \frac{\sqrt{EJ}}{\sqrt{\bar{Q}}\sqrt{\mathrm{r}}}. \tag{4.44}$$

Compare this equation with Equation (3.2) as replicated below:

$$\sigma_r = \frac{K}{\sqrt{2\pi r}}. \tag{4.45}$$

Clearly, Equation (4.44) is the same as the very original Equation (3.2). It shows that for $n = 1$ we have $Q = \sqrt{2\pi}$, and $\sqrt{EJ} = K$. This means for $n = 1$ that $\mathrm{J} = K^2/E = \mathrm{d}U/\mathrm{d}a$. Thus the path-independent contour integral is but the strain energy release rate. It is now apparent why this integral was denoted by J, a symbol already used for the energy release rate.

The above being the case for $n = 1$, it will be true in general, because nowhere above was a restriction made with regard to the shape of the stress-strain curve. Thus, the only significance of the 'path-independent' contour integral is to show that it represents the strain energy release rate, $J = \mathrm{d}U/\mathrm{d}a$, so that the J-integral is indeed the same as J defined previously. Indeed, that is of secondary importance only: the energy release rate can be defined in a much simpler way as was shown earlier in this chapter.

Nevertheless, the integral has its use. As it is known now that it equals $\mathrm{d}U/\mathrm{d}a$, it can be used to calculate G, J or K. It can be applied to the results of a finite element analysis (most codes have post-processors to do this). If the analysis is for the elastic case one can obtain K from $K = \sqrt{EJ}$. When the analysis uses the Ramberg–Osgood it can provide the geometry factor H from:

$$H = \frac{FJ}{\sigma^{n+1} a} . \tag{4.46}$$

This is essentially the way in which H (and h; Section 4.9) were obtained.

4.11. Limitations of EPFM

Although the concepts discussed in this chapter are generally referred to as elastic-plastic, they are in fact elastic. The non-linear stress-strain curve used must apply for loading as well as for unloading. This is illustrated in Figure 4.15. Truly elastic-plastic behavior is shown for comparison. In the latter case linear unloading occurs, and there will be a remaining plastic strain when the stress is reduced to zero. The stress-strain relation used in EPFM, upon unloading, must produce zero strain at zero stress. The curve may be non-linear, but not elastic-plastic. The return to zero strain means that the material is merely non-linear-ELASTIC.

This does not put any restrictions on the use of J as long as there is no unloading. Without any unloading it would never be known that the curve for unloading is different from the loading curve. However, if there is unloading anywhere, the assumption of non-linear elasticity will cause errors. Consider for example Figures 3.15 and 4.16a. It was shown in Chapter 3 on the basis of Figure 3.15 that for linear elasticity the energy release during an infinitesimal extension of a by da is always equal to the change in strain energy regardless of whether there is constant load or constant displacement. In one case the load does work which is twice the increase in strain energy, so that the remaining part of the work by the load (the energy available for fracture) is exactly equal to the change in strain energy. In the case of constant displacement the load does no work, but the strain energy decreases, so that the released energy is again equal to the change in strain energy.

The above is still true if the material is non-linear-elastic (Figure 4.16a), but not if the material is elastic-plastic. During an extension da of a under constant

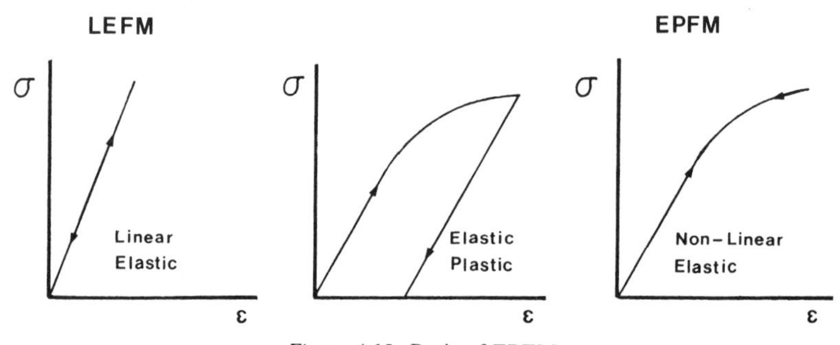

Figure 4.15. Basis of EPFM.

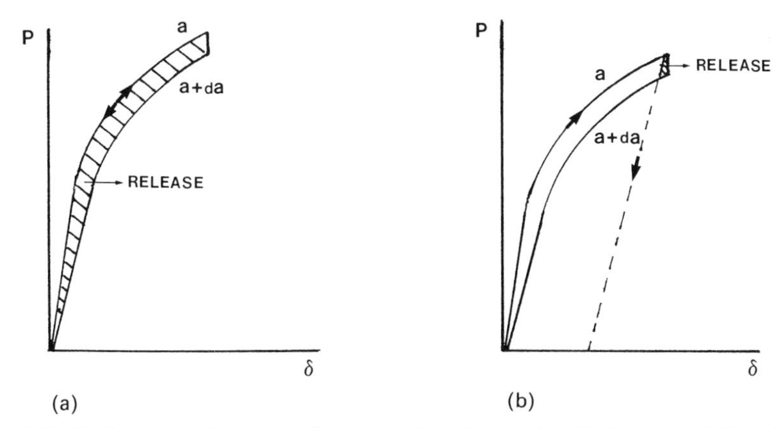

Figure 4.16. Strain energy release upon fracture by da under constant displacement. (a) Non-linear elastic; (b) Elastic plastic.

load the energy available for fracture is equal to the work done by the load minus the increase in (plastic) strain energy. However, during linear elastic unloading the plastic strain energy remains in the material. Only the small elastic part of the strain energy is indeed released: the available energy is much less during constant displacement than during constant load (Figure 4.16b). Hence, strictly speaking, the resulting equations are valid only in load control, so that fracture analysis is meaningful only up to the point of maximim load: this is not a severe restriction, because in most engineering analysis one will be interested only in the maximum stress a cracked structure can sustain.

Yet, even the latter is somewhat questionable. The material at the crack tip is highly stressed. When stable fracture is occurring (up to maximum load), it is this highly stressed crack tip material that is unloading after the fracture has passed through. Hence, the errors due to the unloading assumptions are felt most strongly where it counts most: at the crack tip. The errors will be small initially, but increase with increasing Δa. This sheds doubt on the analysis of stable growth and instability. As a matter of fact the analysis will be only meaningful when the J curve rises very steeply, so that there is very little stable fracture before maximum load (Figure 4.2). It has been suggested [12] that stable fracture should be limited to just a few percent of the remaining ligament, otherwise the errors become considerable.

Using the path-independence of the J-integral it was shown that the crack tip stress field in EPFM is described by Equation (4.44). For an ideally plastic material ($n \rightarrow \infty$) the equation leads to a finite crack tip stress, but for all other values of n the stress is still infinite at $r = 0$. It was pointed out already that the use of Equation (4.14) leads to an infinite fracture stress for $a \rightarrow 0$, and therefore will be increasingly in error for smaller cracks; approximations for small cracks will be necessary just as in the case of LEFM. For this and other reasons, as discussed, collapse still must be evaluated separately as a competing condition.

Apparently, EPFM has cured none of the ills of LEFM; it is a mere extension of LEFM for $n \neq 1$. A host of modifications to J have been proposed [13]; most of these belong in the category of 'patch work' as much as do plastic zone corrections to K [6]; they provide few new insights, complicate the procedure and lead to only marginal improvements.

Nevertheless, EPFM is still very useful and has a definite place in conjunction with LEFM. It has extended the use of fracture mechanics to non-linear materials at least up to a point. Judicious use will provide meaningful engineering answers, provided collapse analysis is done as well, just as in the case of LEFM, and provided appropriate approximations are made for short cracks (Section 3.8). It may be noted again that the accuracy of fracture analysis need only be as good as that of general accuracy of engineering procedures (Chapters 10, 12, 14). With the emergence of geometry factors [4, 5], and the possibility for simple estimates of these (Chapter 8), EPFM has become a useful engineering tool.

4.12. CTOD measurements

An alternative approach to EPFM has been based upon the Crack Opening Displacement (COD). Although referred to as COD, the method actually employs the Crack-Tip Opening Displacement (CTOD).

Consider a crack tip in a stressed body as in Figure 4.17. Let forces be applied over a distance da behind the crack tip in such a manner that the crack just closes over a distance da. The crack is now shorter by da and, therefore, the required closing forces must be equal to the stresses normally present when da is uncracked. For the time being, assume that these stresses are approximately equal to the yield strength.

During their closing action the forces travel over the distance v. Therefore, they do work to the amount of $\mathrm{d}F = 2 \times 0.5F_{ty}v\,\mathrm{d}a$ (plate of unit thickness).

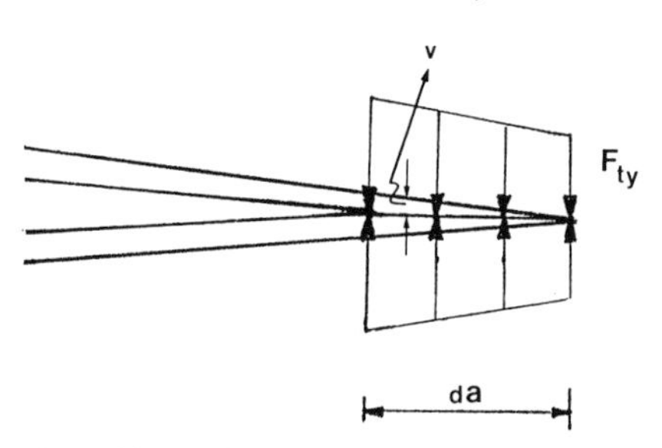

Figure 4.17. Closing forces to close the crack tip over da.

Since v is related to CTOD, the work will be $dF = \alpha \text{CTOD} F_{ty} da$. Upon release of these forces the same amount of energy is released, and the crack will 'grow' again by da. This energy release, $dF/da = \alpha \text{CTOD} F_{ty}$, is what has been called the strain energy release rate G or J. Therefore:

$$G = \alpha F_{ty} \text{CTOD} \quad \text{or} \quad J = \alpha F_{ty} \text{CTOD}. \tag{4.47}$$

The first of these expression would be applicable for LEFM, the second for EPFM.

Naturally, the stresses over the future da are not uniformly equal to F_{ty} as assumed, but if they are not, only the dimensionless factor α will be affected. It turns out [6] that α is approximately equal to unity, but various interpretations would put it between $\pi/4$ and $4/\pi$ in LEFM. However in the case of J the value of α depends upon n [4]. In any case, the above equations lead to:

$$\left.\begin{aligned} \text{CTOD} &\approx \frac{G}{F_{ty}} = \frac{K^2}{EF_{ty}} \quad \text{(LEFM)} \\ \text{CTOD} &\approx \frac{J}{F_{ty}} \quad \text{(EPFM)} \end{aligned}\right\} . \tag{4.48}$$

Fracture occurs at a critical value of G (or K) or a critical value of J. Then, according to Equations (4.48) fracture takes place at a critical value of CTOD, defined as CTOD_c. Consequently, CTOD_c should be a material property characterizing fracture resistance; as such it is a descriptor of 'toughness', and the measurement of CTOD in a test would provide the material's propensity to fracture.

A test for CTOD measurements was standardized first in Great Britain in British Standard BS-5762. Essentially, the test is performed on a small three-point bend specimen, (Figure 4.18). As in other toughness tests a record is made of load versus crack mouth opening displacement. The critical crack tip opening displacement CTOD_c, usually referred to as COD, can be obtained as follows.

The ligament is assumed fully plastic so that all specimen deflection can be considered to be due to rotation around a plastic hinge, the specimen limbs rotating by rigid-body motion. If it is assumed that the center of the plastic hinge coincides with the center of the ligament, then CTOD can be obtained from the crack mouth opening in the manner shown in Figure 4.18.

It is not necessary to make the assumption that the center of the hinge coincides with the center of the ligament. One possibility is to determine its location experimentally. Other options are open [14, 6].

Knowledge of the critical CTOD per se, is of no use for damage tolerance analysis. The number will have significance for engineering only if it can be used to predict fracture in a structure. For this to be possible CTOD must be expressable in terms of the stress acting on a fracture, so that the (fracture) stress

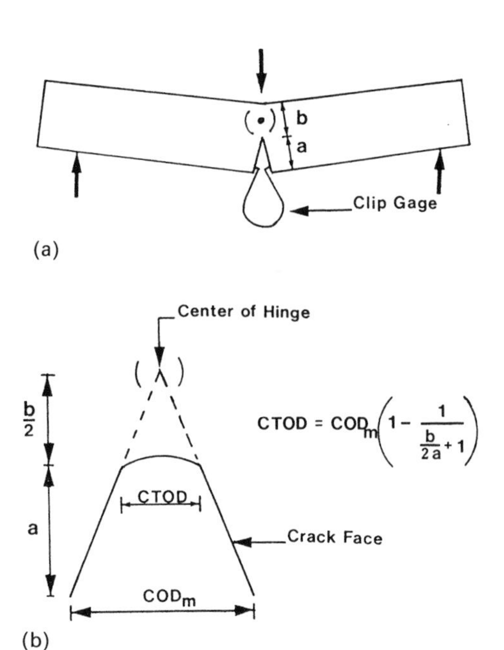

Figure 4.18. Crack opening displacement test. (a) COD specimen; (b) Measured COD_m and inferred CTOD.

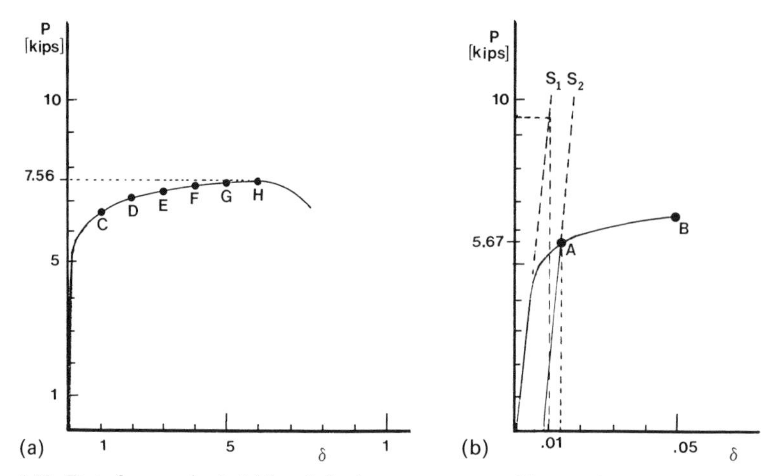

Figure 4.19. Data for exercise 1. (a) Load-displacement measured in test; (b) Enlarged view of small δ regime.

can be calculated as that stress at which the CTOD of the structural crack reaches the critical value. So far only empirical relations between stress (strain) and CTOD have been developed. There is no objection against their use, as long as they are reasonably general. However, with the aid of Equations (4.47) measured CTOD can be converted into J_R and then the result can be used in accordance with Equation (4.16). Certainly, Equation (4.47) is an approximation, but α can be obtained as a function of n [4]. Besides, an approximate general equation is more likely to be useful than a specific empirical one.

The development of the simple EPFM procedures discussed in this chapter have made the semi-empirical CTOD approach somewhat obsolete for damage tolerance analysis. However, the CTOD test is a useful extension of EPFM, as it can provide J_R or K_{Ic} through Equation (4.47) from tests on small specimens.

4.13. Exercises

1. For a test bar of a certain steel one measures a load displacement curve as in Figure 4.19. The original diameter of the cylindrical bar is 0.4 inch; the original length is 4 inch. The final thickness in the neck is 0.28 inch.
   (a) Determine the engineering stress-strain curve, F_{ty} and F_{tu}.
   (b) Determine E, F and n.
   (c) Determine α, σ_0 and ε_0 for $\sigma_0 = 100$ ksi.
   (d) Do the same for $\sigma_0 = 50$ ksi.
   (e) Check whether (a) and (b) lead to the same strain for at least two different stress values, e.g. 50 and 55 ksi.

2. Determine the true stress–true strain curve for the problem in Exercise 1 up to the point of maximum load. Repeat questions (b), (c), (d) of Exercise 1.

3. The bar of Exercise 1 is unloaded at $P = 7.3$ kips and tested as a new bar in a new test. Determine the new load-displacement curve. For this cold-worked material, assuming that F_{ty} coincides with the load at unloading, calculate F_{ty} and F_{tu}.

4. A test on a center cracked panel with $2a = 4$ inch, $W = 32$ inch shows a fracture stress of 50 ksi. The collapse strength is 55 ksi. Given that $W = 20$ inch (500 mm), $B = 0.4$ inch and assuming E, F and n as in Exercise 1, calculate J_r at fracture. Or did failure occur by collapse? (Neglect J_{el}). $H = 8.72$.

5. Using the results of Exercises 1 and 4, calculate the failure stress of a panel 24 inch wide, with a center crack of $2a = 6$ inch. $H = 13.5$; assume that the J_{el} can be neglected. Does failure occur by fracture or by collapse?

6. Given that Equation (4.30) for a center cracked panel is: $J = \alpha\sigma_0\varepsilon_0(1 - (22/W))ah_1(P/P_0)^{n+1}$ with $P = \sigma WB$ and $P_0 = \sigma_0(W - 2a)B$, and that for $a/W = 0.125$ is $h_1 = 4.13$. Calculate the fracture stress with the information obtained in Exercise 1c ($\sigma_0 = 100$ ksi) and in Exercise 1d ($\sigma_0 = 50$ ksi) and show that the results are the same, independent of the arbitrarily selected σ_0. Assume $J = 2$ kips/in and neglect J_{el}.

7. In a J_R test on a CT specimen the load rises almost linearly to 5 kip upon which fracture begins, and upon which the load-displacement diagram becomes essentially horizontal. The initial crack size is one inch; the displacement is 0.15 inch when fracturing begins. The crack (fracture) size reaches 1.05 inch when $\delta = 0.20$ inch and 1.15 inch when $\delta = 0.27$ inch. Calculate the J_R-curve. Thickness is 0.5 inch, $W = 2$ inch.

References

[1] D.P. Rooke and D.J. Cartwright, *Compendium of Stress intensity factors*, H.M. Stationery Office, London (1976).
[2] G.C. Sih, *Handbook of stress intensity factors*, Inst. Fract. Sol. Mech, Lehigh Un (1973).
[3] H. Tada et al., *The stress analysis of cracks handbook*, Del Res. Corp (1973).
[4] V. Kumar et al., *An engineering approach for elastic-plastic fracture analysis*, Electric Power Res. Inst., Rep NP-1931 (1981).
[5] V. Kumar et al., *Advances in elastic-plastic fracture analysis*, Electric Power Res. Inst., Rep NP 3607 (1984).
[6] D. Broek, *Elementary engineering fracture mechanics*, 4th Edition, Nijhoff (1986).
[7] Anon., *Standard method for the determination of J, a measure of fracture toughness*, ASTM Standard E-813.
[8] C.F. Ghih and M.D. German, *Requirements for one-parameter characterization of crack tip fields by the HRR singularity*, GE Tech Rep (1978).
[9] A.S. Tetelman and A.J. McEvily, *Fracture of structural materials*, John Wiley (1967).
[10] J.D. Eshelby, *Calculation of energy release rate prospects of fracture mechanics*, Sih et al. (eds), Noordhoff (1974) pp. 69–84.
[11] J.R. Rice, A Path independent integral and the approximate analysis of strain concentration by notches and cracks, *J Appl. Mach* (1968) pp. 379–386h.
[12] J.W. Hutchinson and P.C. Paris, Stability of J-controlled crack growth, *ASTM STP* **668** (1979) pp. 37–64.
[13] M.F. Kanninen and C.H. Popelar, *Advanced fracture mechanics*, Oxford Un. Press (1985).
[14] C.C. Veerman and T. Muller, The location of the apparent rotational axis in notched bend testing, *Eng. Fract. Mech.* **4** (1972) pp. 25–32.
[15] D. Broek, J. astray and back to normalcy, ECF6-Fracture control of structures, Vol. II EMAS (1986) pp. 745–760.

CHAPTER 5

Crack growth analysis concepts

5.1. Scope

In this chapter the concepts and procedures for crack growth analysis are discussed. Fatigue, being technically the most important crack growth mechanism covers most of the chapter (environmentally assisted growth or combined stress corrosion and fatigue is integrated into this discussion). Stress corrosion cracking by itself is covered in but one section; this is not because it is not considered important, but because stress corrosion cracking is practically covered by prevention and not by control, while analysis procedures are essentially similar to those either for residual strength (Chapter 3) or fatigue. Fatigue crack growth on the other hand can hardly be prevented in many structures; it must be controlled.

The discussions cover the concepts of crack growth analysis, retardation and special effects, as well as the analysis procedure. Examples of analysis are given. However, fatigue crack growth analysis is a complicated subject, and the discussions in Chapters 6 and 7 should be read as well, before attempts to analysize crack growth are made.

5.2. The concept underlying fatigue crack growth

Cyclic stresses resulting from constant or variable amplitude loading can be described by two of a number of alternative parameters, as shown in Figure 5.1. Constant amplitude cyclic stresses are defined by three parameters, namely a mean stress, σ_m, a stress amplitude, σ_a, and a frequency ω or v. The frequency is not needed to describe the magnitude of the stresses. Only two parameters are sufficient to describe the stresses in a constant amplitude loading cycle. It is possible to use other parameters; for example, the miniumum stress, σ_{min}, and the maximum stress, σ_{max}, describe the stresses completely, and so does the stress range, $\Delta\sigma = \sigma_{max} - \sigma_{min}$, in combination with any of the others, except σ_a. Almost any combination of two of the above parameters can completely define

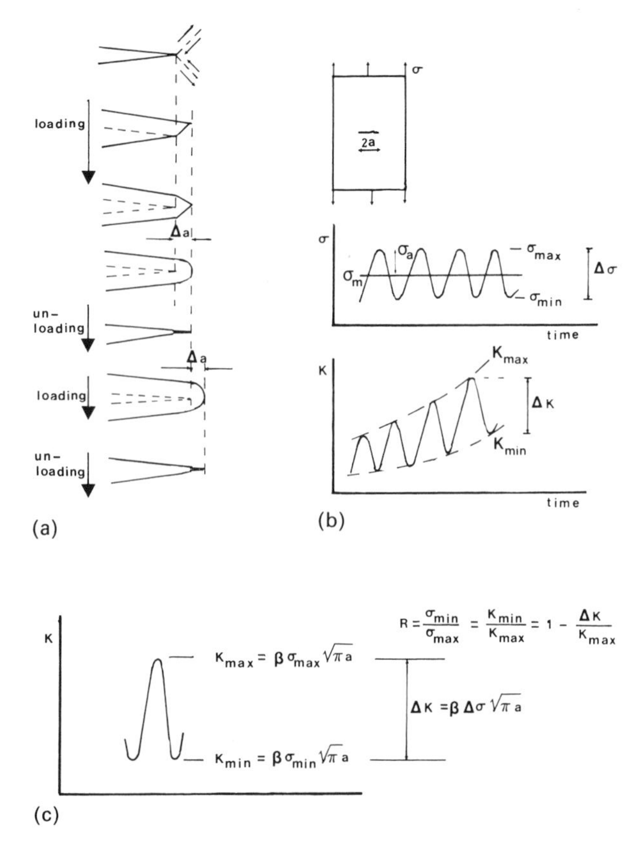

Figure 5.1. Parameters for fatigue crack growth. (a) Blunting and resharpening; (b) $\Delta\sigma$ and ΔK; (c) Stress ratio.

the cycle. Note that in the above the Greek letter Δ is used to indicate a RANGE of the stress. This is not in accordance with the normal use of Δ in mathematics, where Δ indicates a small change. In this case Δ stands for the total range of stress in a cycle: $\Delta\sigma = \sigma_{max} - \sigma_{min}$, which need not be small at all. It would be better to use e.g. σ_r for the stress range, but since the denotation $\Delta\sigma$ has become common practice, it will be used here as well.

Another parameter is often convenient. This is the so-called stress ratio, R, defined as $R = \sigma_{min}/\sigma_{max}$. One of the above parameters can be replaced by R to define the cycling. For instance, any of the following combinations fully defines the stresses: $\Delta\sigma$ and R, σ_{min} and R, σ_{max} and R, σ_a and R, σ_m and R. The case of $R = 0$ defines a situation in which the stress always rises from, and returns to 0. When $R = -1$, the stress cycles around zero as a mean, which is called fully reversed loading. Note that $R = \sigma_{min}/\sigma_{max} = (\sigma_{max} - \Delta\sigma)/\sigma_{max} = 1 - \Delta\sigma/\sigma_{max}$, so that $\Delta\sigma = (1 - R)\sigma_{max}$ and conversely: $\sigma_{max} = \Delta\sigma/(1 - R)$.

Crack growth life is expressed as the number of cycles to grow a fatigue crack over a certain distance. The number of cycles is denoted by N.

The crack growth mechanism, as discussed in Chapter 1, shows that a fatigue crack grows by a minute amount in every load cycle; the mechanism is shown schematically again in Figure 5.1a. Growth is the geometrical consequence of slip and crack tip blunting. Resharpening of the crack tip upon unloading, sets the stage for growth in the next cycle. It can be concluded from this mechanism that the crack growth per cycle, Δa, will be larger if the maximum stress in the cycle is higher (more opening) and if the minimum stress is lower (more re-sharpening), so that:

$$\Delta a_{\text{per cycle}}\uparrow \quad \text{for} \quad \sigma_{\max l}\uparrow \quad \text{and/or} \quad \sigma_{\min l}\downarrow. \tag{5.1}$$

The subscript l indicates the local stresses, at the crack tip, $\uparrow$ stands for larger, $\downarrow$ for lower. Note that in this case the Greek letter Δ is used in its 'normal' sense, meaning that it indicates a small change: growth from a to $a + \Delta a$. In the previous paragraphs the stress range, $\Delta\sigma$, was defined as $\sigma_{max} - \sigma_{min}$. The stress range will be larger when σ_{min} is less, so that the above equation can also be written as:

$$\Delta a_{\text{per cycle}}\uparrow \quad \text{for} \quad \sigma_{\max l}\uparrow \quad \text{and/or} \quad \Delta\sigma_l\uparrow. \tag{5.2}$$

The local stresses at the crack tip can be described in terms of the stress intensity factor K, discussed in Chapter 3, where $K = \beta\sigma\sqrt{\pi a}$, if σ is the nominal applied stress. In a cycle, the applied stress varies from σ_{min} to σ_{max} over a range $\Delta\sigma$. Therefore, the local stresses vary in accordance with:

$$\left.\begin{aligned} K_{\min} &= \beta\sigma_{\min}\sqrt{\pi a} \\ K_{\max} &= \beta\sigma_{\max}\sqrt{\pi a} \\ \Delta K &= \beta\Delta\sigma\sqrt{\pi a} \end{aligned}\right\}. \tag{5.3}$$

Again, the Greek letter Δ stands for range and not for a small increment; the denotation K_r would be better and less confusing, but ΔK is used here in conformance to general practice (see also Figure 5.1b).

With the use of Eqs (5.3) the Equation (5.2) for crack growth becomes:

$$\Delta a_{\text{per cycle}}\uparrow \quad \text{for} \quad K_{\max}\uparrow \quad \text{and/or} \quad \Delta K\uparrow. \tag{5.4}$$

Further, the stress ratio is defined as $R = \sigma_{min}/\sigma_{max}$. It appears from Eqs (5.3) that at any given crack size a, the stress ratio is also equal to K_{min}/K_{max}, since $\beta\sigma_{min} = \sqrt{\pi a}/\beta\sigma_{max}\sqrt{\pi a} = \sigma_{min}/\sigma_{max} = R$, so that:

$$R = \frac{K_{\min}}{K_{\max}} = \frac{K_{\max} - \Delta K}{K_{\max}} \quad \text{or} \quad K_{\max} = \frac{\Delta K}{1 - R}. \tag{5.5}$$

According to Equation (5.4) there is more crack growth when K_{max} is higher. It follows from Equation (5.5) that this is the case when ΔK is larger and/or R is higher, so that Equation (5.4) can be written as:

$$\Delta a_{\text{per cycle}}\uparrow \quad \text{for} \quad \Delta K\uparrow \quad \text{and/or} \quad R\uparrow. \tag{5.6}$$

In this equation Δa is the amount of crack growth in one cycle, which would be expressed in inch/cycle or mm/cycle. If growth were measured over e.g. $\Delta N = 10\,000$ cycles, the average growth per cycle would be $\Delta a/\Delta N$, which is the rate of crack propagation. In the limit where $N \to 1$, this rate can be expressed as the differential $\mathrm{d}a/\mathrm{d}N$. Equation (5.6) indicates that the rate is a rising function of ΔK and R, so that the proper mathematical form of Equation (5.6) is:

$$\frac{\mathrm{d}a}{\mathrm{d}N} = f(\Delta K, R). \tag{5.7}$$

As shown, Equation (5.7) derives directly from the model of crack growth discussed in Chapter 1 and shown in Figure 1.

5.3. Measurement of the rate function

According to Equation (5.7) the rate of crack growth will increase for larger ΔK and higher R. The actual functional form of Equation (5.7) might be derived from the crack growth model in Chapter 1. However, this model – although qualitatively in order – is a two-dimensional simplification of a three-dimensional process that is extremely complicated due to the presence of grains with different orientations, grain boundaries, particles, etc. As a consequence, a rigorous mathematical description of the model is not well possible, and a reliable functional form of Equation (5.7) cannot be obtained from theoretical analysis. This leaves only one possibility to obtain the function: interrogation of the material in a test. Although this might seem objectionable to theoreticians, it should be noted that ALL material data are obtained from tests, such as F_{tu} and F_{ty}, and even the modulus of elasticity E.

Crack growth data are obtained by subjecting a laboratory specimen to cyclic loading. The specimen may be of any kind as long as β is known, so that the stress intensity factors can be evaluated. Most commonly used are center cracked panels and compact tension specimens. The following examples are for a center cracked panel. As long as cracks are small with respect to panel size (e.g. $a/W < 0.4$) the geometry factor, β is approximately equal to one, so that $K = \sigma\sqrt{\pi a}$.

A panel as in Figure 5.2 is provided with as small but sharp central notch, so that cracks at both sides will start almost immediately. The specimen is subjected to a cyclic stress of constant amplitude in a fatigue machine. First

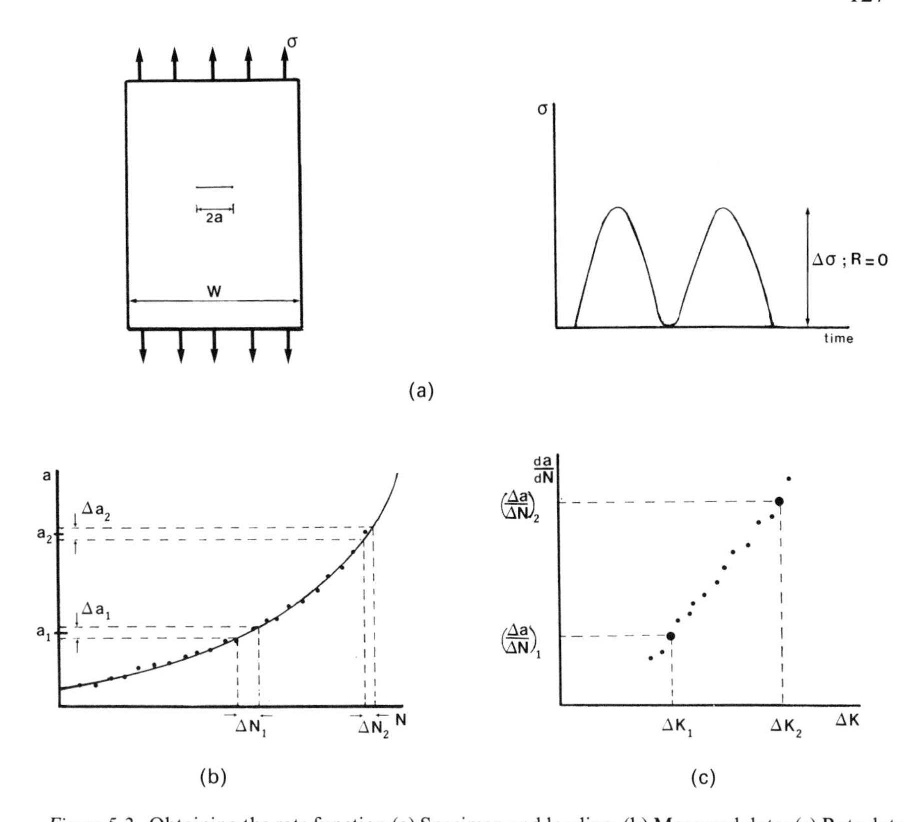

Figure 5.2. Obtaining the rate function (a) Specimen and loading; (b) Measured data; (c) Rate data

consider a stress cycle with $\sigma_{min} = 0$, so that $R = 0$ throughout the test. Also, $\sigma_{max} = \Delta\sigma$ in that case.

Crack growth is monitored throughout the test by measuring the length of the crack at intervals of e.g. 10 000 cycles. The results are plotted to obtain the crack growth curve as in Figure 5.2b. This is all the information that can be extracted directly from the test. It must be interpreted for the determination of the form of Equation (5.7).

Consider a small crack increment, Δa_1, on the curve. (Figure 5.2b) According to the measured curve, it took ΔN_1 cycles for the crack to grow over Δa_1. Thus, the rate of growth is $(\Delta a/\Delta N)_1$. For example, if the crack increment is 0.1 inch and this growth took $\Delta N = 10\,000$ cycles, then the rate is $(\Delta a/\Delta N) = 0.1/10\,000 = 1 \times 10^{-5}$ in/cycle.

The objective is to obtain the growth rate dependence upon ΔK, which requires determination of the stress intensity range. The average crack size at Δa_1 is a_1. The stress range is $\Delta\sigma$, so that $\Delta K_1 = \beta_1 \Delta\sigma\sqrt{\pi a_1}$. Apparently, a value of $\Delta K = \Delta K_1$, produced crack growth at a rate of $(\Delta a/\Delta N)_1$. This result is plotted as a data point in a diagram with $\mathrm{d}a/\mathrm{d}N(=\Delta a/\Delta N)$ and ΔK along the axes, as shown in Fig. 5.2c.

The above procedure is repeated for a number of points along the crack growth curve. At a larger crack size a_2, an amount of growth Δa_2 takes only ΔN_2 cycles. Because the curve is steeper, the rate is higher: as $a_2 > a_1$, also $\Delta K_2 > \Delta K_1$. Hence, a larger ΔK indeed produces a higher rate of growth. A plot of the data points as in Figure 5.2c confirms this.

Because differentiation is a very inaccurate procedure, large 'scatter' may occur in $\mathrm{d}a/\mathrm{d}N$. This problem is discussed in Chapter 7. It is the reason why in practice $\mathrm{d}a/\mathrm{d}N$ is obtained as a running average of 5–7 points along the crack growth curve.

Figure 5.2c provides the growth rate for any given ΔK. In Chapter 3 it was shown that the crack tip stress distribution is unique and depends only upon the stress intensity factor. If at two different cracks in the same material have equal stress intensity then the two crack tip stress fields are identical; there is similitude. Hence, if the stress intensities are equal the response of the cracks must be the same. This means that the crack growth rate will always be the same, if ΔK is the same. Thus, Figure 5.2c is the material's rate response in all cases. It can be used to analyze crack growth in a structure built of this material.

The validity of this similitude argument can be checked by performing a second test on a similar (or different) panel, but with a different $\Delta\sigma$. The crack growth is measured (Figure 5.3), the results analysed in the same manner as before, and the data of both tests plotted in the same rate diagram (Figure 5.3). The rate data of the second test will fall on the same line as the data of the first test. This confirms that the same rate was obtained at the same ΔK in both tests. For example, take a point on crack growth curve 1 at a crack size of $a_1 = 0.2$ in.

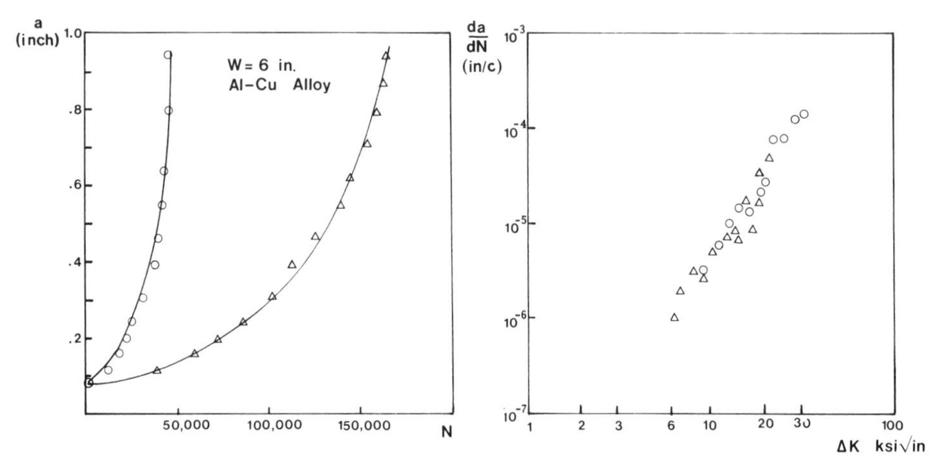

Figure 5.3. Data for two tests. Tests at $\Delta\sigma = 17.6$ and 11.4 ksi; $R = 0$ on center cracked panels; left: measured data; right: reduced data.

With a stress range of $\Delta\sigma = 17.6$ ksi, the value of the stress intensity range is $\Delta K = 17.6\sqrt{0.2 \times \pi} = 14\,\text{ksi}\sqrt{\text{in}}$. In the second test the stress range was $\Delta\sigma = 11.4$ ksi. This would produce a stress intensity of $14\,\text{ksi}\sqrt{\text{in}}$, if $a = 2.38 \times 0.2 = 0.48$ in (assuming $\beta = 1$) in the second test, i.e. $\Delta K = 11.4\sqrt{0.48 \times \pi} = 14.0\,\text{ksi}\sqrt{\text{in}}$, still assuming $\beta = 1$. This means that at a crack size 0.2-in in test 1 and 0.48-in in test 2, the stress intensities were the same, so that the rates should be the same. Figure 5.3a shows that the slopes (rate) of the curves at these two crack sizes are indeed equal (naturally, this followed immediately from the fact that the two tests led to the same $\mathrm{d}a/\mathrm{d}N - \Delta K$ diagram). Similitude in behaviour is hereby established. The results can be used to analyze crack growth in a structure.

The tests discussed so far were all at the same stress ratio, namely $R = 0$. According to Equation (5.7) the rates also depend upon R. This dependence can be assessed by performing tests at different values of R. The results are plotted versus ΔK, to obtain data such as in Figure 5.4. Indeed, higher R produces higher growth rates as should be anticipated on the basis of Equation (5.7). The data in Figure 5.4 show that the effect of R is smaller than that of ΔK, but that is simply the way it comes out.

Data are always plotted on logarithmic scales of $\log(\Delta K)$ and $\log(\mathrm{d}a/\mathrm{d}N)$,

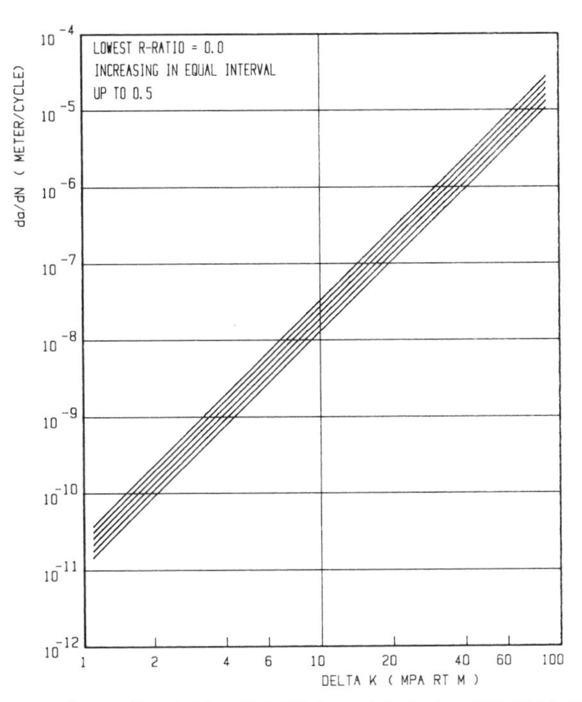

Figure 5.4. Crack growth rates for Ti-6AL-4V; Mill Anneal; Lab. Air; 70F (21C). Effect of R-Ratio.

because the rates vary over several orders of magnitude. A log scale for ΔK is not strictly necessary, but it has become standard practice to use a logarithmic scale for ΔK as well.

Crack growth properties of a number of structural alloys are compared in Figure 5.5. Environment, loading frequency, and temperature may have a significant effect on growth rates. Examples of some of these effects are shown in Figure 5.6. For a discussion of crack growth at negative R see Chapter 7.

5.4. Rate equations

The form of Equation (5.7) follows from the test data; it cannot be obtained from a theoretical model. Naturally, a functional form can be established by fitting a curve through the test data. The resulting equations are sometimes useful as they eliminate the necessity of using a graph.

From Figures 5.3 through 5.6 it appears that the rate data for one particular R-ratio fall more or less on a straight line in a logarithmic plot. The equation for a straight line is $y = mx + b$. In the present case $y = \log(\mathrm{d}a/\mathrm{d}N)$ and $x = \log(\Delta K)$, so that:

$$\log\left(\frac{\mathrm{d}a}{\mathrm{d}N}\right) = m_P \log(\Delta K) + \log(C_P). \tag{5.8}$$

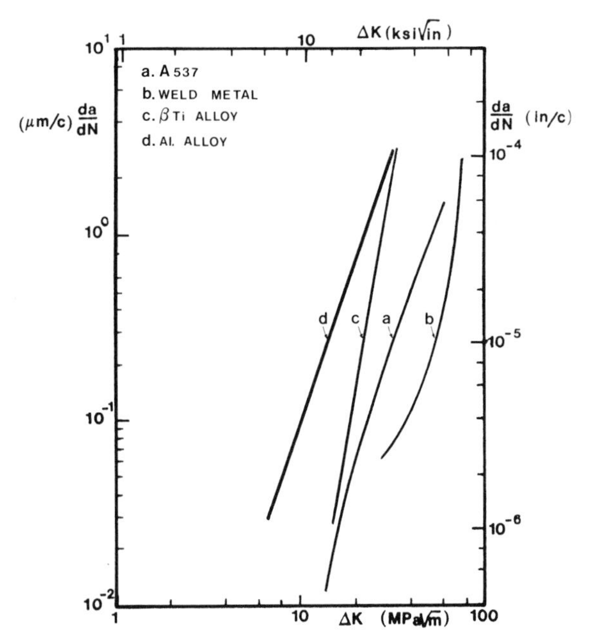

Figure 5.5. Typical rate properties of different alloys.

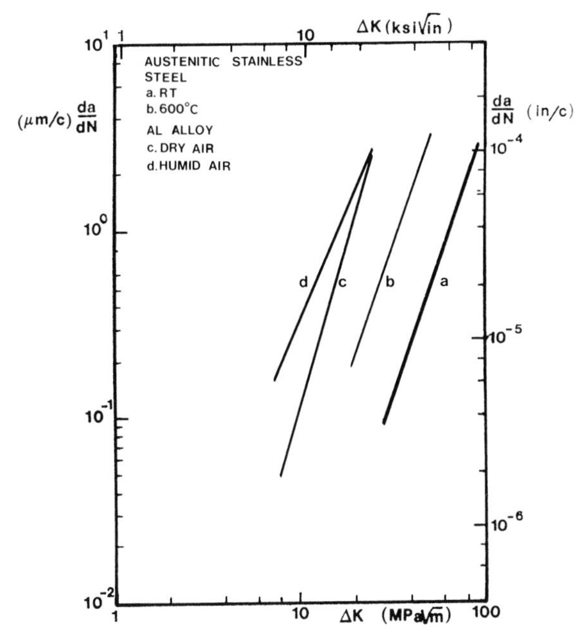

Figure 5.6. Typical effects of environment and temperature.

Taking the anti-log provides:

$$\frac{\mathrm{d}a}{\mathrm{d}N} = C_P(\Delta K)^{m_P}. \tag{5.9}$$

This equation is generally known as the Paris equation.

The parameters C_P and m_P can be easily determined. For example, using the two points A and B in Figure 5.7, yields:

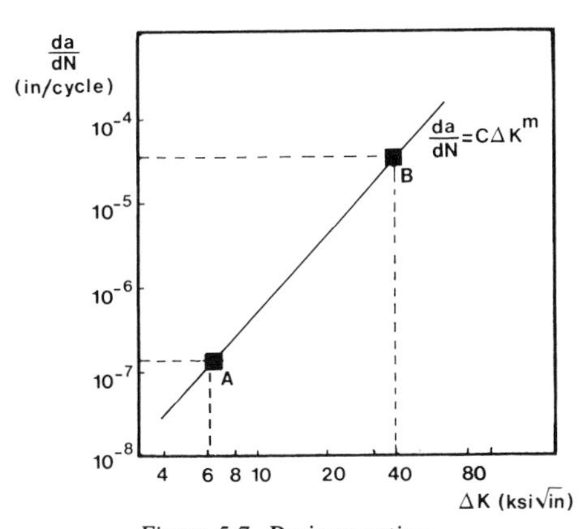

Figure 5.7. Paris equation.

$$\begin{aligned} \text{point } A\text{: } \log(1.6 \times 10^{-7}) &= m_P \log 6.3 + \log C_P \\ \text{point } B\text{: } \log(4 \times 10^{-5}) &= m_P \log 40 + \log C_P. \end{aligned} \tag{5.10}$$

Taking the logarithms provides:

$$\left.\begin{aligned} -6.8 &= 0.8m_P + \log C_P \\ -4.4 &= 1.6m_P + \log C_P \quad - \\ \hline 2.4 &= 0.8m_P \end{aligned}\right\}. \tag{5.11}$$

This provides $m_P = 3$; substitution of which in one of the equations leads to $\log C_P = -9.2$ or $C_P = 6.3 \times 10^{-10}$. The rate equation becomes:

$$\frac{da}{dN} = 6.3 \times 10^{-10} \Delta K^3 \tag{5.12}$$

for this particular material. For most materials the value of m_P is between 3 and 5. The value of C_P is more strongly material dependent; it has also largely different values in different unit systems (see Chapter 7).

The Paris equation covers only one R-value. The lines for different R are often parallel, i.e. have equal slope as e.g. in Figure 5.4. Thus all these lines would have the same m-value, but different C; the latter depending upon R as $C(R)$. Hence, the following equation could cover all R-values:

$$\frac{da}{dN} = C(R) \Delta K^{m_R}. \tag{5.13}$$

For many materials, the dependence of C on R can be described in a simple manner, as e.g.

$$\frac{da}{dN} = \frac{C_w}{(1 - R)^{n_w}} \Delta K^{m_R}. \tag{5.14}$$

Where C_w is the value of C when $R = 0$. Equation (5.14) can be used as is. Often it is further modified by substituting $K_{max}^{n_w} = (\Delta K/(1 - R))^{n_w}$, so that:

$$\frac{da}{dN} = C_w \Delta K^{m_R - n_w} K_{max}^{n_w} = C_w \Delta K^{m_w} K_{max}^{n_w}, \tag{5.15}$$

where $m_w = m_R - n_w$, which is known as the Walker equation. Note that Equation (5.15) essentially reverts back to the original Equation (5.4); of course Equations (5.14) and (5.15) are equivalent, and both are in use.

One may argue that fracture occurs when the maximum stress intensity in a cycle equals the toughness, i.e. if $K_{max} = K_c$ or K_{Ic}. Since $K_{max} = \Delta K/(1 - R)$, this would happen when $\Delta K = (1 - R) K_c$. At fracture the growth rate would tend to infinity. A functional value can be made to go to infinity through division by zero:

$$\frac{\mathrm{d}a}{\mathrm{d}N} = C_F \frac{\Delta K^{m_F}}{(1 - R)K_c - \Delta K}. \tag{5.16}$$

At fracture, where $\Delta K = (1 - R)K_c$, the above equation indeed provides an infinite growth rate. This equation is known as the Forman equation. It shows the growth rate to depend upon R and should therefore apply for all R-values: the equation 'pretends' to 'know' how $\mathrm{d}a/\mathrm{d}N$ depends upon R. A strong objection against the equation is that in many cases fracture is not governed by the toughness, because of collapse (Chapters 2 and 3).

In addition to Equations (5.9) through (5.16) many different curve fitting equations can be developed to describe the test data. As a matter of fact, there are probably as many equations as there are researchers in the field. Several others are in common use. But none of these, nor the above equations, have any physical significance; they are merely curve fitting equations. If they do fit the data properly, there is no objection against their use. But no equation can fit all data, so that religious adherence to one equation is not advisable. One should use the equation providing the best fit in a particular case. An equation may be used if convenient, but direct graphical use of the rate diagram is just as reliable. Most crack growth analysis is done by computer, which can be supplied the rate diagram in tabular form. It makes little difference to a computer whether it interpolates in a table or uses an equation. It should be noted that the parameters for the various equations are different, even if they cover the same data set. For this reason the coefficients C_P, C_w, and C_F and exponents m_P, m_w, and m_F are used to indicate that they are specifically for a certain equation (Paris, Forman, Walker). Use of the parameters of one equation for another – even for the same material – may lead to dramatic errors. Conversion of parameters to other unit systems requires great care (Chapter 7).

5.5. Constant amplitude crack growth in a structure

Most structures experience some form of variable amplitude loading in which case the crack growth analysis is considerably more complicated than for constant amplitude, as will appear later in this chapter. However, in the few cases of constant amplitude loading the analysis can be readily performed with or without the use of a computer. A crack in a structure will grow at the rates indicated by the rate diagram because of the similitude discussed. Analysis of structural crack growth can be carried out if the geometry factor is known (Chapter 8) for the structural configuration at hand. The crack growth (curve) in the structure follows from an integration of the rates:

$$\frac{\mathrm{d}a}{\mathrm{d}N} = f(\Delta K, R) \quad \text{or} \quad \mathrm{d}N = \frac{\mathrm{d}a}{f(\Delta K, R)}. \tag{5.17}$$

Integration provides:

$$N = \int_{a_0}^{a_p} \frac{\mathrm{d}a}{f(\Delta K, R)}. \tag{5.18}$$

Generally, the integration is done numerically; it can seldom be performed in closed form, because of the complexity of the functions f and β in ΔK, and of the stress history. The function f might be as simple as the Paris equation:

$$N = \frac{1}{C_p} \int_{a_0}^{a_p} \frac{\mathrm{d}a}{\{\beta(a/w)\,\Delta\sigma\sqrt{\pi a}\}^{m_p}}. \tag{5.19}$$

The β for a structural crack is usually a lengthy polynomial in a/W or known only in tabular form, so that numerical integration is indicated even if f is simple and $\Delta\sigma$ is constant (independent of a). Integration is performed most easily through the use of a computer, but in the case of constant amplitude loading a hand computation is very well possible. The principle for a simple numerical integration in the case of constant amplitude loading is shown below.

If the loading is of constant amplitude, the integration can be done in small steps with little error; the step size might be taken as e.g. a crack increment of one percent of the current crack size. Assume for example (Figure 5.8) a case of constant amplitude loading at e.g. $\Delta\sigma = 20$ ksi; further assume that a Paris equation applies with $\mathrm{d}a/\mathrm{d}N = 6.17E - 10\,\Delta K^3$. Let β for the structural crack be given as $\beta = 1.12 + (a/W)^2$ (a hypothetical case) and let $W = 4$ inches (Figure 5.8). The first two steps of a calculation starting at a crack size of 0.75 inch are shown below:

Initially:

$a = 0.75$ in; $N = 0$ cycles;

$\Delta a = 0.01 \times 0.75 = 0.0075$ in (one percent increase);

$\Delta K = [1.12 + (0.75/4)^2] \times 20 \times \sqrt{0.75\pi} = 35.5\,\text{ksi}\sqrt{\text{in}}$;

$\mathrm{d}a/\mathrm{d}N = 6.17 \times 10^{-10} \times 35.5^3 = 2.75 \times 10^{-5}$ in/cycle;

$\Delta N = \Delta a/(\mathrm{d}a/\mathrm{d}N) = 0.0075/(2.75 \times 10^{-5}) = 273$ cycles;

$N = N + \Delta N = 0 + 273 = 273$ cycles;

$a = a + \Delta a = 0.75 + 0.0075 = 0.7575$ in;

$\Delta a = 0.01 \times 0.7575 = 0.007575$ in (one percent);

$\Delta K = [1.12 + (0.7575/4^2] \times 20 \times \sqrt{0.7575\pi} = 35.7\,\text{ksi}\sqrt{\text{in}}$;

$\mathrm{d}a/\mathrm{d}N = 6.17 \times 10^{-10} \times 35.7^3 = 2.81 \times 10^{-5}$ in/cycle;

$\Delta N = 0.007575/(2.81 \times 10^{-5}) = 269$ cycles;

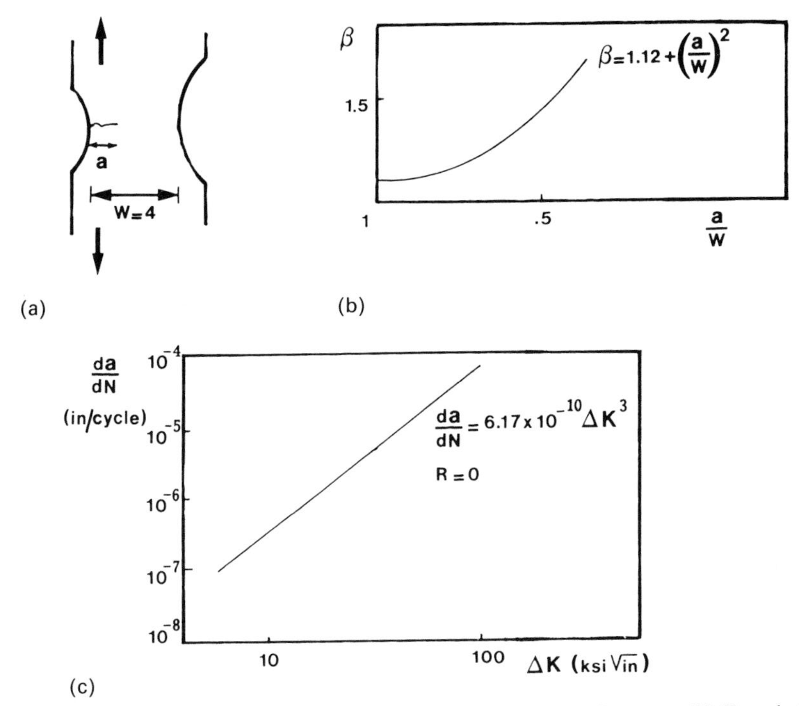

Figure 5.8. Crack growth analysis; Example for constant amplitude. (a) Structure; (b) Hypothetical β-curve; (c) Material's rate data.

$$N = 273 + 269 = 542 \text{ cycles;}$$

$$a = 0.7575 + 0.0076 = 0.76510.$$

This process is continued until the crack size a is reached. It can be programmed for execution by a computer:

$$a_j = a_0;\ N = 0;$$

$$a_i = a_j;$$

$$\beta_i = f(a_i/w);$$

$$\Delta K_i = \beta_i \Delta\sigma\sqrt{\pi a_i};$$

$$\frac{\mathrm{d}a}{\mathrm{d}N} = f(\Delta K_i,\ R)$$

$$\Delta a_i = \delta a_i \text{ (e.g. } \delta = 0.01);$$

$$\Delta N_i = \Delta a_i/(\mathrm{d}a/\mathrm{d}N)_i$$

$$N_j = N_i + \Delta N_i;$$

$$a_j = a_i + \Delta a_i;$$

if $a_j < a_P$ then return to beginning.

With the steps on the order of one percent of the current crack size (or fixed step sizes if so desired) good accuracy can be obtained. If large steps are taken an integration rule such as Simpson or Runge–Kutta should be used. These integration rules were divised for numerical integration prior to the computer era, when large steps had to be taken for hand calculations. In true integration where in the limit $\mathrm{d}a \to 0$ no such rule is required. The computer can indeed let the step size go to zero, so that the integration rules are not necessary.

In an emergency constant amplitude analysis as above can be done by hand in rather large steps as shown in Table 5.1. Even without the use of an integration rule, reasonable accuracy is obtained when the steps are not made too large, especially at the beginning (Table 5.1). The results of Table 5.1 are compared to Figure 5.9 with a computer analysis of the same problem, showing that the hand calculation gives a reasonable approximation. The procedure illustrated in Table 5.1 lends itself very well for execution with a spread sheet program, or a dedicated program such as shown above can be used. Integration is a forgiving procedure in contrast to differentiation. The latter problem is discussed in Chapter 7.

Most structures being subjected to variable amplitude loading (Chapter 6), the above procedure for constant amplitude initegration is seldom applicable. Crack growth analysis for variable amplitude loading will be discussed later in this chapter. Such analysis is much complicated by the problem of load interaction and retardation as discussed in the following section.

5.6. Load interaction: retardation

When one single high stress is interspersed in a constant amplitude history, the crack growth immediately after the 'overload' is much slower than before the overload. Figure 5.10 shows how three single overloads increase the crack growth life by almost a factor of five (compare A and B). After a period of very slow growth immediately following the overload, gradually the original growth rates are resumed. This phenomenon is known as retardation. A negative load following the overload reduces retardation but does not eliminate it (compare *B* and *C* in Figure 5.10). Crack growth analysis for variable amplitude loading, is not very well possible without an account of retardation effects. Before such an account can be made, retardation must be explained.

Consider (Figure 5.11) a crack subjected to constant amplitude loading at

Table 5.1. Hand calculation. Case = Center cracked panel; W = 6 inch; $da/dN = 4 \times 10^{-9} \Delta K^{3.5}$; $\Delta\sigma$ = 12 ksi; R = 0.

a (in)	Δa (in)	Average a in step a_{ave} (in)	$\beta = \sqrt{\sec(\pi a_{ave}/W)}$	$\Delta K = \beta \Delta\sigma \sqrt{\pi a_{ave}}$ (ksi $\sqrt{\text{in}}$)	$da/dN = 4 \times 10^{-9} \Delta K^{3.5}$ (in/cycle)	$\Delta N = \Delta a / da/dN$ (cycles)	$N = N + \Delta N$ cycles
0.1							0
	0.4	0.30	1.006	11.72	2.20×10^{-5}	18182	
0.5							18182
	0.5	0.75	1.040	19.16	1.23×10^{-4}	4065	
1							22247
	0.5	1.25	1.123	26.70	3.93×10^{-4}	1272	
1.5							23519
	0.5	1.75	1.282	36.07	1.13×10^{-3}	442	
2							23961
	0.8	2.40	1.799	59.28	6.42×10^{-3}	125	
2.8							24085
Same calculation with finer steps during early growth							
0.1							0
	0.1	0.15	1.002	8.25	6.45×10^{-6}	15504	
0.2							15504
	0.2	0.30	1.006	11.72	2.20×10^{-5}	9091	
0.4							24595
	0.3	0.55	1.021	16.11	6.71×10^{-5}	4471	
0.7							29066
	0.3	0.85	1.053	20.65	1.60×10^{-4}	1875	
1							30941
	0.5	1.25	1.123	26.70	3.93×10^{-4}	1272	
1.5							32213
	0.5	1.75	1.282	36.07	1.13×10^{-3}	442	
2							32655
	0.8	2.40	1.799	59.28	6.42×10^{-3}	125	
2.8							32780

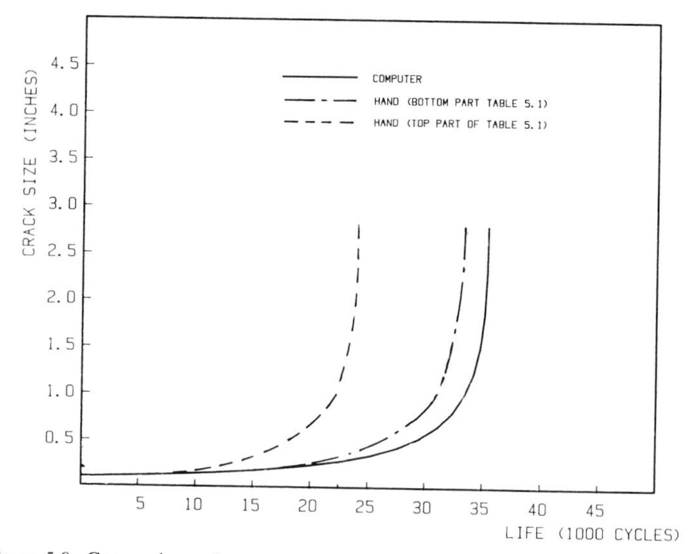

Figure 5.9. Comparison of computer analysis and the hand analysis of Table 5.1.

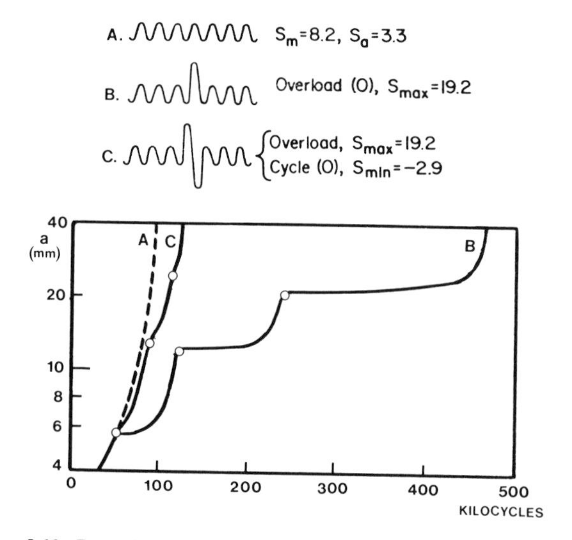

Figure 5.10. Retardation after overload (Courtesy Aircraft Engineering).

$R = 0$; during the first cycle the load varies from A to C through B. Before the loading starts, imagine a little (dashed) circle (Figure 5.11a) at the crack tip, indicating the material that will undergo plastic deformation in the future plastic zone (Chapter 3). The plate is then loaded to B. Imagine that one could now

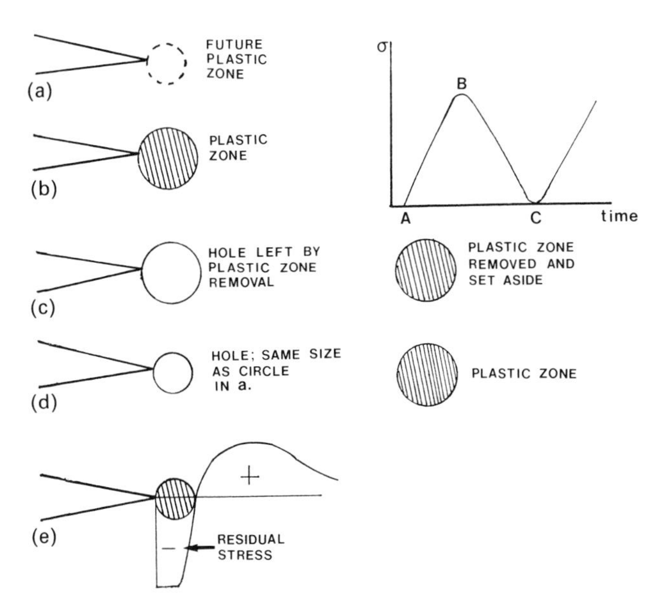

Figure 5.11. Residual stress at crack tip (a) Load at A; (b) Load at B; (c) Load at B; plastic zone removed; (d) Load at C; plastic zone still out; (e) Load at C; plastic zone back in.

remove the plastic zone and put it aside (Figure 5.11c). After unloading to C the situation of Figure 5.11d is reached: all material is elastic – the plastic material having been cut out – so that all strains and displacements are zero after unloading. Hence, after unloading the hole at the crack tip in Figure 5.11d is equal in size to the dashed circle in Figure 5.11a. The plastic zone has become permanently deformed; it is larger than before the loading started. Thus it will not fit in the hole of Figure 5.11d. In order to make it fit it must be squeezed back to its original size for which a stress at least equal to the yield strength is needed (it must be deformed plastically again in compression to be squeezed back to its original size; this requires stresses at least equal to the yield).

The plastic zone in fatigue loading is very small. Most of the fatigue crack growth takes place at low values of ΔK as can be appreciated from the data in previous figures. At high ΔK the rates are so high that little crack growth life is left; most of the life is covered at low ΔK. If for example, $\Delta K = 10\,\text{ksi}\sqrt{\text{in}}$ in a material with a yield strength $F_{ty} = 50\,\text{ksi}$, the plastic zone size (Chapter 3) would be $r_p = 10^2/(6\pi 50^2) = 0.0021$ inch. As the remainder of the plate is elastic and returns to zero strain after unloading, this small plastic zone indeed will be squeezed back to its original size and made to fit in its surroundings. In order to squeeze the permanently deformed material back to its original size reverse yielding is necessary: the compressive stress must be at least equal to the yield strength. Whether the plastic zone is taken out as hypothesized above, or whether it remains in place as it normally will, the end result will be the same.

It follows that after unloading, there is a compressive stress at least equal to the yield strength at the crack tip. This is a residual stress (no external load) which must be internally equilibrated by tension further away. The residual stress distribution is as shown in Figure 5.11e. During subsequent cycling this residual stress system will be present upon each unloading and it will have to be added to (act together with) the applied stress. The crack growth response by the material automatically accounts for this residual stress system; the data in e.g. Figures 5.4–5.6 already reflect its effect, because the material 'knows' about these residual stresses and shows growth rates in accordance with their presence.

If an overload occurs, a much larger plastic zone is formed (Figure 5.12). After the overload a more extensive residual stress system is present than before the overload. This more extensive system, acting against the applied stress,

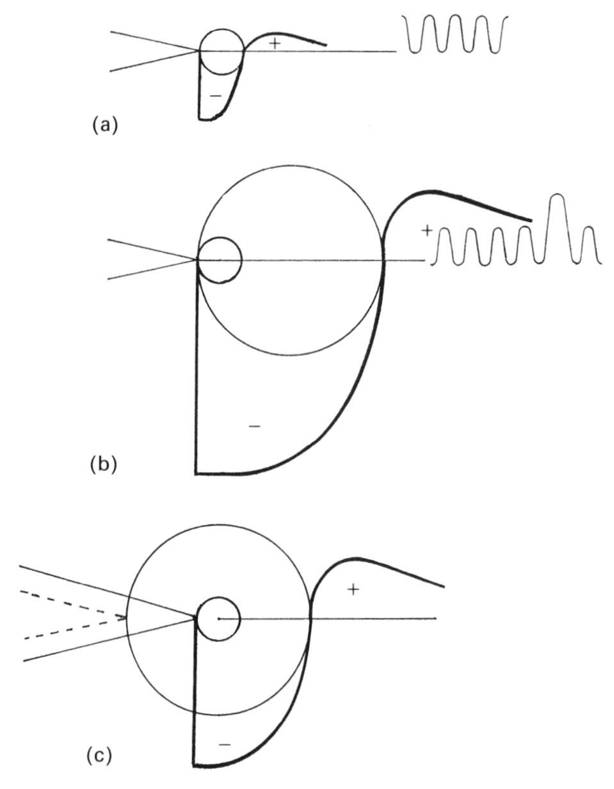

Figure 5.12. Residual stresses before and after overload. (a) Before; (b) After; (c) Situation after overload.

causes subsequent growth to be slower (retarded). Once the crack has grown through the overload residual stress field ($\approx$ overload plastic zone), the original residual stress field is restored (Figure 5.12) and the 'normal' growth is resumed. (In reality the compressive stresses do not extend throughout the plastic zone, because the compressive yield zone is smaller. However, the principle is still maintained).

Another way of looking at this problem is illustrative. It was shown in Chapter 1 that crack growth occurs by plastic deformation (slip). As a consequence, the crack tip plastic strain range is the best measure of crack growth. The stress and strain being singular at the crack tip, it is somewhat hard to envisage how they vary. However, the singularity occurs only in an infinitesimally small material element, while it will effectively disappear due to crack tip blunting. Therefore, crack growth phenomena can be discussed on the basis of finite values of stress and strain, their values being bounded by the (cyclic) stress–strain curve.

First, consider a crack with no previous history (no previous plastic zones) with a very small plastic zone, cyclically loaded at $R = 0$ (Figure 5.13). At maximum load, the stress and strain are defined by point A. Assuming an extremely small plastic zone in a large plastic plate, the elastic deformations of the plate will completely dominate the problem. Hence, all deformations will come back to almost zero upon load release: the elastic plate will squeeze the permanently deformed material in the (extremely small) plastic zone at zero load. This implies that the crack tip strain will be reduced to almost zero (point B in Figure 5.13).

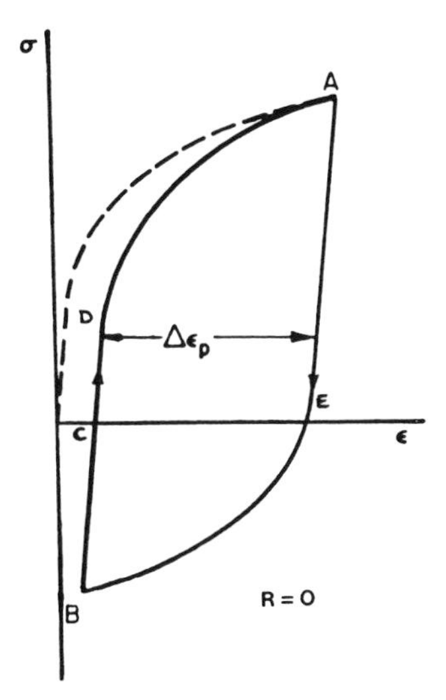

Figure 5.13. Stress-strain loop at crack tip. [7]. Copyright ASTM reprinted with permission.

Upon reloading, the stress and strain must conform to point A, because that is the condition dictated by the surrounding elastic material. Obviously, the crack tip is subjected to a plastic strain range $\Delta\varepsilon_p$ with $R_\varepsilon \approx 0$ and R_σ close to -1. The material experiences a stress–strain cycle BCDAEB.

The crack tip strain at $R = 0$ will return to almost zero, but not exactly so, because there must be equilibrium of residual stresses, as illustrated in Figure 5.14. The compressive stresses in the plastic zone have to be equilibrated by tension stresses around the plastic zone, which give cause to a small remaining positive strain. Note however that the extent of residual tension stresses is actually less than shown in Figure 5.12. This is due to the fact that upon return from point A in Figure 5.13 yielding begins earlier, because $AE = BD$ in Figure 5.13 due to the Bauschinger effect.

The return of the crack tip strain to almost zero is a result of the action of the large elastic plate, or rather of the remaining ligament. Note that the stiffness of the plastic zone is very small compared to the stiffness of the ligament.

Now consider a real fatigue crack with a previous history (Figure 5.15); in its wake along the crack edges is a strip of material representing the accumulation of all previous crack tip plastic zones through which crack has progressed. This material is no longer loaded, but at one time it underwent plastic deformation. Closure stresses arise because the permanent elongation of the crack lips will

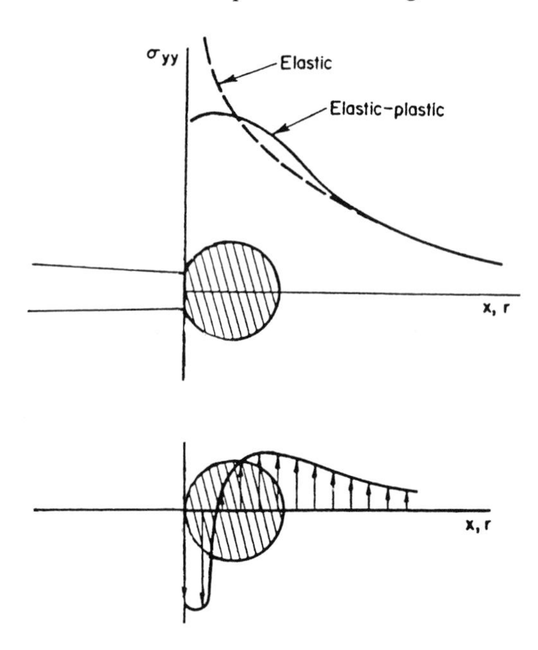

Figure 5.14. Crack tip stresses at $R = 0$ loading. top: at max load. Bottom: at zero load. Compare with Figure 5.13. [13]. Courtesy EMAS.

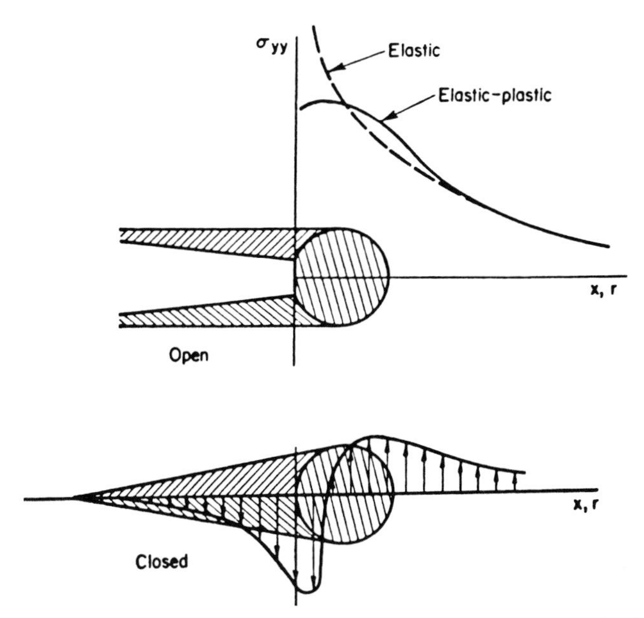

Figure 5.15. Crack tip stresses at $R = 0$ loading while accounting for previous plastic deformation in wake of the crack, causing closure of crack tip at zero load. [13]. Courtesy EMAS.

close the crack before the load is zero. As such, they are similar to the compressive stresses built up in the plastic zone. As a matter of fact, both stress systems result from the same action of the surrounding elastic field. Therefore, Figure 5.14 can be redrawn by including the crack lips as shown in Figure 5.15.

The closure stresses at the crack tip never exceed the yield. Thus, it requires only an elastic strain to accommodate the remaining strain at the crack tip (in case of complete closure), which means that the remaining crack tip strain is elastic and the return point in Figure 5.13 is still B. Upon reapplication of the load, the crack will remain closed until the closure stresses and the compressive stresses are relaxed. But, the crack tip material is already straining. Its stress-strain condition moves from B to C in Figure 5.13 while the crack is still closed.

Figure 5.16 shows the consequences of an overload. Depending upon the minimum stress in the cycle and upon the relative stiffness of previously plastic material with respect to the elastic material, the remaining strain after the overload will be larger or smaller (point F in Figure 5.16a). Therefore, depending upon return to F_1 or F_2, the straining during the subsequent cycles will be as in Figure 5.16b. In any case, the cyclic strain range (i.e. the opening of the σ–ε loop) is considerably reduced and the crack growth rate will decrease accordingly (retardation).

According to some retardation models (see Section 5.7) the growth rate reduction is proportional to the ratio between the overload plastic zone and the

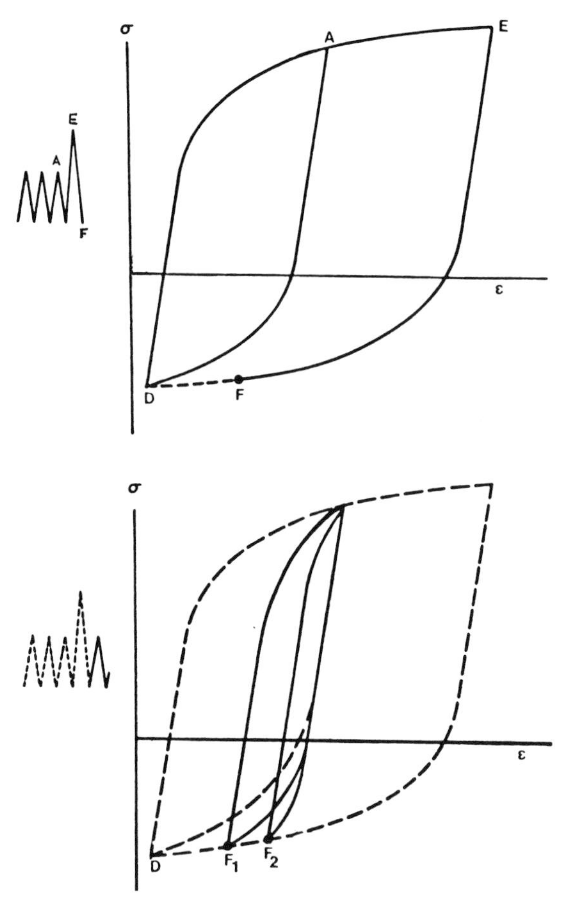

Figure 5.16. Consequences of overload for crack tip plastic strain loop. Top: during overload; Bottom: after overload. [7]; Copyright ASTM. Reprinted with permission.

current plastic zone. Although the retardation models may be somewhat artificial, they do contain the relevant parameters. With increasing crack size (and plastic zone size) it becomes increasingly difficult for the elastic material to restore the zero strain field after unloading (the stiffness ratio between ligament and plastic zone decreases). Hence, at small crack sizes the return point may be F_1 (Figure 5.16), but at large crack sizes, it shifts to F_2, so that retardation becomes more pronounced (see also Figure 5.10). This may explain why various investigators find different retardation effects in the same material: retardation is crack size dependent and panel size (ligament) dependent.

The above discussion explains why a compressive stress following the overload will reduce the retardation, but cannot completely annihilate it (compare C and B in Figure 5.10); during compression the closed crack is no stress raiser and therefore all compressive stresses and strains are elastic (elastic strains are negligible as compared to plastic strains).

5.7. Retardation models

The complexities of the retardation phenomenon so far have precluded the development of an all-encompassing mathematical-physical treatment of the problem in a retardation model. More than half a dozen models have been proposed [1–7], none of which covers all aspects of the problem.

A number of models are based on crack closure. As was explained in the previous section, crack tip closure occurs even in constant amplitude loading. Some closure models consider only that part of the cycle effective over which the crack is open. However, even during that part of the cycle the crack tip material is straining (*B–C* in Figure 5.13) and this straining is just as much part of the crack-tip strain loop as is the part associated with the open crack. Hence, considering only that part of the cycle during which the crack is open, is physically incorrect. Certainly, closure changes after an overload and this affects the general residual stress field at the crack tip, but this residual stress field is changed more due to the larger plastic zone. In view of this, pure closure models do not cover all aspects of the problem. There is no doubt that overloads affect closure; hence there will be a correlation between closure and retardation. But retardation and changes in closure are both consequences of the overload. Closure changes are not the cause of retardation. Both are symptoms, and one symptom cannot explain another; both are caused by the 'disease'. A proper treatment of the model must consider the effect of overloads upon the total residual stress field in the wake of the crack and ahead of the crack, and in particular upon the plastic strain range.

Other retardation models attempt to account for the residual stress field directly by superposing it on the stress field due to the external load. The residual stresses themselves cause a certain stress intensity which can be added to the stress intensity due to the applied loads. Such models have several draw-backs and short-comings. In the first place, although the residual stress field can be assessed qualitatively in an easy manner, the quantitative evaluation (and thus evaluation of the resulting K) is difficult. A second problem is, that a residual stress field already exists even in constant amplitude loading. Also this field should be accounted for if the model is to be based on residual stresses. (Normally, it will be automatically accounted for in the data base).

Despite the mathematical complexity of some, all models are two-dimensional over-simplifications of a complicated three-dimensional problem, and full of assumptions. For example, all models must consider plastic deformation, but even plastic deformation is not easily treated analytically, further the yield strength, F_{ty}, is arbitrarily defined (e.g. 0.2% plastic strain), while it is an essential number in the calculation of the plastic zone. Plastic zone size equations by themselves are subject to doubt, especially in the case of a changing state of stress. Some models use empirical equations for changes of the closure stress.

All of this leads to the conclusion that any model must contain one or several unknown parameters which must be obtained from tests (i.e. adjusted empirically). In one model such a parameter was later included to make empirical adjustment possible in the first place. There is no practical objection against the use of empirical parameters (E, v and F_{ty} follow from experiments and so does after all $\mathrm{d}a/\mathrm{d}N$), provided the use of these empirical parameters leads to useful crack growth predictions for structural applications.

If all models are simplifications, none can be preferred over another. Moreover, if they all contain adjustable parameters, they can all be made to work if the parameters are adjusted appropriately. All claims that one model is better than another are improper. Each model can be made to work if empirically adjusted; if it does not work, it was not adjusted properly. Generality of the adjustment may be a problem. In that respect some models may be somewhat better than others. Clearly, the adjustment parameters will be material dependent. But should they also depend strongly upon the stress history, as they do (see Chapters 6, 7) then they cannot be used generally. Attempt to make general use of these parameters then lead to false claims with regard to a model's adequacy.

There are so many retardation models that a discussion of all would be beyond the scope of this book. Review of just a few might suggest that these would be better than others. The interested reader is therefore referred to the relevant literature. The general reader probably has no use for such a review at all, because crack growth analysis with retardation requires the use of a computer anyway and the general user will employ existing software. Such software should have the option for the use of several retardation models. If just one model is available generality may be hampered, because one model may be somewhat more appropriate for certain applications and vice versa. Which models are included in the software is rather secondary, because all can be made to work if properly calibrated.

More important than which retardation model is used, may be the following:

(a) As demonstrated in the previous section retardation will depend upon the relative stiffness of plastic zone and elastic material. This may become important for larger cracks in small components. No known retardation model considers this problem but some account could be made in the computer code.

(b) All models must consider crack tip plasticity and therefore will use arbitrary numbers. As was explained in Chapters 2 and 3, the state of stress has a great influence on plasticity. The plastic zone is $r_p = K^2_{\max}/\alpha\pi F^2_{ty}$ where α is commonly considered to be 2 for plane stress and 6 for plane strain. Further, the condition for plane strain is $B > 2.5K_{max}^2/F_{ty}^2$, in which 2.5 is a somewhat arbitrary number (Chapter 3). 'Normal' stress cycles at low K_{max} may give plane strain, but an overload may cause plane stress due to its larger K_{max}. This also causes a more extensive residual stress field and more retardation. Thus, in accounting for retardation the computer code should assess the state of stress in each cycle. If it does not, even the 'so-called' sophisticated retardation model may give large errors. For example, the larger retardation at longer cracks in Figure 5.10 may be caused by a change from plane strain to plane stress during the overloads. But even if the state of stress is evaluated, the value of α will be arbitrary, the factor of 2.5 is arbitrary and F_{ty} is more or less arbitrary. Some models use different plastic zone formulations than above, but these still contain arbitrary numbers.

In the following crack growth analysis for variable amplitude loading will be illustrated on the basis of the Wheeler model. This model is used here not because it is believed to be better than any other, but because it is very simple, so that it can be used easily for illustrations. It is worth mentioning however that if all models are simplifications anyway, the simplest certainly is the most appealing. If all models must be calibrated for general use, even the simplest model can be made to work by calibration.

Wheeler introduces a retardation parameter ϕ_R. It is based on the ratio of the current plastic zone size and the size of the plastic enclave formed by an overload (Figure 5.17a). An overload occurring at a crack, size a_0 will cause a crack tip plastic zone of size

$$r_{P0} = \frac{K_0^2}{\alpha\pi F_{ty}^2} = \frac{\beta^2\sigma_0^2 a_0}{\alpha F_{ty}^2}, \tag{5.20}$$

where σ_0 is the overload stress. When the crack has propagated further to a length a_i the current plastic zone size will be

$$r_{Pc} = \frac{K_{i\max}^2}{\alpha\pi F_{ty}^2} = \frac{\beta^2\sigma_{i\max} a_i}{\alpha F_{ty}^2}, \tag{5.21}$$

where $\sigma_{i\max}$ is the maximum stress in the i-th cycle. This plastic zone is still embedded in the plastic enclave of the overload: the latter proceeds over a distance ϱ in front of the current crack (Figure 5.17b). Wheeler assumes that the retardation factor ϕ_R will be a power function of r_{Pc}/ϱ. Since $\varrho = a_0 + r_{P0} - a_i$, the assumption amounts to:

$$\left.\begin{aligned} \frac{da}{dN} &= \phi(\Delta K, R) \\ \text{with} \quad \phi_R &= \left(\frac{a_0 + r_{P0} - a_i}{r_{Pc}}\right)^{\gamma} \end{aligned}\right\} \tag{5.22}$$

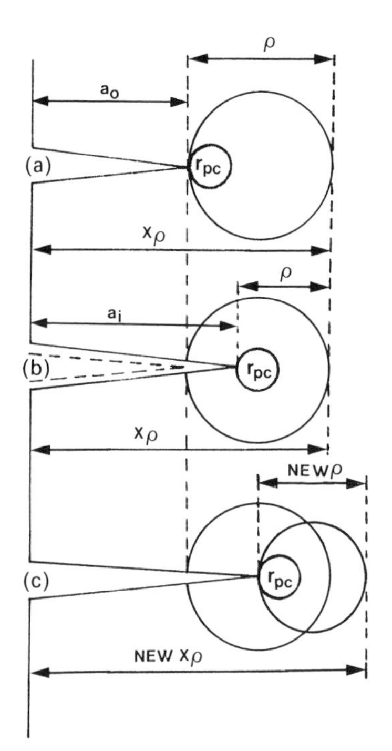

Figure 5.17. The model of Wheeler. (a) Situation immediately after overload; (b) After some crack growth; (c) Situation after second overload.

If $r_{P_c} = \varrho$, the crack has grown through the overload plastic zone, and the retardation factor becomes $\phi_R = 1$ by definition. The exponent in Equation (5.22) has to be determined empirically. This is the adjustable calibration factor. Note that if $\gamma = 0$ there will be no retardation at all under any circumstances. Hence, the minimum value of γ is zero. Typically for variable amplitude loading $0 < \gamma < 2$, depending upon the material but also upon the spectrum.

For the case of a single overload in a constant amplitude test the retardation factor gradually decreases to unity while the crack progresses through the plastic enclave (Figure 15.17b). If a second high load occurs, producing a plastic zone extending beyond the border of the existing plastic enclave, the boundary of this new plastic zone will have to be used in the equation (Figure 5.17c), and the instantaneous crack length will then become the new a_0.

Calibration of the above model (and all other models) proceeds as follows. A test is performed under variable amplitude loading. The test result is then 're-predicted' several times using the proper $\mathrm{d}a/\mathrm{d}N - \Delta K$ data and the proper β, but with different values of the adjustable parameters; (in the case of Wheeler γ-values taken are e.g. 0, 0.5, 1.15 etc.). The parameter value(s) that produce(s) the best coverage of the test data, is (are) the values to be used in analysis. An example of such a calibration [8] is shown in Figure 5.18. Clearly, in this case $\gamma = 1.4$ is the parameter value to be used.

Unfortunately, the parameter calibration is not general. It depends upon the load-history and spectrum (Chapters 6 and 7). A different spectrum with a different mixture of high and low loads requires a different calibration factor. E.g. the non-linear man-induced exceedance diagram (Chapter 6) requires different calibration parameter(s) than a nature-induced log-linear exceedance diagram; the calibration parameters are suitable for one type of spectrum, but they cannot be generalized for all spectra. Failure to perform this re-calibration and subsequent general use of calibration factors, is the main cause of claims that one retardation model is better than another. If proper calibration is performed for each spectrum type, any model can be as good as any other. Calibration for a certain spectrum type and material generally gives good results [8] for all variations of the same type of spectrum (Chapter 6), as shown in Figure 5.19. This figure shows results of about 70 predictions for random loading with the spectrum and calibration as used in Figure 5.18. More information on model calibration can be found in Chapter 7.

5.8. Crack growth analysis for variable amplitude loading

Most structures are subjected to variable amplitude loading, i.e. of the type shown in Figure 1.2 (For a detailed discussion of load histories see Chapter 6). In such cases the crack growth rate da/dN varies from cycle to cycle, depending upon ΔK and R of the cycle involved, and upon retardation (any cycle can be

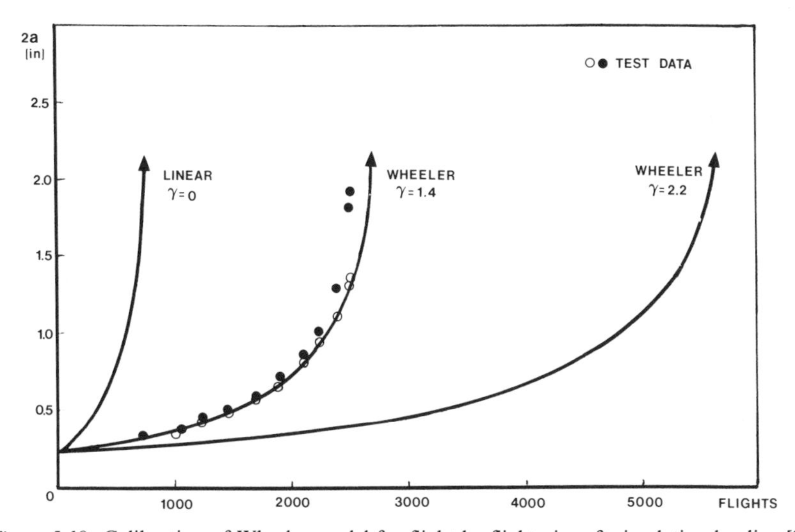

Figure 5.18. Calibration of Wheeler model for flight-by-flight aircraft simulation loading [8].

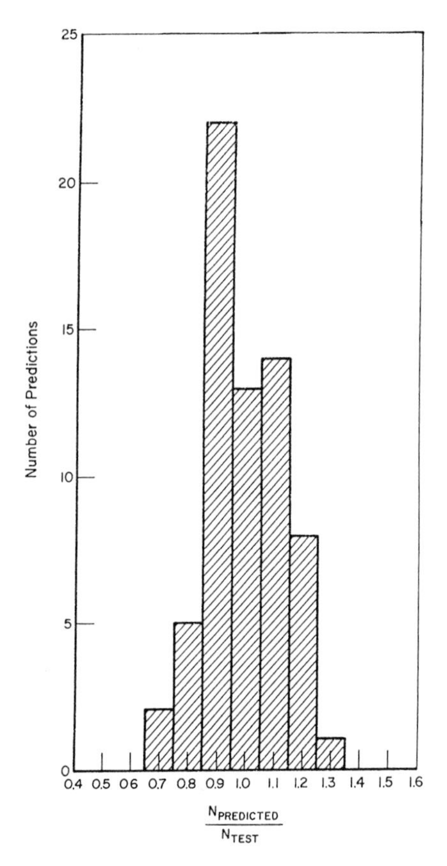

Figure 5.19. Accuracy of 70 predictions as compared to tests [8], Courtesy Engineering Fracture Mechanics.

an overload cycle with respect to the subsequent cycle). Therefore, the computer program must evaluate and add crack growth on a cycle-by-cycle basis. For example if crack growth takes 200 000 cycles, the computer must perform the 'operation' 200 000 times. This may take considerable computer time. Typically, a mainframe computer can perform at about 50 000 cycles a minute, so that the above computation would take about four minutes. A personal computer might take as much as two to three hours for the same job (1987).

A logic diagram for the integration is shown in Figure 5.20. This is again based on the Wheeler model because of the latter's simplicity, but it is not essentially different for other retardation models. Naturally, in order to perform the calculation for a structural crack, the computer must be provided with applicable da/dN data for the material at hand (Chapter 7), and last but not least, the stress history for the structure (Chapter 6).

A more detailed representation of the algorithm involved is shown in Table

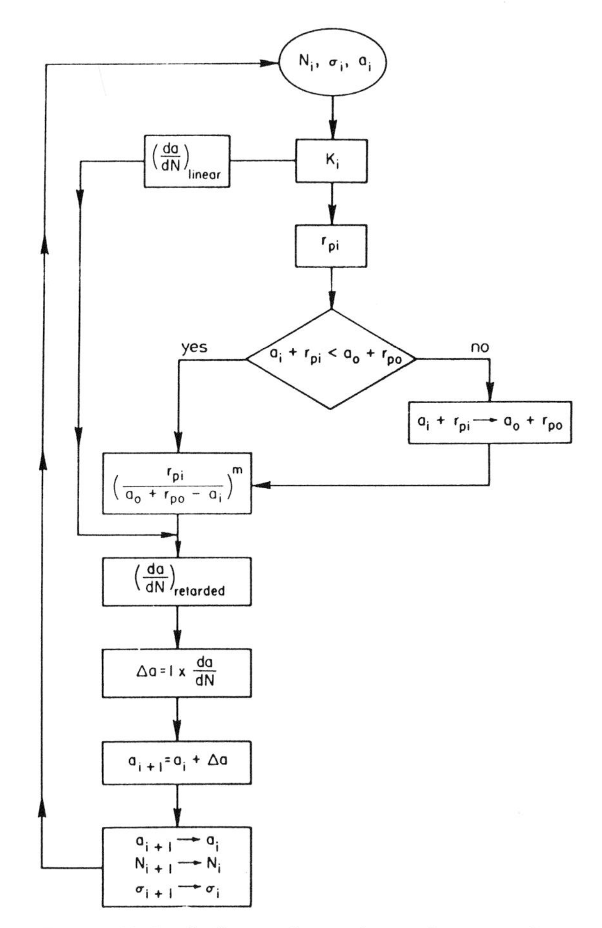

Figure 5.20. Logic diagram for crack-growth computation.

5.2. A hand-calculation for a few successive cycles is shown in Table 5.3. These tables show the basic algorithm which is quite simple; a useful computer code with many options, especially for the complicated book keeping for stress histories, is rather involved and will contain approximately 3000 statements (lines).

The accuracy of the computation depends somewhat on the retardation model, but useful results can be obtained with any well-calibrated model. Most influential to the accuracy are the input of the stress sequence (Chapter 6) and material data (Chapter 7). A general discussion of accuracy and errors is presented in Chapter 12. The accuracy problem involved in the simple algorithm of Table 5.2 is in the addition of a very small da to a relatively large a. For example, 8-bit personal computers evaluate a number into 8 digits. If $a = 1$ inch and da in a given cycle is 0.000001 inch, the new crack size will become

Table 5.2. Crack growth analysis in variable amplitude

$a_j = a_0;\quad N = 0$		
$a_i = a_j$		
$\beta_i = f\left(\frac{a_i}{L}\right)$	←	Subroutine or preprocessor; β-Library or methods of Chapter 8
$\Delta\sigma_i,\ R_i$	←	Subroutine; generation of stress history (Chapter 6), randomization, counting if necessary (Chapter 6).
$\Delta K_i = \beta_i \Delta\sigma_i \sqrt{\pi a_i}$		
$K_{max} = \Delta K_i/(1 - R_i)$		
$r_{pi} = K^2_{max}/\alpha\pi F^2_{ty}$	←	Subroutine state of stress for α.
$X_{\varrho new} = a_i + r_{pi}$		
if $X_{\varrho new} > X_{\varrho old}$	then	$X_\varrho = X_{\varrho new}$
$\varrho = X_\varrho - a_i$		
ϕ_{Ri}	←	Subroutine various retardation models. Only Wheeler shown $\phi_{Ri} = \varrho/r_{pi}$
$\frac{da}{dN} = f(\Delta K, R)$	←	Subroutine or data library
$\frac{da}{dN} = \phi_{Ri}\frac{da}{dN}$ (retarded)		
$\Delta N = 1$		
$\Delta a = 1 \times da/dN$		
$a_j = a_i + \Delta a$		
$N = N + 1$		
if $a_j < a_p$ then return to line 2		
a, N		
a (years, voyages, flights)	←	Subroutine conversion
Output	←	Subroutines
Plots	←	Subroutines
End		

$a = 1.0000001$. However, if $da = 0.000{,}0001$, the computer evaluates $1 + 0.000{,}00001 = 1$, i.e. the crack has not grown due to a computer rounding error. This means that a and da must be evaluated in double precision which in an 8-bit personal computer provides 16 digits. Hence $a = 1$ and $da = 1E - 15$

Table 5.3. Example of retarded crack growth by hand calculation: on the basis of Table 5.2.
Note a: β assumed equal to 1 throughout; b: $\alpha = 2$ (plane strain assumed); c: $da/dN = 2E-9\ \Delta K^2\ K_{max}$ assumed: γ assumed as 1.5; $F_{ty} = 50$ ksi; d: Stress history assumed as in columns 3 and 4; e: Table must be worked horizontally.

1	2	3	4	5	6	7	8	9	10	11	12	13	14	15	16
Cycle	a	$\Delta\sigma$	R	ΔK	K_{max}	r_{pi}	$X_{\varrho new}$	$X_{\varrho old}$	X_{ϱ}	ϱ	ϕ	(da/dN) Retarded $= \phi\ da/dN$	N	Δa	$a = a + \Delta a$ new a for next line
1	0.1	10	0	5.61	5.61	0.0020	0.10020	–	0.10020	0.0020	1	3.53 E–7	1	3.53 E–7	0.100000353
2	0.1^+	10	0	5.61	5.61	0.0020	0.10020	0.10020	0.10020	0.0020	1	3.53 E–7	2	3.53 E–7	0.100000706
3	0.1^+	15	0.1	8.42	9.35	0.0056	0.10056	0.10020	0.10056	0.0056	1	1.32 E–6	3	1.32 E–6	0.100002026
4	0.1^+	10	0	5.61	5.61	0.0020	0.10020	0.10056	0.10056	0.0056	0.21	7.41 E–8	4	7.41 E–8	0.100002100
5	0.1^+	12	0	6.73	6.73	0.0029	0.10029	0.10056	0.10056	0.0056	0.37	2.25 E–7	5	2.25 E–7	0.10002325
6	0.1^+	10	0	5.61	5.61	0.0020	0.10020	0.10056	0.10056	0.0056	0.21	7.41 E–8	6	7.41 E–8	0.100002399
7	0.1^+	19	0	10.66	10.66	0.0072	0.10072	0.10056	0.10072	0.0072	1	2.42 .E–6	7	2.42 E–6	0.100004822
8	0.1^+	15	0.1	8.42	9.35	0.0056	0.10056	0.10072	0.10072	0.0072	0.69	9.11 E–7	8	9.11 E–7	0.100005733
9	0.1^+	10	0	5.61	5.61	0.0020	0.10020	0.10072	0.10072	0.0072	0.15	5.17 E–8	9	5.17 E–8	0.100005785
10	0.1^+	12	0	6.73	6.73	0.0029	0.10029	0.10072	0.10072	0.0072	0.26	1.59 E–7	10	1.59 E–7	0.100005944

will be evaluated correctly, but if $da = 1E - 16$ and $a = 1$, the computer will still not recognize growth. There is nothing that can be done about this rounding problem. Usually it is not serious, but it may become a problem in evaluating retardation. Computer programs working in single precision may be one cause of claims regarding the accuracy of retardation models.

It should be noted, that the above does not change when a is evaluated in meters. For example if $a = 0.01$ m and $da =$ 1-E16 m, the addition will be performed properly in double precision because leading zeros do not count. Mainframe computers already carry 16 decimals in single precision; they carry 32 in double precision. Even in that case rounding errors may occur, but they are even less significant.

Although the algorithm in Table 5.2 is simple, computer codes are generally rather complicated [e.g. 9], because there must be

(a) preprocessors for β, or a β library.
(b) options for various rate equations, da/dN table and/or library of data;
(c) options for various retardation models;
(d) accounting for state of stress;
(e) options for random loading;
(f) accounting procedures for stress history;
(g) options for cycle counting.

If all of the above are included, the main code of 3000 statements as mentioned above can easily triple or quadruple in size. Of the above (e), (f) and (g) may be the most important and most involved; they are discussed separately in Chapter 6. Further discussions of the subject are found in Chapter 7 (data and calibration) and Chapter 12 (errors and accuracy).

It is well-known that fatigue predictions, in general, have a low accuracy. In the case of crack propagation, a linear integration (without interaction effects or retardation) will generally yield results which are on the safe side. As shown in Figure 5.10 negative loads reduce the retardation caused by positive loads, but the net effect is usually a deceleration of crack growth, so that retardation models must be used.

Figure 5.21 shows results of crack growth in rail steel [10] under simulated train-by-train (Chapter 6) loading. Retardation hardly plays a role in rail steels, therefore predictions were made by means of linear integration. The figure shows that they are within a factor 2 of the experimental data.

Better accuracy can be obtained in general, provided the retardation model is adjusted. Predicted crack growth for a titanium alloy subjected to aircraft service loading [8] are shown in Figure 5.22 together with experimental data. Generally, part of the discrepancy between computation and test may be caused by scatter in crack growth properties. Most retardation models can be empirically adjusted. In this respect, the Wheeler model is attractive, because it contains only one adjustable constant.

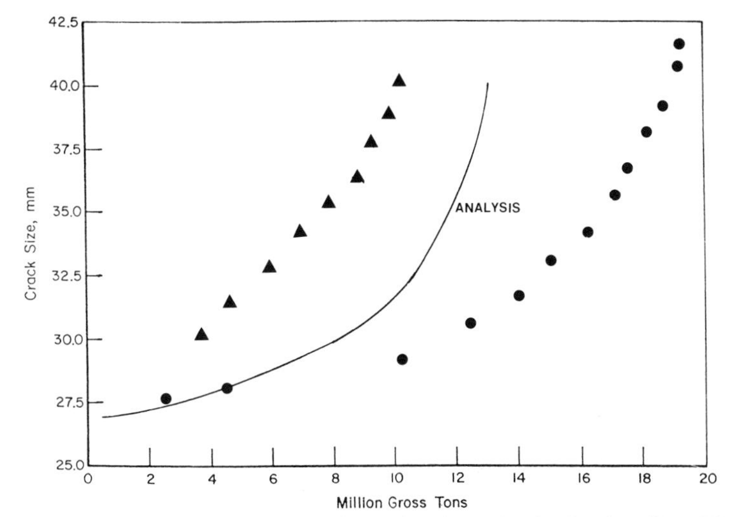

Figure 5.21. Predictions and test data for service simulation loading in rail steel [10].

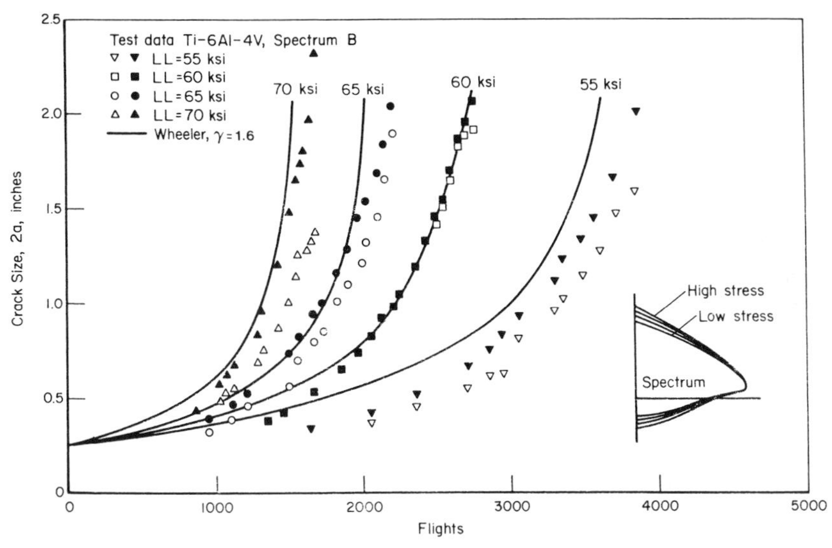

Figure 5.22. Predicted crack growth and test data for aircraft spectrum [8]. Courtesy Engineering Fracture Mech.

Apparently, a crack growth prediction can be substantially more accurate than a fatigue life prediction. Admittedly, a few experiments are necessary with a spectrum of certain shape to empirically adjust the retardation model parameters. From then on, predictions can be made for the same general spectrum shape and variations thereof, for structural parts subjected to lower and higher stresses and for cracks of different types (different β). Crack growth properties of most materials show considerable scatter. The rail steels that are the subject of Figure 5.21 showed variations of almost a factor of ten in constant amplitude crack growth (see Chapter 14). Therefore, discrepancies between predicted and experimental crack growth are not a shortcoming of the predictive method per se, but are due to anomalies in material behaviour.

Fortunately, most materials are well-behaved in comparison, by showing less scatter in crack growth. Nevertheless, there is enough scatter that predictions will always have some uncertainty. This would still be the case if better retardation models were developed.

However, the prediction procedure in general contains many more uncertainties, which may be just as detrimental to the final results as are the shortcomings of the retardation model. These are:

(a) Uncertainty in the local stress level.
(b) Uncertainty in the stress intensity calculation.
(c) Insufficient knowledge of the load spectrum.
(d) Possible environmental effects.

Consider first the uncertainty in stress level and stress intensity. In the case of a complex structure consisting of many elements, an error of five percent in the stress analysis would be quite normal. The subsequent determination of the stress intensity can easily add another five percent, especially in the case of corner cracks or surface flaws. Thus, the final inaccuracy of the stress intensity may be in the order of 10%. If the crack growth rate is roughly proportional to the fourth power of ΔK, the error in the crack growth prediction will be on the order of $(1.1)^4 = 1.45$ (45%); see also Chapter 12.

Despite extensive load measurements, the prediction of the load spectrum is still an uncertain projection in the future. Slight misjudgements of the spectrum can have a large effect on crack growth.

Even if possible environmental effects are disregarded, the errors in crack growth prediction due to uncertainties in stress analysis and loads analysis can be just as large or larger than the errors due to the crack growth integration. Development of better crack growth integration techniques will not improve this situation. Therefore, the shortcomings of the retardation models can hardly be used as an argument against crack growth predictions.

Taking into account all errors that can enter throughout the analysis, it is obvious that a safety factor should be used. This safety factor should not be taken on loads or stresses or da/dN data. Doing this would make some predictions more conservative than others. The complexity of crack growth behaviour does not permit an easy assessment of the degree of conservatism attained through the application of such safety factors. A safety factor should rather be applied to the final result, i.e. to the crack growth curve, by dividing the number of cycles to any given crack size by a constant factor. The problem of accuracy and sources of error is discussed further in Chapter 12.

5.9. Parameters affecting fatigue crack growth rates

When predictions of crack propagation have to be made, data should be available relevant to the conditions prevailing in service. Such data may be hard to find (for pragmatic solutions see Chapter 7). Fatigue crack propagation is affected by an endless number of parameters, and the circumstances during the test will seldom be the same as in service. The influence of the environment is the most conspicuous.

The effect of environment on crack growth rates has been the subject of many investigations on a variety of materials; the rate of fatigue crack propagation in wet air may be an order of magnitude higher than in vacuum, the effect being attributed to water vapour. The influence of salt water (seawater) is of particular interest to marine structures. An example of its effect will be shown in Chapter 7. It is generally accounted for in the crack growth analysis by submitting the computer program with the actual data in tabular form. It is then assumed that during each cycle in a variable amplitude sequence the rate will immediately adjust to the one found in the constant amplitude test data at the same ΔK. This is assuming that chemical/load equilibrium will be immediately attained. More elaborate accounting can be implemented however. No single model can explain the influence of the environment on the rate of propagation of fatigue cracks. Different explanations apply to different materials. The effect is certainly a result of corrosive action and as such it is time-dependent. Therefore, the environmental effect is dependent upon the cycling frequency.

Among the many factors that affect crack propagation, the following should be taken into consideration for crack growth predictions:

(a) thickness;
(b) type of product;
(c) heat treatment;
(d) cold deformation;
(e) temperature;
(f) manufacturer;
(g) batch-to-batch variation;
(h) environment and frequency.

For the factors lower in this list it is less likely that they can be properly accounted for. No attempt will be made to illustrate the effects of all these factors with data, because some have greatly different effects on different materials.

Many of these effects cannot be accounted for properly in the analysis of structural cracks, primarily because the data are simply not available. A pragmatic approach to solve the problem is discussed in Chapter 7. At this point it is sufficient to note that the necessary use of estimated data may be of considerable influence on the accuracy of the analysis. With this in mind, the acclaimed inaccuracies of e.g. retardation models may well become secondary (see also Chapter 12).

In sheets there is a systematic, effect of thickness on crack propagation, especially before the fracture mode transition. Fatigue cracks in sheets start perpendicular to the sheet surface. When the crack grows the size of the plastic zone increases and plane stress develops. This causes the fatigue crack to change to single or double shear, as depicted in Figure 5.23. Plane stress develops when the size of the plastic zone is in the order of the sheet thickness (Chapter 3). In thicker sheets the transition will require a large plastic zone and occur at a greater length of crack. The data suggest that crack growth is slower in plane stress than in plane strain at the same stress intensity.

Although the effect of thickness on crack growth has been recognized for over 20 years, little effort has been expended in developing a useful model for everyday damage tolerance analyses. Figure 5.24 emphasizes the necessity to include this effect in the crack growth analysis. A tentative semi-empirical model has been proposed [11], but the best way to account for thickness effects is probably to submit the proper data to the computer code. A factor of two error (due to thickness) in $\mathrm{d}a/\mathrm{d}N$ data may overshadow any effects of 'inaccurate' retardation models.

Many investigators hold that there is a threshold for fatigue cracking: below a certain ΔK the rate $\mathrm{d}a/\mathrm{d}N$ is supposed to be essentially zero. At least one conference, resulting in a two-volume book [12] was devoted to this subject. The threshold would be reflected in a vertical $\mathrm{d}a/\mathrm{d}N - \Delta K$ curve at low ΔK, as shown in Figure 5.25. The threshold is usually determined by gradually decreasing the stress in a test until crack growth comes to a halt. In view of possible retardation this procedure is subject to some doubt. Besides, the threshold is definitely crack size dependent; it is not unique. However, if one accepts the presence of a threshold, the practical question is "what is the effect on predicted crack growth". Figure 5.26 shows what it may amount to: the two curves are indistinguishable. Generally speaking, the effect is hardly worthwhile considering, but of course, in each case one would have to use judgement. If for example the initial ΔK is below threshold no growth occurs at all. In this respect Figure 5.26 is somewhat deceiving; one could select a case where the effect is larger. However in random loading many cycles will be above threshold and the effect on life (in years or hours) is small, especially when there is retardation. Most computer codes provide an option to use a threshold. Referring back to Section 5.8 it should be noted that the computer uses a threshold automatically because it rounds any data smaller than a certain value, depending upon the current crack size.

There is often much concern regarding the anomalous behaviour of very small

(a)

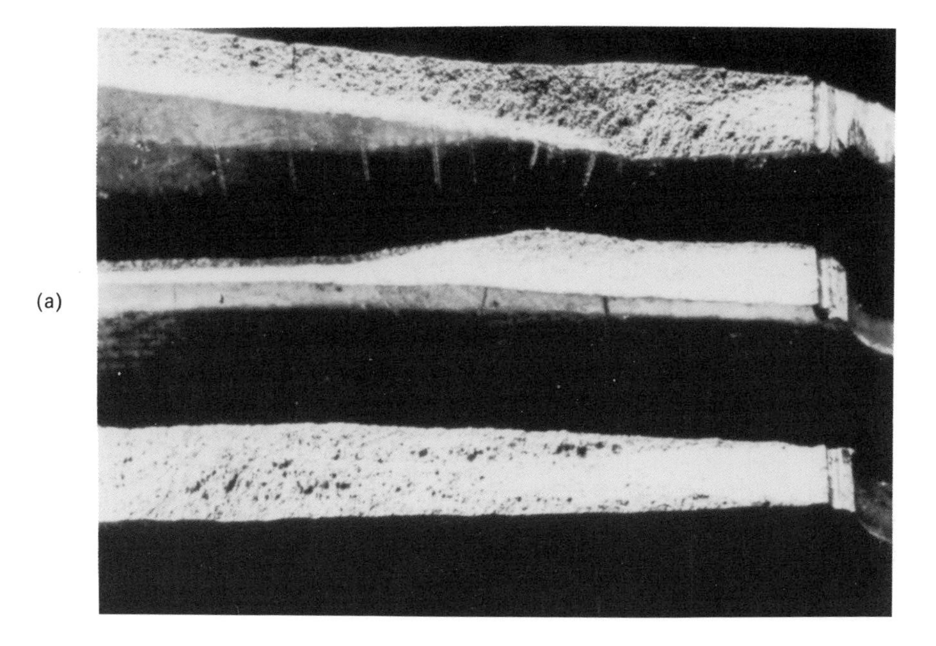

(b)

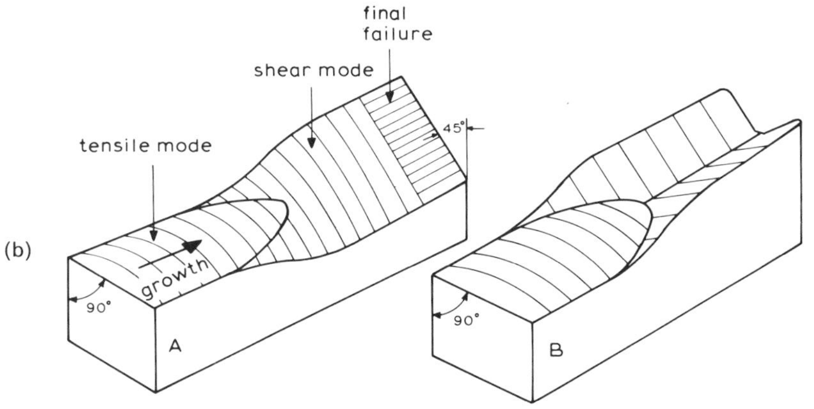

Figure 5.23. The transition of a fatigue crack in sheet. (a) Transition of fatigue cracks to double shear (top) and single shear (center and bottom) in Al-alloy specimens; (b) Single shear (A) and double shear (B).

cracks [13]. As shown in Figure 5.27 small cracks tend to show growth rates much higher than would be expected on the basis of the acting ΔK. Various explanations have been put forward but the 'cure' proposed is mainly artificial use of an apparent crack size $a + \lambda$ where λ is a fixed quantity determined empirically.

Most of the short crack data stem from strain control fatigue tests at $R = -1$

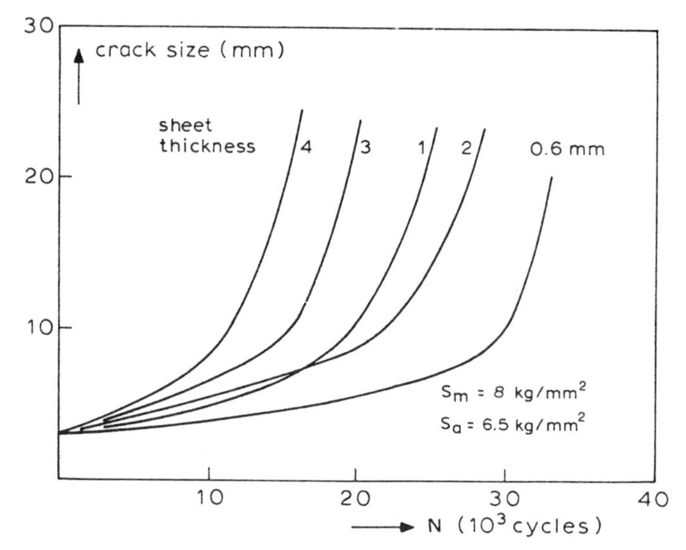

Figure 5.24. Effect of sheet thickness on crack growth.

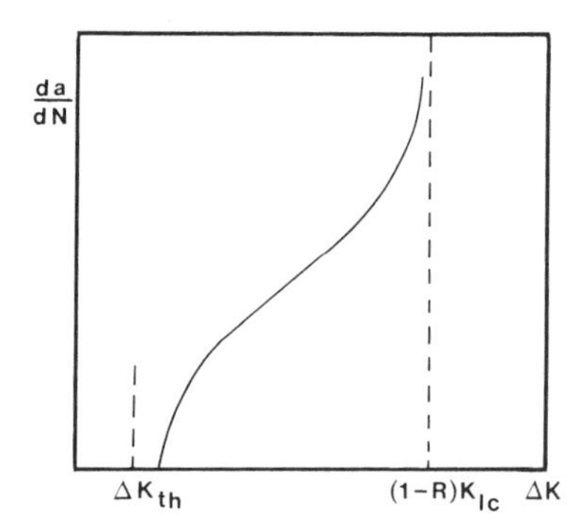

Figure 5.25. Threshold in rates (schematic).

on small notched coupons (usually with central holes). Consider a short crack at such a hole as in Figure 5.28a and compare it with a long crack at the same ΔK (Figure 5.28b). By the nature of the test, the plastic zone at the hole is much larger than the crack tip plastic zone. Since completely reversed plastic strain is enforced in most of these fatigue tests, the crack tip will also be subjected to completely reversed strain ($R_\varepsilon = -1$). A larger crack at $R = -1$ will close during compressive loading, so that its strain range will still be as if $R_\varepsilon = 0$. The small crack at the hole is experiencing a strain range twice as large as a regular crack at $R_\sigma = 0$. Hence, the small crack should show a rate of growth as if its ΔK were approximately twice as large as the calculated value. This is indeed the

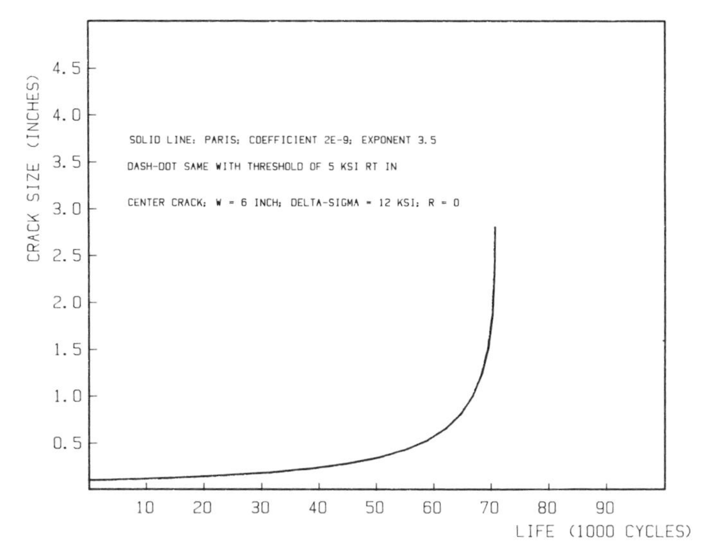

Figure 5.26. Effect of using threshold in computation; threshold is 5 ksi; Curves indistinguishable.

case as can be judged from Figure 5.27. When the small crack grows, it will gradually move away from completely reversed straining so that the growth rate decreases. Once the crack tip is outside the plastic zone of the hole, the small crack behaves as a normal crack.

The small crack behaviour depends upon the loading and upon the ratio between ligament and hole diameter (notch depth). A hole, unlike a crack, does not close in tension and it is equally as much a stress raiser in compression as in tension, so that even under load control reverse plastic strains can occur. Therefore, similar, but smaller, effects should be anticipated under general loading conditions. However, if the hole is filled with a fastener, it will essentially 'close' and little or no small crack effects should be anticipated. In technical problems many cracks start from holes, but these are often "filled" fastener holes, so that the 'small crack problem' may be of little relevance. In most other cases these so-called 'small cracks' are well below the detection limit in practical inspections. They are then irrelevant because crack growth below the detectable crack size is hardly of interest. Naturally, one can always give examples of cases where the problem might appear, but generally speaking, it is an interesting research subject, but its technical relevance is small.

As a final note in this section consider the use of J (Chapter 4) as opposed to K for representing $\mathrm{d}a/\mathrm{d}N$ data. By far the larger part of fatigue crack growth lives is spent at low ΔK. Let e.g. $\Delta K = 10\,\text{ksi}\sqrt{\text{in}}$ and $R = 0$. Taking a (low) yield strength of $F_{tSy} = 40\,\text{ksi}$, the plastic size would be $r = 10/(6 \times \pi \times 40^2) = 0.003\,\text{in}$. If this is not a small enough plastic zone, none will ever be.

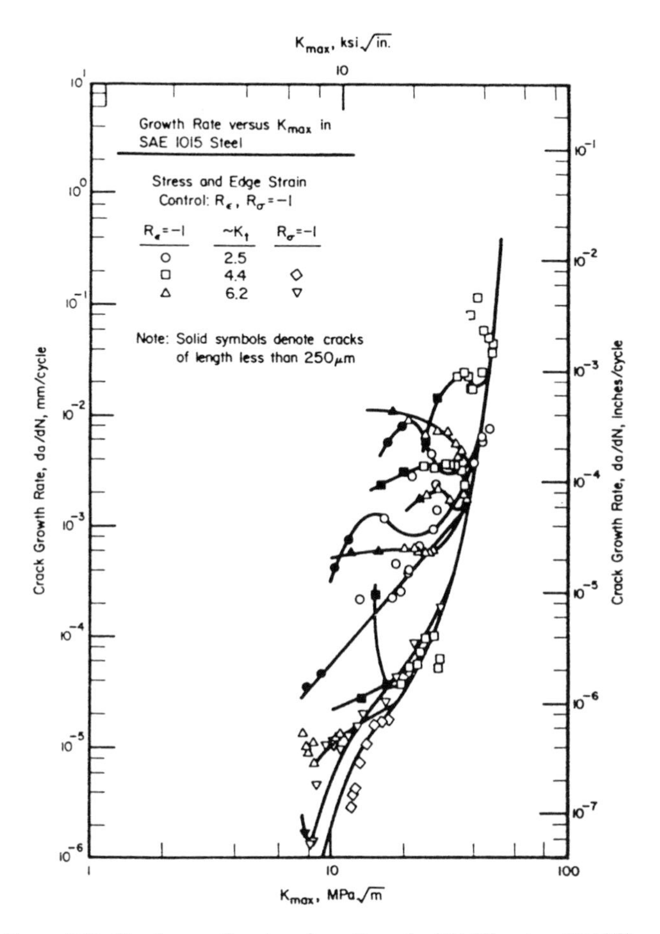

Figure 5.27. Crack growth rates of small cracks [13] (Courtesy EMAS).

Naturally, at large crack sizes the plastic zone will be larger, but does this justify the use of EPFM. Consider Figure 5.29. The crack growth life might be either N_1, N_2 or N_3, but practically this makes little difference on the total. The difference is only caused at the high ΔK. During most of the crack growth life, the stress intensity is well below K_{Ic} (fortunately). Thus, the use of K for the analysis is well justified. Finally, the champions of using J for representing da/dN data have always used approximations of J rather than actual J values obtained properly (Chapter 4). By using 'appropriate' approximations, one can always make data look better. However, should the data at high ΔK indeed be better represented by J, the differences in analysis results (Figure 5.29) would still not justify the use of a more complicated parameter; a parameter moreover that has more drawbacks than advantages (see Chapter 4). In short, the use of

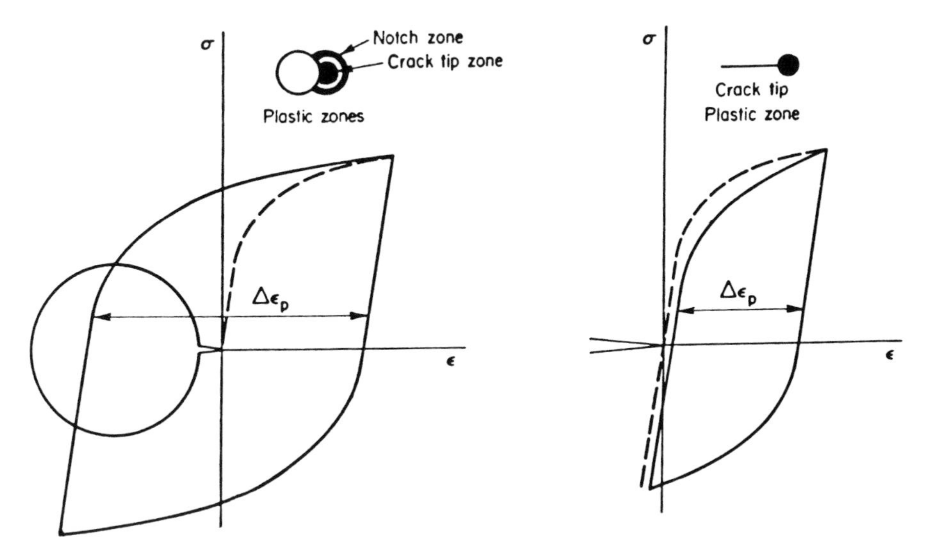

Figure 5.28. Stress-strain loop for small crack at notch. (left) as compared to regular (large crack; right); both $R = 0$ loading. [7, 13]. Copyright ASTM. Reprinted with permission.

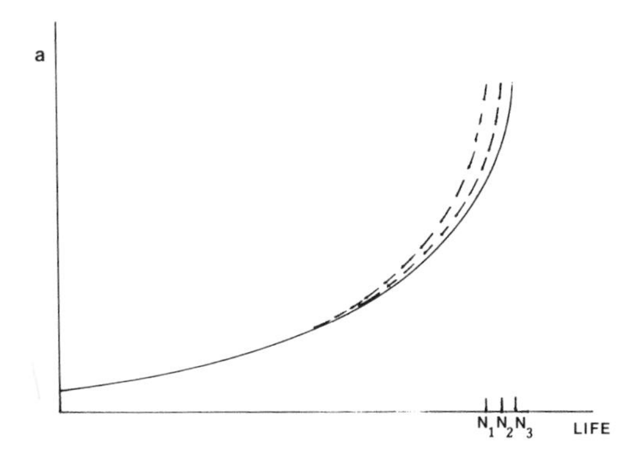

Figure 5.29. Effect on life of using different parameter in high rates regime.

J for fatigue crack growth analysis is another interesting research subject, but not of practical interest.

5.10. Stress corrosion cracking

Crack growth can occur by other mechanisms than fatigue (Chapter 1). The most important one is stress corrosion cracking. Given a specific material-environment interaction the stress corrosion cracking rate is governed by the stress intensity factor. Specimens with the same initial crack but loaded at different levels (different initial K-values) show different times to failure as shown diagrammatically in Figure 5.30. The specimen initially loaded to K_{Ic} fails immediately. Specimens subjected to K values below a certain threshold level never fail. This threshold level is denoted as K_{Iscc}. The subscript *scc* standing for stress corrosion cracking.

During the stress corrosion cracking test the load is kept constant. Since the crack extends, the stress intensity gradually increases. As a result the crack growth rate per unit of time, $\mathrm{d}a/\mathrm{d}t$, increases according to:

$$\frac{\mathrm{d}a}{\mathrm{d}t} = f(K). \tag{5.23}$$

This is shown in Figure 5.31. When the crack has grown to such a size that K becomes equal to K_{Ic}, final failure occurs, as indicated in Figure 5.30b. Obviously, if failure does not occur (infinite time to failure) the crack did not grow at all, because if it did, K would increase causing more growth, etc. Thus, the K giving an 'infinite' life is indeed a threshold.

The stress corrosion cracking threshold, K_{Iscc}, and the rate of crack growth depend upon the material and the environment. From Figure 5.30 it follows that a component with a certain size of crack loaded such that $K = K_{Ic}$, fails immediately. Components loaded to K values at or above K_{Iscc} will show crack growth to failure. No failure (no crack growth) occurs when $K < K_{Iscc}$.

Fracture mechanics can deal with stress corrosion crack growth in the same manner as with fatigue crack growth. Obtaining the crack growth curve as a function of time amounts to integration of $\mathrm{d}a/\mathrm{d}t$ in much the same manner as integration of $\mathrm{d}a/\mathrm{d}N$ (see previous sections). However, the total times to failure in stress corrosion are usually from 1 to 1000 hours or they are infinite (Figure 5.30). If the crack growth life is less than 1000 hours there is hardly time to

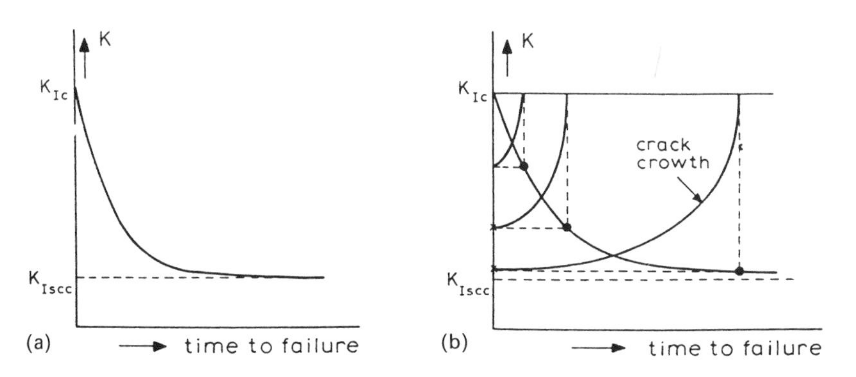

Figure 5.30. Stress corrosion test data. (a) Stress corrosion time to failure, upon loading to initial K-level; (b) Cracking during test.

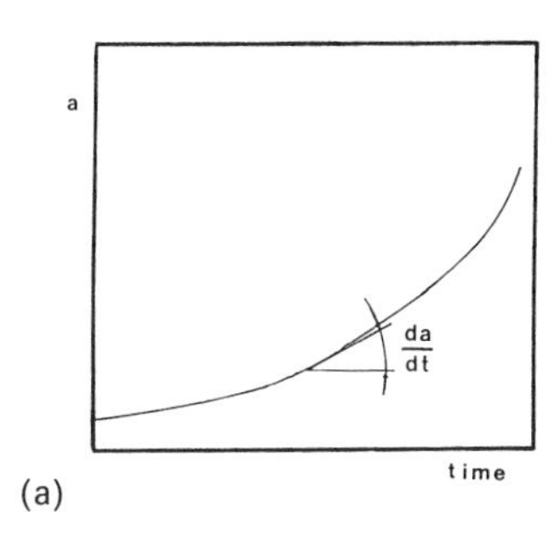

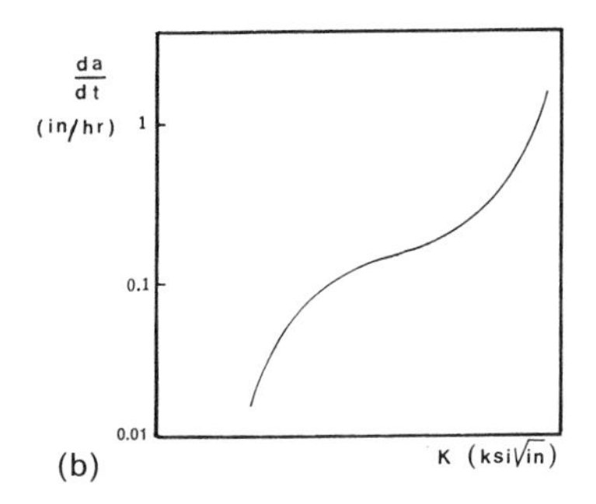

Figure 5.31. Stress Corrosion Crack growth. (a) Crack growth as a function of time; (b) Growth rate as function of K.

inspect or deal with cracks otherwise. Thus crack growth analysis is seldom worthwhile.

Stress corrosion cracking should be prevented rather than 'controlled'. Crack growth is prevented by keeping $K < K_{Iscc}$. Mathematically, this problem is identical to keeping $K < K_{Ic}$ or K_c (i.e. the fracture problem as discussed in Chapter 3). Thus prevention of stress corrosion cracking can be dealt with in damage tolerance analysis in the same manner as fracture.

The discussions in Chapters 3 and 10 apply if K_{Ic} is replaced by K_{Iscc}. By the same token any computer software that can solve fracture problems (residual strength) can solve stress-corrosion prevention problems.

For example if $K_{Iscc} = 15\,\text{ksi}\sqrt{\text{in}}$, would an edge crack of $a = 0.5$ inch in a structure subjected to $\sigma = 18\,\text{ksi}$ present any danger? Assuming $\beta = 1.15$ (Chapter 3) the stress intensity is $K = \beta\sigma\sqrt{\pi a} = 1.15 \times 8 \times \sqrt{0.5 \times \pi} = 11.5\,\text{ksi}\sqrt{\text{in}}$. This stress intensity is below K_{Iscc} and therefore crack growth by stress corrosion should not occur.

Exercises

1. A fatigue crack grows in constant amplitude from $a = 1$ to $a = 1.05$ inch in 7000 cycles. What is the rate of growth?

2. The crack growth properties of a certain material can be described by $\mathrm{d}a/\mathrm{d}N = 1 \times 10^{-8}\Delta K^2 K_{max}^{1.5}$. If $\beta = 1$, what is the rate of growth of a crack of length $a = 0.5$ inch if $\sigma_{max} = 12\,\text{ksi}$ and $R = 0.2$. How many cycles does it take for this crack to grow to $a = 0.51$ inch?

3. If $\mathrm{d}a/\mathrm{d}N = 1 \times 10^{-8}$ m/cycle at $\Delta K = 8\,\text{MPa}\sqrt{\text{m}}$ and $\mathrm{d}a/\mathrm{d}N = 4 \times 10^{-6}$ m/cycle at $\Delta K = 30\,\text{MPa}\sqrt{\text{m}}$, determine the parameters of the Paris equation. Convert the data to in/cycle and $\text{ksi}\sqrt{\text{in}}$ and then determine the parameters again.

4. The following data are obtained from crack growth tests at constant amplitude; center crack with $W = 20$ inches.

a (inch)	N (cycles)	
	$\Delta\sigma = 16$ ksi; R = 0	$\Delta\sigma = 10$ ksi; $R = 0.5$
0.1	0	0
0.105	1100	2000
1.5	i	k
1.55	$i + 100$	$k + 170$

Establish the rate diagram using two data points for each R-value; assume straight lines between data points.

5. For the data in Exercise 4 determine a Paris equation (calculate C_P and m_P) for each of these R-values.

6. Using the results of Exercise 4 determine the Walker equation.

7. Assuming $K_c = 80\,\text{ksi}\sqrt{\text{in}}$, determine the parameters of the Forman equation for the data in Exercise 4. Hint: first replot the data as $\{[(1 - R)K_c - \Delta K]\mathrm{d}a/\mathrm{d}N\}$ versus ΔK.

8. Using each of the equations of Exercises 5, 6 and 7 calculate the growth rates for 10 values of ΔK and 2R-values. Plot the calculated results in three separate rate diagrams and compare the results with the original test data.

9. Using the equation obtained in Exercise 6 calculate the crack growth curve of a center crack, starting at $a = 0.5$ up to $a = 3$ inch; $W = 100$ inches. Use 5 integration steps (0.5–0.7, 0.7–1.0, 1.0–1.5, 1.5–2.0, 2.0–3). $\sigma_{max} = 15$ ksi; $R = 0.3$; constant amplitude. Plot the crack growth curve. Hint: use a similar procedure as in Table 5.1.

10. Repeat Exercise 9 using the Forman equation of Exercise 7. Compare with results of Exercise 9 by plotting in the same graph.

11. For a structural crack of $a = 0.1$ inch we find $\beta = 3$. At $a = 0.1$, $\sigma_{max} = 11$ ksi and $R = 0.2$ what is the crack growth rate predicted by the equations of problems 6 and 7?

12. Assuming that γ for Wheeler retardation is $\gamma = 1.2$, calculate the rates for the case of Exercise 11 with the Walker equation of Exercise 6 if the given cycle was preceded by an overload cycle with $\sigma_{max} = 17$ ksi at $R = 0$. Also calculate the rates after the crack has grown by 0.003 inch after the same overload. What would these rates be if $\gamma = 0$? Assume plane strain plastic zone as $K_{max}^2/6\pi F_{ty}^2$; $F_{ty} = 70$ ksi.

13. If $K_{Iscc} = 15\,\text{ksi}\sqrt{\text{in}}$, what is the highest permissible stress for the configuration of Exercise 11 for stress corrosion cracking to be prevented?

14. A material with $K_{Ic} = 30\,\text{ksi}\sqrt{\text{in}}$ is subjected to stress corrosion tests in salt water. Three specimens with a single edge crack are loaded in uniform tension at loads of 18, 24 and 30 kips respectively. All cracks are 0.5 inch long; $W = 3$ inch and $B = 0.5$ inch; $F_{ty} = 70$ ksi. The times to failure of the three specimens are respectively: 5000, 100 and 3 hours. Estimate K_{Iscc} and calculate the amount of crack growth that occurred in each of the tests. Ignore collapse. Assume $\beta = 1.12$ throughout.

References

[1] B.J. Habibie, *Eine Berechnungsmethode zum Voraussagen des Fortschritten von Rissen*, Messerschmidt-Bolkow-Blohm Rep. UH-03-71 (1971).

[2] J. Willenborg et al., *A crack growth retardation model using an effective stress concept*, AFFDL-TM-71-1-FBR (1971).

[3] P.D. Bell and A. Wolfman, Mathematical modeling of crack growth interaction effects. *ASTM STP* **595** (1976), pp. 157–171.

[4] J.C. Newman, A crack closure model for predicting fatigue crack growth under aircraft spectrum loading, *ASTM STP* **748** (1981) pp. 53–84.

[5] A.U. de Koning, A simple crack closure model for prediction of fatigue crack growth, *ASTM STP***743** (1981) pp. 63–85.

[6] H. Fuhring and T. Seeger, Structural memory of cracked components under irregular loading. *ASTM STP***667** (1979) pp. 144–167.

[7] D. Broek, A similitude criterion for fatigue crack growth modeling, *ASTM STP* **868** (1985), pp. 347–360.

[8] D. Broek and S.H. Smith, Fatigue crack growth prediction under aircraft spectrum loading, *Eng. Fract. Mech.***11** (1979) pp. 123–142.

[9] D. Broek, *Fracture Mechanics Software*, FracturREsearch (1987).

[10] D. Broek and R.C. Rice, *Prediction of fatigue crack growth in railroad rails*, SAMPE Nat. Symp. Vol. 9 (1977) pp. 392–408 11. A. Nathan et al., *The effect of thickness on crack growth rate*, paper presented at 14th symposium on aeronautical fatigue (ICAF) (1987).

[12] Various Authors, *Fatigue thresholds*, Eng. Mat. Adv. Serv. (EMAD), (1982), 2 volumes.

[13] D. Broek and B.N. Leis, *Similitude and anomalies in crack growth rates*, Mat. Exp. and Design in Fatigue, Westbury (1981) pp. 129–146.

CHAPTER 6

Load spectra and stress histories

6.1. Scope

In this chapter the word spectrum will be used to mean any statistical representation of loads or stresses. A stress history (or load history) is the particular sequence of stresses (or loads) experienced by a structure or component in service or as used in crack growth analysis.

For the application of damage tolerance concepts, a reliable prediction must be made of the number of load cycles that will propagate a crack from a certain starting size to the permissible size. Inspection intervals or any other fracture control measures envisaged will be based on this prediction (chapters 11, 12). The prediction of fatigue crack propagation rates and propagation time of a structural crack requires the input of relevant crack propagation data, geometry factors and stress history. Although often great efforts are expended in obtaining geometry factors (to great accuracy) and material data, and much significance is attached to the applicability of retardation models, the stress history frequently is the stepchild in the analysis. Yet, the effects of stress history are so significant, that slight misinterpretations may lead to errors in crack growth life far overshadowing those in data and geometry factors.

This chapter presents a general discussion on the generation of stress histories for use in crack growth analysis. Most load spectra can be represented by exceedance diagrams; therefore the latter will be used as the basis for the stress history development for structures subjected to random or semi-random loading, but the essentials of the discussions apply just the same to load spectra of different form and/or different formats. The representation by an exceedance diagram has many advantages for random and semi-random loading, not in the least because of its simplicity. Examples of exceedance diagrams will be given for many types of structures.

In most cases the stress history used in the crack growth analysis is a simplified version of the stress history anticipated to be experienced in service. The simplification is necessary because (1) the actual stress cycles cannot be

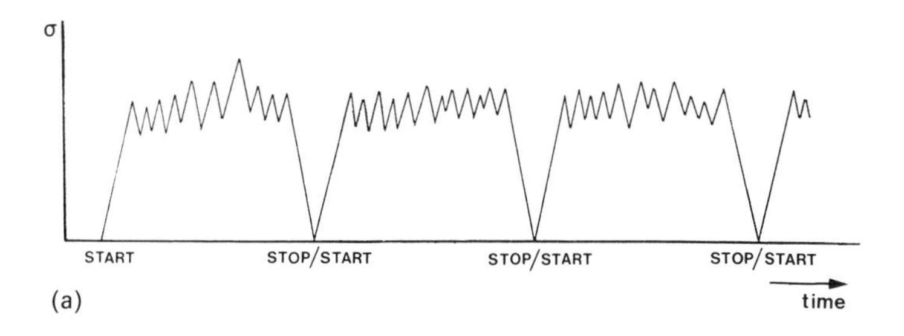

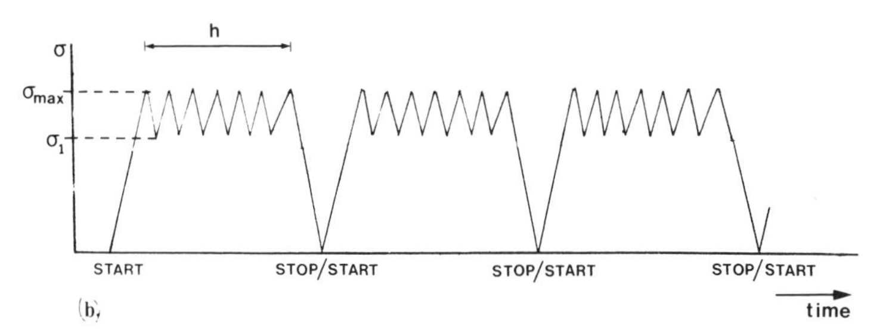

Figure 6.1. Loading of Rotating machinery. (a) Typical; (b) Idealization.

known in advance, especially in the case of random loading, and because (2) the stress history must be derived as an interpretation of past experience on other, similar structures, obtained either from load/stress measurements or from calculations.

All major considerations in the development of a representative stress history, including selection of clipping and truncation levels, will be reviewed and examples will be given to illustrate the effects on the results of the crack growth analysis. The use of a representative stress history is extremely important -probably more so than any other ingredient of the analysis- in particular when there is retardation.

6.2. Types of stress histories

Few structures or components are subjected to constant amplitude loading. An example of these few is perhaps a buried pipeline subjected to occasional de- and re-pressurization cycles; but even in that case loadings due to hydrostatic tests, possible seismic events, and perhaps frost heave, thaw settlement and land slides, give cause to a stress history of variable amplitude. For those structures subjected to loading of constant amplitude, there is hardly an interpretation problem, provided the magnitude of the loads and consequent stresses are known with sufficient accuracy.

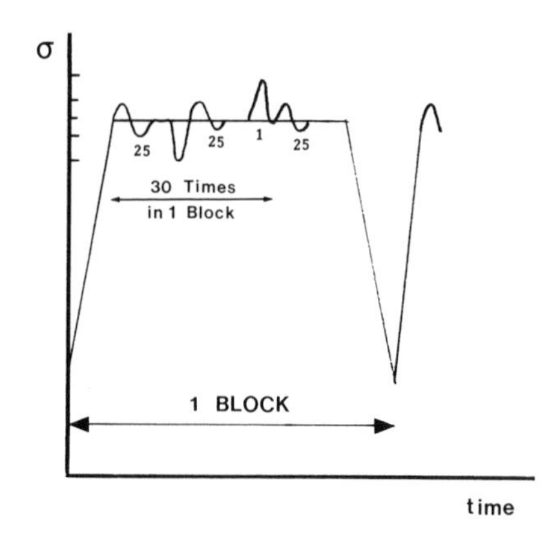

Figure 6.2. Proposed stress history for nuclear power generation components [2].

A class of structures with relatively simple load histories is rotating machinery, among which one can reasonably categorize all turbomachinery, propellers, helicopter blades and so on. The major load cycles consist of the start-up-shut-down cycle (mostly centrifugal force), upon which small cycles are superposed as shown in Figure 6.1. The latter cycles may be due to hydrodynamic/aerodynamic bending loads on the blades of a turbine, or propeller, and/or vibratory loads. The superposed small cycles are usually not of constant amplitude (because of changing power input or output). Neither is the start-stop cycle always of the same magnitude because it depends upon the r.p.m. In some cases however, one may be justified in using a constant amplitude start-stop cycles and superposed constant amplitude small cycles as shown in Figure 6.1b. The input to the computer program then could be a simple table of stresses e.g.:

$0 \quad - \sigma_{max} \quad 1$ time;

$\sigma_1 - \sigma_{max} \quad n$ times.

This sequence can be repeated throughout the crack growth analysis, provided the computer program offers the facility to take tabular stress input and interpret its meaning.

Should the above simplification not be justified, then a longer table, with more values of the superposed stress cycle (and number of occurrences) can be used. The sequence as given in the table could be applied as is and repeated, or if the computer code offers this possibility, the sequence could be randomized.

Piping components of nuclear power structures are subjected to a stress history of very similar nature. A more or less standardized (the meaning of the word here being: agreed upon by a group of people) form of the history [1, 2] is shown in Figure 6.2, which has been used for tests and analysis development.

Although the above histories are simple, there still may be interpretation problems. The problems of counting, clipping, trunction, number of levels, etc.,

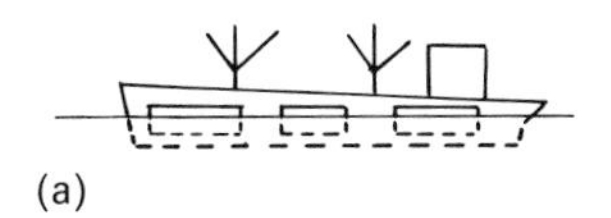

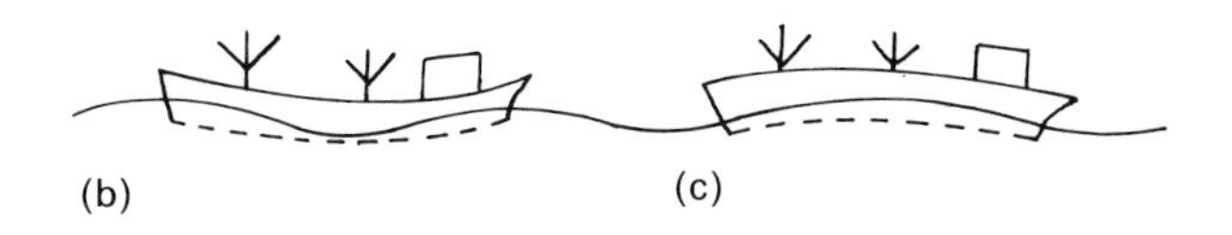

Figure 6.3. Loads on ship. (a) Still water; (b) Sagging; (c) Hogging.

discussed in this chapter still apply. Also the discussion on 'errors' in Chapter 12 are of relevance.

Many structures experience loads (stresses) of variable amplitude in 'random' sequence. Actually, the load sequence most often is semi-random, and sometimes interspersed with deterministic loads (stresses). Typical examples are airplanes, ships, offshore structures, railroad rails etc. The load sequence on a ship is not the same during every voyage, because storms and high seas are encountered only occasionally. Similarly, the loads on an offshore structure depend upon winds and waves, and are different from month to month. Such

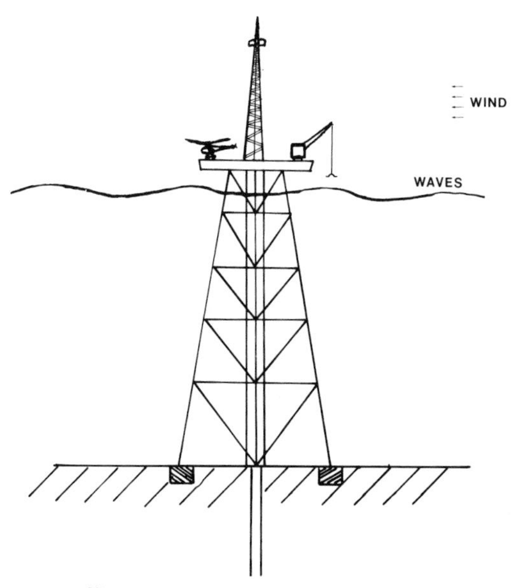

Figure 6.4. Loads on offshore platform.

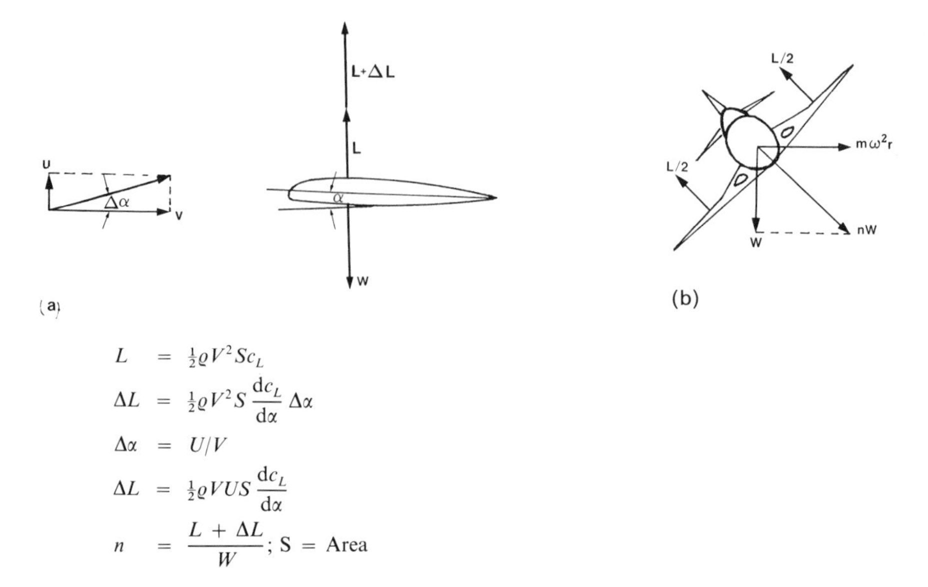

Figure 6.5. Loads on aircraft wing. (a) Gust loads on commercial aircraft; (b) Maneuver loads on fighter aircraft.

sequences are not truly random; high loads are clustered in certain periods (storms). The word semi-random is used here to describe such sequences.

The loads on e.g. a crane derive from the anticipated weights of various magnitudes lifted. Loads on a ship arise from hogging and sagging across the waves. In still water the buoyancy forces on a ship are more or less evenly distributed (apart from bow and stern). The resulting bending moment then depends upon the distribution of masses such as engine, fuel and cargo (see Figure 6.3). In waves the buoyancy forces are unevenly distributed, causing bending of the ship, and leading to tension in the deck during hogging and compression during sagging, and to tension in the bottom during sagging.

Loads on off-shore structures arise from winds and waves (Figure 6.4), on automobile structures from uneven roads, on railroad rails from different wheel loads and different car weights, as well as from impacts due to wheel flats and gaps (bolted rails and points).

Aircraft wings and tails experience loads due to maneuvers and gusts (turbulence), as shown in Figure 6.5. Commercial airliners are subject to some maneuver loading, but gust loading is the most important. Fighters on the other hand do experience gusts, but maneuver loading is the most important. In upward wing bending the lower surface of the wing is in tension and vice versa. The variable bending due to gust and maneuvers is superposed on the steady bending of stationary horizontal flight (1g), where the wing just carries the weight of the aircraft.

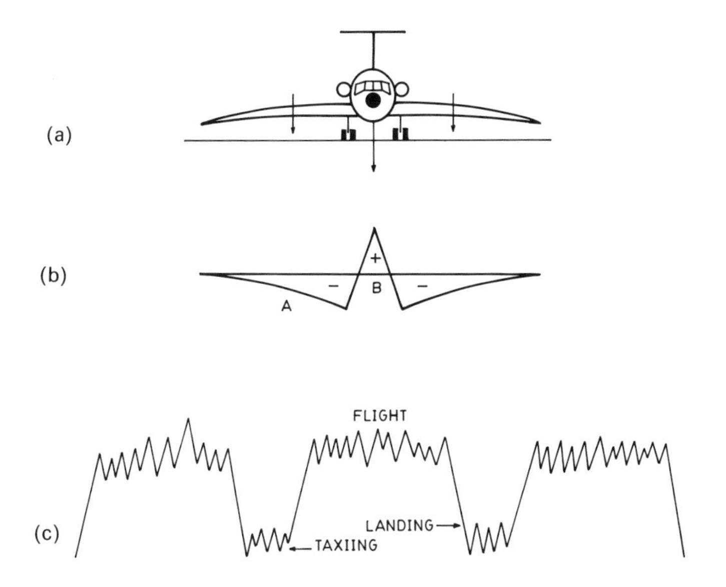

Figure 6.6. Landing loads on aircraft. (a) Ground loads; (b) Moment-line; (c) Stress in lower wing skin during flight and landing.

Following each flight is a ground-air-ground cycle (g.a.g.), sometimes combined with taxiing loads (Figure 6.6). Of course the g.a.g. cycle depends upon the airplane configuration, size, landing gear location and is different at different locations. For example, in a wing (Figure 6.6) the ground loads may cause compressive stresses at A, and tension at B depending upon weight distribution and location of the landing gear. The g.a.g. cycle is a deterministic event in an otherwise semi-random stress history. It is known from experiments that the g.a.g. cycle causes faster crack growth, in particular when the stresses are compressive. This is largely due to the annihilation of residual compressive stresses introduced during tension overloads, and consequently, the elimination of retardation (Chapter 5). Some retardation models attempt to account for the effect of compressive stress and thus for g.a.g. Because of its sometimes large influence, the g.a.g. has to be accounted for properly.

As it is very important, it is emphasized again, that the loading on most of these structures is not truly random. High loads are clustered (in periods of bad weather); this must be accounted for if retardation is significant, because complete randomization would spread out all high loads throughout the history so that they would cause much more retardation than when occurring in clusters. Moreover the history may be interspersed with deterministic loads, such as g.a.g. cycles in aircraft, unloading and refourageing of ships after each voyage, etc. If these deterministic loads are significant, they must be included in the history at the proper intervals (not at random).

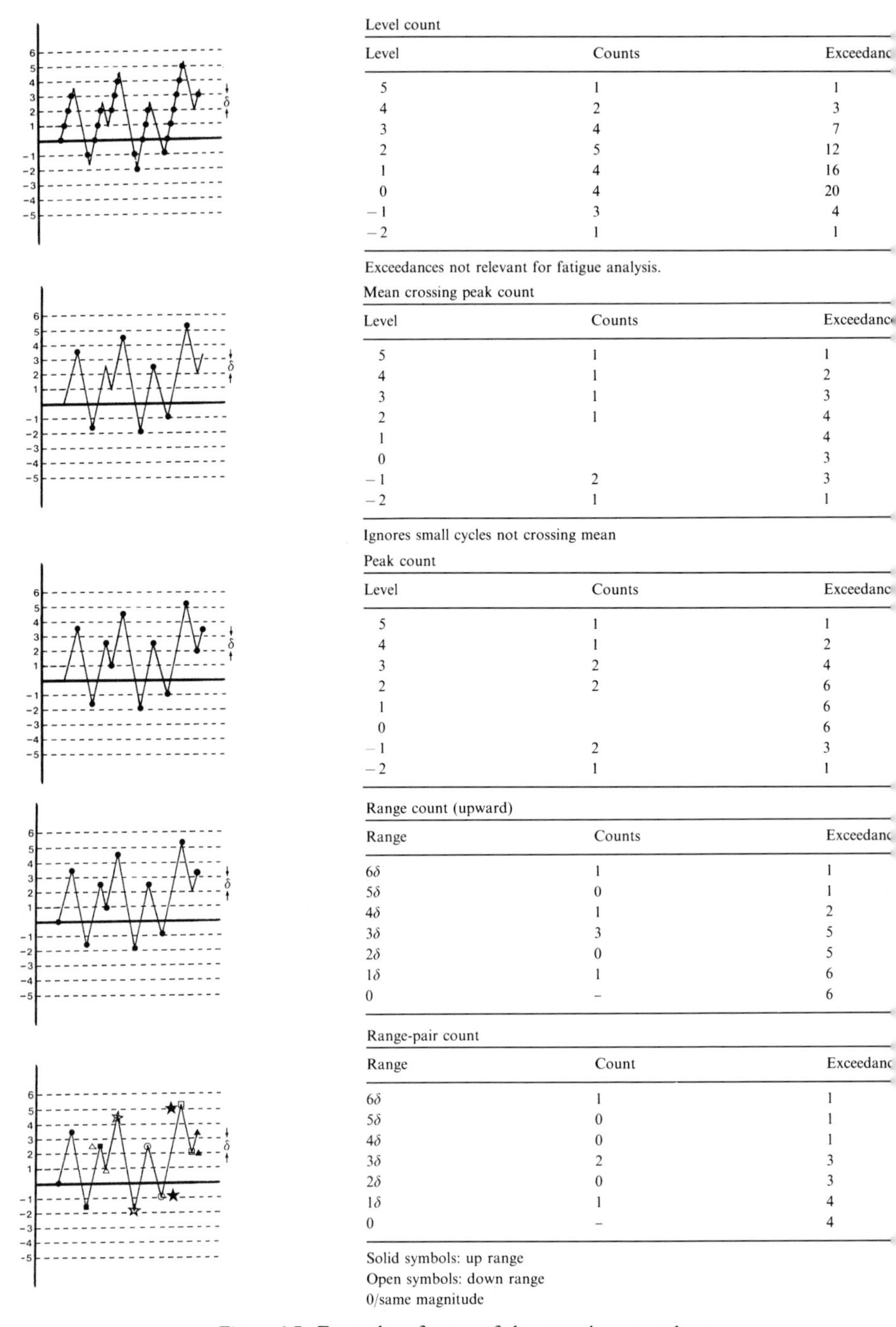

Level count

Level	Counts	Exceedanc
5	1	1
4	2	3
3	4	7
2	5	12
1	4	16
0	4	20
−1	3	4
−2	1	1

Exceedances not relevant for fatigue analysis.

Mean crossing peak count

Level	Counts	Exceedanc
5	1	1
4	1	2
3	1	3
2	1	4
1		4
0		3
−1	2	3
−2	1	1

Ignores small cycles not crossing mean

Peak count

Level	Counts	Exceedanc
5	1	1
4	1	2
3	2	4
2	2	6
1		6
0		6
−1	2	3
−2	1	1

Range count (upward)

Range	Counts	Exceedanc
6δ	1	1
5δ	0	1
4δ	1	2
3δ	3	5
2δ	0	5
1δ	1	6
0	–	6

Range-pair count

Range	Count	Exceedanc
6δ	1	1
5δ	0	1
4δ	0	1
3δ	2	3
2δ	0	3
1δ	1	4
0	–	4

Solid symbols: up range
Open symbols: down range
0/same magnitude

Figure 6.7. Examples of some of the counting procedures.

6.3. Obtaining load spectra

If the loading is man-induced the load spectrum often can be calculated on the basis of (anticipated) usage. This would be the case for cranes, fighter aircraft (maneuvers) and rotating machinery. If, on the other hand, loading is induced by nature (waves, winds, gusts, rough roads), the load spectrum is usually based upon measurements, from which the loads must be inferred. Subsequently, the stress history must be obtained from the load spectrum.

Load spectra can be measured in service by indirect means only, for example from continuous strain gage records over long periods of time. The measurement is indirect because the loads must be inferred from the measured strains or from e.g. measured wave heights or wind force. As such records are very extensive a representation in some concise form is desirable: they must be interpreted so that an envelope can be established. The envelope is called the spectrum. Power spectrum density analysis can be used or one can establish exceedance diagrams from counts of the records.

Many counting methods have been developed (peak count, mean-crossing peak count, range pair count, range-pair-mean count or rainflow count, etc.), a few of which are shown in Figure 6.7. At present, it is generally agreed that the latter method gives the best representation for fatigue, but in many cases the differences among the better counting methods are small [3]. It should be mentioned here that it is rather immaterial whether different methods produce somewhat different spectra; the only criterion is whether such a spectrum when used in a crack growth analysis produces the correct crack growth result. This brings about the major problem of the counting procedures, namely what is important for crack growth: the maxima and minima or the ranges and means.

On the basis of counts of past measurements, the envelope of the future load history can be estimated, but the actual load experience of a new structure cannot be known until the service life is expired. All counting methods have shortcomings and the tendency to neglect certain small load reversals. Their usefulness depends upon their purpose. When the data are to be used for design, the usefulness may be judged by how well the counting method has described the actual loads, in particular the highest loads. For fatigue crack growth calculations the usefulness depends on how well the methods describe those loads which are the most relevant to the crack growth process. Although the question regarding the use of maxima and minima instead of ranges is important, the most important difficulty in the use of spectra is that after the counting all information about the sequence is lost. A new sequence must be generated for use in damage tolerance analysis.

The interpretation (counting) of measured load experience is a very interesting problem by itself, but a detailed discussion is beyond the scope of this book. The reader is referred to the excellent review of the problem by Schijve [3]. It should be noted here that the counting procedure presently known as the

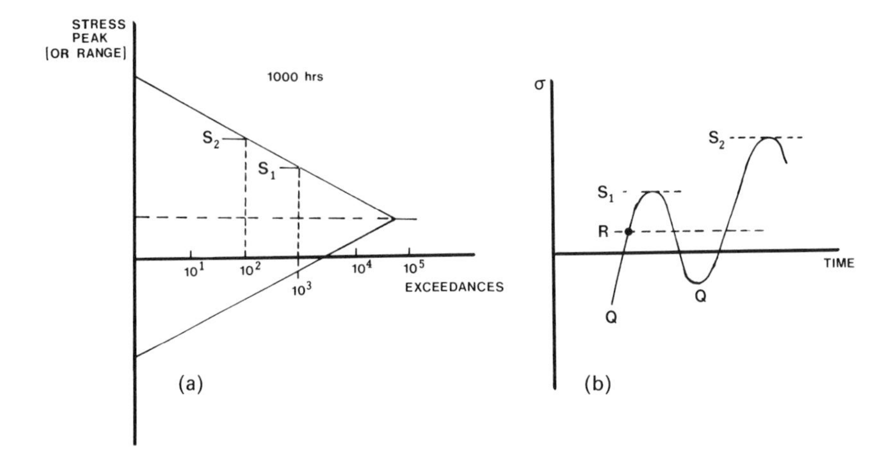

Figure 6.8. Significant counts for fatigue analysis. (a) Exceedances of maxima (or ranges: not levels); (b) Levels and maxima.

'rainflow' method is virtually identical to the much older range-pair-mean count method [3]. For the purpose of this book, it is sufficient to start with the end-result of the measurements, namely the load spectrum. Nevertheless, the problem of counting will arise again later in this chapter

6.4. Exceedance diagrams

Although power spectra are useful, in particular when there are environmental and frequency effects, exceedance diagrams are more convenient because they lend themselves easily for the (re-) generation of stress histories. A typical exceedance diagram is shown in Figure 6.8. The diagram shows how many times a certain stress (or load) level is exceeded. This by itself is a 'loaded' statement. More properly expressed, the diagram gives the number of times certain maxima and minima (or ranges) are exceeded. For example in Figure 6.8b the level R is exceeded twice when the load varies from Q through R to S_1 and S_2. These exceedances are NOT and should not be represented in the diagram, because they would be of no use for crack growth analysis; only Q and S are of interest. Rather, e.g. point S_1 in Figure 6.8a indicates that the maximum stress in A CYCLE exceeded level S_1 a certain number (1000) times during the time (1000 hours) for which the diagram was established.

Thus, the exceedances must be interpreted as exceedances of peaks or RANGES, depending upon the counting method, and not as level-crossings. A level-crossing diagram (first case in Figure 6.7) might look very different and would be rather useless for fatigue crack growth analysis without further interpretation. On the other hand, level crossings may provide some informa-

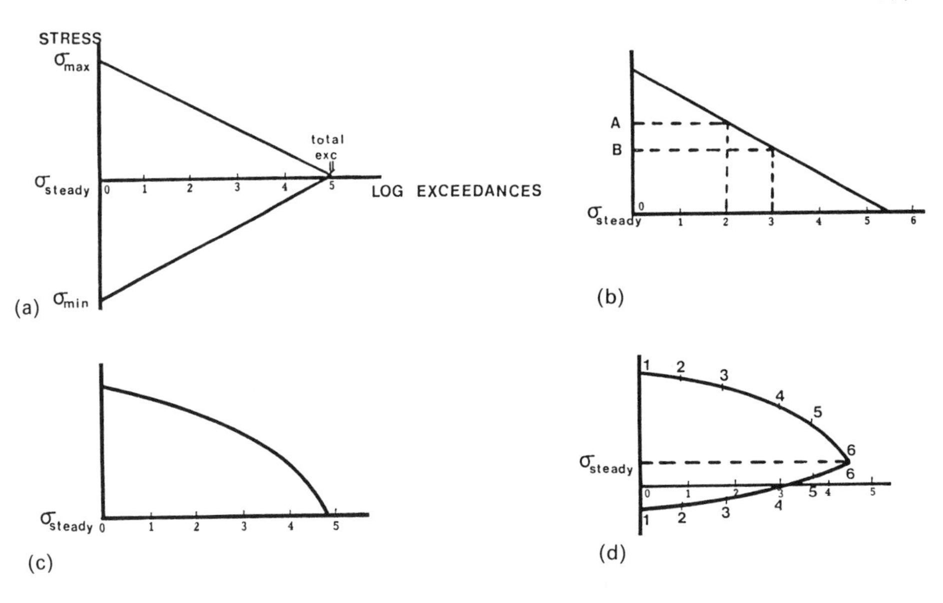

Figure 6.9. Typical exceedance diagrams. (a) typical for weather induced; ships, aircraft; earthquakes; (b) Typical weather induced; offshore structures; (c) Typical for mechanically induced; railroad rails; (d) Typical for mechanically induced; fighter aircraft.

tion about the time spent at or above certain load levels, which maybe of interest if there are significant environmental (time-dependent) effects. Note that exceedances are generally presented on a logarithmic-scale, the loads (stresses) on a linear scale.

Exceedance diagrams can be categorized roughly in two classes, namely (almost) semi-log linear and non-linear. These types are shown in Figure 6.9. The almost semi-log-linear diagram is typical for nature induced (e.g. weather) loading: very similar diagrams apply to ships, offshore structures and commercial airplanes. The diagram may be asymmetric, or symmetric around a non-zero 'steady' load or stress. The non-linear diagram is typical for mechanically induced loads. This type of diagram applies to fighter airplanes, cranes and e.g. railroad rails. It is almost always asymmetric. For lack of a better word, the level at which top and bottom part of the diagram meet, will be called the 'steady stress' (load). The exceedance diagram may be expressed in loads, g-levels, wave heights, gust velocity or any measure of load. The diagram gives the load experience for a certain time interval, number of flights or voyages, amount of traffic (rails or bridges), number of years or hours; it is a statistical average for that time interval. It is usually assumed that for each such time interval the same load envelope applies, be it that the load sequence may be different in each interval.

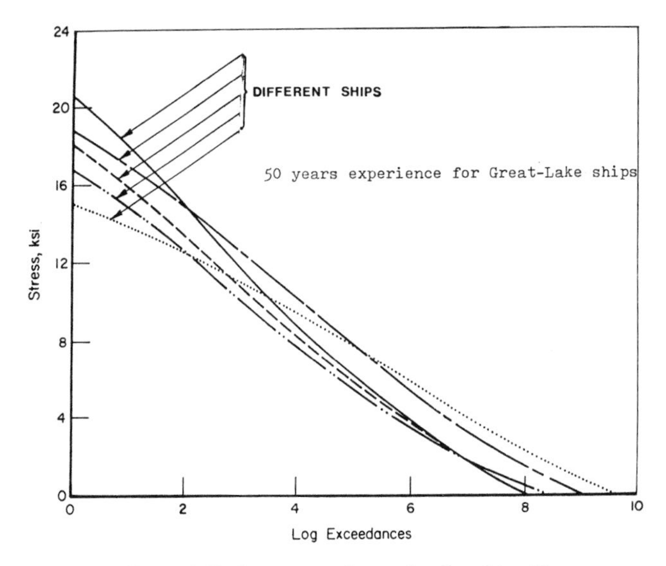

Figure 6.10. Stress exceedances for five ships [6].

Typical exceedance diagrams for ships are shown in Figure 6.10. These are induced by nature and are indeed nearly semi-log-linear (note that they are symmetric, but only the top half is shown). Spectra for off-shore structures are very similar, but often asymmetric (Figure 6.11). An aircraft gust spectrum (nature induced) is shown in Figure 6.12 (only top half is shown; diagram is symmetric); indeed it is almost semi-log-linear, while the man-induced fighter

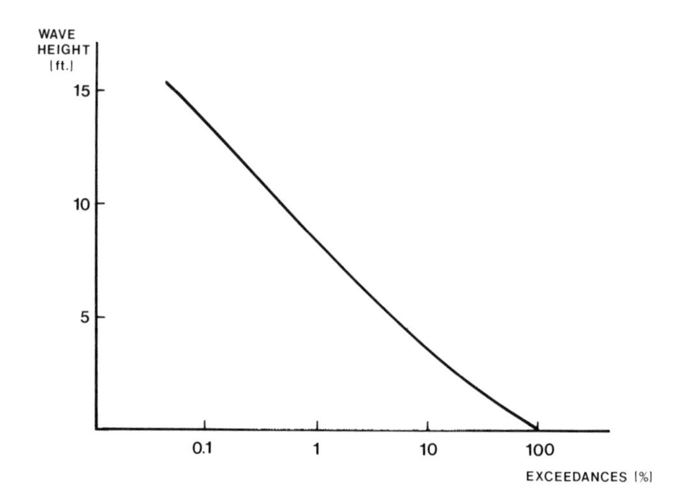

Figure 6.11. Offshore spectrum.

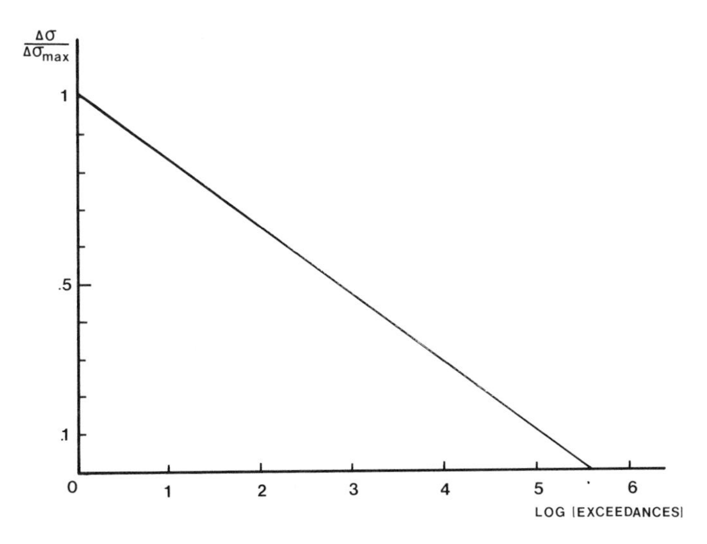

Figure 6.12. Aircraft gust spectrum [8].

maneuver spectrum in Figure 6.13 is non-linear. Similarly, the man-induced spectrum for railroad rails in Figure 6.14 is non-linear (and asymmetric). It must be emphasized that the spectra in Figures 6.10–6.14 are examples only; they may differ from structure to structure and should not be used as the general spectrum for the above types of structures; they show the general trend.

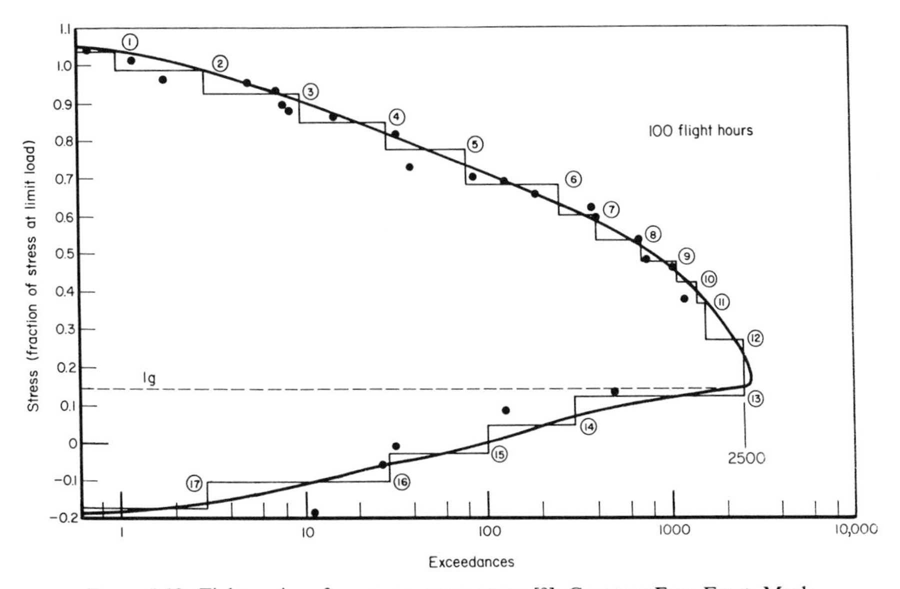

Figure 6.13. Fighter aircraft manoeuver spectrum [9]. Courtesy Eng. Fract. Mech.

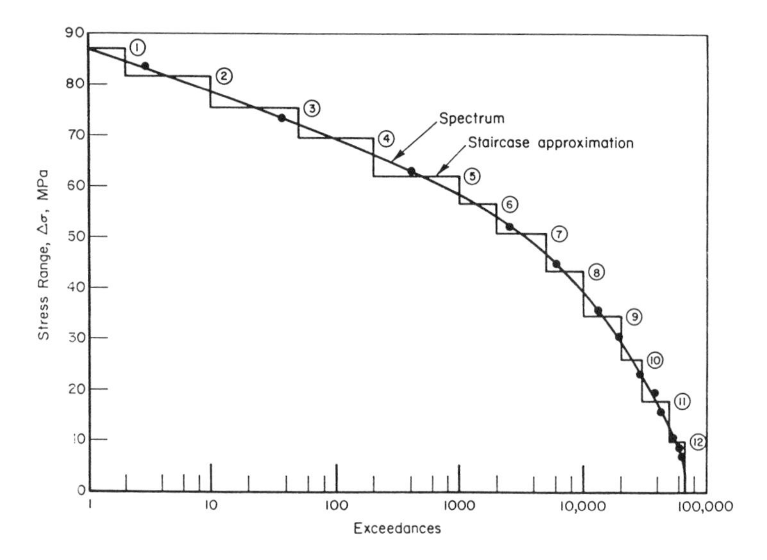

Figure 6.14. Spectrum for railroad rails [10]. Courtesy SAMPE.

No load experience is available for a new structure. Hence, the fatigue crack growth analysis must be based on past experience and be a projection in the future. In some cases the loads (or stresses) in a structure are continously monitered, so that the analysis can be updated from time to time depending upon the actual service experience. (The United States Air Force 'tracks' many of its aircraft for this specific purpose). The projection of load experience in the future introduces an uncertainty in the crack growth prediction. Nothing can be done about this problem other than performing multiple computer runs, using various alternative load spectra, to bound the fatigue performance for best, worse and average scenarios.

6.5. Stress history generation

As fatigue crack growth analyses is based upon stresses, the spectrum, whatever its basis, must be converted into stresses. If the stresses are proportional to the quantity given in the exceedance diagram (load, wave height, gust velocity and so on), the diagram can be converted easily into a stress exceedance diagram through a multiplication factor. This is often the case or nearly so, because the nominal stresses during fatigue are generally elastic. In some cases more complicated transfer functions will be involved.

The exceedance diagram gives the number of times that a certain stress level is exceeded during a given time interval, the level being a maximum, minimum or range, not a level crossing (see previous section). In the example in Figure 6.15, the exceedances are for one year of operation. This means that the stress

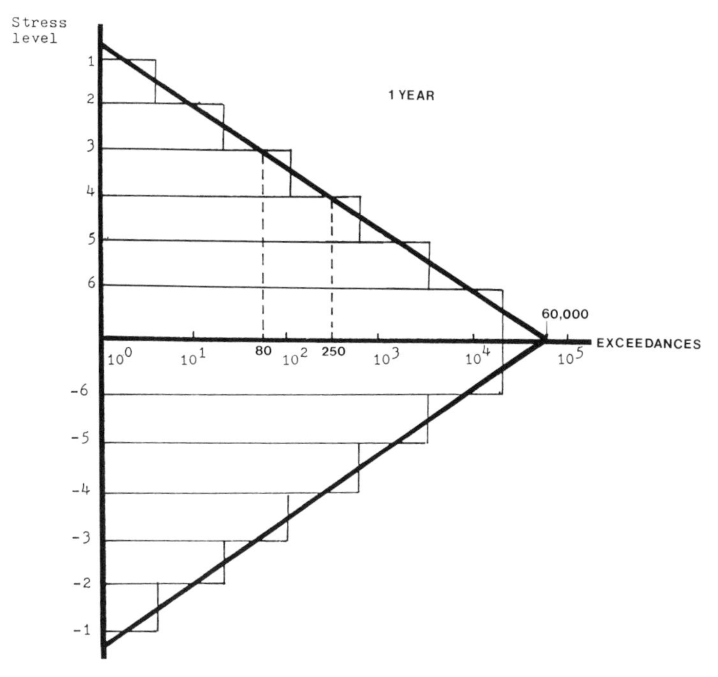

Figure 6.15. Determination of stress levels from exceedance diagram.

Level	Exceedances	Occurences
1	4	4
2	20	16
3	100	80
4	800	700
5	3000	2200
6	20000	17000

level 4 will be exceeded 250 times per year, and level 3 will be exceeded 80 times per year. As a result, there will be 250–80 = 170 events in which the maximum stress reaches a level somewhere between 3 and 4.

One can identify a stress level i that is exceeded 99 times, and a level j that is exceeded 98 times. Thus, there would be one stress excursion to j. Similarly in the case of Figure 6.15, one could identify 60 000 different stress levels, each of which would occur just once per year. In general, one could define as many stress levels as there are exceedances in the diagram. Obviously, it is impractical to consider so many different stress levels. Not only would it be impractical, it would also ignore the fact that the spectrum is a statistical representation of past experience and that the analysis is a prediction of the future. Accounting for so many stress levels would be presuming that the stresses are known to occur in the future exactly as they did in the past, which they will not.

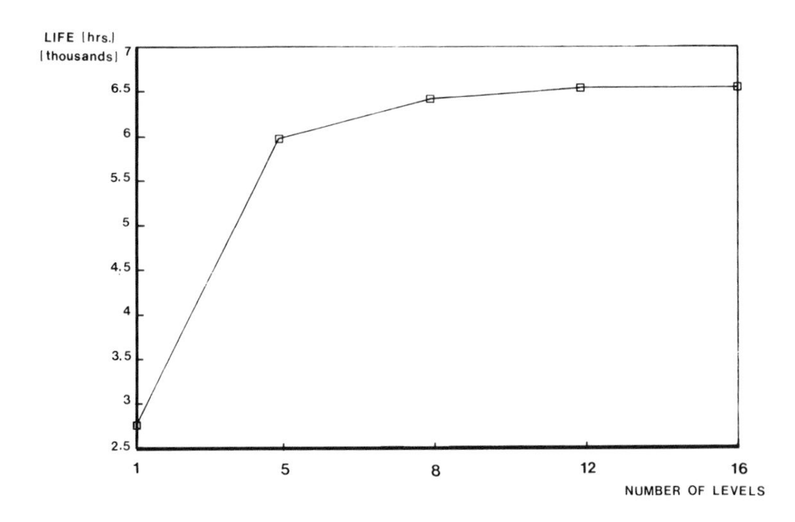

Figure 6.16. Effect of levels in exceedance diagram approximation; computed number of hours for crack growth as a function of number of levels. One level is constant amplitude.

Instead, the exceedance diagram is approximated by a limited number of discrete levels; six such levels are shown in Figure 6.15. The stress levels do not have to be evenly spaced, but they usually are selected that way. Alternatively, one may select a number of exceedances and start constructing the levels from the abscissa. This has the advantage that for a change of the load-stress conversion factor always the same number of cycles would occur; only the stress values would change. Experience shows that 10 to 12 levels (each for positive and negative excursions) are usually sufficient to give the desired accuracy; the use of more than 12 levels hardly changes the results. This can be appreciated from the results of actual fatigue crack growth analyses for one particular exceedance diagram, shown in Figure 6.16.

For clarity only 6 levels (6 positive and 6 negative) are shown in the example in Figure 6.17. At each level a line is drawn that intersects the exceedance curve. Usually the steps are completed by placing the vertical AB in such a manner that the shaded areas above and below the exceedance curve are equal, as shown in Figure 6.17. If the exceedance diagram is semi-log-linear, this vertical AB will be at the logarithmic average of C and D, i.e midway between C and D on the log scale of exceedances. Other procedures have been suggested, but these do not lead to essentially different results, as can be inferred from Figure 6.16: if more and more levels are selected, the approximation has less and less significance. Besides, great sophistication (complication) is not warranted, because the exceedance diagram of the past will not be repeated exactly in the future anyway. Figures 6.15 and 6.17 shows how the exceedances and from these the number occurrences of each level are obtained.

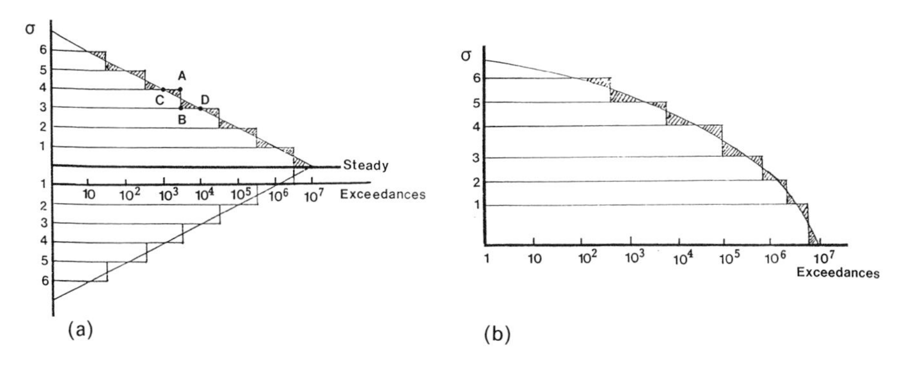

Figure 6.17. Example of level approximation (only 6 shown for clarity).

Level	Exceedances	Occurences
(a) Symmetric and log linear		
6-6	30	30
5-5	300	270
4-4	3000	2700
3-3	30000	27000
2-2	300000	270000
1-1	3000000	2700000
(b) Asymmetric		
0-6	60	60
0-5	8000	7940
0-4	100000	92000
0-3	900000	800000
0-2	2000000	1100000
0-1	5000000	3000000

If the exceedance diagram is for ranges of stress, the approximation is now essentially complete but a problem remains as shown below. If the diagram is for peaks (maxima and minima), positive and negative excursions still have to be combined. One might be tempted to select positive and negative excursions in random combinations. This could introduce a new and serious problem, as shown on the basis of Figure 6.18.

Four cyclic sequences A, B, C and D are shown in Figure 6.18. Let it be assumed that the material follows a Paris equation for crack propagation with a power of 4 (Chapter 5), merely to facilitate the discussion. Sequence A has a small reversal which certainly will not contribute to the crack growth. Hence, only the large excursion of 8δ would have to be accounted for. The growth during this cycle is then $C(8\delta)^4 = 4096\,C\delta^4$. The little reversal in sequence B cannot be neglected. If this sequence were interpreted as two stress ranges of 5δ each, the total growth would be $2 \times C(5\delta)^4 = 1250\ C\delta^4$. Similarly, for sequence C the result would be $2 \times C(6\delta)^4 = 2592\ C\delta^4$. With this interpreta-

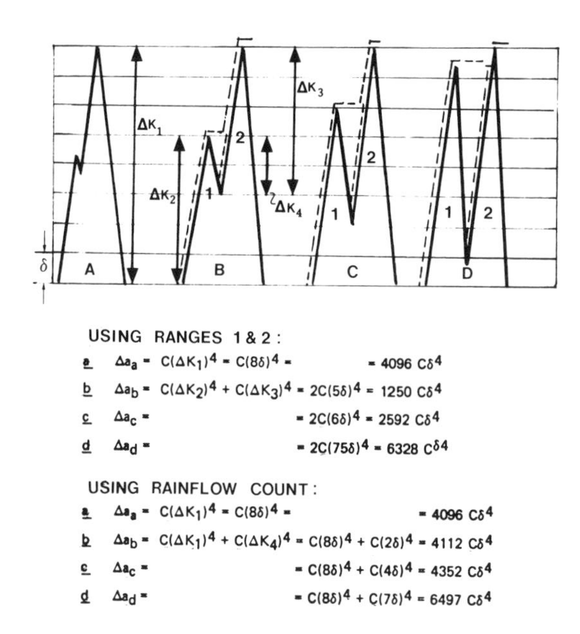

Figure 6.18. Two interpretations of ranges for crack growth analysis.

tion sequence B would cause only about 1/3 of the growth of sequence A, while even sequence C would cause less growth than A. Clearly, B and C are more severe than A, and this should be reflected in the growth, otherwise erroneous answers will be obtained. The solution to this problem is in the interpretation, as shown below.

In all sequences the material is subjected to the large excursions of 8δ, causing a growth of $C(8\delta)^4 = 4096\ C\delta^4$ in ALL FOUR cases. In addition there is a small reversal with a range of 2δ in sequence B causing a growth of $C(2\delta)^4 = 16\ C\delta_4$, and bringing the total to 4112 $C\delta^4$. Similarly, the small range of 4δ in sequence C causes an additional growth of $C(4\delta) = 256\ C\delta^4$ to bring the total at 4352 Cs. With this correct interpretation sequences B and C indeed produce more growth than A. The conclusion is, that in an arbitrary sequence of peaks and valleys, the identification of significant stress ranges requires a new interpretation. A counting procedure such as the rainflow count will accomplish this (Figure 6.18 bottom); effectively the example above is the consequence of a rainflow count. Should an arbitrary sequence of peaks and valleys be used in the analysis, then the computer code must have a facility to count the stress history again. Besides, a new problem of sequencing arises: should the large range in sequences B and C be applied first and then the small range, or vice versa?

Positive and negative excursions following from the exceedance diagram must be combined and sequenced. If this is done randomly, rainflow counting of the history will again be necessary as shown in Figure 6.18 to determine what constitutes a range. Although this is a legitimate approach, a simpler procedure is often employed. Because the spectrum was developed from a counted history in the first place, why would it be necessary to disarrange it again, and then count it again, if the result was counted already a priori. Besides, the effective result of the second counting will be that the largest positive peak will be combined with the lowest valley. Foreseeing this, one may combine positive and negative excursions of equal frequency of occurence immediately. The ranges are then easily established as shown in Figure 6.17 top, and these can now be applied randomly because they are already pre-counted and interpreted. This leads to the largest possible load cycles, which is conservative; the computer code does not need a counting routine. Sea and air are continous. A down-gust must soon be followed by an up-gust of approximately equal magnitude otherwise air would disappear in space. Similarly, when the waves are high, a ship is likely to experience hogging and sagging of the same magnitude in close succession. Hence, a priori combination of positive and negative excursions of the same frequency of occurrence is realistic; it avoids the necessity of renewed counting and sequencing.

The content of the stress history is now known, but the sequence must still be determined. If retardation is not an issue, sequencing of stresses is rather irrelevant. Fatigue crack growth analysis without load interaction is virtually insensitive to load sequence, provided the stress history is more or less random. If the material exhibits little or no retardation, the problem of sequencing as discussed below does not apply. The same holds, if for reasons of conservatism, retardation is not considered. In the latter case it should be recognized that the ANALYSIS is indeed insensitive to the sequencing, but if load interaction actually occurs, the analysis without retardation does not give answers that have any bearing upon reality.

On the other hand, if load interaction is (or must be) considered, stress sequencing becomes of eminent importance. In many computer analyses the loads are simply applied in random order. However, because load interaction is indeed important, a random sequence is certainly not providing correct answers when actual service loading is semi-random. Not all voyages of a ship are of the same severity; a ship experiences storms only occasionally, and so does an offshore structure. A jet liner experiences many smooth flights and occasionally a rough flight. This means that the loading is not truly random, but clusters of high loads occur (Figure 6.19). Were these high loads randomly distributed throughout, they would cause large retardation continually. Because of the clustering, the retardation will be much less. The fatigue crack growth analysis must account for these effects: a mixture of PERIODS or flights or voyages of different severity must be applied. This is defined here as semi-random loading.

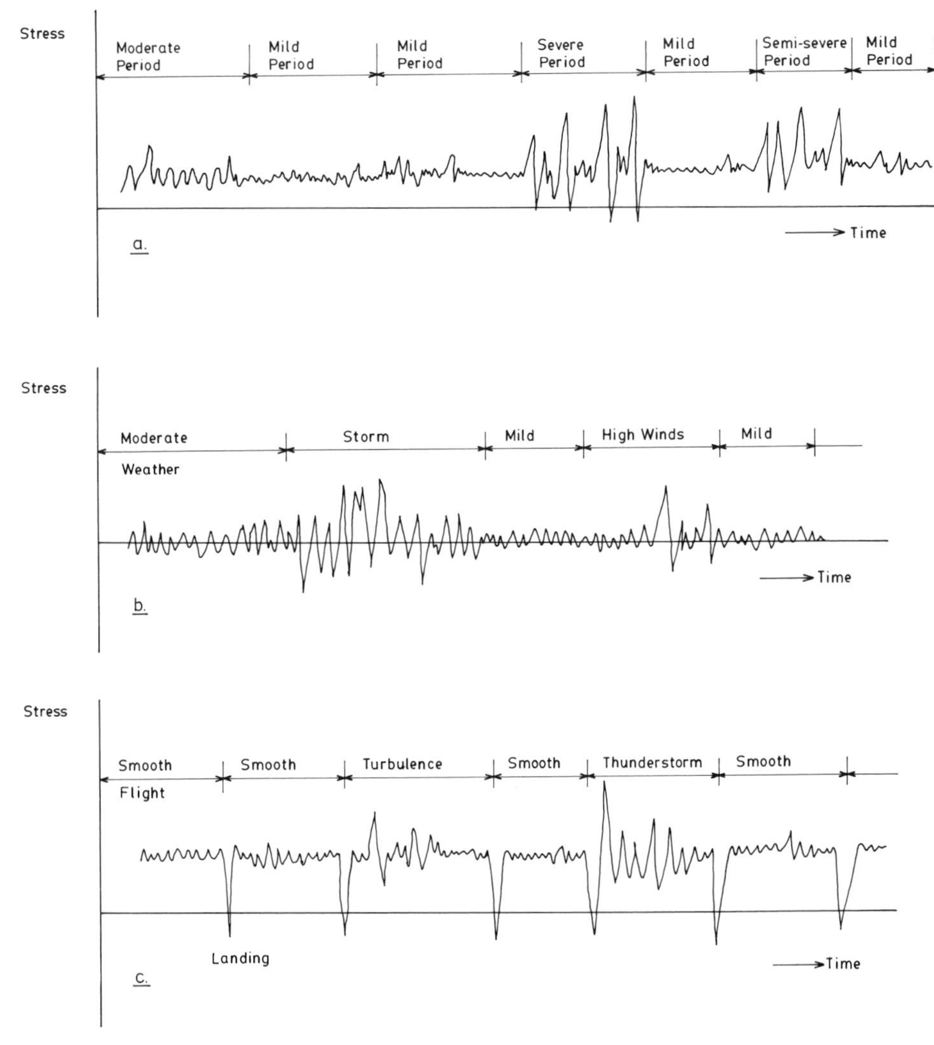

Figure 6.19. Semi-random loading in different periods (voyages, months, flights). (a) General case; (b) ship or offshore structure; (c) Commercial aircraft.

Of course there will be fewer severe periods than mild periods, as shown in the example in Table 6.1 and in Figure 6.19. In the computer analysis periods of different severity are to be applied in random sequence, the cycles in each period to be applied randomly, but different from the random sequence in previous applications of the same period.

A semi-random sequence can be developed in many ways. A simple algorithm is shown below. The mild and severe periods are constructed by recognizing that

Table 6.1. Periods of different severity
(a) Exceedance diagram
Stress conversion: 1 unit = 1 ksi.

Max. stress	Min. stress	Steady stress	Exceedances
15.000	−5.000	5.000	1
13.000	−3.000	5.000	5
11.000	−1.000	5.000	30
9.000	1.000	5.000	166
7.000	3.000	5.000	910
5.000	5.000	5.000	5000

Exceedances for 120 blocks.

(b) Stresses for 120 periods
1 unit of stress is 1 ksi

Exceedances Cycles	Occurr. Cycles	Min. str.	Max. str.	Periods type/number Total 120 periods 1/96	2/19	3/4	4/1
1	1	−5.000	15.000	0	0	0	1
3	2	−4.000	14.000	0	0	0	2
8	5	−3.000	13.000	0	0	1	1
19	11	−2.000	12.000	0	0	2	3
46	27	−1.000	11.000	0	1	1	4
108	62	0.000	10.000	0	2	4	8
253	145	1.000	9.000	1	2	2	3
594	341	2.000	8.000	2	6	7	7
1393	799	3.000	7.000	6	9	10	12
3266	1873	4.000	6.000	15	18	18	19

the exceedance diagrams for the individual periods are of the same shape [11], but with a different severity factor (slope), their total making up the diagram of total exceedances. The example will be based upon an exceedance diagram for 100 periods (be it voyages, hours, months or flights) as shown in Figure 6.20, and again only six levels are used, but the same procedure can be followed with exceedance diagrams for fewer or more periods, and/or more levels.

The different periods are constructed as illustrated in Figure 6.20 and Table 6.2. The total number of exceedances is 100 000, so that the average number of exceedances per period 1 000 000/100 = 1000. The highest level occurs 3 times (Table 6.2, column 8). Naturally, it will occur only in the severest period denoted as period A. Letting this level occur once in period A: the exceedance diagram for period A is established as shown in figure 6.20b. Note that the total exceedances are 1000 (per period) and that level 6 must occur once (step at level 6 must be at 1), so that the exceedance diagram of period A can indeed be constructed as in Figure 6.20b; then all levels can be drawn as

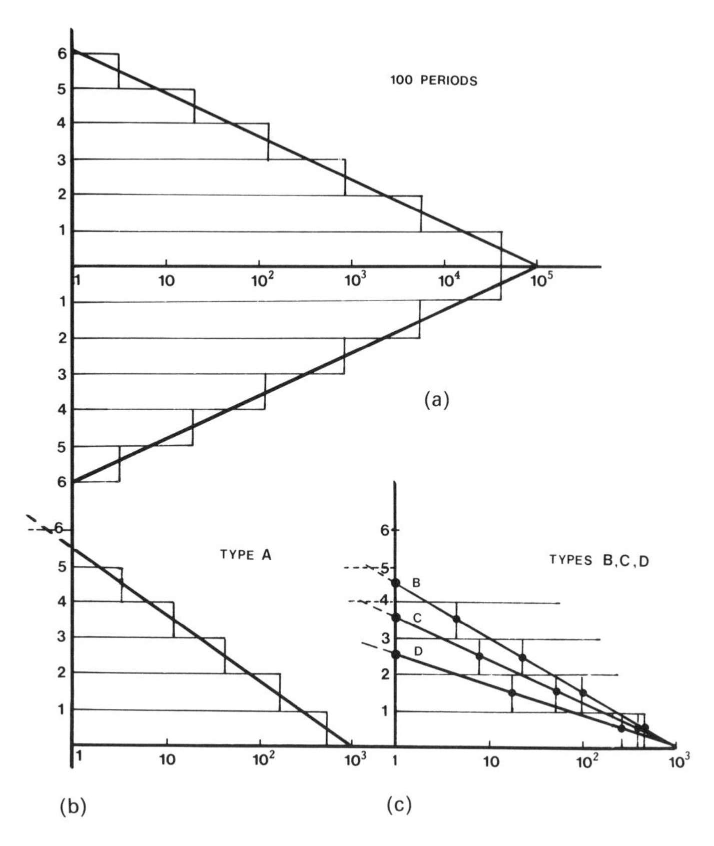

Figure 6.20. Stress history generation (see also Table 6.2). (a) Total exceedances; (b) Exceedances for period A: only top half shown; (c) Exceedances for other periods: only top half shown.

shown. Period A can occur only three times, because then the cycles of level 6 will be exhausted. The exceedances for period A are read from the exceedance diagram of A (Figure 6.20b), and from these the occurrences (number of cycles) are determined as in Table 6.2 columns 4 and 5. There being three periods A, the total cycles for all A are as in column 6. These cycles are subtracted from the total (column 3) so that the remainder for all other 97 periods is as shown in column 7. The next most severe period is B. Its highest will be level 5 which will occur once. This information permits construction of the exceedance diagram for B in the same way as for A as shown in Figure 6.20c. The exceedances and occurrences are determined as in columns 8 and 9 in Table 6.2. As there were only 12 cycles of level 5 left after subtraction of three periods A (column 7), there can be 12 periods B. These 12 periods will use the numbers of cycles shown in column 10, which must be subtracted from those in column 7

Table 6.2. Generation of stress history with different periods based in Figure 6.20.

1	2	3	4	5	6 3 × 5	7 3–6	8	9	10 12 × 9	11 6–10
Level	Exceed-ances Fig. 6.20a	Occurrences	Type A exceed Fig. 6.20b	Occurr in A	Occur in 3 types A	Remainder	Type B exceed Fig. 6.20c	Occur in B	Occur in 12 types B	Remainder
6	3	3	1	1	3	–	–	–	–	–
5	21	18	3	2	6	12	1	1	12	–
7	132	111	12	9	27	84	5	4	48	36
3	830	698	48	36	108	590	20	15	180	410
2	5750	4920	158	110	330	4590	100	80	960	3630
1	43650	37900	575	417	1251	36649	480	380	4560	32089
			Total number of periods		3				12 + 3 = 15	

If considered necessary →

12	13	14 36 × 13	15 11–14	16 15/49	17 16 × 49	18 15–17	19 18/12	20 9 + 19	21	22
Type C exceed Fig. 6.20c	Occur in C	Occur in 36 types C	Remainder for 49 type D	Occur in type D	Occur in 49 type D	Remains	Distributed in 12 type B	New type B	Exceed of D	According to diagram Fig. 6.20c
–	–	–	–	–	–	–	–	–	–	–
–	–	–	–	–	–	–	–	1	–	–
1	1	36	–	–	–	–	–	4	–	–
8	7	252	158	3	147	11	1	16	3	1
52	44	1584	2046	42	2058	−12	−1	79	45	20
400	348	12528	19561	399	19551	10	1	381	444	300
		15 + 36 = 51			51 + 49 = 100					

Same for negative levels if applicable. Periods: 3A + 12B + 36C + 49D = 100 total.

to leave the remaining cycles in column 11. Period C is constructed in the same manner. There can be 36 periods C and then the cycles of level 4 are exhausted.

One could go on in this manner, but since there now are only 49 periods left, it is better to divide the remaining cycles in column 15 by 49 in order to distribute them evenly over 49 types D. This is done in columns 15–17. There are some cycles unaccounted for, and also a few too many cycles were used as shown in column 18. These are cycles of lower magnitude contributing little to crack growth, and since the diagram is only a statistical average – and the procedure an approximate one – this little discrepancy could be left as is. However, if one would want to be precise, they could be accounted for by a little change in the content of period C, as shown in columns 18–20.

By means of the manipulation in column 16 the exceedance diagram of type D is determined as shown in column 21. Were the exceedance diagram for D determined in accordance with the above principle, it would be as shown in Figure 6.20c and the exceedances would be as in column 22. The latter are certainly somewhat different from those in column 21, but since they concern only the lower stresses, the effect on crack growth analysis will be minor (note again that the procedure is based upon statistical averages).

The above is an example only. If more than 6 levels are used, more (and different) types of periods can be generated. However, there is no need to go to extremes as long as a semi-random history is obtained, recognizing that periods of different severety do occur and that the higher loads are clustered in those periods. No matter how refined the procedure, the actual load sequence in practice will be different anyway. No mater how many levels are used and how many different types of periods are generated, there comes a point where the remaining cycles must be lumped into the remaining periods. Thus the mildest period will never be entirely representative; but since this concerns only the lowest stresses, no great harm is done.

Clearly, the total number of cycles in the stress history as generated above is less than the total number of exceedances of the steady level. This is entirely acceptable, because the very smallest cycles certainly will have no effect on crack growth, the procedure being for crack growth with retardation. The larger the number of levels used, the more (small) cycles there will be. But as shown in Figure 6.16 the result of the analysis will be the same (see also section 6.7 on truncation).

In accordance with the nature of the loading, there are only three severe periods A and 12 periods B in the total of 100 in Table 6.2. The majority consists of mild periods D (49) and C (36). Regardless of the number of levels chosen and the number of periods concerned, the above procedure will reflect this reality. Naturally, other procedures can be devised, but the above is certainly a rational one and easy to implement; besides it is based upon reality [11].

In the crack growth analysis the various periods should be applied in random order and the cycles within each period should be applied randomly. Thus the

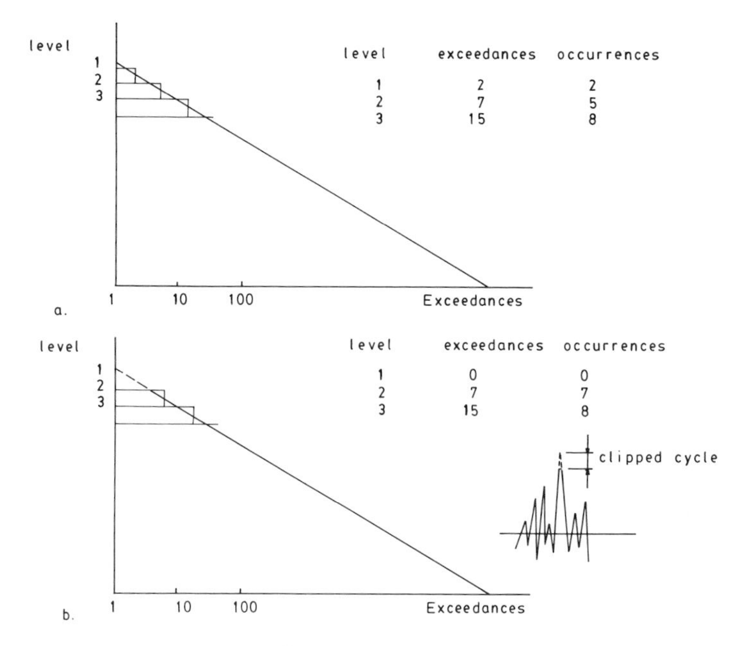

Figure 6.21. Clipping.

second occurrence of e.g. period C would apply the stresses in a different sequence than the first occurence, but the total cycle content of period C would be the same. If the 'basket' with 100 periods is empty, it is 'refilled', and the process starts anew; yet because of the randomization the periods appear in different order.

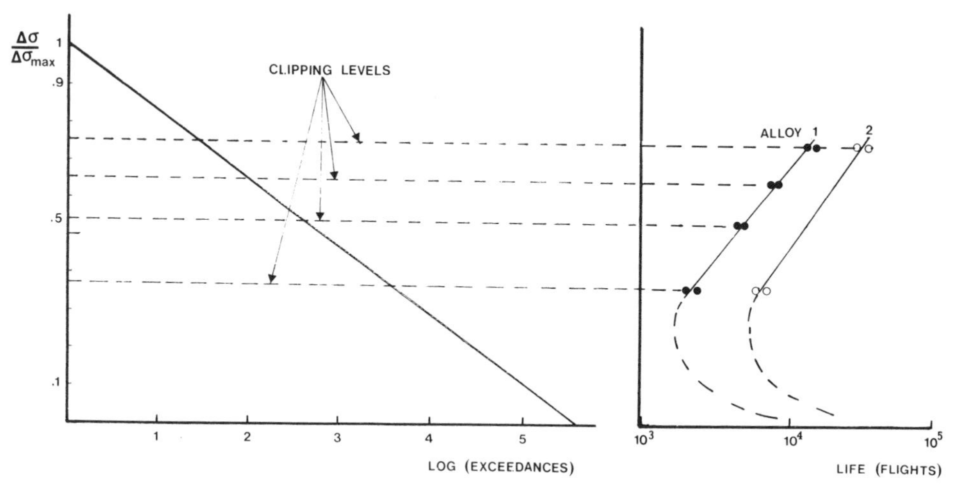

Figure 6.22. Effect of clipping on life; test data [8].

The above procedure can be implemented by hand calculations, and by submitting a resulting table of stresses and occurrences for use in a computer analysis. Aternatively, it can be programmed and included in the computer code. If this is done, the input is very simple: only a few data describing the exceedance diagram need be submitted; the computer does the rest.

If there are deterministic loads these must be interspersed in the above sequence. For example, in case of aircraft there will be taxing cycles and ground-air-ground cycles after each period (flight). They should not be applied randomly among the other cycles.

6.6. Clipping

The spectrum is a statistical average of previous load experiences. A level j may appear to occur e.g. 10 000 times. It would cause no surprise if it would actually occur 9900 or 10 100 times in a future case. There will be a certain level in the stress history that occurs only once. Since this is an extreme statistical number, it is very well possible that the level is never reached in service. For example, it would be quite natural if level 6 in Figure 6.20 and Table 6.2 were never reached. Two high loads might still occur, but only go as high as level 5. In that case there would be twenty-one cycles to level 5 instead of eighteen to level 5 plus three to level 6. The total number of cycles would not change, but the level 6 cycles would be 'clipped' to level 5. Also the exceedance diagram would be clipped as shown in Figure 6.21.

Since the highest stresses cause most of the retardation, clipping can cause dramatic effects on crack growth [8]. In the older literature, clipping is often called 'trucation', but presently the word truncation is reserved for the procedure discussed in the following section. Note that clipping merely reduces the magnitude of the highest loads to the clipping level; no loads are omitted, as shown in Figure 6.21b, insert.

The first tests on the effect of clipping were performed by Schijve [8] using aircraft flight simulation loading. Figure 6.22 presents a summary of his test results. Various clipping levels were used by reducing the size of the largest cycles–which are small in number–to the size of the next highest level (no cycles omitted). In tests with lower clipping levels all stresses above the clipping level were reduced in magnitude to the clipping levels. The exceedance diagram in Figure 6.22 shows how many cycles would be affected (up to about 800 at the very lowest clipping level, out of a total of over 300 000). According to Figure 6.22 clipping of the two highest levels (affecting only 80 of the 200 000 cycles), already reduced the crack propagation lives by almost a factor of about 2. It appears that crack propagation may be faster than expected if the structure encounters less severe service loading than was foreseen. The effect of clipping on crack growth can be demonstrated also in crack growth analysis as shown

Symbol			Spectrum	Linear Analysis (Flights)
a	▲	Willenborg	Fighter	270
	△	Wheeler, 2.3		
b	●	Willenborg	Trainer	460
	○	Wheeler, 2.3		
c	■	Willenborg	B-I class bomber	140
	□	Wheeler, 2.3		
d	▼	Willenborg	C-transport	1270
	▽	Wheeler, 2.3		

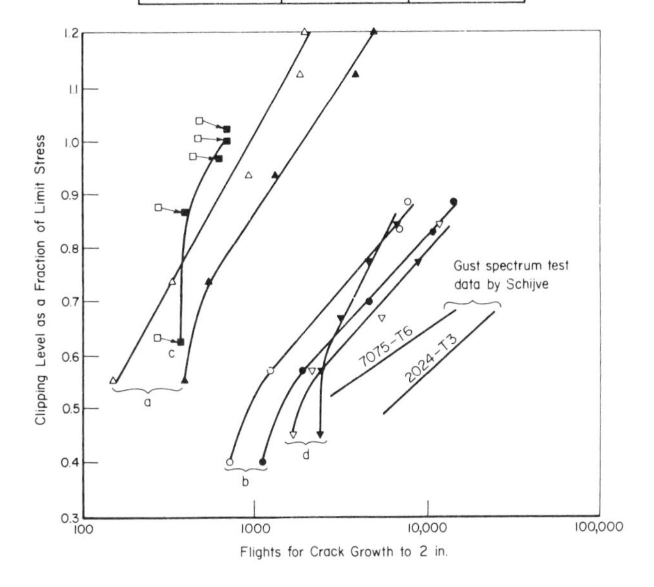

Figure 6.23. Effect of clipping [9], Courtesy Eng. Fract. Mech.

in Figure 6.23. These are computational results [9] for four different spectra. Clearly, clipping has an effect only if there is retardation; in an analysis without retardation the differences would not be noticeable.

As discussed above, it should be expected that clipping occurs in service. The exceedance diagram is a statistical average, and loads that are occurring only a few times might reach to a slightly lower level only. Should this occur then crack growth would be faster; if the analysis did not account for clipping the crack growth in service might be much faster than predicted. It is sometimes argued that clipping is unrealistic and that all those load levels should be included that may be anticipated to occur in service. The latter part of the argument is crucial; if they indeed occur, they should be included. However, whether they will occur is questionable. The spectrum is only a conjecture or, at best, an interpretation of measured data. Slight variations of the spectrum may be unimportant in the lower part, but they are very significant in the upper part.

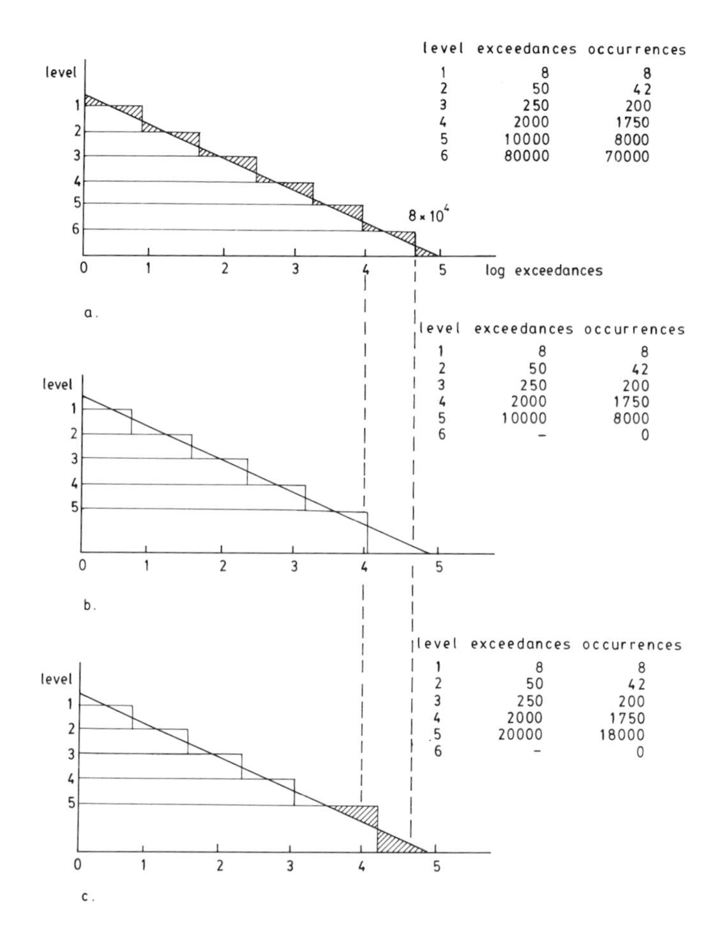

Figure 6.24. Truncation. (a) Complete spectrum; (b) Improper truncation (omission of lower level); (c) Proper truncation at level 5.

Clipping of the spectrum should be a factor for serious consideration in crack growth analysis when retardation is taken into account. It is easy to produce optimistic crack growth curves if high enough stresses are included (Figures 6.22 and 6.23), but the objective of the analysis is to obtain realistic information. Engineering judgement is the only guideline for the selection of an appropriate clipping level. A reasonable level might be the one that is exceeded ten times. Although this level is often selected, it is too arbitrary. Depending upon the clipping level, and upon the retardation properties of the material and the spectrum shape, the life may vary by a factor of two or three as shown in Figures 6.22 and 6.23) so that a categorical selection of ten exceedances would still be giving disputable results for some spectra and some materials.

The best solution to the problem is to perform multiple computer runs. A well-designed computer code will have options for automatic clipping. Once the preliminary work is done, multiple runs do not require any further labor other than by the computer. Multiple runs using different clipping levels will establish the sensitivity to clipping in any particular case. If the effect is small, no special problems arise, but if it is large the upper and lower bound of the crack growth curve can be established, as well as an average.

6.7. Truncation

It takes as much computer time (and testing time) to deal with one small cycle as with one large cycle. Thus the small cycles at the lower end of the exceedance diagram consume most of the time (cost) while their effect may be very small. Therefore, it would be advantageous if their number could be reduced. This is called truncation.

In essence the spectrum approximation by discrete levels already causes truncation of lower stressess. Figure 6.24a shows truncation from 100 000 exceedances to 80 000 due to the selection of the lower level. Further truncation could be achieved by raising the lower level to level 5. This is sometimes understood to mean elimination of all levels 6 as shown in Figure 6.24b. In that case 70 000 cycles would be simply thrown out without any account of their effect. This is an improper procedure.

True truncation involves reconstruction of the lower step as shown in Figure 6.24c. This is in accordance with the stepwise approximation of the idagram which is known to be legitimate. In this way the 70 000 cycles of level 6 are replaced by an additional 10 000 cycles of level 5, and as such they are still accounted for in a manner consistent with the entire procedure. Yet there is a savings of 60 000 cycels. The total cycles are reduced from 80 000 to 20 000 and the computation time is reduced accordingly by 75%. A proper truncation procedure (not elimination) must be included in software.

Truncation requires judgement and it is recommended that the effect of truncation on analysis results be evaluated by making different computer runs to determine whether truncation is permissible (giving the same results as the full history). The larger the number of stress levels the less the effect of truncation on crack growth. Hence, truncation is better justifiable if the user selects e.g. 12–16 levels instead of the smallest possible number of levels. Figure 6.25 and Table 6.3 show that proper truncation is indeed permissible (conservative), improper truncation leading to unconservative results.

If truncation is simply understood to mean the elimination of small cycles, the effect on the analysis can be unconservative, as shown in Figure 6.25b. Then concerns about the effect of truncation are legitimate, and a sensitivity analysis should be performed. On the other hand, if done properly, true truncation is

Table 6.3. Stress levels in case of truncation (compare Figure 6.24. Results potted in Figure 6.25). Unit of stress is ksi

Case 1; no truncation; 10 levels

Exceedances cycles	Occurr. cycles	Min. str.	Max. str.
1	1	−5.000	25.000
4	3	−3.500	23.500
10	6	−2.000	22.000
25	15	−0.500	20.500
63	38	1.000	19.000
158	95	2.500	17.500
398	240	4.000	16.000
999	601	5.500	14.500
2511	1512	7.000	13.000
6309	3798	8.500	11.500

Case 2; proper truncation; 9 levels

Exceedances cycles	Occurr. cycles	Min. str.	Max. str.
1	1	−5.000	25.000
4	3	−3.500	23.500
10	6	−2.000	22.000
25	15	−0.500	20.500
63	38	1.000	19.000
158	95	2.500	17.500
398	240	4.000	16.000
999	601	5.500	14.500
3980	2981	7.000	13.000

Case 3; improper truncation; 9 levels

	Occurr. cycles	Min. str. /load	Max. str. /load
	1	−5.000	25.000
	3	−3.500	23.500
	6	−2.000	22.000
	15	−0.500	20.500
	38	1.000	19.000
	95	2.500	17.500
	240	4.000	16.000
	601	5.500	14.500
	1512	7.000	13.000

All cycles below 13 ksi omitted (compare case 1).

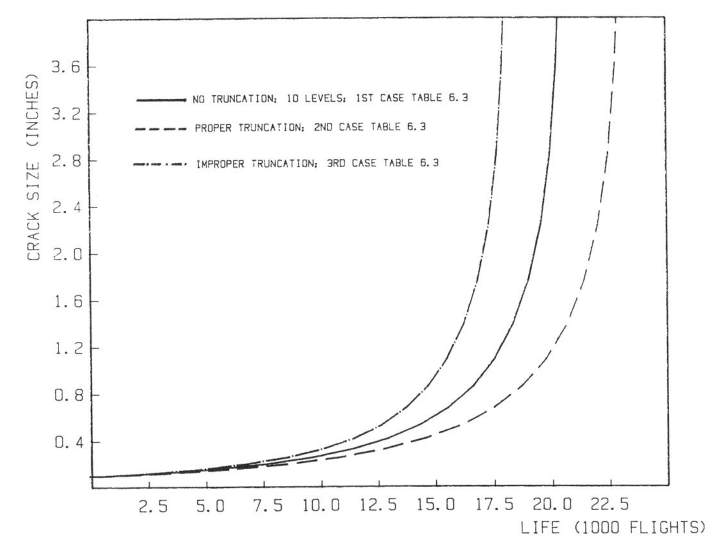

Figure 6.25. Effect of proper and improper truncation (unconservative) on predicted crack growth. Warning: curves for proper and improper truncation are inversed in above figure.

perfectly legitimate as can be seen from Figure 6.25. A well designed computer code includes provisions for proper truncation, the truncation being performed automatically upon user specification of the truncation level. The issue is then of little importance. If the computer code understands truncation to be the elimination of the lower levels multiple runs will always be necessary to prove that this 'improper' truncation is justifiable. Moreover, multiple runs must then be made in every case, because the effect of the 'improper' truncation depends upon spectrum shape, material and geometry.

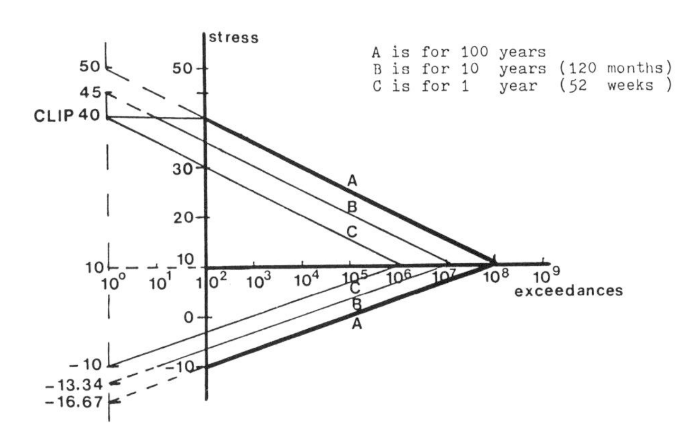

Figure 6.26. Manipulation of Spectrum.

Table 6.4. Stress history for 100 years

Max. stress	Min. stress	Steady stress	Exceedances
50.00	−16.67	10.00	1
42.00	−11.34	10.00	40
34.00	−6.00	10.00	1 535
26.00	−0.67	10.00	63 096
18.00	4.67	10.00	2 511 876
10.00	10.00	10.00	100 000 000

1 unit of stress is 1.00 ksi.

History for 100 years 1 unit of stress = 1.00 ksi

Max. str.	Min. str.	Occurr	Years 79	Years 16	Years 3	Years 1	Years 1
40.0	−16.7	1	0	0	0	0	1
40.0	−14.4	9	0	0	1	2	4
40.0	−12.2	36	0	1	3	4	7
40.0	−10.0	169	1	4	4	5	8
36.7	−7.8	785	7	10	13	12	20
33.3	−5.6	3 642	35	40	46	38	60
30.0	−3.3	16 903	164	181	204	170	268
26.7	−1.1	78 455	764	830	939	780	1 222
23.3	1.1	364 156	3 550	3 840	4 341	3 604	5 639
20.0	3.3	1 690 264	16 480	17 818	20 134	16 712	26 141
16.7	5.6	7 845 510	76 493	82 705	93 460	77 572	121 331
13.3	7.8	36 415 624	355 052	383 875	433 780	360 038	563 138

6.8. Manipulation of stress history

The stress history generated in the manner discussed in Section 6.5 may become a little awkward if the number of exceedances is very small or the number of flights, time, voyages etc., covered by the diagram is very large. The best results are obtained when the total exceedances are on the order of 2 000 to 100 000, and the number of periods (voyages, months, etc.) on the order 50–1000. Therefore it may be advantageous to reduce exceedance diagrams for smaller or larger numbers to the above ranges.

For the following discussion all exceedance diagrams are presumed to start at one exceedance. If this requires extrapolation to undesireable higher stresses, the spectrum should be clipped at the desired level. Examples of how the exceedance diagram can be manipulated to obtain a realistic stress history follow below.

Let the exceedance diagram A in Figure 6.26 be for 100 years of usage (1200 months or 5200 weeks). The diagram starts at 100 exceedances and ends at 100 000 000 exceedances. Extrapolation to one exceedance brings the highest stress level to 50, which cannot be justified. The problem can be rectified by

Table 6.5. Stress history 1200 months
Exceedance table for 1200 months

Max. stress	Min. stress	Steady stress	Exceedances
50.00	−16.67	10.00	1
42.00	−11.34	10.00	40
34.00	−6.00	10.00	1 585
26.00	−0.67	10.00	63 096
18.00	4.67	10.00	2 511 876
10.00	10.00	10.00	100 000 000

1 unit of stress is 1.00 ksi

History for 1200 months 1 unit of stress = 1.00 ksi

Max. str.	Min. str.	Occurr	Months 959	Months 192	Months 39	Months 9	Months 1
40.0	−16.7	1	0	0	0	0	1
40.0	−14.4	9	0	0	0	0	9
40.0	−12.2	36	0	0	0	2	18
40.0	−10.0	169	0	0	3	3	24
36.7	−7.8	785	0	3	4	4	16
33.3	−5.6	3 642	2	6	11	11	43
30.0	−3.3	16 903	13	17	22	24	97
26.7	−1.1	78 455	63	72	82	79	305
23.3	1.1	364 156	295	328	356	342	1 313
20.0	3.3	1 690 264	1 373	1 509	1 634	1 567	5 999
16.7	5.6	7 845 510	6 374	7 003	7 571	7 253	27 722
13.3	7.8	36 415 624	29 587	32 500	35 132	33 660	128 603

clipping the spectrum at 40. Using a stress history with different periods of years (i.e. years with different cycle content) leads to the results in Table 6.4. The period 'year' may be too long, and it may be more realistic to construct 1200 different months, or even 5200 different weeks. This would result in stress histories as shown in Tables 6.5 and 6.6. Obviously, the latter two stress histories are not realistic, because too many cycles are concentrated in one period (month or week). A much better result can be obtained by using a spectrum for 10 years (i.e. 120 months), which is spectrum B in Figure 6.26. Note that B and A are still the same spectrum; a factor of ten in exceedances is made up for by a ten times smaller time interval. As can be seen from Table 6.7 a more realistic stress history is obtained. If weeks are used as a period, a spectrum for one year (52 weeks) can be used, which is spectrum C in Figure 6.26. The resulting stress history is shown in Table 6.8. Note that the highest stress is now 40 without clipping.

It is advisable to reduce the spectrum (or stress table) to one that covers 50–1000 periods in the above manner, in order to arrive at a realistic stress history. All of the above may seem a little artificial, but it should be realized that

Table 6.6. Stress history for 5200 weeks
Exceedance table for 5200 weeks

Max. stress	Min. stress	Steady stress	Exceedances
50.00	−16.67	10.00	1
42.00	−11.34	10.00	40
34.00	−6.00	10.00	1 585
26.00	−0.67	10.00	63 096
18.00	4.67	10.00	2 511 876
10.00	10.00	10.00	100 000 000

1 unit of stress is 1.00 ksi

History for 5200 weeks 1 unit of stress = 1.00 ksi.

Max. str.	Min. str.	Occurr	Weeks 4159	Weeks 832	Weeks 167	Weeks 41	Weeks 1
40.0	−16.7	1	0	0	0	0	1
40.0	−14.4	9	0	0	0	0	9
40.0	−12.2	36	0	0	0	0	36
40.0	−10.0	169	0	0	0	3	45
36.7	−7.8	785	0	0	3	5	78
33.3	−5.6	3 642	0	3	5	5	105
30.0	−3.3	16 903	3	4	5	4	98
26.7	−1.1	78 455	14	18	24	23	302
23.3	1.1	364 156	68	76	84	75	1 009
20.0	3.3	1 690 264	316	352	386	347	4 466
16.7	5.6	7 845 510	1471	1617	1774	1597	20 542
13.3	7.8	36 415 624	6827	7511	8231	7398	95 084

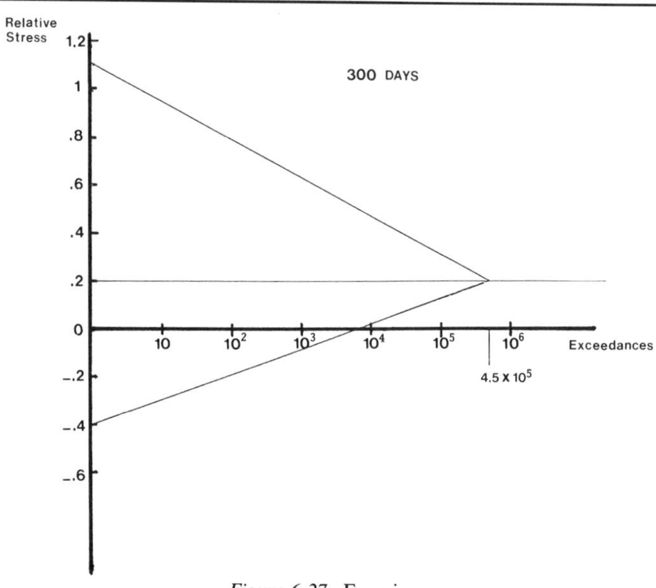

Figure 6.27. Exercises.

Table 6.7. Stress history for 120 months
Exceedance table for 120 months

Max. stress	Min. stress	Steady stress	Exceedances
45.00	– 13.34	10.00	1
38.00	– 8.67	10.00	25
31.00	– 4.00	10.00	631
24.00	0.66	10.00	15 849
17.00	5.33	10.00	398 106
10.00	10.00	10.00	10 000 000

1 unit of stress is 1.00 ksi

History for 120 months 1 unit of stress = 1.00 ksi

Max. str.	Min. str.	Occurr	Months 95	Months 19	Months 4	Months 1	Months 1
40.0	– 13.3	1	0	0	0	0	1
40.0	– 11.4	7	0	0	0	2	4
39.2	– 9.5	21	0	0	3	3	6
36.2	– 7.5	81	0	3	3	4	8
33.3	– 5.6	312	2	4	7	6	11
30.4	– 3.6	1 194	9	13	14	13	22
27.5	– 1.7	4 574	37	41	43	42	66
24.6	0.3	17 524	142	157	163	155	243
21.7	2.2	67 138	545	599	621	583	1 914
18.7	4.2	257 216	2 089	2 291	2 376	2 233	3 495
15.8	6.1	985 446	8 006	8 770	9 086	8 541	13 361
12.9	8.1	3 775 436	30 675	33 591	34 800	32 713	51 169

the purpose of crack growth analysis is to make a projection in the future, so as to exercise fracture control (Chapter 11). As the future is not known in detail, any reasonable projection that accounts for the salient features of the load history is as good or better than any other. Furthermore, such seemingly trivial things as clipping may have a greater influence on the predictions than all other painstaking efforts for accuracy and realism. Clipping levels are often selected on the basis of a mere assumption, yet unduly 'sophisticated' (read time consuming and complex) procedures are often used to generate e.g. flight-by-flight stress histories. All of the latter's 'sophistication' can be overridden by one 'simple' assumption with regard to clipping. More often than not, the 'sophisticated' stress history is than applied randomly with total disregard for the fact that high loads are clustered as discussed extensively in this chapter.

It should be clear from the foregoing discussions that there are only a few issues that count in the generation of a stress history. These are:

(a) Different periods of severity (flights, voyages, etc.) must be used and applied semi-randomly if in practice the loading is semi-random. Random

Table 6.8. Stress history for 52 weeks
Exceedance table for 52 weeks

Max. stress	Min. stress	Steady stress	Exceedances
40.00	−10.00	10.00	1
34.00	−6.00	10.00	16
28.00	−2.00	10.00	251
22.00	2.00	10.00	3 981
16.00	6.00	10.00	63 095
10.00	10.00	10.00	1 000 000

1 unit of stress is 1.00 ksi

History for 52 weeks 1 unit of stress = 1.00 ksi

Max. str.	Min. str.	Occurr	Weeks 41	Weeks 8	Weeks 2	Weeks 1
40.0	−10.00	1	0	0	0	1
37.5	−8.3	5	0	0	1	2
35.0	−6.7	12	0	1	1	2
32.5	−5.0	38	0	3	3	8
30.0	−3.3	122	2	3	3	9
27.5	−1.7	385	7	8	8	17
25.0	0	1 216	22	27	25	47
22.5	1.7	3 845	71	80	76	142
20.0	3.3	12 159	225	254	234	434
17.5	5.0	38 451	714	795	732	1 353
15.0	6.7	121 593	2258	2514	2314	4 275
12.5	8.3	384 511	7140	7953	7318	13 511

application of stresses derived by complicated means will negate all complicated efforts to determine stress histories.

(b) Deterministic loads must be applied at the point where they occur. For example g.a.g. cycles must occur between flights; random application may defy all other sophisticated procedures.

(c) A reasonable number of stress levels (10–12 positive and negative) must be selected. More levels will complicate the procedure without improving the usefulness of the results.

(d) Largest positive and negative excursions must be combined, and so on. Random combinations will require subsequent counting, the result of which can be foreseen, while the stress history was based on an already counted history in the first place.

(e) The total number of periods and cycles must be in accordance with the exceedance diagram.

The above criteria account for what may be called the signature of the loading. Small changes in these, including clipping, will usually have more effect on crack growth than any complex means of establishing stress levels.

Hypothesize for a moment that it would be necessary to use a complicated procedure to establish e.g. the loads or stresses in every segment of a flight (ascend, cruise, descend etc.) or of a voyage of a ship (river, port, locks, cruise, etc.). Any such procedure would be full of assumptions, and it needs no explanation that no airplane or ship would ever encounter the assumed circumstances. If it were necessary to go through this 'sophisticated' procedure, it would only be because the predicted crack growth would be different if other assumptions were made. But in that case the prediced crack growth would have no relevance to the service behavior in the first place. Actual service loading would be different from the assumptions made, and if the results were that critical to the assumptions the predictions would be useless, regardless of all sophisticated procedures used. The simple fact that one would consider the result usefull nevertheless, implies that one makes the assumption that it does not really matter whether the loads are as derived from this sophisticated procedure. If this is the case, why would the 'sophisticated' procedure be necessary in the first place. Surprizingly, if such complicated procedures are used, the stresses are subsequently applied in random sequence, while they should be semi-random and clipping is ignored. All apparent sophistication is then thrown overboard.

It is very easy to perform many computer runs with different stress levels, simple and sophisticated stress histories, etc. If the results do come out about the same (which they do), the above point is proven. One might ask whether tests would prove the same. Not enough test data are available to conclude that they do. But, after all, crack growth prediction are made by analysis, and hence, only the analysis results count. If the computer cannot distinguish the difference, the results apply.

Next hypothesize that there would be a difference, depending upon the sophistication of the stress history generation. Then of course, also the behavior in service would be different. In other words: no matter how sophisticated the stress history generation, it would have no bearing upon the service behavior, and all sophistication would be a waste. The mere fact that it is assumed that the stress history generated will provide data of relevance to service behavior implies the assumption that stress history used is not overly important. That being the case, any such sophistication is implicityly making the assumption that the sophistication is not necessary in the first place: and if the sophistication were necessary, no level of complication would produce a useful answer.

The above discussion shows that procedures to derive stress histories in a complicated way are defying themselves. If they were necessary, they would be useless by implication. If they are not necessary, they are useless from the start.

It must be concluded that only those effects count that could make a significant difference in the calculated crack growth. Those are the effects listed above concerning the signature of the loading and clipping. They have to be accounted for as was demonstrated by test and analysis results. All others are secondary; they complicate procedures without adding to their usefulness.

A final problem in the generation of the stress history is in the definition of stress. Should one use nominal stress, local stress, or hot spot (welded structures) stress? Should secondary stresses, residual stresses and dead stresses be accounted for, and if yes, how? To a certain degree, this can be dealt with through the baseline data used in the analysis. However, there is more here than meets the eye, the main problem being that local plasticity during the highest load cycles tends to make even the most sophisticated load-interaction models go awry. This problem can be solved only by pragmatic engineering judgement.

6.9. Environmental effects

If there is load- or load-environment interaction it is especially important that clustering of high loads in storms is recognized by comparing analyses for fully random and clustered loading (mild weather or storms). Computer runs with similar clustering, but with a totally different sequence, should provide essentially the same result.

For marine structures especially, the effect of environment should receive ample consideration. Quite obviously, the main environments to consider are 'salt air', sea water, and the 'splash zone'. The question arises: if going from one cycle to the next (e.g. high ΔK – low ΔK or vice versa) in successive cycles, does the new rate immediately fall in place with the (constant amplitude) baseline data. (In other words: is there no environment interaction).

Models for environment interaction have been proposed. As the environmental effect is time dependent, these models bring in the element of time, or the frequency of the loading. They are certainly of scientific interst, but the practical problem is one of load (retardation) and environment interaction. As the latter problem has not been addressed at all, the complication of accounting for environment equilibrium is hardly worthwhile for practical crack growth analysis. Thus, the models mentioned above remain in the realm of research and 'theory' (hypothesis) and are not too relevant to practical applications as discussed in this book at this time. A pragmatic approach calls for the assumption that chemical equilibrium is indeed established immediately. In that case, the problem is solved by using crack growth rate data (da/dN-ΔK) for the relevant environment. Retardation effects are accounted for as discussed. Retardation parameters must be determined empirically, and therefore, these will automatically account for any chemical interaction effects, included in the test data used for calibration.

6.10. Standard spectra

So-called 'standard' spectra have been developed in Europe for general use for a variety of structures. The word standard is to be interpreted to mean a general norm based upon a very great number of measurements; it is not meant to be a design standard or specification. These standard spectra were intended primarily for use in tests, so that results of various experimental investigations might be better comparible. There are standard spectra [12–19] for airplanes (Falstaff and Twist), for helicopters (Helix and Felix), for offshore structures (Wash), etc.

Stress histories for these spectra were obtained by using algorithms very similar to those discussed in the previous sections using semi random loading with periods of different severity. In essence, not the exceedance diagrams but the stress histories derived in this manner are considered to be the standard. A limitation of the standard spectra is that they essentially always perform in the same way; this is useful for testing and data comparisons, but may be too severe a limitation for practical use.

Using the standard stress histories in a crack growth analysis, would necessitate input of all stress cycles in the sequence generated, and thus require special input facilities. Fortunately, the algorithm as discussed here (which is somewhat simpler than the one used for the standard histories), can be rather easily incorporated in a computer code for crack growth analysis. In that case only a few data points describing the exceedance diagram would be needed as input, upon which the analysis code would generate a load history, perform clipping and truncation as prescribed, and subsequently the crack growth calculation. In this manner, the standard spectra would be more useful and more easy to use. Their exceedance diagrams could be applied for crack growth analysis if a specific exceedance diagram were lacking. Most stresses are proportional to loads, so that the stress levels in different parts of the structure can be obtained through a multiplication factor. The stress levels are given as relative numbers; a conversion factor is sufficient to determine all stress levels. Hence, judiciously used, the standard exceedance diagrams may provide spectrum information for damage tolerance analysis where none is available. Suitable crack growth analysis software only needs input of the exceedance diagram.

6.11. Exercises

1. Approximate the exceedance diagram of Figure 6.27 by 8 equally spaced positive and 8 negative stress levels for equal exceedances as the positive levels. Determine the number of occurrences of each level. Assuming that the 100% level represents a stress of 15 ksi, determine the stress ranges for each level by combining positive and negative excursions of equal frequency of occurrence.

2. Repeat exercise 1 by constructing 6 levels starting with the exceedances: 2, 10, 100, 1000, 10 000, 100 000. If the stresses of exercise 1 and 2 were used in a crack growth analysis, would there be a difference in results?

3. Using the results of Exercise 1, generate a stress history with five 'periods' of different severity.

4. Truncate the spectrum of Exercise 1 properly at the lowest level but one. Generate a new stress history with five different 'periods'. Next use the improper truncation procedure discussed in the text and compare the results of the two procedures (Note that for the latter part the results of Exercise 1 can be used directly; only for the first part a new history must be generated.)

5. How would clipping to the highest level but one change the stress histories of exercise 4?

6. Repeat Exercises 1, 3, 4, 5 using the spectrum of Figure 6.27 and by selecting 7 levels.

7. Change the stress history developed in Exercise 3 for the case that the maximum level represents a stress of 21.5 ksi.

8. Why is clipping not important if there is no retardation?

References

[1] M.E. Mayfield et al., *Cold leg integrety evaluation*, USNRC Report NUREG/CR-1319, February 1980.

[2] International group on Crack Growth in Nuclear, *Structures* (ICCGR).

[3] J. Schijve, *The analysis of random load-time histories with relation to fatigue tests an life calculations*, Fatigue of Aircraft Structures, p. 115, Pergamon (1963).

[4] J.B. de Jonge, *The monitoring of fatigue loads*, ICAS congress, Rome (1970), Paper 70–31.

[5] G.M. van Dijk, *Statistical load data processing*, ICAF symposium Miami (1971).

[6] D. Broek et al., *Fatigure strength of great-lakes ships*, Battelle Rept to Am. Bur. Shipping (1979).

[7] Private communication.

[8] J. Schijve, Cummulative damage problems in aircraft structures and materials, *The aeron. J.* **74** (1970) pp. 517–532.

[9] D. Broek and S.H. Smith, Fatigue crack growth prediction under aircraft spectrum loading, *Eng. Fract. Mech.* 11 (1979) pp. 122–142.

[10] D. Broek and R.C. Rice, Prediction of fatigue crack growth in railroad rails, *SAMPE Nat. Symposia*, **9** (1977) pp. 392–408.

[11] N.I. Bullen, *The chance of a rough flight*, Royal Aircraft Est. TR 65039 (1965).

[12] G.M. van Dijk and J.B. de Jonge, *Introduction to a fighter aircraft loading standard for fatigue evaluation*, FALSTAFF, NLR-Report MP 75017 (1975).

[13] J.B. de Jonge, D. Schuetz, H. Nowack, and J. Schijve, *A standard load sequence for flight simulation testing*, NLR- Report TR 73029, or LBF-Report FB – 106, or RAE-Report TR 73183 (1973).

[14] J.J. Gerharz, *Standardized environmental fatigue sequence for the evaluation of composite components in conbat aircraft* (*ENSTAFF*), Lab. für Betriebsfestigkeit, Fraunhofer Inst. für Betriebsfestigkeit FB-179 (1987).

components in conbat aircraft (ENSTAFF), Lab. für Betriebsfestigkeit, Fraunhofer Inst. für Betriebsfestigkeit FB-179 (1987).

[15] G.E. Breithopf, *Basic approach in the development of TURBISTAN, a loading standard for fighter aircraft engine disks*, ASTM conference Cincinati OH (1987).

[16] M. Huck and W. Schutz, *A standard load sequence of Gaussian type recommended for general application in fatigue testing*, Lab. für Betriebsfestigisellschaft IABG rept TF-570 (1976).

[17] J. Darts and W. Schutz, *Helicopter fatigue life assessment*, AGARD-CP-297, (1981), pp. 16.1 - 16.38.

[18] P.R. Edwards and J. Darts, *Standardized fatigue loading sequence for helicopter rotors (HELIX and FELIX)*, Royal Aircraft Establishment RAE TR 84084, Part 1 and 2, Augutst 1984.

[19] W. Schutz *Standardized stress-time histories–An overview*. ASTM conference, Cincinnati OH (1987).

CHAPTER 7

Data interpretation and use

7.1. Scope

Material data are an essential input to all fracture and crack growth analysis; without applicable data analysis is not possible. Misinterpretation and misuse of data are major contributors to the acclaimed shortcomings of fracture mechanics, because the data interpretation problem is not a trivial one, especially where it concerns fatigue crack growth. The phrase: 'data are data, and cannot be argued with', is commonly misapplied. The statement may be true for the raw data, i.e. a load-COD curve, or a measured crack growth curve, but cannot be applied to the derived data, K_c, K_{Ic}, J_R, and $da/dN - \Delta K$. The latter are obtained from the raw data through an interpretation process full of assumptions such as the data reduction procedures stipulated in the relevant ASTM specifications [1–5]. Although these reduction procedures are probably the best available, they are not indisputable.

This chapter is not intended to argue the shortcomings of data reduction procedures. Rather, it is concerned with how these data are subsequently interpreted and used in the analysis. Questions arise regarding constraint, scatter, equation fitting, data errors and inaccuracies, retardation parameters, mixed or changing environments, etc. These are the type of problems addressed here, because they may affect the accuracy of the damage tolerance analysis, more than the shortcomings of fracture mechanics.

Problems in the use of toughness data, both in terms of K and J, will be briefly addressed first. As the toughness affects the calculated permissible crack size, a_p, inaccuracies change a_p only. From the point of view of fracture control this is often not very important as will be discussed in Chapter 11. A small change in a_p does not affect H much in general, unless a_p is already small – especially when in rare cases it is below the detection limit, or where it affects arrest and leak-before break (Chapter 9). Errors in rate data on the other hand, affect H directly. A factor of two difference in rates (which is not uncommon) changes the period H by a factor of two, which is significant. For this reason, the main emphasis in this chapter is on the interpretation and use (estimation sometimes) of fatigue rate data.

In many cases the analysis must make use of data provided in handbooks [6] or the general literature [7], or of data in magnetic files in software libraries. These are almost always reduced data, as opposed to raw data: they have been manipulated. Often some interpretation has been done as well, such as averaging or curve fitting. The user of such data should not overlook this 'triviality', and if possible, check whether reduction and interpretation was appropriate for the application envisaged. In practically all cases a decision must be made on how to deal with scatter, in particular for rate data. Data for the exact condition or alloy at hand (regarding temperature, environment, material direction, R-ratios, etc.) often are not available. They then must be estimated on the basis of available data for similar alloys or circumstances. These kinds of problems are addressed in this chapter as well. Recipes cannot be given; only guidelines can be provided. In the end the user must exercise engineering judgement, possibly honed by the following discussions.

No attempt will be made to provide any material data here as this is not a materials handbook. Considering that the DT-handbook [6] consists of thousands of pages, any attempt to present data here would be inadequate and selective. Real data will be used for illustrative purposes; in some cases hypothetical data will be used to better demonstrate a specific point.

7.2. Plane strain fracture toughness

The plane strain fracture toughness, K_{Ic}, is commonly obtained from a standard test [1] on a compact tension (CT) specimen. Standardization can be defended on many grounds (also the tensile test and the hardness test are standardized). Does this mean that data obtained from non-standard tests are inadmissable? If this question had to be answered affirmatively, the result of the test would be of no use in the first place. It is assumed that the test result, K_{Ic}, can be used for the prediction of fracture in a structure, on the basis of the similitude argument that fracture takes place at the same value of K as in the laboratory test. Hence, if a specimen other than the standard could not be used, the implied assumption would be that the other configuration does not fracture at K_{Ic}; ergo, it would be impossible to predict fracture in a structure on the basis of the K_{Ic} from a standard test, which would render the standard test useless as well.

The concept of fracture mechanics is that fracture in plane strain occurs when $K = K_{Ic}$ regardless of the configuration. Therefore, K_{Ic} is obtainable from a test on any configuration, provided one knows the expression for K, or rather for β. Similarly, fracture can be predicted for any configuration for which the expression for β is known. The main justification for the standard test is that β for the CT specimen has been calculated to great accuracy, for straight-front through-the-thickness cracks. The standard declares a test invalid if the fatigue crack front is not straight. This is because the β-expression provided is not valid for such a crack, but fracture still took place at $K = K_{Ic}$, and the test result could still be used if β were known for the actual crack shape.

The test also is declared invalid when the thickness is less than 2.5 $(K_{Ic}/F_{ty})^2$. As was shown already in Chapter 3, the factor of 2.5 was determined by committee agreement on the basis of test data, but it appears from Figure 3.8 that a factor of 2 or 4 would be just as defendable. Yet, the number of 2.5 appears in the standard, which does not make the number indisputable and certainly not sanctimonious. The 'candidate toughness' obtained for a test where the factor is 2.4 is not by definition worse than the one obtained for a test where the factor is 2.7. As constraint depends upon the yield strength and K, the factor in reality will depend upon F_{ty} and K_{Ic} and upon configuration (i.e. it is material dependent and configuration dependent).

It should be pointed out also, that the toughness obtained when the factor is e.g. 2 may not be declared a valid plane strain toughness by the standard, but it is the toughness for the thickness at hand. For the given thickness the number is at least as reliable as K_{Ic} obtained in a valid test. Thus, if one is not interested in K_{Ic} per se, but in the toughness for the chosen thickness, the result is useful. If one insists on knowing the plane strain toughness the test must be 'valid', but one may want to exert caution with regard to the factor 2.5.

As pointed out the required factor for plane strain is likely to be material dependent. Almost certainly, it is also geometry dependent. This can be appreciated from the data [6] for large center cracked panels shown in Figure 7.1. The

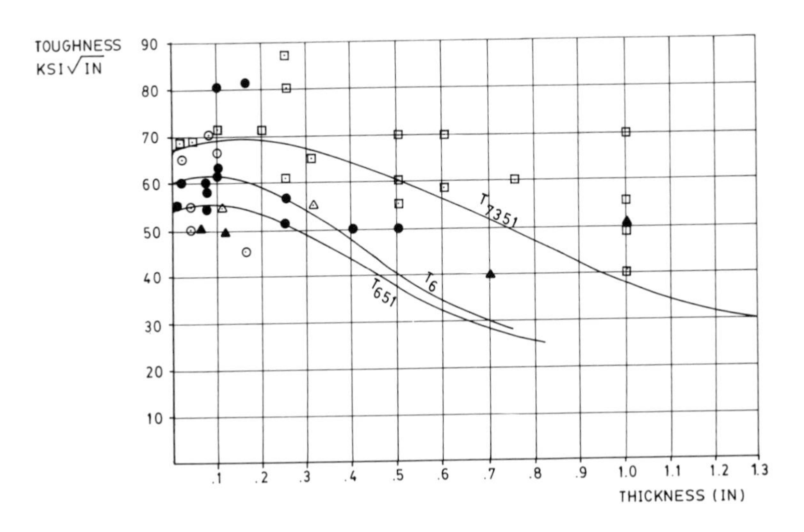

Figure 7.1. Toughness data [6] for 7075 Aluminum alloy (large center cracked panels).

same reference [6] quotes $K_{Ic} \approx 30\,\text{ksi}\sqrt{\text{in}}$ for the same material, obtained from a standard test. As the alloy's yield strength is $F_{ty} \approx 80\,\text{ksi}$, the standard would predict plane strain at a thickness of $2.5\,(30/80)^2 = 0.35$ inch. Yet, the center cracked panels of 0.35 inch thickness have a toughness much higher than $30\,\text{ksi}\sqrt{\text{in}}$. Similar discrepancies are found for other materials. Constraint in CT specimens is higher than in some other configurations. This is convenient because it permits the use of small CT specimens and tends to lead to conservative toughness values. But clearly, if the factor 2.5 is applied to the center cracked panel, a toughness of only $30\,\text{ksi}\sqrt{\text{in}}$ would be counted on for a plate of 0.35 inch thickness, while the actual toughness is as high as $50\text{–}60\,\text{ksi}\sqrt{\text{in}}$. For a given crack size this would be underestimating the residual strength by almost 50%, certainly not a negligible error.

Clearly, plane strain in center cracked panels has not been reached yet at a thickness equal to $2.5\,(K_{Ic}/F_{ty})^2$. At this thickness the toughness is higher (transitional). Plane strain occurs at a thickness of 4–5 times $(K_{Ic}/F_{ty})^2$. Besides, there is another reason why the center-cracked panel behaves differently, namely the rate of change of $K(a)$ as explained below.

Every material, except a truly brittle one, exhibits a rising R-curve (Chapter 3), as shown in Figure 7.2. This is usually not a problem in a standard test on a CT specimen, because $G(a)$ and $K(a)$ rise very sharply with crack size in this specimen ($G = K^2/E$; Chapter 3), so that the beginning of fracture usually coincides with the instability. The standard prohibits excessive non-linearity of the load displacement diagram in order to exclude cases in which instability is not immediate [8]. In a center cracked panel $G(a)$ or $K(a)$ increase only moderately with a. Hence, fracture is first stable, and instability occurs at higher values of K, especially for larger cracks (Figure 7.2). Usually, the question of interest is how much load a structure can carry and at which stress it breaks. This is determined by the instability, and thus the K at instability is the relevant number, called K_c or K_{eff} (see also chapter 3).

Which cases warrant the use of a valid K_{Ic}? It is obvious from the above arguments that there is no categoric answer to this question. Engineering

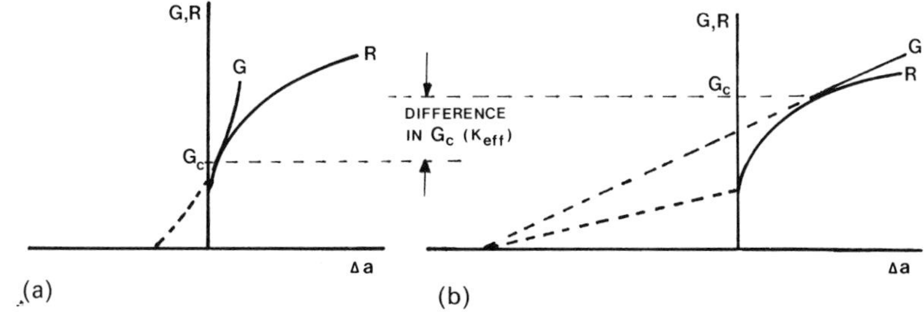

Figure 7.2. Different behavior of compact tension specimen and center cracked panel. (a) CT specimen; (b) Center crack.

judgement must be exercised unless conservatism is desirable. Cases where the use of K_{Ic} is always indicated concern surface flaws and corner cracks. It is emphasized that in these cases the thickness plays no role (Chapter 3). For the analysis of part-through cracks refer to Chapter 9.

Although possibly superfluous, it is pointed out again that even in plane strain, and regardless how low the toughness, the predicted residual strength for small cracks is too high, the predicted permissible (or critical) crack size for high stresses too large. For $a \rightarrow 0$, the predicted stress tends to infinity. Using EPFM in this regime (for an otherwise LEFM material) would not solve this problem either. An approximate solution must be used. This is amply discussed in Chapters 3, 4 and 10.

Finally, in using toughness data from handbooks, the relevant conditions (temperature and direction) should apply. Like most other material properties, toughness depends upon temperature. Engineers are used to the material properties being temperature dependent, but the (often) strong dependence upon direction of fracture and crack growth properties is sometimes not recognized. The going nomenclature for crack direction and material direction is shown in Figure 7.3. The first letter, L, T or S indicates the direction of the loading, the second letter (again L, T or S) the direction of the crack. Toughness values for the ST-direction may be 30–60% lower than for the LT direction. Not accounting for such effects may lead to errors. As the problem is most often associated with the growth of part-through cracks, the reader is referred to the more detailed discussion of the matter in Section 7.6.

Many times, toughness data for the temperature and direction of interest are not available. Section 7.5 provides some guidelines for estimating the toughness in such cases.

7.3. Plane stress and transitional toughness, *R*-curve

Data on plane-stress and transitional toughness should always be assessed with extreme caution. There are many ways in which raw data can be erroneously interpreted and derived data misquoted. There are equally many ways to misuse them.

Many K_c data were derived from panels of insufficient size, which renders them useless. As was shown in Chapter 3, when the toughness is high – which it usually is in plane stress – small panels will fail by net section collapse, i.e. at $K < K_c$. Thus the K_c derived from such a test is too low. It is sometimes quoted as an apparent toughness, K_{app}, but this practice is suggesting more than it should. Fracture due to collapse should be treated differently as discussed in Chapters 3, 4 and 10, and not on the basis of K_c nor K_{Ic}. A test on a panel of insufficient size yields no useful result other than F_{col}, and the knowledge that K_c is higher than the 'apparent K_{app}'. A good data handbook [6] quotes the panel

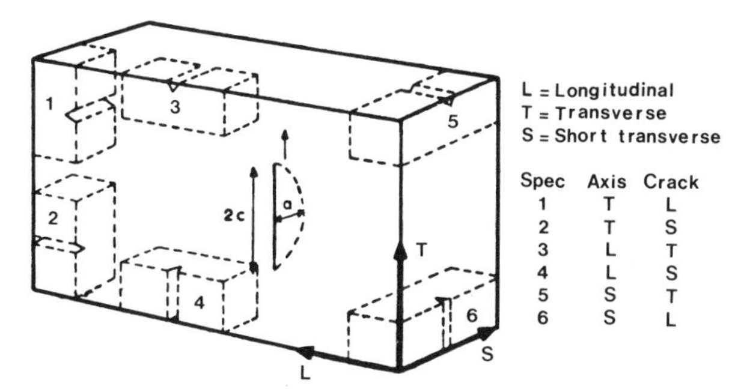

Figure 7.3. Nomenclature for loading and Crack Directions.

size as well as the raw data so that the user can readily check whether collapse or net section collapse (yield) occurred, and then make a judgement. If no panel size is quoted the numbers are suspect.

For example if the toughness provided is $80\,\mathrm{ksi}\sqrt{\mathrm{in}}$ for a 6-inch wide center cracked panel of a material with $F_{ty} = 60\,\mathrm{ksi}$, the number is useless. A 2-inch crack (a = 1 inch) in such a panel would provide $\sigma_{fr} = 80/1.07\sqrt{\pi a} = 42\,\mathrm{ksi}$, but failure by net section yield is at $\sigma_{fc} = (6 - 2)/6) \times 60 = 40\,\mathrm{ksi}$. Obviously, the test panel failed at net section yield. The value of this to the user is only that it is certain that $K_c \geqslant 80\,\mathrm{ksi}\sqrt{\mathrm{in}}$ and that all center cracked panels smaller than 6-inch will fail by collapse.

Another problem arises because of the definition of toughness. The value quoted may be the actual K_c or the effective toughness, K_{eff}, which is sometimes called K_c. As explained in Section 3.12 neither of these is actually a material property; both depend upon crack size and configuration, because they are derived for fracture instability. In principle, the actual K_c is not a very useful number: if the amount of stable fracture is unknown, the 'crack' size itself is unknown; the residual strength cannot be calculated. Conversely the calculated 'critical' crack size is always too large (Chapter 3). If some information on the amount of stable fracture is available K_{eff} can be estimated from K_c. Without such information, a safe assumption is that $K_{\mathrm{eff}} = 0.9\text{–}0.8\,K_c$. The use of K_{eff} is not without problems either, but it is the most rational solution short of using the *R*-curve. The criteria for panel size apply.

Using the *R*-curve and an instability analysis may seem to solve the difficulties with K_{eff}, but in practice it does not. First of all, *R*-curve data are scarce. More important, accurate measurement of *R*-curves are difficult. As a consequence, the inaccuracy of the data usually negates the advantages of the more refined procedure. Furthermore, collapse considerations apply to *R*-curves as well.

Like the plane strain toughness, K_c and K_{eff} depend upon temperature. Directional dependence is usually small (TL versus LT), because plane stress and transitional toughness always apply to through-the-thickness cracks and these do seldom, if ever, occur in ST direction. The problem of estimating the toughness where no data are available, is discussed in Section 7.5.

7.4. Toughness in terms of *J* and *J*

Toughness data for use with EPFM are scarce. Available data pertain mostly to materials used in nuclear power structures such as 304 SS, A-533B etc., many of which can be found in a series of reports issued by the U.S. Nuclear Regulatory Commission (NRC – NUREGs). Data for other materials are scattered throughout the literature. No systematic data compilation is as yet available.

Practically all data have been obtained from CT specimens and in almost all cases through the *J*-estimation procedure prescribed in the test standard [2]. The latter is based on the assumption of full net section yield, which by itself is subject to some doubt (Chapter 4). Although many estimation schemes for *J* have been devised, the only viable general procedure for the application of EPFM in engineering fracture analysis is through the use of $J = H\sigma^{n+1}a/F$, or equivalent, as discussed in Chapter 4 (excluding finite elements for routine fracture assessments). Since H is known for CT specimens, it would be better to obtain J_{Ic} and J_R data through the use of the above equation rather than from a doubtful approximation, especially if the data are to be used in this manner. Fortunately EPFM equations are so forgiving that the discrepancies seldom are of great relevance in the application (see Chapter 4, Figure 4.11). This has the additional advantage that the user need not seriously worry if the data appear to show large scatter; the scatter is logical and rather immaterial in the fracture analysis (Chapter 4).

Apart from scarcity there remain two significant problems with the toughness data. The first and most important is the constraint. All data are obtained from through the thickness cracks in relatively thin specimens. Often constraint is not made an issue, but in most tests there will not be plane strain. This does not matter if the data are to be used for through-the-thickness cracks in the same thickness. But most cracks are surface flaws or corner cracks where plane strain prevails. In such cases the toughness J_{Ic} and J_R will be less than for the test specimens. One can only speculate on the magnitude of the difference, but it would be reasonable to expect similar differences as in LEFM. Constraint criteria for plane strain have been proposed [9], but (in comparison with the LEFM plane strain conditions) these seem to be too optimistic (favoring plane strain in small thickness, as discussed in Chapter 4).

Another problem is that the J_R data are obtained from small specimens and hence, usually are only for small values of Δa (generally for Δa up to 0.2 inches at best). In large structures the amount of stable fracture will be considerably more than 0.2 inches (Figure 7.2 applies just as well to the J_R-curve). Analysis then can be performed only if the J_R curve is extrapolated. In view of the fact that the analysis is very forgiving, such extrapolations can be made rather easily without sacrificing much accuracy. Nevertheless, this is an unfortunate situation which should be corrected through a thorough revision of the standard test procedure with regard to specimen size requirements.

7.5. Estimates of toughness

As shown in the foregoing sections, estimates of toughness are often necessary; even if data are available they may not be for exactly the circumstances prevailing in the structure. Considerations upon which such estimates might be based will be given, but the actual estimate remains the responsibility of the user. If no data at all are available, the problem of estimating a reasonable toughness value, J_R or R-curve becomes more difficult; the success of the estimate depends upon the availability of comparable data. The procedures discussed below lead to ROUGH estimates; they are not recommended for general use. When there is a lack of data, tests are always preferable.

Consider a structure built of alloy Y, and suppose that a handbook provides data for alloy X for LT and TL direction, and for alloy Y for LT, while the value for the TL direction is sought. If it can be ascertained that alloy Y is very similar to X and that yield strength, tensile strength, grain size and production process are similar, the estimate might be based on:

$$\text{Toughness}\ (Y_{\text{TL}}) = \text{Toughness}\ (X_{\text{TL}}) \times \text{Toughness}\ (Y_{\text{LT}})\ /\text{Toughness}\ (X_{\text{LT}})$$

following common rules for estimates based on reference values.

If no data at all are available for alloy Y, the estimate is more precarious. Usually, if the alloys are similar

$$\text{Toughness}\ (Y) = F_{tyX} \times \text{Toughness}\ (X)\ /\ F_{tyY}$$

should provide a safe (conservative) estimate, but only if $F_{tyY} > F_{tyX}$. This should not be used if $F_{tyY} < F_{tyX}$. In that case it would be safer to use the same toughness as for X. For dissimilar alloys such estimates should not be attempted at all.

Estimating the effect of temperature is more dangerous because temperature affects the yield strength. One may try:

$$\text{Toughness}\ (Y_{T2}) = \frac{\text{Toughness}\ (X_{T2})}{\text{Toughness}\ (X_{T1})} \times \frac{F_{ty}\,YT_1}{F_{ty}\,YT_2} \times \text{Toughness}\ (Y_{T1})$$

but it is advisable to base the estimate on more extensive comparisons of available data.

Data handbooks (e.g. [6]) provide an abundance of toughness data for high strength alloys. But even for those materials estimates are often necessary; because of the enormous number of parameters not all cases are covered. The situation is worse for the most widely used materials: common structural steels. On the other hand for those materials one often knows some Charpy data. The lower shelf Charpy-value is essentially a fracture energy and can thus be expected to correlate with K_{Ic}. The Charpy test measures the total fracture energy of the specimen, which is essentially the integral of $R(a)$ over the ligament. If the R-curve would be horizontal the value of this integral divided by the ligament would indeed be R and $K = \sqrt{ER}$. For the low toughness at the lower shelf the R-curve will be nearly horizontal, so that a correlation between K_{Ic} and Charpy energy is indeed likely. However, there are some essential differences between a Charpy test and a toughness test. The most important of these are the difference in loading rate (affecting the yield strength) and the difference in notch acuity (affecting the state of stress at the notch root, and as such the stress at yield).

From empirical comparisons [10, 11, 12], it appears that

$$K_{Ic} = 12\sqrt{Cv} \quad \text{for K in ksi} \sqrt{\text{in}} \text{ and } Cv \text{ in ftlbs;}$$

$$K_{Ic} = 11.4\sqrt{Cv} \quad \text{for K in MPa} \sqrt{\text{m}} \text{ and } Cv \text{ in Joules.}$$

A conservative lower bound [11] is claimed to be:

$$K = 22.5\,(Cv)^{0.17} \quad \text{for ksi} \sqrt{\text{in}} \text{ and ftlbs;}$$

$$K = 21.6\,(Cv)^{0.17} \quad \text{for MPa} \sqrt{\text{m}} \text{ and Joules.}$$

Toughness values so obtained would be for the same high loading rates (impact) as prevalent in the Charpy test.

The toughness for slower loading rates may be obtained from the same equations, while accounting for a transition temperature shift given by:

$$\Delta T = 215 - 1.5\,F_{ty} \text{ for } °F \text{ and ksi and } 36 < F_{ty} < 140 \text{ ksi;}$$

$$\Delta T = 0 \text{ if } F_{ty} > 140 \text{ ksi;}$$

$$\Delta T = 119 - 0.12\,F_{ty} \text{ for } °\text{C and MPa and } 250 < F_{ty} < 965 \text{ MPa;}$$

$$\Delta T = 0 \text{ if } F_{ty} > 965 \text{ MPa.}$$

Note that the slower loading will cause the transition temperature to be lower (lower F_{ty}), so that the estimated toughness values will be useful – if the loading rate is low – even at these lower temperatures $T - \Delta T$. Other empirical correlations have been derived [12].

If for example the Charpy value is 20 ftlbs, the yield strength $F_{ty} = 60$ ksi at a temperature of 65F, the toughness for high loading rates would be estimated as $K_{Ic} = 12\sqrt{20} = 54\,\text{ksi}\sqrt{\text{in}}$. This would be a safe value to use, because the toughness would be higher for slower loading. One could probably use this toughness value for temperature as $65 - (215 - 1.5 \times 60) = -60$F (see above equations for ΔT).

Empirical correlations to estimate the toughness at the upper shelf from Charpy data have been developed as well [10]. The generality of these is less certain, because the upper shelf Charpy energy is not directly related to toughness. If the R-curve rises steeply – which it does at the upper shelf – the integral of $R(a)$ over the ligament is not uniquely related to R. Besides, much of the Charpy energy on the upper shelf is energy used for general specimen deformation rather than for fracture: in the extreme case Charpy specimens do not fracture at all, but are simply folded double, so that only deformation energy is measured. Nevertheless, in some cases the empirical correlations may be the only way to arrive at an estimate.

Toughness estimates can also be based upon the results of COD tests. Correlations between CTOD and toughness were discussed in Chapter 4.

Estimating K_c, or the effect of thickness is usually somewhat easier. Assuming that true plane stress occurs when the thickness is equal to the plastic zone, plane stress will develop during loading when $B = (K_{Ic}/F_{ty})^2/2\pi$. Note that K_{Ic} is used in this equation, because at very low stress (low K) there will be plane strain, but plane stress must have developed when $K = K_{Ic}$, otherwise K cannot be increased to the full plane stress toughness K_c. Further assume that plane strain is reached at a thickness of $B = 2.5\,(K_{Ic}/F_{ty})^2$. These two points can be identified in the diagram of toughness versus thickness as shown in Figure 7.4. A straight line approximation between the full plane stress and full plane strain value is usually a conservative one (compare Figure 7.1). The straight line can then be used as the basis for the estimate, provided the plane strain toughness and the

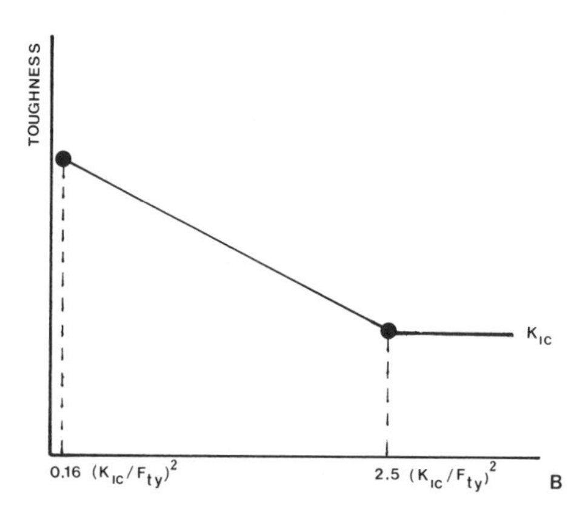

Figure 7.4. Conservative estimate of transitional toughness on the basis of two data points (compare with Figure 7.1.).

toughness for at least one other thickness are known. If only K_{Ic} is known, one may estimate that K_{eff} for full plane stress is between 2 and 2.5 times K_{Ic} (K_c is another 10–20% higher), and then follow the above procedure. (For examples see solutions to Exercises.)

Actual and reliable data are preferable above estimates. In the lack of handbook data a test on a specimen of sufficient size is always preferable. However, most engineers, unlike researchers, do not have easy access to a laboratory; even if they do, economic conditions and/or pressures for immediate answers often preclude obtaining data for each and every alloy, condition, heat treatment, temperature, material direction and thickness. Thus, estimates are often necessary. Used judiciously, the above procedures should provide conservative answers.

7.6. General remarks on fatigue rate data

Usually da/dN data obtained from handbooks or the literature need further interpretation and so, naturally, do test data. A typical data set as might appear in a handbook [6] is shown in Figure 7.5. The set shown is rather complete as it covers several R-ratios. Not always are such complete sets available; in general $R = 0$ is covered (or $R = 0.05$ or 0.1), but often this is the only R-value for which data are available.

There may be a question about the effect of negative R. Some schools hold that negative stresses hardly affect crack growth, while others maintain that data clearly demonstrate a substantial effect. Both schools are essentially right, the dichotomy being a matter of interpretation.

Consider Figure 7.6. A small negative stress may still have some effect until the crack is fully closed. However, after complete closure there is no longer a stress concentration: the compressive stress can be carried through and does not have to by-pass the crack. In tension the crack forms a load-path interruption which must be bypassed and which causes the high crack tip stresses, crack growth and fracture in the first place. In compression the crack faces carry through the load, as a loose pile of bricks can carry compressive loads. Thus indeed the compressive part of the cycle has no effect (low elastic crack tip stresses, as opposed to a very high stress concentration and yielding in tension), and e.g. crack growth curves obtained for $R = 0$ and $R = -1$ with the loading as in Figure 7.7a, would be almost identical as shown in Figure 7.7b. But the da/dN data can be represented in two ways.

By accounting for the above argument the negative part of the cycle could be ignored and the stress ratio be defined as $R = 0$, with $\Delta\sigma = \Delta\sigma_1$ and $\Delta K_1 = \beta\Delta\sigma_1\sqrt{\pi a}$. The da/dN data will then be as shown in Figure 7.7c (school 1), and there will no apparent effect of negative R. On the other hand, in a 'formal' interpretation, $R = -1$, the stress range $\Delta\sigma = \Delta\sigma_t$, and $\Delta K_t = \beta\Delta\sigma_t$

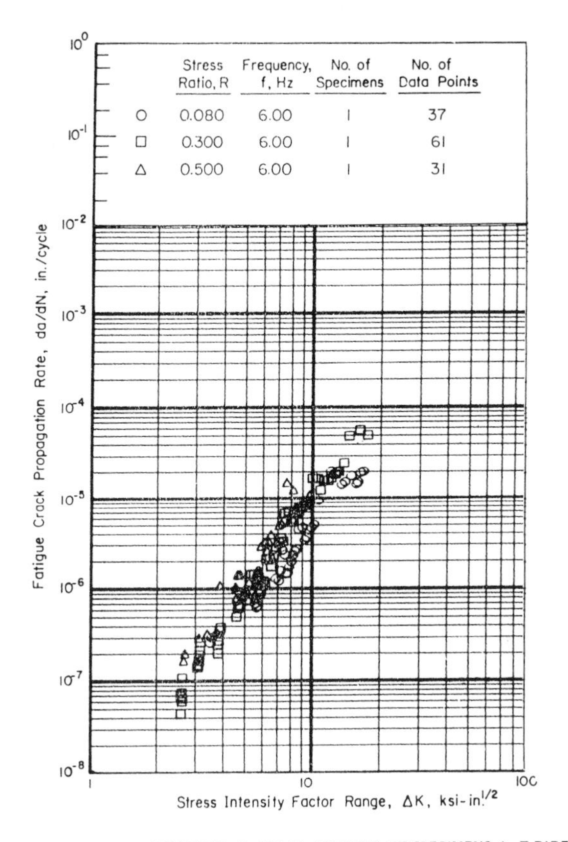

Figure 7.5. Example of rate data in MCIC Handbook [6]; Courtesy MCIC.

$\sqrt{\pi a}$ would be employed. Since the rates are obviously the same ones as in the previous case (same data as in Figure 7.7b) the same rate data are plotted at twice as large a ΔK, namely $\Delta K_t = 2\Delta K_1$, as in Figure 7.7c (school 2). Now there is a considerable effect of negative R; nevertheless the crack growth curves are identical; both rate diagrams in Figure 7.7c are based on the same data of Figure 7.7b.

Both interpretations are tenable as long as each is interpreted in the same manner in a crack growth analysis. In a loading case such as in Figure 7.7d school 1 must take $R = 0$, $\Delta\sigma_3$ and $\Delta K_3 = \beta\Delta\sigma_3\sqrt{\pi a}$, and use data set 1 to find $\mathrm{d}a/\mathrm{d}N$, but school 2 must take the larger $\Delta\sigma_4$ and $\Delta K_4 = \beta\Delta\sigma_4\sqrt{\pi a}$ with negative R, to find $\mathrm{d}a/\mathrm{d}N$ from data set 2. (Note that both will find the same $\mathrm{d}a/\mathrm{d}N$ as they should). Consistent interpretation and usage will prevent errors.

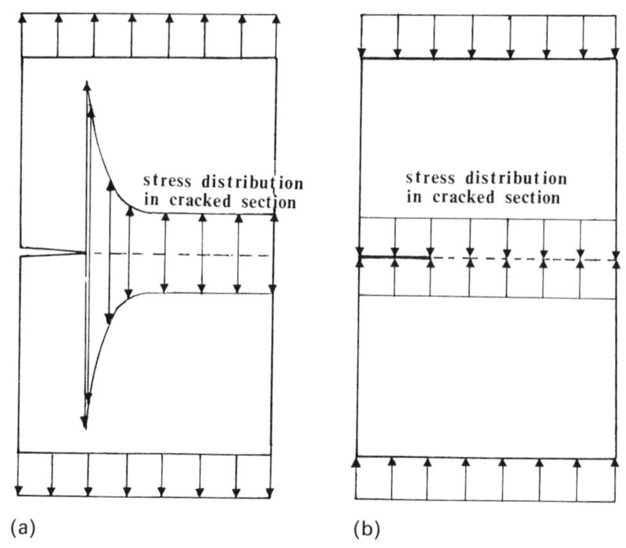

Figure 7.6. Crack Tip Stresses in Tension and Compression. (a) Tension; (b) Compression.

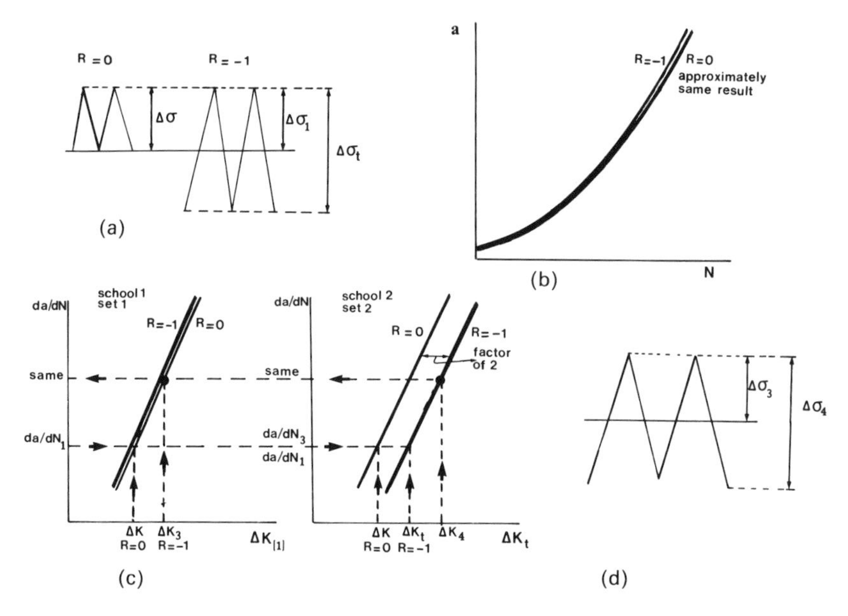

Figure 7.7. Effect of Negative *R*. (a) Loading, (b) Test data (hypothetical); (c) Rate plots for two schools; (d) Use of data in analysis.

One should be aware however, of the 'school' built into computer software. Some programs use the school 1 interpretation, others use school 2, some provide the option for either one.

Fatigue crack growth properties depend upon direction *T-L*, *L-T* etc. (Figure 7.3). In particular in the case of surface flaws one must be cautious in the selection of the appropriate data set, especially in forgings where due to the different grain flow the identification of *L*, *T* and *S* direction is not always trivial. Figure 7.8 shows a part produced by two different methods. Machining to provide for the seat of the bolt head will expose grain endings in different ways. The crack will select the weakest path along the long exposed grain boundaries and tend to grow in the *ST* (or *SL*) direction, so that *ST*-data should be used for the analysis. The analyst must be cognizant of the production procedure and use the appropriate data. A crack assumed in the wrong direction (i.e. reversing the directions in the forged and machined-out-of-plate parts), results in erroneous predictions, because *ST* toughness values and rates are usually much lower than those for *LT* or *TL* (cross grain growth).

It is crucial that the proper crack direction and data set are selected for fracture and crack growth predictions. The analyst must be aware of how the part is made in order to identify crack location and direction of growth. Arbitrarily choosing the crack in the same direction in all cases (and using the wrong data) will lead to predictions with no bearing upon practice. 'The computer says so', is no excuse; neither the computer code nor fracture mechanics can be blamed if wrong assumptions are made.

When the environment to be considered is different from air, the rates may

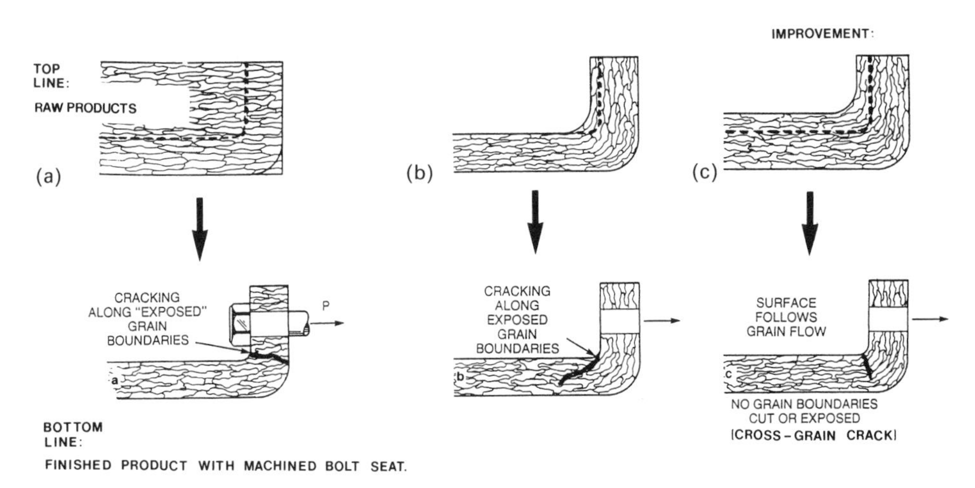

Figure 7.8. Effect of Production Procedure on Crack direction. (a) Rolled plate; (b) Forging; (c) Oversize Forging.

be substantially different and also the effect of R-ratio. A typical data set for a pipeline steel in seawater [13] is shown in Figure 7.9. It needs no emphasis that the data used, must be for the proper environment. The damage tolerance requirements may prescribe a data set for environmental effects (ASME). But in general, an estimate must be made for the 'average' environmental effect as will be discussed later.

7.7. Fitting the da/dN data

Data sets such as shown in Figure 7.5 still must be interpreted before they can be used in analysis. Clearly, the 'scattered' data points cannot be used directly.

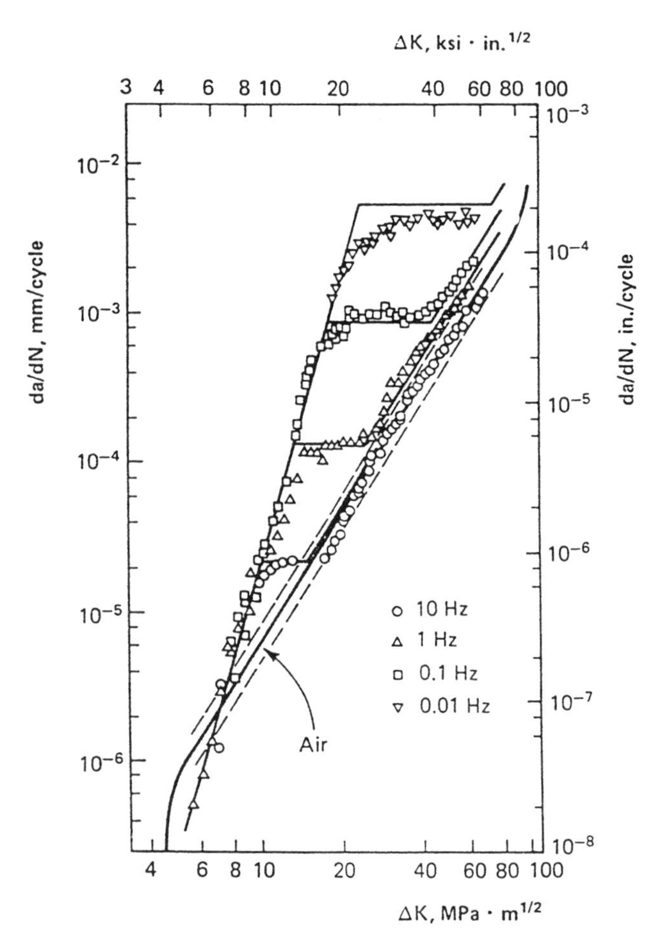

Figure 7.9. Effect of Sea Water on Crack Growth in Pipe Steel [13] (Courtesy ASME).

The problem of how to deal with scatter per se will be discussed in Section 7.8. This section concerns the first step of the interpretation.

In some cases a growth rate equation may be desirable; in particular if the data appear to fall on straight lines, the Paris or Walker equations may be convenient. Most computer programs have options for the use of a number of equations (Walker, Forman and some threshold equation are the most common), but accept tabular data as well. The latter eliminates the need for force-fitted (sometimes poorly-fitted) equations. Whatever the shape of the rate curves, they can be represented in tabular form. This is especially convenient if the data can be read from a permanent magnetic data file that can be called by the program. However, the program does not interpret the data, so that one cannot submit the data points as they appear in the $da/dN - \Delta K$ data plot. First lines must be drawn through the data for different R and points of these lines must appear in the table, be it that this line may have any form without being fixed by an equation.

Whether equations or tabular data are used, interpretations must be made. All equations derive from 'curve' fitting and have no physical basis. None of them is fundamentally better than any other; none is more universally useful than any other. The most appropriate equation is the one giving the best fit for the case at hand, which in turn depends upon the material, environment etc. Also for the use of tabular data, the best fitting line must be drawn through the data.

Commonly fits are obtained by using e.g. a least squares fit of the $da/dN - \Delta K$ data. This is all that can be done if the original raw test data (a versus N) are not available; the original crack growth data usually are not reported in the handbooks or the literature. The best fit through the $da/dN - \Delta K$ data must then be used, although this 'best' fit may not give the best predictions as shown below.

A measured crack growth curve is shown in Figure 7.10a, the da/dN data in Figure 7.10b. In this particular case a straight line is appropriate (Paris equation). A least-squares fit of these data provides $C_P = 6.496\text{E-}11$, and $m_P = 3.43$. When these parameters are used in a crack growth analysis routine to re-predict the original crack growth curve, the result is as shown in Figure 7.10a, which is certainly not the best fit to the actual crack growth curve. In this case the curvature of the predicted curve seems appropriate (which means that the value of m_P is correct), but there is a more-or-less proportional error, which can be corrected by adjusting C_P. The predicted life is too long as compared to the test life by a factor of $1\,937\,483/1\,871\,080 = 1.035$. Multiplication of C_P by this factor to obtain a new $C_P = 1.035 \times 6.496\text{E-}11 = 6.72\text{E-}11$, will result in a better prediction as shown in Figure 7.10a.

Apparently the regression fit of the da/dN data is not necessarily the best fit for analysis. The reason for this is that in all curve fitting procedures every data

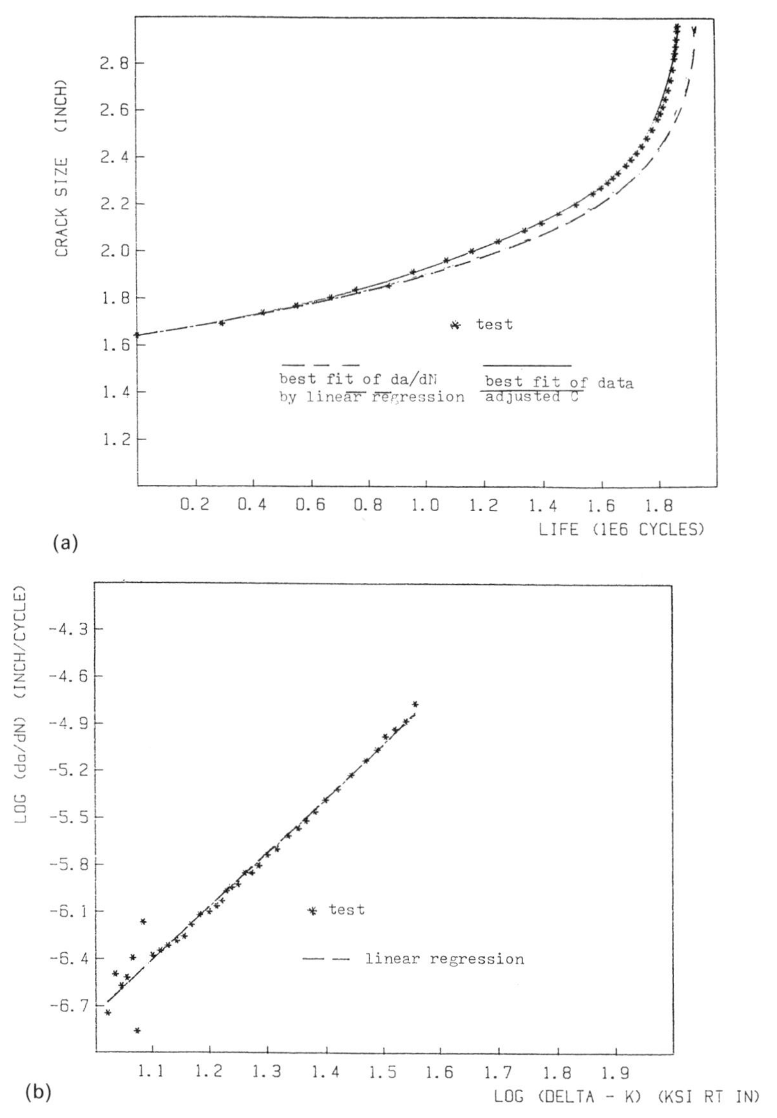

Figure 7.10. Finding the best representation of rate data. Material: A-533 B at 550 F; $R = 0.1$. (a) Test Data and Predicted curves (see text); (b) Regression Fit of Rate data.

point gets equal weight. But the points for high da/dN affect only a small portion of the life, while those for low da/dN are of much more influence on the life (they are the relevant data during most of the life); the latter should weigh more in the curve fitting.

In the above examples, the curvature of the predicted curves was appropriate, and only C_P needed adjustment. If the curvature is found to be incorrect, m_P must be adjusted first by trial and error, as shown in Figure 7.11. This usually causes wild gyrations in the predicted curve, but these can be ignored. The objective is to arrive at a line with the proper curvature, which will have the proper m_P. Once this m_P-value is found (TRY 2), the curve can be adjusted by adjusting C_P in the manner discussed above (Figure 7.11).

Clearly, fitting an equation is not trivial; it requires judgement, and re-prediction of the original data using a predictive computer code. If the original crack growth curve is not available, there is no other option than to fit the da/dN data as well as possible. This can be done by regression, but this not necessarily the best; regression analysis may be a refinement and not necessarily an improvement over a hand-drawn best fit line. With the latter the parameters for some equations can be derived very easily by hand.
The Paris equation for example is a straight line on a log scale, with the equation $y = m_P x + b$, where $y = \log(da/dN)$, $x = \log(\Delta K)$, and $b = \log C_P$. By taking two points (as far apart as possible) on the hand-drawn fit, the values of da/dN and ΔK can be substituted in the equation. This provides 2 equations with two unknowns (C_P and m_P), which can be solved to obtain the values of the parameters. An example was given in Chapter 5.

If data are available for more than one R-ratio (Figure 7.12), the Walker equation can be derived in an equally simple manner. All data sets are fitted with

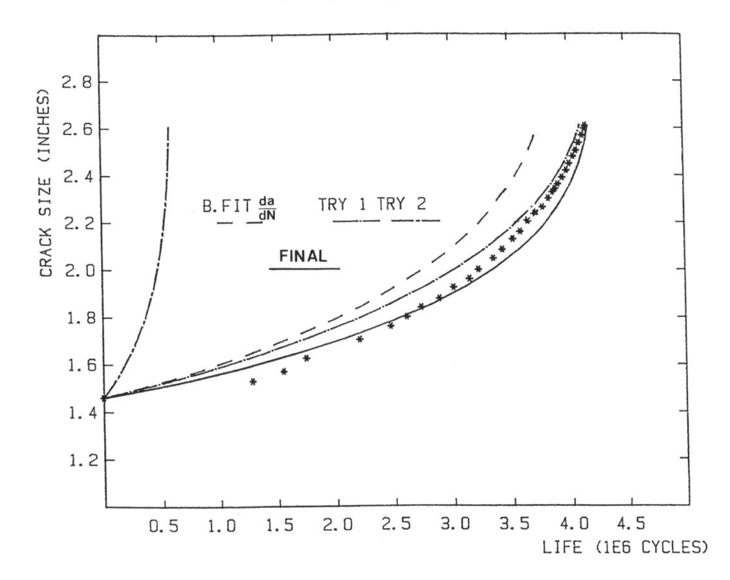

Figure 7.11. Adjustment of data for best predictions. Try 1 provides best m; adjustment of C as in Figure 7.10 (see text) Material: A-533 B at 550 F; $R = 0.7$.

parallel straight lines (same m). For each line m and C_R are determined as for the Paris equation, C_R being the value of C for each particular R-ratio and C_W is the value of C_R for $R = 0$. The Walker equation reads:

$$\frac{\mathrm{d}a}{\mathrm{d}N} = \frac{C_w}{(1 - R)^{n_w}} (\Delta K)^{m_R} = C_R (\Delta K)^{m_R} \tag{7.1}$$

or with $K_{\max} = \Delta K/(1 - R)$:

$$\frac{\mathrm{d}a}{\mathrm{d}N} = C_w \frac{\Delta K^{n_w}}{(1 - R)^{n_w}} (\Delta K)^{m_R - n_w} = C_w K_{\max}^{n_w} \Delta K^{m_R - n_w} \tag{7.2}$$

so that:

$$\frac{\mathrm{d}a}{\mathrm{d}N} = C_w \Delta K^{m_w} K_{\max}^{n_w}. \tag{7.3}$$

By plotting the calculated C_R values versus the corresponding values of $(1 - R)$ on a double-logarithmic scale and drawing a straight line, one obtains C_W and n_W in the same manner as above. Table 7.1 and Figure 7.13 show the derivation for the data set in Figure 7.12. With these parameters the Walker equation can

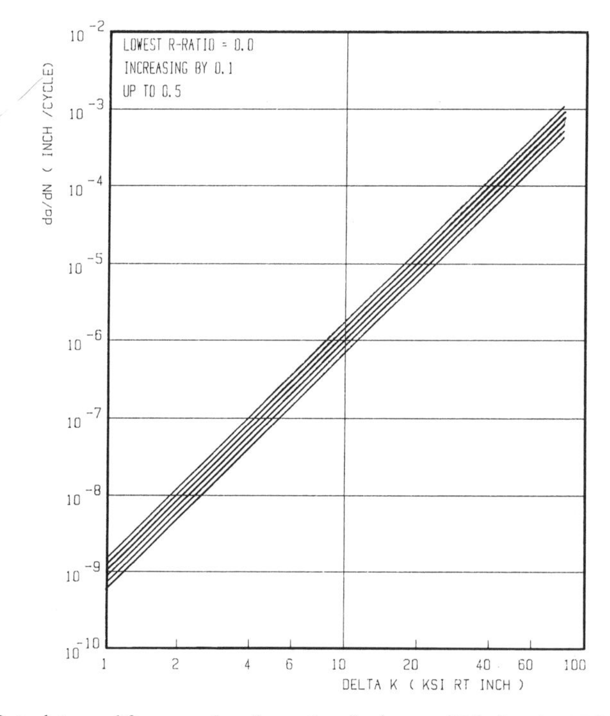

Figure 7.12. Rate data used for examples of equation fits (see text) Ti-alloy (see Tables 7.1 and 7.2).

Table 7.1. Derivation of Walker equation for data of Figure 7.12

Paris equation for $R = 0$, using $\Delta K = 2\,\text{ksi}\,\sqrt{\text{in}}$ and $20\,\text{ksi}\,\sqrt{\text{in}}$.

$$\text{Log}\,\frac{da}{dN} = m \log \Delta K + \log C$$

$$\left.\begin{array}{l}\text{Log } 4.7\text{ E-9} = m \log 2 + \log C \\ \text{Log } 5.6\text{ E-6} = m \log 20 + \log C\end{array}\right\} \qquad \begin{array}{rl} -8.32 = & 0.301\text{ m} + \log C \\ -5.25 = & 1.301\text{ m} + \log C \\ \hline 3.07 = & m \end{array} -$$

$\text{Log } C = -8.32 - 0.301 \times 3.07 = -9.24 \qquad C = 5.7\,E\text{-}10\ (R = 0).$

Same for other R-ratio's or C_R can be read directly from Figure at $\Delta K = 1$ ($\log \Delta K = 0$).

Results

R	C_R	$1 - R$	log (1 − R	$\log C_R$	Predicted C_R from solution below	Ratio C_R pred/ C_R figure
0	5.7 *E*-10	1	0	−9.24	6.30 *E*-10	1.10
0.1	7.6 *E*-10	0.9	−0.046	−9.12	7.56 *E*-10	0.99
0.2	8.9 *E*-10	0.8	−0.097	−9.05	9.27 *E*-10	1.04
0.3	1.1 *E*-9	0.7	−0.155	−8.96	1.17 *E*-10	1.06
0.4	1.3 *E*-9	0.6	−0.222	−8.88	1.52 *E*-10	1.17
0.5	1.6 *E*-9	0.5	−0.301	−8.79	2.08 *E*9	1.30
					Average	1.11

log (1 − R) versus $\log C_R$ plotted in Figure 7.13. From Figure 7.13 and Equation (7.1):

$$\log C_R = \log C_W - n_w \log (1 - R); \qquad \begin{array}{rl} -8.78 = & \log C_w + 0.3\,n_w \\ -9.20 = & \log C_w + 0 \\ \hline 0.52 = & 0.3\,n_w \end{array} -$$

$n_w = 0.52/0.3 = 1.73; \quad \log C_w = -9.20; \quad C_w = 6.3\,E\text{-}10$

$$C_R = \frac{6.3\,E\text{-}20}{(1 - R)^{1.73}} \quad \text{For } R = 0.5\text{: } C_R = \frac{6.3\,E\text{-}10}{0.5^{1.73}} = 2.08\,E\text{-}9$$

for other predicted values see last column in table.

$m_w = m_R - n_w = 3.07 - 1.73 = 1.34$

$$\text{Equation 7.1: } \frac{da}{dN} = \frac{6.3\,E\text{-}10}{(1 - R)^{1.73}}\,\Delta K^{3.07}$$

$$\text{Equation 7.3: } \frac{da}{dN} = 6.3\,E\text{-}10\,\Delta K^{1.34}\,K_{\max}^{1.73}.$$

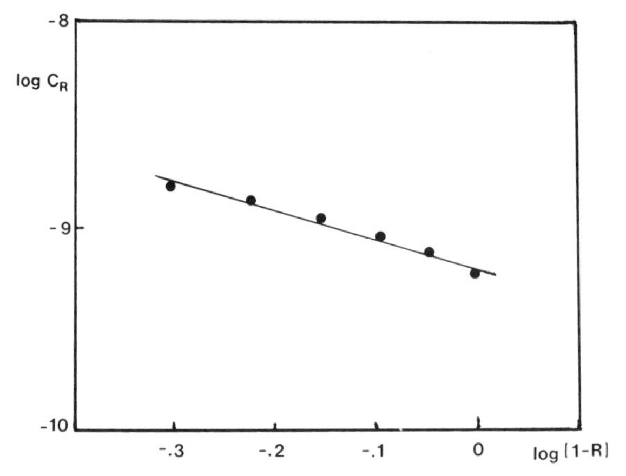

Figure 7.13. Fit of Walker equation for data of Figure 7.12 (see Table 7.1).

then be used in the form Equation (7.1), or in the alternative form of Equation (7.3) with $m_W = m_R - n_W$. Some computer codes use Equation (7.1), others use Equation (7.3), so that the required input may be different (either C_W, m_R, n_W or C_W, m_W, n_W).

Also the parameters of the Forman equation can be obtained in a simple manner. The Forman equation is:

$$\frac{\mathrm{d}a}{\mathrm{d}N} = C_F \frac{\Delta K^{m_F}}{(1 - R)\, K_c - \Delta K} \tag{7.4}$$

which can be written as:

$$\left\{(1 - R)\, K_c - \Delta K\right\} \frac{\mathrm{d}a}{\mathrm{d}N} = C_F\, \Delta K^{m_F}. \tag{7.5}$$

Hence, the Forman equation ASSUMES that in a plot of log $[(1 - R)\, K_C - \Delta K)\mathrm{d}a/\mathrm{d}N]$ *versus log* (ΔK) all data for all R ratios consolidate to one line and that this line is straight. Figure 7.14 and Table 7.2 demonstrate this. From this line C_F and m_F can be determined in the same manner as for the Paris equation. Should the data (for more than one R-ratio) not fall on a single line as e.g. in Figure 7.15 then the Forman equation is not a good fit, at least not with the value of K_c that was used. Note that K_c is the only adjustable parameter in the equation, the effect of R-ratio is implied. Since fracture of the crack growth specimens may have occurred by collapse, K_{Ic} or K_c would not be appropriate to fit the data and the value used becomes an adjustment parameter indeed, although the generality of the equation becomes doubtful as it would apply to the specimens only. Yet, it can then be tried to consolidate the data using a different K_c value.

If the data do not provide a single straight line as is the case in Figure 7.15

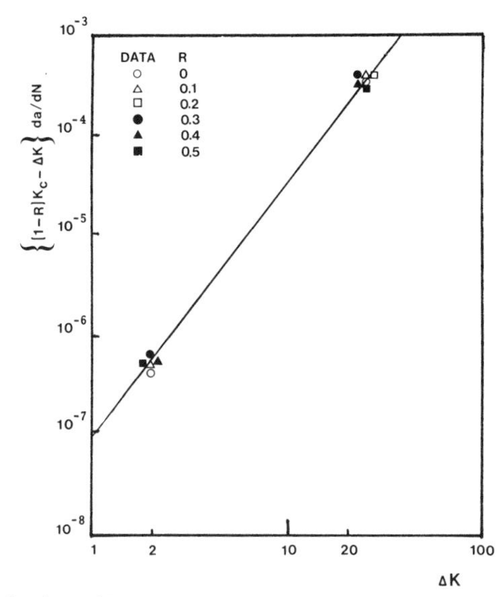

Figure 7.14. Derivation of Forman equation for data set of Figure 7.12 (see Table 7.2).

then the Forman equation is not a good fit at all. Similarly, if $\log(C_R)$ versus $\log(1 - R)$ data do not fall on a straight line, the Walker equation is not a good fit. In that case a more complicated curve fit could be used, but curve fitting is appealing only if the equation is a simple one. If not, it is much more convenient to use tabular data inputs instead of an equation. A computer works just as easily with a data table (ΔK, $\mathrm{d}a/\mathrm{d}N$ for various R) as with an equation.

This section would not be complete without a warning with regard to conversion of equation parameters to other equations or to other unit systems. It appears that C_P, m_P, C_F, m_F, are NOT transferable, nor are C_W, m_W, n_W. A little exercise deriving all parameters for all equations from the same data set (regardless of fit) will easily demonstrate the difference. For example, Tables 7.1 and 7.2 show the Walker and Forman equations for the same data sets. Note that $C_F \neq C_W$, and $m_W \neq m_F$.

Unit conversions can be made, but they are treacherous. Fortunately the n and m values do not change for different unit systems, because the exponents are dimensionless by nature. But C does change, because C is not dimensionless.

For example, in a Paris equation with ΔK in ksi$\sqrt{\text{in}}$ and $\mathrm{d}a/\mathrm{d}N$ in in/c, the units for C_P follow from (note that 'cycle' is dimensionless):

$$\text{in/cycle} = \text{Unit}\,(C_p)\,(\text{ksi}\,\sqrt{\text{in}})^{m_p} \quad \text{or} \quad \text{Unit}\,(C_p) = \frac{\text{in}}{(\text{ksi}\,\sqrt{\text{in}})^{m_p}}. \tag{7.6}$$

Table 7.2. Derivation of Forman equation from data set of Figure 7.12

Assumption: $K_C = 100$ ksi $\sqrt{\text{in}}$ (other values may be assumed to obtain better fit).

R	da/dN from Figure 7.12 (in/c)		$K_c = 100$ ksi $\sqrt{\text{in}}$ $\{(1-R)K_c - \Delta K\}\frac{da}{dN}$	
	$\Delta K = 2$ ksi $\sqrt{\text{in}}$	$\Delta K = 20$ ksi $\sqrt{\text{in}}$	$\Delta K = 2$ ksi $\sqrt{\text{in}}$	$\Delta K = 20$ ksi $\sqrt{\text{in}}$
1	2	3	4	5
0	4.55 *E*-9	4.81 *E*-6	4.46 *E*-7	3.85 *E*-4
0.1	5.39 *E*-9	6.03 *E*-6	4.74 *E*-7	4.22 *E*-4
0.2	6.75 *E*-9	7.14 *E*-6	5.27 *E*-7	4.28 *E*-4
0.3	8.94 *E*-9	8.64 *E*-6	6.08 *E*-7	4.32 *E*-4
0.4	9.45 *E*-9	1.00 *E*-5	5.48 *E*-7	4.00 *E*-4
0.5	1.12 *E*-8	1.25 *E*-5	5.38 *E*-7	3.75 *E*-4
		Average	5.24 *E*-7	4.07 *E*-4

Data of columns 4 and 5 are plotted in Figure 7.17. On a log-log-plot, these cannot be properly separated and must be plotted next to each other instead of at proper ΔK.

Equation of straight line in Figure 7.14:

$$\log\left[\left\{(1-R)K_c - \Delta K\right\}\frac{da}{dN}\right] = C_F\,\Delta K^{m_F}$$

2 points:

$$\left.\begin{aligned}\log(5.24\,E{-}7) &= m_F \log 2 + \log C_F\\ \log(4.07\,E{-}4) &= m_F \log 20 + \log C_F\end{aligned}\right\}\qquad \begin{aligned}-6.28 &= 0.301\,m_F + \log C_F\\ -3.39 &= 1.301\,m_F + \log C_F\end{aligned}\;-$$

$$2.89 = m_F$$

$$C_F = 7.08\,E\text{-}8$$

Equation (7.4): $$\frac{da}{dN} = 7.08\,E\text{-}8\,\frac{\Delta K^{2.89}}{(1-R)\times 100 - \Delta K}$$

This strange unit comes about by the curve fit, and because C_P therefore has no physical significance. Conversion of the units to psi$\sqrt{\text{in}}$ and in/c would require a conversion factor of:

$$1\,\frac{\text{in}}{(\text{ksi}\sqrt{\text{in}})^{m_p}} = \frac{\text{in}}{(1000\text{ psi}\sqrt{\text{in}})^{m_p}} = \left(\frac{1}{1000}\right)^{m_p}\frac{\text{in}}{(\text{psi}\sqrt{\text{in}})^{m_p}} \tag{7.7}$$

i.e. the conversion depends upon m_P, and C_P should be multiplied by $(1000)^{-m_P}$ for use of the same equation with psi $\sqrt{\text{in}}$. If $C_P = 1\text{E-9}$ and $m_P = 3.2$ for ksi $\sqrt{\text{in}}$, its value becomes $1\text{E-9}/(1000)^{3.2} = 2.5\text{E-19}$ for psi$\sqrt{\text{in}}$.

Conversion to ΔK in MPa $\sqrt{m}$ da/dN in m/c would require a conversion factor of:

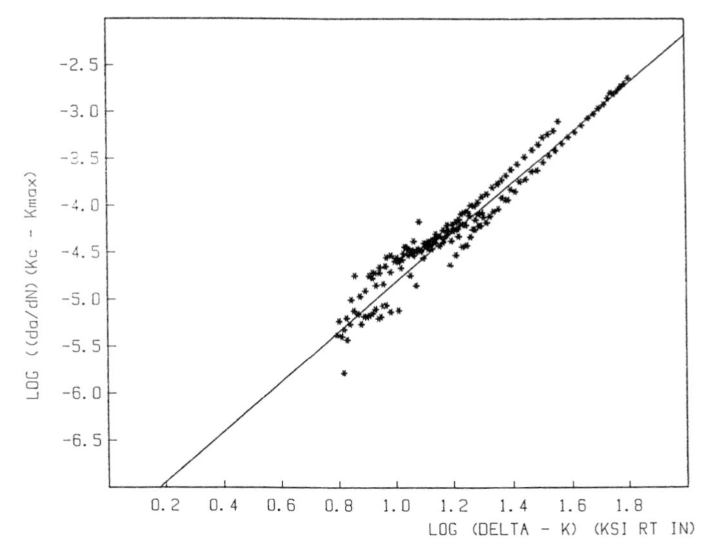

Figure 7.15. Data for A-533 B steel for various R-ratios plotted to fit Forman equation. Data do not consolidate to one line and do not fit equation.

$$1 \frac{\text{in}}{(\text{ksi}\sqrt{\text{in}})^{m_p}} = \frac{0.0254\ m}{(6.86\ \text{MPa}\sqrt{0.0254\ m})^{m_p}} = \frac{0.0254}{(1.09)^{m_p}} \frac{m}{(\text{MPa}\sqrt{\text{m}})^{m_p}}. \tag{7.8}$$

In the above example the conversion to MPa $\sqrt{m}$ and m/cycle, C_p would become $1E\text{-}9 \times 0.0254/(1.09)^{3.2} = 1.93E\text{-}11$.

Conversions of equations other than the Paris equation are even more tricky. They can be made, but usually it is easier to first convert the da/dN and ΔK data and then derive the new equation parameters from the converted data set. It is advisable to make it a common practice to specify the unit system used when quoting equation parameters, a practice not often followed in the literature. Equation parameters without specification of the unit system are useless; the effects of the unit system are such that indiscriminate use of the parameters will be disastrous.

In the use of computer software some caution is needed when working with MPa $\sqrt{m}$ for ΔK. In that case the crack size, a, must be given in meters (instead of the more customary mm). Since 1 MPa = 1N/mm^2, it is more rational to use the unit MPa $\sqrt{\text{mm}}$, as some codes do. This may require unit conversion for the coefficients in the rate equations or even for the ΔK values in tabular input. Although these seem trivial matters, little slips can lead to such large errors in predictions that the problem is worth mentioning.

It may be obvious from this section that indeed the fitting of rate data is not trivial. As was shown in Table 7.1 and Figure 7.13, the Walker equation is a good fit to the data set in Figure 7.12. Yet, the resulting C_W-values are off by a factor of 1.11 on the average (Table 7.1). Thus, all rates will be 11% higher, and therefore predicted lives will be 11% too low (conservative). Naturally, this is largely due to the difficulty of reading data from a log–log–plot. However, it is not really the reading of data that counts; after all rate data vary as strongly as they do and direct use of data would not improve the situation. The reader is challenged to derive more accurate equation parameters (Exercise 11). It should be noted that these inaccuracies are not due to shortcomings of fracture mechanics; they are caused by material behavior. The latter is the 'real world'; it is 'the way it is', theory or not. No better 'theory' would improve the situation (see also Chapter 12 on accuracy).

As may be clear from the examples, the Forman equation often will be a poor fit (Figure 7.15). This is because the equation implicity accounts for the effect of R, which is usually more complicated as can be seen from Table 7.1.

Any equation fitting leads to errors, even if it may seem (on a log-scale) that the fit is very good. On the other hand, the use of tabular data may not be much better, because the reading of data from a log-plot is full of error also. It should be realized however, that this is not due to the log-plot, but due to the material behaving in the manner shown. No theory or equation can accurately account for material behavior if it depends as strongly as shown upon ΔK. Any small error in ΔK will cause large errors in $\mathrm{d}a/\mathrm{d}N$.

7.8. Dealing with scatter in rate data

Despite regular practice a regression fit is seldom the appropriate fit of rate data, as was explained in the previous section. Similarily, use of statistical procedures to account for the scatter are seldom in accord with physical reality. Most statistical procedures ignore the physics and mechanics of the problem. Commonly e.g. the 90% confidence curves are determined by using the individual $\mathrm{d}a/\mathrm{d}N$ data points as the statistical population sample. However, applying the correct mathematics does not lend credence to the physical result.

First consider the nature of the scatter. The three main sources of scatter are:

(a) Consistent difference between heats A and B of the same alloy.

(b) Local differences due to inhomogeneities and 'weak spots'.

(c) Errors in measurement.

Essentially, only the first source of scatter is relevant; the other two have little bearing on the problem. The three sources of scatter will be considered independently as if the others were absent. Consistent differences between various heats or batches of materials, and to a certain degree differences from location to location in one batch or plate, will be reflected in a more-or-less consistent difference in crack growth curves. As a consequence, the rate also will be consistently different as shown in Figure 7.16. This is true material scatter and it must be accounted for in an analysis because it is not known what the exact properties are of the batch used in the structure.

Next consider the scatter due to inhomogeneities and 'weak spots'. These may occur throughout the material, but only locally. At some locations the crack will accelerate, but soon afterward it resumes normal behavior. Another crack (in a different specimen) will encounter such 'weak' spots at other locations and will speed up locally (at different stages in life than the first one), and then resume normal growth; similarly, it slows down sometimes locally. On the whole, the two crack growth curves will be essentially identical as shown in Figure 7.17a. Also the rate data will be identical, except that each test provides a few outlying data points due to local higher or lower rates as indicated in Figure 7.17b. The outlying data points occur at different locations in the two tests. If instead of two, many tests are performed on specimens from the same plate, more and more outlying data points appear and the scatter seems to become well established (Figure 7.17c).

Taking the upper (or 90%) and lower bound of this scatter (Figure 7.17c), and reconstructing crack growth curves on this basis, results in Figure 7.17d. Clearly, the upper and lower bound data lead to unrealistic results, as all measured crack growth curves are essentially identical. This is caused by the implicit assumption that it is possible that in some cases all crack growth could be through a continuous string of weak spots (note that this scatter is caused by local weak spots). This is an untenable assumption. Weak spots are local; in each case only a few will be encountered; the entire bulk will not be one great weak spot. Hence, the average curve is the relevant one; the 'scatter' is apparent only and not real.

The third type of scatter is due to measurement errors, shown in Figure 7.18.

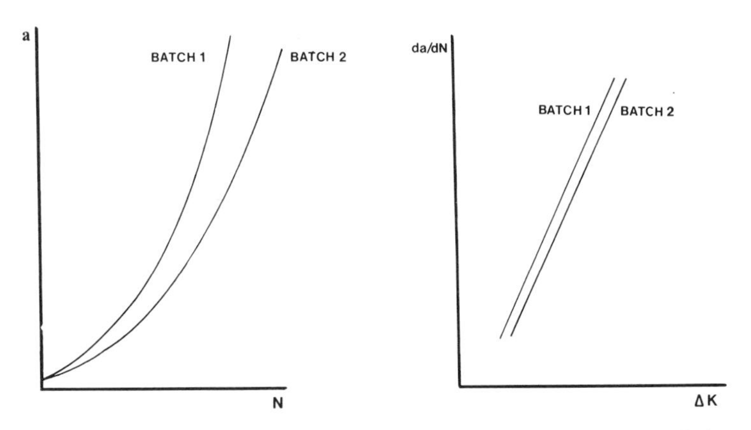

Figure 7.16. Typical scatter due to batch-to-batch and heat-to-heat variations.

In general the crack size will not be consistently over-measured or consistently under-measured; both over- and under-measurements occur. Even if the error would always be to one side, it would result only in a shift of the crack growth curve; the rates would not be much wrong:

$$\frac{a_2 + \delta - (a_1 + \delta)}{\Delta N} = \frac{a_2 - a_1}{\Delta N} = \frac{\Delta a}{\Delta N}. \tag{7.9}$$

Actual crack growth measurements, sometimes even show the crack to become smaller. This indicates that the measurement interval is too small: it is erroneous to believe that more measurements are always better. If the crack appears to become smaller the data point is useless (negative da/dN). Even if the crack 'appears' not to grow the data point is useless: zero cannot be plotted on a logarithmic scale. The above shows that wild gyrations can occur if measurements are made too close and scatter can actually get worse because too many data are taken, each of them having an error.

The main reason for the problem is the differentiation (obtaining da), an inherently inaccurate procedure. It tends to exaggerate measurement inaccuracies. For example consider a case with a measurement accuracy of 0.005 inch, which is about as good an accuracy as can be attained by any

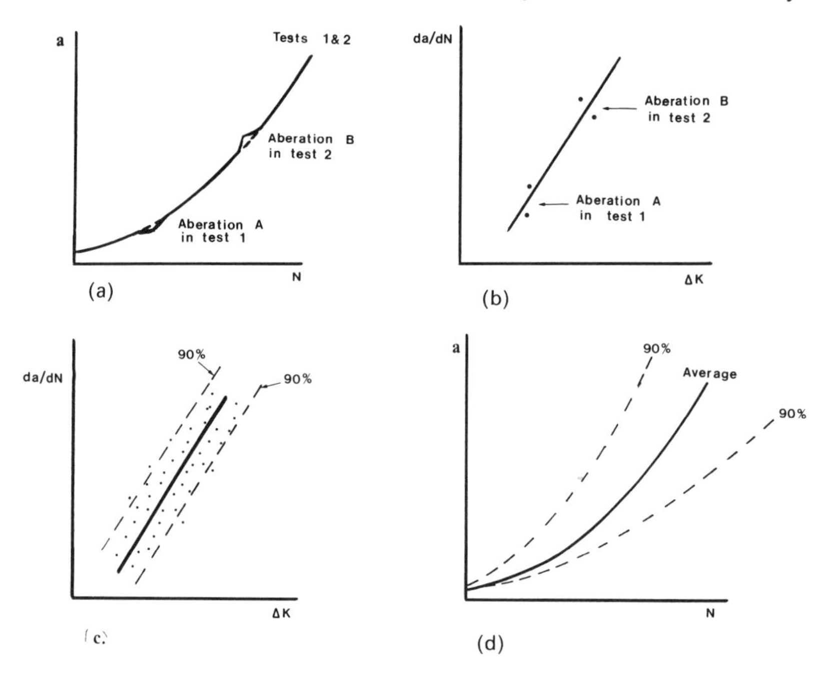

Figure 7.17. Scatter due to local inhomogenieties. (a) Measured data for two tests; (b) Rate data for two tests; (c) Multiple tests; (d) Predictions.

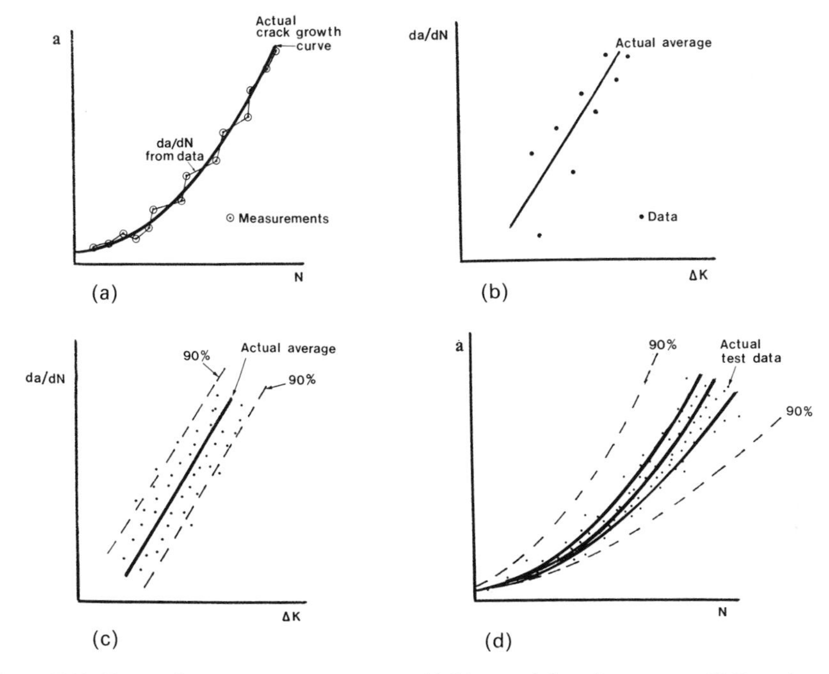

Figure 7.18. Scatter due to measurement errors. (a) Measured data for one test; (b) Rate data for one test; (c) Multiple tests; (d) Predictions.

measurement of crack size. If $a = 0.5$ inch nominally, its value is between 0.495 and 0.505 inch; the possible error is only one percent. Let the next measurement be 0.52 inch, indicating a crack size between 0.515 and 0.525, again with an accuracy of one percent. The value of Δa is then between 0.525–0.495 = 0.030 and 0.515–0.505 = 0.010, with an expected value of 0.52–0.50 = 0.020. Hence, the error in rate will be as large as 50% (0.020 ± 0.010). This problem is not unique for crack growth; it always occurs where (numerical) differentiation is performed.

To counteract this problem, the ASTM procedure recommends to take a moving average for the rate determination. Yet the differentiation will still exaggarate measurement inaccuracies, which will appear as exaggarated 'scatter' in the rate diagram. If a sufficient number of such data is accumulated the scatterband becomes impressive, even if the other sources of scatter were completely absent. The resulting problem is the same as was discussed on the basis of Figure 7.17. Upper and lower bound would give crack growth curves bearing no relation to the tests (Figure 7.18d), since all tests essentially showed the same crack growth curves. Using the errors of all measurements and assuming they all worked in one direction, ignores the mechanics and physics of what causes this 'apparent' scatter. Determining a 90% band with a statistical treatment that does not account for this reality is inappropriate.

The question now arises what to do about the scatter in practical damage tolerance analysis. Most of the scatter observed in data plots can be ignored, because it is apparent scatter only. The line representing the average of the data is the only realistic one (Figures 7.17 and 7.18). However, it is prudent to account for the batch-to-batch, heat-to-heat, and manufacturer-to-manufacturer variation (Figure 7.16), because it is not known a priori which batch of material will be used in the structure. As a rule of thumb, these effects can cause a difference of about a factor of 2 in crack growth rates between worst and best. Thus the worst would be about a factor of 1.41 higher than the average, the best a factor of 1.41 lower than the average ($1.41 \times 1.41 = 2$). On the logarithmic scale of the rate plot this will only be a slight shift. If a Walker, Paris or Forman equation is used, the correction can be effected by multiplication of C_t by a factor of 1.4, after the coefficient has been determined (see previous section) from the average of the data. If a threshold equation is used, an assumption must be made on how the threshold is affected. The threshold being a disputable property, no general rule can be given. The damage tolerance analyst will have to use judgement. In the case of tabular data, a shift of about 1.4 could be used as well, and if a threshold is used the same remarks apply as above.

With regard to the threshold in tabular data it is worthwhile noting that most computer codes cannot handle $da/dN = 0$, because log (0) is undefined. Therefore the rate at the threshold should be given as a small but non-zero number, e.g. as 10^{-20}, otherwise the computer may declare an input error and the run may be terminated prematurely.

In conclusion, it is worthwhile mentioning that integration is as forgiving as differentiation is inaccurate. Integration, being the inverse of differentiation, tends to 'cover-up' (i.e. compensate for) the errors of differentiation. Examples of this were given in Chapter 5, comparing hand analysis with computer analysis.

7.9. Accounting for the environmental effect

Should the environmental effect result in data of a form such as [13] in Figure 7.9, then none of the common equations is applicable. It might well be possible to use a more complex equation if the computer code is equipped to deal with these, but most general computer codes can deal only with the common equations or with tabular data. Thus a tabular representation of such data would be indicated.

In the case of variable amplitude loading ΔK varies from one cycle to the next. In one cycle the ΔK may call for rates at the 'plateau' level of Figure 7.9, in the next cycle ΔK may be low and call for rates close to the threshold. The baseline data were obtained at continually increasing ΔK. As the environmental effect is time dependent and measured at 'chemical equilibrium' at the crack tip, it is questionable whether in variable amplitude loading the effect is the same; the equilibrium condition for the given ΔK cannot be reached immediately in one cycle. Although procedures have been proposed to deal with the problem of time-dependence, they do not appear in general computer codes, nor do most of them consider the problem of variable amplitude loading. Thus, the user may take a risk when indiscriminately using the data for variable amplitude without the benefit of some variable amplitude test data. The latter can be used as a check. If the analysis can reasonably reproduce the test result, one may have some confidence that it can deal with the structural problems to be analysed.

In many cases the environmental effect is less complicated and common equations can still be used if so desired. The questions about equilibrium in variable amplitude loading and of time dependence, as mentioned above, remain however.

Should the environment change during crack growth, a new and major problem arises. This occurs in many structures exposed to weather. In winter the environment is cold, dry air, in summer it is warm, wet air. It should be noted here that there may be 100% relative humidity in winter, but this cold air (even if saturated) does contain much less moisture than the summer's warm air; for example air with 100% relative humidity at room temperature contains approximately 30 000 ppm of water, but air with 100% relative humidity at $-55°C$ contains only about 200 ppm of water (relative humidity is the fraction of the possible moisture content).

The problem occurs for transport vehicles, bridges, airplanes, and many other structures. Thus, while the following example is for an airplane, it applies in the same manner to other structures. A wing full of fuel may warm up considerably when the airplane is serviced, standing on the tarmac. During ascend when many of the gust and maneuver loadings are encountered, the material is warm (heat content of fuel), the air is warm and the moisture content high; at this temperature saturated air contains 30 000 ppm of water. When flying in the stratosphere the structure and fuel cools to $-55°C$. Cyclic loading still occurs, but at this temperature even saturated air contains only 200 ppm of water. If the flight is over the ocean, the air will contain a certain amount of salt. Upon descend, the structure is cold (due to cold fuel) but the air warm with much more water. On top of this, the air may contain sulphuric acid (pollution). One side of the crack (if it is a through crack in the skin) is exposed to air, the other side to jet fuel. Clearly, such an environment and the changes thereof, defies any theory and any modeling. Not even tests can conceivably provide a solution to this problem. Only pragmatic engineering is of use.

Pragmatism can be exercised in many ways, depending upon goals, outlook, and desirable conservatism. Assumptions must be made and they can have far reaching consequences. No categoric recipes can be given, only an example. Consider a case where the data set can be represented by a Paris or Walker equation, all with the same exponent(s). Consequently all environmental effects would be only in the coefficient C. Estimating the relative times spent in each environment, one could take a weighted average of C as shown in Table 7.3. The final equation then will be as shown (note: the numbers in Table 7.3 are fictitious).

Naturally, these assumptions are disputable, but so are all other assumptions. Clearly, the time spent in the different environments can be estimated only, so that more 'refined' assumptions are as good as the above estimates. Will different assumptions and refinements do any more than complicate the procedure while not giving better results? The reader be the judge.

The above procedure is an example only, but it shows roughly how the problem is handled in some cases. Every analyst encountering similar situations must make such approximations and judgements, but the problem cannot be ignored. Obviously, a problem such as this is not likely to be solved by theoretical considerations. Fracture mechanics cannot be blamed, since this is an engineering problem. With the potential inaccuracies following from this inescapable engineering approach, it is hardly realistic to require extreme accuracy in stress and geometry factors.

7.10. Obtaining retardation parameters

All retardation models contain 'unknown' parameters; if they do not, too many assumptions were made, e.g. about the accuracy of the plastic zone size and constraint equations, the yield strength (arbitrarily defined), the strain

Table 7.3. Pragmatic estimate of da/dN for changing environment for the case of equal m_p for all curves

m_p = 3 (Paris)
C_p = 1 E-8 (salt air)
C_p = 3 E-9 (air at RT)
C_p = 5 E-10 (cold air)

All data fictitious.

Experience (estimated) % time	Environment	C_p	Weighted C_p
10	Salt air	1 E-8	0.1 × 1 E-8 = 1.0 E-9
30	Air RT	3 E-9	0.3 × 3 E-9 = 0.9 E-9
60	Cold air	5 E-10	0.6 × 5 E-10 = 0.3 E-9
Total		Weighted average:	2.2 E-9

Average C_p = 2.2 E-9; m_p = 3.

hardening etc. Models without any unknown parameters can hardly be expected to have general applicability other than by happenstance. Any model that does contain such parameters can be made to work because the parameters can be obtained (adjusted) empirically. (Chapter 5).

From an engineering point of view there is no objection against empirical adjustment, provided the result is reasonably general. Objections against empericism are somewhat exaggerated.

ALL material properties used in engineering are obtained from tests. Crack growth data (da/dN) must be obtained from tests as well. Then it is hardly objectionable to use empirical retardation parameters. The only objection can be that retardation models are primitive and may not reflect the actual physics of retardation, so that the empirical parameter may not be applicable for general usage.

For the latter reason a model with one and only one (adjustable) parameter is the most attractive. Without any such parameters, the model can hardly be expected to be general; with too many parameters, it will always be possible to fit a specific case, but the generality becomes more dubious.

Regardless of the number of parameters, an experimental determination of their values is necessary. This is called calibration. For any such calibration at least one test is needed for the type of variable amplitude loading relevant to the structure. It must contain the signature of that loading as discussed in Chapter 6. The test can be done on a simple specimen (β is not important as it is not the β-analysis that is at issue, but the retardation model). CT specimens should not be employed for this purpose unless the stress history contains no compressive stresses at all. In the case of compressive loading, the stress distribution (load path) in a CT specimen bears no resemblance to that around structural cracks in compression. (The specimen can be used in tension because of the uniqueness of the near crack tip stress field.) Although one test is necessary, results of several tests are preferable for a reliable calibration.

Performing the calibration is a simple matter, especially if there is only one parameter in the model. Predictive crack growth analysis is performed, using the proper loading spectrum with the appropriate da/dN data, and assumed parameters for the retardation model. The analysis is repeated several times with different parameter value(s), and the results compared with the test data. The parameter value giving the best reproduction of the test data is the sought value.

An example for an aircraft spectrum loading case using the Wheeler retardation model was shown in Figure 5.18. Note that the case with a Wheeler exponent, $\gamma = 0$, represents the case without retardation, called the linear case. The results in Figure 5.18 show that $\gamma = 1.4$ gives the best representation of the test.

The generality of this calibration must now be brought to trial. In practice it must assumed to be rather general, but in a research project generality can be investigated by performing analysis for many different stress histories and by comparing the results with data obtained from tests using similar stress histories. Examples of such a generality check were presented in Figures 5.19 and 5.22. Data from tests with considerably different spectra (causing largely different crack growth) were all covered very well by the analysis. It should be noted that an analytical result within 20% of the test life is very good, in comparison with errors caused by the data – interpretation (see previous spectrum) and those arising from inaccuracies in stress history prediction and clipping (Chapter 6).

Unfortunately, true generality cannot be claimed. First of all, retardation depends upon the yield strength, and therefore the parameters are material specific. They must be determined for each alloy separately (as must the da/dN data). Second, and more important they are spectrum dependent. They can be used for variations of the same type of spectrum as shown in Figures 5.22, but they are not applicable to a spectrum of altogether different shape (Chapter 6): retardation parameters determined for an off-shore spectrum cannot be used for a pipe-line (even if the material were the same). Retardation depends upon the mixture of high and low loads. A different spectrum shape (Chapter 6) will give a different mixture of high and low loads. Its retardation parameters must be determined separately, because after all, retardation models are simplifications. Hence, retardation parameters are spectrum specific. Nature induced log-linear spectra need different parameters than man-induced spectra (Chapter 6), and so do spectra for e.g. rotating machinery. Carrying over retardation parameters from one type of spectrum to another is not permitted; it is one of the main reasons why some retardation models are acclaimed to be better than others. Each model should be calibrated for the spectrum relevant to the application. If it is, any model will perform satisfactorily, if the provisions stipulated in Chapter 6 are implemented.

Naturally, one can ignore retardation altogether, which is almost always conservative. Doing this is also making an assumption, namely that there is no retardation. It certainly will not provide accurate analysis results (though conservative). Ignoring retardation on the basis of the argument that it cannot be properly accounted for, is naively believing that the implied crude assumption of no retardation is better than an engineering approximation. Calibration of the retardation model has several additional advantages. For example, the problem of data interpretation is partially resolved. In the calibration analysis one already uses the interpreted data, so that any misinterpretations will be automatically compensated for in the calibration parameter(s). Also the problem of rainflow counting (Chapter 6) is largely resolved. If the stress history is not 'counted' in the calibration analysis, potential errors due to counting are automatically compensated for in the parameters derived from the calibration analysis. Hence, in subsequent predictions the problem can be ignored as well. Should one elect to use counting in the calibration analysis the parameters may be found to have different values if there happens to be an effect of counting at all (Chapter 6). Then of course, one must use counting in subsequent analysis as well.

A problem often not recognized is that the calibration is also specific to the computer code used. Some computer codes account for changes in state of stress, others do not. Some use different equations for the plastic zone size than others. Consequently, the calibration parameters obtained with different computer codes will be different as well; they depend upon the equations used. Every retardation model must be calibrated with the same computer code as used for analysis. Transfer of calibration parameters from one code to another is not permitted, unless the codes use exactly the same equations for plastic zone, state of stress, etc., and use the same definitions of F_{ty}. This may seem cumbersome and unscientific, but the reason is that the above equations and definitions are arbitrary. If one specific computer code is used however, for calibration as well as for predictions, this is only a one-time problem. The transfer of calibration factors may be another reason for acclaimed inaccuracy of some models.

Despite the arguments in this section, objections may still be raised against retardation handling and calibration. Unfortunately then there is no other alternative than using linear analysis, which is not very accurate either. In the extreme one might forego all analysis; this would provide no information at all. Waiting for the perfect retardation model is no way out. Even the perfect model must work with interpreted estimates of future stress histories. In that respect, they are hardly different from weather predictions; the reader be the judge of the latter's accuracy. Again, the resulting inaccuracies are not due to fracture mechanics, but due to assumptions made for input.

7.11. Exercises

1. Estimate K_c for a material with $F_{ty} = 80$ ksi and $B = 0.3$ inch, if the quoted plane strain toughness is $K_{Ic} = 40$ ksi $\sqrt{\text{in}}$, and the plane stress toughness 90 ksi $\sqrt{\text{in}}$.

2. A center cracked panel test shows *a* $K_c = 75$ ksi $\sqrt{\text{in}}$. The panel was 16 inch wide, and the fatigue crack size was eight inches. Estimate K_{eff}, if there was a stable fracture causing $\Delta a = 0.3$ inch, and $F_{ty} = 80$ ksi.

3. Supposing $F_{ty} = 35$ ksi for the material in Exercise 3, estimate K_c.

4. If the Charpy energy is 16 ft/lbs at room temperature and the yield strength $F_{ty} = 70$ ksi, estimate the toughness. For which temperature can this toughness still be used if the loading rate is low?

5. The LT toughness for material A is 50 ksi $\sqrt{\text{in}}$. Estimate the toughness of material B, assuming both materials have the same composition and if material A has a yield strength of 60 ksi and B a yield strength of 75 ksi. Estimate the toughness for B if the yield strength values are reversed.

6. The transitional toughness for a thickness of 0.5 inch of a material with $F_{ty} = 70$ ksi has been determined as $K_c = 60$ ksi $\sqrt{\text{in}}$, the plane strain toughness as $K_{Ic} = 40$ ksi $\sqrt{\text{in}}$. Estimate the toughness for a through-crack in a plate of 0.3 inch thickness.

7. Determine the parameters for the Walker equation for the data in Figure 7.12 in your own way by reading the data yourself. Then calculate the data lines with your equation and draw your own conclusions.

8. Determine the parameters for the Forman equation for the data in Figure' 7.12 in your own way by reading the data and draw your own conclusions for recalculating the data with your own equation.

9. For the cases in Exercises 7 convert the parameters to units of psi $\sqrt{\text{in}}$.

10. Rederive the parameters in Exercises 9 by first converting the scales. Compare with the results.

11. Using the data of Table 7.3 obtain the best constant for an equation governing five percent usage in salt air, 25% in room temperature air, remainder in cold air.

References

[1] Standard test method for plane-strain fracture toughness of metallic materials ASTM standard E-399.

[2] A standard method for the determination of J, a measure of fracture toughness, ASTM standard E-813.

[3] Methods for crack opening displacement (COD) testing, British Standards Institution BS.5762.

[4] Standard recommended practice for R-curve determination ASTM standard C-561.

[5] Tentative test method for constant-load-amplitude fatigue crack growth rates above 10^{-8} m/cycle ASTM standard E-647.

[6] Anon., *Damage tolerant design handbook*. Mat & Ceramics Info Center, (Columbus) MCIC HB-01; yearly updates.

[7] J.E. Campbell et al. (ed.), *Application of fracture mechanics for selection of metallic structural materials*, Am. Soc. Metals (1982).

[8] D. Broek, *Elementary engineering fracture mechanics*, 4th Ed, Nijhoff (1986).

[9] M.F. Kannenin and C.H. Popelar, *Advanced fracture mechanics*, Oxford Un. Press (1985).

[10] S.T. Rolfe and J.M. Barson, *Fracture and fatigue control in structures*, Prentice-Hall (1977).

[11] R. Roberts and C. Newton, Interpretive report in small scale test correlation with K_{Ic} data, *Welding Res. Council. Bulletin* **265** (1981).

[12] B. Marandet and G. Sanz, Evaluation of the toughness of thick medium strength steels by using LEFM and correlations, *ASTM STP* **631** (1977) pp. 72–84.

[13] O. Vosikovsky, Fatigue crack growth in X-65 line pipe steel at low frequencies in aquous environments, ASME trans **H97** (1975) pp. 298–305.

CHAPTER 8

Geometry factors

8.1. Scope

For the solution to any fracture or crack growth problem the analyst must know the geometry factors for either K, J, or both, for the structural crack of interest. Geometry factors for many generic configurations already have been obtained and compiled in handbooks [1, 2, 3]. This can be done a priori for generic loading and geometries, but actual structural details are often unique so that ready made handbook solutions cannot be expected to be available.

In such a case a formal solution can be obtained in principle but due to the complexity of many structural details such analysis may be too costly or prohibitive for any number of other reasons. Structural cracks usually are of the part-through type so that 3-D analysis would be indicated. If the structure is expensive, the cost of fracture high and only one type of crack is of interest, a finite element analysis may be a solution. If on the other hand one must consider literally hundreds of potential cracks, it is not possible to obtain geometry factors in this manner other than for a few of the most critical cases. Finite element analysis of the uncracked structure may well be performed, but detailed finite element analysis of models with cracks for hundreds of potential crack locations is prohibitive.

There is a great need for simple (be it approximative) methods to obtain geometry factors. Fortunately, there are many practical and easy procedures which can provide geometry factors with good accuracy; they are almost always adequate for engineering applications in view of the general accuracy of damage tolerance analysis (Chapter 12).

Methods to obtain geometry factors are the following:

(a) Direct use of handbook solutions.

(b) Indirect use of handbook solutions through superposition and compounding.

(c) Methods based on insight and engineering judgement, combined with (b) and (a).

(d) Use of Green's functions or weight functions, if necessary in combination with finite element stress analysis of the uncracked structure.

(e) Detailed finite element analysis of models with cracks.

Although all these will be discussed in this chapter, emphasis will be on simple methods. Where possible, the success and accuracy of these will be illustrated through application to cases for which solutions are known.

Methods (a), (b), (c), and (d) are already included in some of the general purpose fracture mechanics software [4], but they can be done easily by hand in a short time. Use of all methods requires access to one of the handbooks [1, 2, 3].

In many cases one must define a reference stress for the stress intensity factor, and this reference stress must be used consistently throughout the analysis. Large errors may occur when this is not done correctly. For this reason the problem of the reference stress will be discussed first in Section 8.2.

8.2. The reference stress

The stress intensity factor is defined as $K = \beta\sigma\sqrt{\pi a}$, in which σ is the nominal stress away from the crack. The geometry factor β accounts for the fact that average stresses in the cracked section are higher, as well as for all free boundaries affecting the crack tip stress as expressed by K. Thus $\beta = \beta(a/W, a/D, a/S, \ldots)$, where W, D, and S are relevant structural dimensions (Chapter 3), or in general $\beta = \beta(a/L)$, L being a generalized length. Determining geometry factors means derivation of the function $\beta(a/L)$ for the specific loading and geometry details relevant to the crack to be analyzed. Simple methods to derive these functions are the subject of this chapter. The way in which the function is to be used in crack growth and fracture analysis depends upon the definition of stress as well, as shown below.

In the case of uniform applied stress there is no problem in the definition of σ in the stress intensity factor. But if the stress distribution is non-uniform it may not be immediately obvious which stress should be used in the expression for K. An example of this was given already in Chapter 3 where the stress intensity for the compact tension specimen was discussed. Commonly the latter is expressed in load instead of stress, because 'the stress' in a compact tension specimen cannot be defined easily. Use of the load presents no problems in the evaluation of a toughness test result, but in structural analysis stresses are used, not loads. Especially in crack growth analysis for complex structures, the use of loads would be awkward; more specifically, computer crack growth analysis codes are based upon stress. It was shown in Chapter 3 that the problem can be mended for the compact tension specimen, by simply defining a reference stress as $\sigma = P/WB$. Although this reference stress has no physical significance, the stress intensity is evaluated correctly, provided β is changed in accordance with

Equation (3.29). All analysis can then be based upon this reference stress.

Before turning to the case of non-uniform stress distributions, it may be worthwhile to consider another example. For a central crack of size $2a$ in a plate of width W under uniform tension it has been shown that $\beta = \sqrt{\sec(\pi a/W)}$, and $K = \beta\sigma\sqrt{\pi a} = \sqrt{\sec(\pi a/W)}\,\pi\sqrt{\pi a}$. Should one insist on using a reference stress different from σ in this expression, one can legitimately do so. For example, one could use the average stress in the cracked section, σ_{net}, which is given by $\sigma = \sigma_{net} W/(W - 2a)$. Then K would become:

$$\left.\begin{aligned} K &= \frac{\sqrt{\sec \pi a/W}}{1 - 2a/W}\,\sigma_{net}\sqrt{\pi a} = \beta\,\sigma_{net}\sqrt{\pi a} \\ \text{with} \\ \beta &= \frac{\sqrt{\sec \pi a/W}}{1 - 2a/W} \end{aligned}\right\}. \qquad (8.1)$$

Obviously, the values of K obtained would be identical to those based upon σ. Consistent use of Equation (8.1) in a residual strength analysis would provide as output the fracture stress in terms of σ_{net} (from which σ could be obtained). Similarly, cyclic stress input in a crack growth analysis should then be in terms of σ_{net}, and the input for β should be in accordance with Equation (8.1). In this case a reference stress other than σ offers no advantage, but the example illustrates the principle. Any reference stress can be used provided β is adjusted accordingly, so that the product $\beta\sigma$ remains the unaffected.

The simplest case of a non-uniform stress distribution is a bending moment (Figure 8.1). The β for this case can be found in handbooks, but different handbooks may provide different β's. This difference is due to the use of a different reference stress, as shown in the figure. In one case the reference is the maximum bending stress in the outer fiber, σ_{max}, while in the other case it is the bending stress at $x = a$ given as $\sigma_{xa} = (1 - 2a/W)\sigma_{max}$. Defining β as β_m and β_{xa} respectively for the two cases, a conversion can be made as follows:

$$\left.\begin{aligned} K &= \beta_m\,\sigma_{max}\sqrt{\pi a} = \beta_{xa}\,\sigma_{xa}\sqrt{\pi a} \\ \beta_{xa} &= \beta_m \cdot \frac{\sigma_{max}}{\sigma_{xa}} = \frac{\beta_m}{(1 - 2a/W)} \end{aligned}\right\}. \qquad (8.2)$$

Clearly β_{xa} goes to infinity for $a = W/2$, because σ_{xa} goes to zero, while the stress intensity is finite. Numerical evaluation of a few cases will readily demonstrate that the two lead to the same value of K. In this case, the maximum bending stress is obviously the easiest to use. In employing handbook solutions one should ascertain which reference stress is used, and make conversions if desirable. It is good practice to define the reference stress when quoting β.

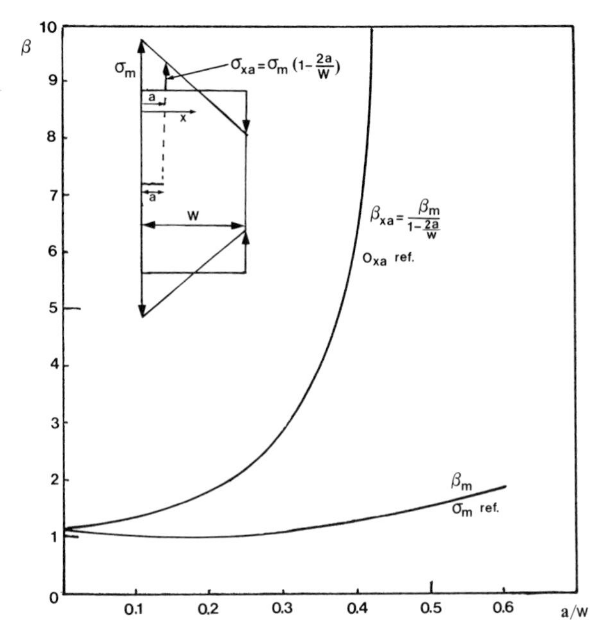

Figure 8.1. Geometry factors for different reference stresses in bending.

The need for a clearly defined reference stress is most obvious in the case of a generally non-uniform stress distribution, such as in Figure 8.2. The highest stress is usually the best choice. But all stresses are proportional to load (elastic), so that the stress at any point is proportional to the highest stress in the distribution. Hence, any other local stress can be used as the reference, by multiplying β by the proportionality factor. With the nomenclature of Figure

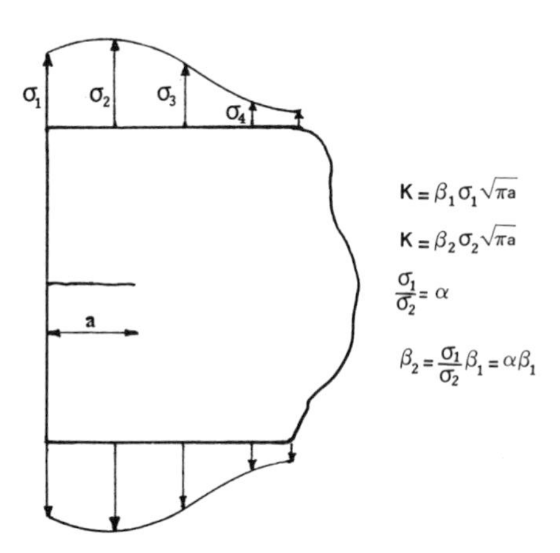

Figure 8.2. Reference stress in case of non-uniform stress distribution.

8.2:

$$\beta_2 = \frac{\sigma_1}{\sigma_2} \beta_1. \tag{8.3}$$

It is emphasized once more that the reference stress must be used consistently. The β-input to the crack growth analysis must be for the proper reference stress, and the input of stress ranges, exceedance diagrams, stress histories and/or stress occurrence tables must be in terms of the reference stress. The output of a residual strength analysis will be in terms of the reference stress and may require further interpretation. For example, let β_2 in Equation (8.3) for a crack size of e.g. two inches be 1.1, and $\sigma_2/\sigma_1 = 2$, so that $\beta_1 = 2.2$. Supposing $K_{Ic} = 60\,\text{ksi}\sqrt{\text{in}}$, the residual strength would be evaluated for the two cases as $\sigma_1 = 60/(2.2\sqrt{\pi.2}) = 10.9\,\text{ksi}$, and $\sigma_2 = 60/(1.1\sqrt{\pi.2}) = 21.8\,\text{ksi}$. The latter solution predicts fracture to occur when the local stress σ_1 has the value 10.9 ksi. This is the case when the maximum stress in the distribution is $2 \times 10.9 = 21.8\,\text{ksi}$. Although this may seem trivial in this example, misinterpretations are easily made when input and results of a lengthy analysis are reconsidered at some later time or reviewed by others. Providing the definition of the reference stress in input and output tables is a good practice.

8.3. Compounding

In the general expression for the stress intensity factor $K = \beta\sigma\sqrt{\kappa a}$, the geometry factor β accounts for the effect of all boundaries: $\beta = f(a/W, a/D \ldots)$, where W, D, etc. are relevant dimensions of the structure. In many cases the individual effects of these boundaries can be found in handbooks; Their composite effect is obtained by compounding, which is multiplication of all individual effects.

Possibly, the most prominent example of compounding is demonstrated by the classical solution (other solutions have since been obtained [5–8]) for the elliptical surface flaw (Figure 8.3). The various boundary effects are due to: back free surface (BFS), front free surface (FFS), width (W), and crack front curvature (CFC), i.e.

$$\left.\begin{aligned} K &= \beta_{\text{BFS}}\,\beta_{\text{FFS}}\,\beta_w\,\beta_{\text{CFC}}\,\sigma\sqrt{\pi a} = \beta\sigma\sqrt{\pi a} \\ \beta &= \beta_{\text{BFS}}\,\beta_{\text{FFS}}\,\beta_w\,\beta_{\text{CFC}} \end{aligned}\right\}. \tag{8.4}$$

If W is large, $\beta_w = 1$, and $\beta_{\text{BFS}} = 1.12$; then the classical solution provides:

$$\text{K} = 1.12\,\beta_{\text{FFS}} \frac{(\sin^2\varphi + a^2/c^2\cos^2\varphi)^{1/4}}{\int_0^{\pi/2} \{1 - (1 - a^2/c^2)\sin^2\varphi\}^{1/2}\, d\varphi}\,\sigma\sqrt{\pi a}. \tag{8.5}$$

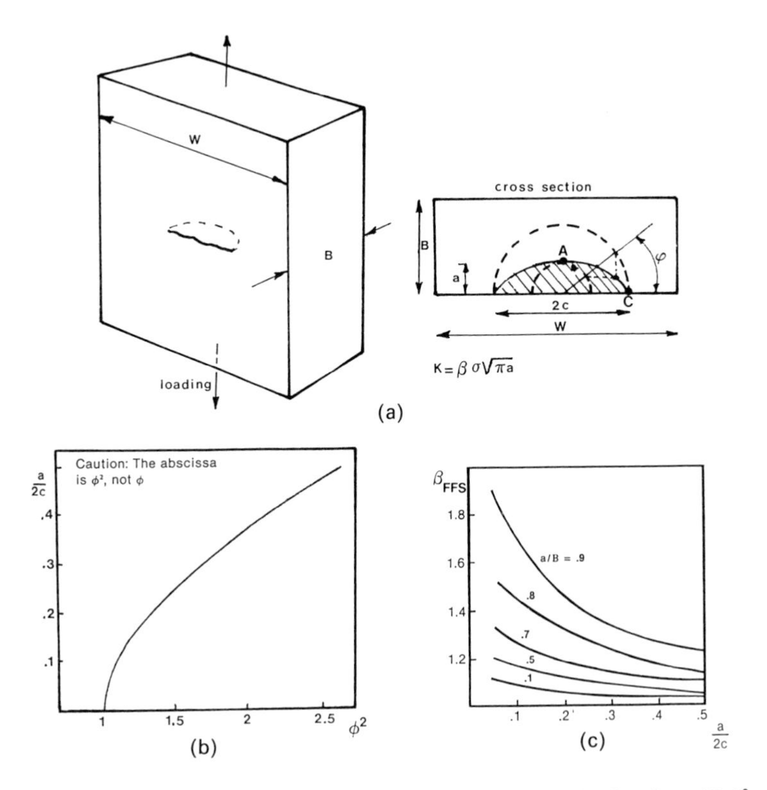

Figure 8.3. Stress Intensity of Surface Flaw under uniform tension. (a) Surface flaw; (b) ϕ^2 versus $a/2c$; (c) β_{FFS} versus $a/2c$.

The effect of the back free surface is simple, and almost always taken as $\beta_{BFS} = 1.12$ regardless of crack type (Section 8.5). The effect of the front free surface depends upon crack shape (Figure 8.3c). Finally, the effect of curvature $\beta_{CFS} = 1/\phi$ is a complicated function of the parametric angle, φ, and the flaw aspect ratio a/c. The elliptical integral in the denominator can be and has been evaluated once and for all for various a/c as is shown in Figure 8.3b. Its value is ϕ, but the common representation is as in Figure 8.3b, providing $Q = \phi^2$ instead of ϕ. For a crack with a certain aspect ratio a/c (or $a/2c$), the value of Q can be read from this diagram and then $\phi = \sqrt{Q}$ is obtained. Alternatively, the curve can be fitted with an equation so that ϕ can be calculated in a simple manner.

The function of φ in the numerator must be evaluated for a certain point at the crack front. Note that Equation (8.5) implies that the stress intensity varies along the crack front. For the deepest point A ($\varphi = \pi/2$; Figure 8.3), the value of $f(\varphi)$ equals 1 because $\cos(\pi/2) = 0$, so that:

$$\left.\begin{aligned} \beta_A &= 1.12\,\beta_{\text{FFS}}\frac{1}{\sqrt{Q}} \\ K^A &= \beta_A\,\sigma\,\sqrt{\pi a} \end{aligned}\right\} \tag{8.6}$$

in which β_{FFS} and Q follow from the graphs in Figure 8.3.

For point C at the surface ($\varphi = 0$; $\sin\varphi = 0$; $\cos\varphi = 1$) the numerator in Equation (8.5) becomes equal to $\sqrt{a/c}$, and therefore:

$$\begin{aligned} \beta_c &= 1.12\,\beta_{\text{FFS}}\frac{1}{\sqrt{Q}}\sqrt{a/c} \\ K^c &= \beta_c\,\sigma\,\sqrt{\pi a}\,. \end{aligned} \tag{8.7}$$

Note that $\beta_c = \beta_A\sqrt{a/c}$, so that with $a < c$, the stress intensity at the deepest point of the flaw is higher than at the surface ($\rho_A > \beta_c$).

There is no objection in expressing K^c in terms of c instead of a; in that case:

$$\left.\begin{aligned} K^c &= 1.12\,\beta_{\text{FFS}}\frac{1}{\sqrt{Q}}\sqrt{a/c}\,\sqrt{a/c}\,\sigma\,\sqrt{\pi c} = \bar{\beta}_c\,\sigma\,\sqrt{\pi a} \\ \bar{\beta}_c &= 1.12\,\beta_{\text{FFS}}\frac{1}{\sqrt{Q}}\,a/c = \beta_c\,\sqrt{a/c} = \beta_a\,\frac{a}{c} \end{aligned}\right\}. \tag{8.8}$$

Equations (8.7) and (8.8) are equivalent, provided β is properly adjusted as in Equation (8.8). In crack growth analysis it is often more convenient to express both K^A and K^c in terms of a by using Equations (8.6) and (8.7). The different stress intensities at a and c have significant consequences for the behavior of surface flaws, as discussed in Chapter 9.

In obtaining β for a structural crack in a complex geometry, the effect of the individual boundaries can often be found in handbooks or determined otherwise (see following sections). By compounding these effects the 'total' b is obtained. More examples of compounding will appear in this chapter. It should be noted that rigorous compounding adheres to slightly different rules [9], but the procedure shown here is generally used and accepted. In view of the general accuracy of damage tolerance analysis (Chapter 12) the above method is acceptable.

8.4. Superposition

While compounding is multiplication of geometry factors, superposition is addition of stress intensity factors due to various mode I loadings. For example (Figure 8.4) in a combination of bending and tension the total crack tip stress from Equation (3.1) is:

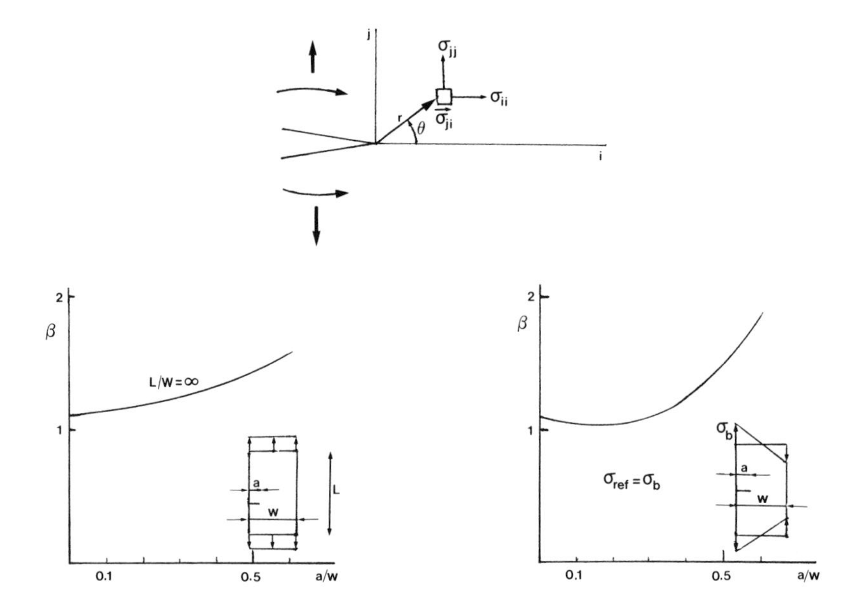

Figure 8.4. combination of tension and bending.

$$\sigma_{ij} = \frac{K_{\text{ben}}}{\sqrt{2\pi r}} f_{ij}(\theta) + \frac{K_{\text{ten}}}{\sqrt{2\pi r}} f_{ij}(\theta). \tag{8.9}$$

Because the solution of the crack tip stress field is universal, the functions $f_{ij}(\theta)$ in both terms of Equation (8.9) are identical for each $i-j$ combination, so that the equation can be written as:

$$\sigma_{ij} = \frac{K_{\text{ben}} + K_{\text{ten}}}{\sqrt{2\pi r}} f_{ij}(\theta) = \frac{K_{\text{tot}}}{\sqrt{2\pi r}} f_{ij}(\theta). \tag{8.10}$$

The total stress intensity is:

$$K_{\text{tot}} = K_{\text{ten}} + K_{\text{ben}}. \tag{8.11}$$

Apparently the total K follows from the addition of stress intensity factors, owing to the fact that both terms in Equation (8.9) for any of the individual stresses contain the same function of θ. It should be emphasized that superposition of stress intensities of different modes of loading is NOT possible (Chapter 9). In such a case different θ-functions apply to the different modes and the the step from Equations (8.9) to (8.10) cannot be made. For a discussion of combined mode loading see Chapter 9.

As long as all loading is mode I superposition is permissible. For example, if there is tension, in and out of plane bending, plus pressure inside the crack, the total stress intensity becomes:

$$K_{\text{tot}} = K_{\text{beno.u.p.}} + K_{\text{beni.p.}} + K_{\text{ten}} + K_{\text{pr}}. \tag{8.12}$$

After the superposition is completed, β must be obtained by selecting a suitable reference stress. Generally, all damage tolerance analysis is based on $K = \beta\sigma\sqrt{\pi a}$, and Equation (8.12) is not very suitable for analysis.

Assume a combined bending and tension case as in Figure 8.4; the geometry factors for the two cases are as shown. The stress intensity is:

$$K_{\text{tot}} = \beta_{\text{ben}}\,\sigma_{\text{ben}}\sqrt{\pi a} + \beta_{\text{ten}}\,\sigma_{\text{ten}}\sqrt{\pi a}. \tag{8.13}$$

In order to obtain β for the combination a reference stress must be selected. This can be any of σ_{ten}, σ_{ben}, $\sigma_{\text{tot}} = \sigma_{\text{ben}} + \sigma_{\text{ten}}$, or any other suitable stress (Section 8.2). Selection of σ_{ten} leads to:

$$\left.\begin{aligned} K &= \left(\beta_{\text{ben}}\frac{\sigma_{\text{ben}}}{\sigma_{\text{ten}}} + \beta_{\text{ten}}\right)\sigma_{\text{ten}}\sqrt{\pi a} = \beta\,\sigma_{\text{ten}}\sqrt{\pi a} \\ \beta &= \beta_{\text{ben}}\frac{\sigma_{\text{ben}}}{\sigma_{\text{ten}}} + \beta_{\text{ten}} \end{aligned}\right\} \tag{8.14}$$

while the use of σ_{tot} provides:

$$\left.\begin{aligned} K &= \left(\beta_{\text{ben}}\frac{\sigma_{\text{ben}}}{\sigma_{\text{tot}}} + \beta_{\text{ten}}\frac{\sigma_{\text{ten}}}{\sigma_{\text{tot}}}\right)\sigma_{\text{tot}}\sqrt{\pi a} = \beta\,\sigma_{\text{tot}}\sqrt{\pi a} \\ \beta &= \beta_{\text{ben}}\frac{\sigma_{\text{ben}}}{\sigma_{\text{tot}}} + \beta_{\text{ten}}\frac{\sigma_{\text{ten}}}{\sigma_{\text{tot}}} \\ \sigma_{\text{tot}} &= \sigma_{\text{ten}} + \sigma_{\text{ben}} \end{aligned}\right\}. \tag{8.15}$$

Equations (8.14) and (8.15) are equivalent and both can be used as long as one is aware that all solutions are obtained in terms of the reference stress. For example, if one wants to find the residual strength for a crack size of $a = 2.4$ inch (in an 8-inch wide panel with $K_c = 70\,\text{ksi}\sqrt{\text{in}}$) the solution is as shown in Table 8.1. The same result is obtained in either case, provided the results are interpreted correctly. The solution in terms of σ_{tot} is least likely to lead to interpretation errors, but careful application of the rules will provide correct answers in all cases.

Equation (8.13) can be used directly, provided one of the stresses is defined in value. The fracture condition $K = K_c$ (or K_{Ic}) leads to: $K_{\text{ben}} + K_{\text{ten}} = K_c$, which represents a straight line in $K_{\text{ten}} - K_{\text{ben}}$ – space, as depicted in Figure 8.5. It shows the combinations of tension and bending that lead to fracture. In the above example of a 2.4-inch crack and $K_c = 70\,\text{ksi}\sqrt{\text{in}}$, one obtains the solutions as in Figure 8.5b. If the tension stress in e.g. 12.62 ksi, an additional bending of 8.45 ksi would cause fracture (compare with results in Table 8.1).

Although Figure 8.5 is illustrative, a solution by means of Equations (8.14)

Table 8.1. Residual strength analysis with different reference stresses. (Case of Figure 8.4). Compare results of columns 7 and 9

Specifics: $W = 8$ inch; $K_c = 70\,\text{ksi}\,\sqrt{\text{in}}$
$\sigma_{ben}/\sigma_{ten} = 0.67$; $\sigma_{ten}/\sigma_{ben} = 1.5$
$\sigma_{tot} = 2.5\,\sigma_{ten}$
$\sigma_{ten}/\sigma_{tot} = 0.6$
$\sigma_{ben}/\sigma_{tot} = 0.4$

Example: $a = 2.4$ inch; $\sigma_{fr\,ten} = 12.62$ ksi
$K_{ten} = 1.27 \times 12.62 \times \sqrt{\pi \times 2.4} = 44\text{ ksi}\sqrt{\text{in}}.$
$K_{ben} = 70 - 44 = 26\text{ ksi}\sqrt{\text{in}}$
$\sigma_{ben} = 26/(1.12\sqrt{\pi \times 2.4}) = 8.45$ ksi; $\sigma_{tot} = 12.62 + 8.45 = 21$ ksi

1 a (in)	2 a/w	3 β_{ten} Figure 8.4	4 β_{ben} Figure 8.4	5 σ_{ten} reference β From Equation (8.14)	6 $\sigma_{fr\,ten} = \frac{70}{\beta\sqrt{\pi a}}$ (ksi)	7 $\sigma_{fr\,tot} = \frac{1}{0.6}\sigma_{fr\,ten}$ (ksi)	8 σ_{tot} reference β From Equation (8.15)	9 $\sigma_{fr\,tot} = \frac{70}{\beta\sqrt{\pi a}}$ (ksi)
0.8	0.1	1.15	1.045	1.850	23.87	39.78	1.108	39.85
1.6	0.2	1.20	1.055	1.907	16.37	27.28	1.142	27.34
2.4	0.3	1.27	1.120	2.020	12.62	21.03	1.210	21.07
3.2	0.4	1.35	1.255	2.191	10.08	16.80	1.312	16.83
4.0	0.5	1.45	1.500	2.455	8.04	13.40	1.470	13.43

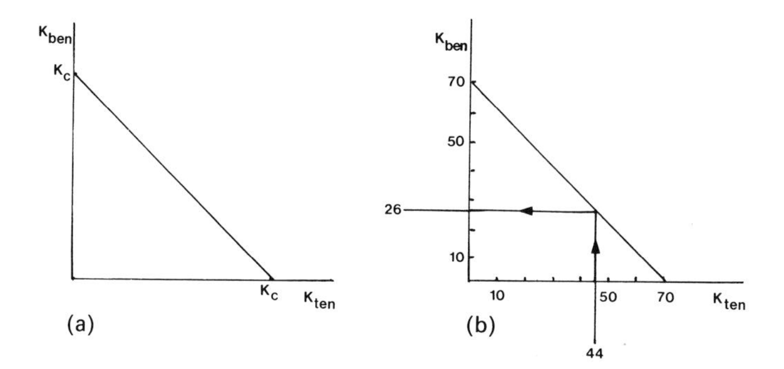

Figure 8.5. Combination of mode I loadings (right diagram shows specific example of Table 8.1).

or 8.15 is usually more convenient. Most software is based on this solution, the user deciding the reference stress. In the case of (fatigue) crack growth analysis Equation (8.13) cannot be used and some form of Equations (8.14) or (8.15) must be applied. All stresses, whether constant amplitude or variable amplitude must be expressed in terms of the reference stress, usually taken as σ_{tot}. In most loading cases, the bending and tension will be in phase (e.g. eccentric load), so that the ratio $\sigma_{ben}/\sigma_{ten}$ is always the same (and so is $\sigma_{ben}/\sigma_{tot}$ or $\sigma_{ten}/\sigma_{tot}$), regardless of the actual stress values. Hence β can be evaluated a priori with either Equations (8.14) or (8.15). As long as the table of stresses or the exceedance diagram and the β-values submitted to the crack growth analysis computer program are compatible for the proper reference stress, correct answers will be obtained.

If there is more than one boundary to be considered, compounding should be performed first, and then the superposition executed. E.g. for a crack at a hole in a plate of width W:

$$\left.\begin{aligned} \beta_{ben} &= \beta_{hole,ben}\,\beta_{w,ben} \\ \beta_{ten} &= \beta_{hole,ten}\,\beta_{w,ten} \\ \beta &= \beta_{hole,ben}\,\beta_{w,ben}\frac{\sigma_{ben}}{\sigma_{tot}} + \beta_{hole,ten}\,\beta_{w,ten}\frac{\sigma_{ten}}{\sigma_{tot}} \\ K &= \beta\,\sigma_{tot}\sqrt{\pi a} \end{aligned}\right\}. \tag{8.16}$$

The analysis then proceeds as above.

The principle of superposition has an important and very interesting consequence as is demonstrated on the basis of Figure 8.6. Consider a plate under uniform tension with no crack (case A). A similar plate with a crack of $2a$ (case B) can be 'fooled' into believing that there is no crack by applying stresses to the faces of the crack. Originally the material at the crack location was carrying a

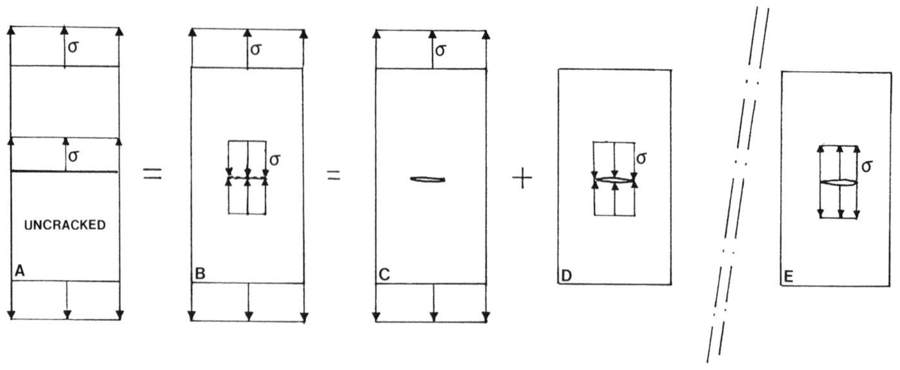

Figure 8.6. Superposition for uncracked plate.

uniform stress (case *A*). Hence, applying this uniform σ to the crack faces as in case *B* will result in the identical situation as case *A*. Since case *B* is the superposition of 2 mode I loading cases, the stress intensity of *B* (and of *A*) is equal to the sum of the stress intensities of cases *C* and *D*:

$$K^A = K^C + K^D. \tag{8.17}$$

The stress intensity of case *A* is zero ($K^A = 0$), the case of no crack, so that:

$$K^D = -K^C. \tag{8.18}$$

Case *D* cannot exist by itself as the crack faces would interfere, but if *C* is applied first then *D* can be superposed.

If we reverse the applied stresses in case *D*, the sign of the cracks tip stresses (and hence of *K*) also will change (case *E*), and hence:

$$K^E = -K^D. \tag{8.19}$$

Combination of Equation (8.18) and (8.19) yields:

$$K^E = K^C. \tag{8.20}$$

Equation (8.20) provides a new stress intensity factor for a crack loaded by internal pressure. This is useful for surface flaws along the internal wall of pressure vessels, where the pressurized medium enters the crack (Figure 8.7). However Equation (8.20) has more important implications. Note first that the picture of Figure 8.6 can be redrawn for any geometry and stress distribution, as long as the proper stresses are applied to the crack faces (Figure 8.8). The following rule emerges:

The stress intensity for any loading case is equal to the stress intensity obtained by applying to the faces of the crack the stresses that used to be there when there was no crack ($K^c = K^E$).

This rule can be used to obtain stress intensities and geometry factors in many

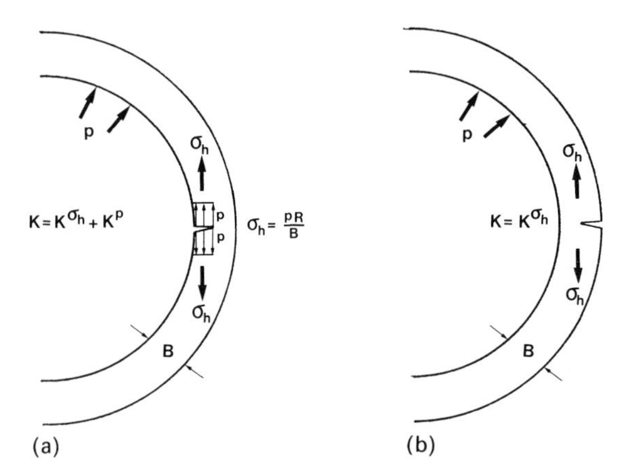

Figure 8.7. Cracks in pressurized container. (a) Internal: pressure inside crack; (b) External.

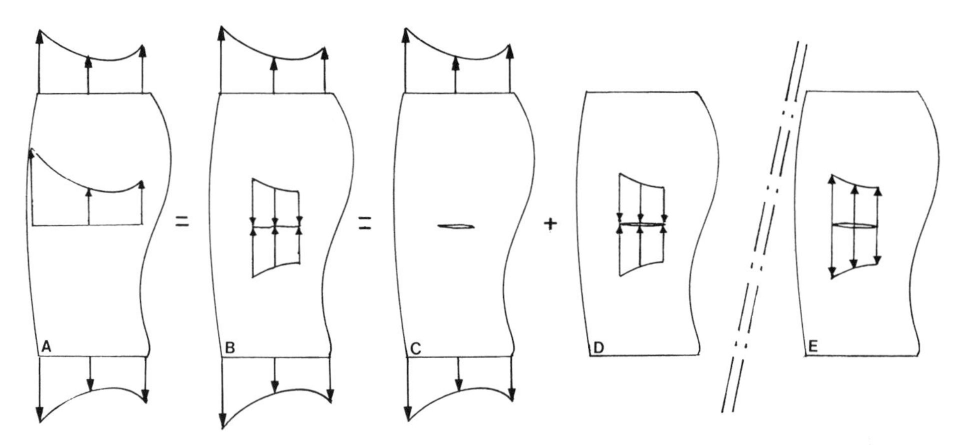

Figure 8.8. 'Uncracked stress distribution' rule.

cases by simple means as will be demonstrated at various places in the following sections. In the following the rule will be referred to as the uncracked stress distribution rule; it means that K^c can be obtained from the solution to K^E.

8.5. A simple method for asymmetric loading cases

Pictorial superposition as used in Figures 8.6 and 8.8 is often very helpful in obtaining geometry factors, especially in the case of asymmetric loading. The most prominent asymmetric loading case is the lug (Figure 8.9). This case can be built up [10] by superposition of two symmetric loading cases (*B* and *C*). Then of course the downward load *P* and the upward stress system σ are

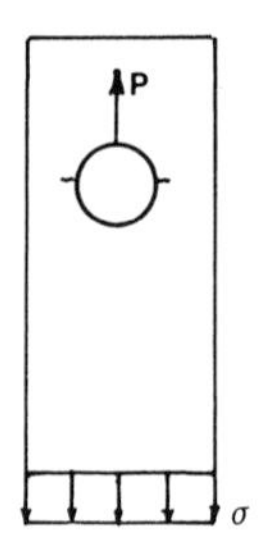

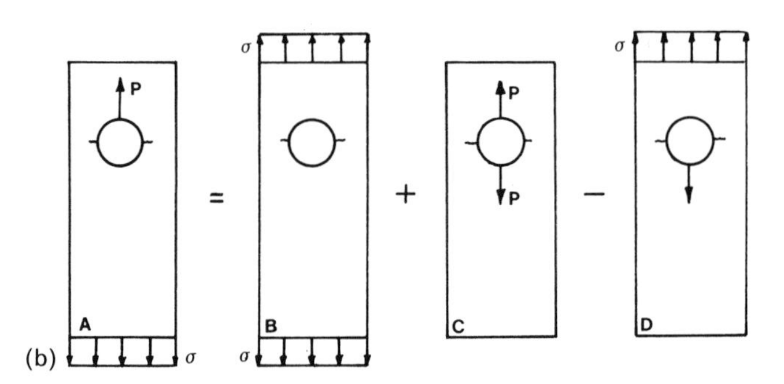

Figure 8.9. Superposition to Derive K for Asymmetric case. (a) Basic configuration; (b) Inverse superposition.

superfluous and must be subtracted (case D). Superposition yields:

$$K^A = K^B + K^C - K^D. \tag{8.21}$$

It is obvious that $K^D = K^A$ (inversing the picture does not change the stresses). Then it follows that the stress intensity of the lug crack is:

$$2\,K^A = K^B + K^C \quad \text{or} \quad K^A = \frac{K^B + K^C}{2}. \tag{8.22}$$

The geometry factors for both K^B and K^C can be found in handbooks, so that indeed K can be derived. Should the hole be negligably small, and W large, then Equation (8.22) leads to:

$$K^A = \left(\frac{P}{\sqrt{\pi a}} + \sigma\sqrt{\pi a}\right)\Big/ 2, \tag{8.23}$$

where P is the load per UNIT thickness, Equilibrium requires that $P = \sigma W$, so that:

$$\left.\begin{aligned} K^A &= \left(\frac{\sigma W}{\sqrt{\pi a}} + \sigma\sqrt{\pi a}\right)\Big/2 = \left(\frac{W}{2\pi a} + \frac{1}{2}\right)\sigma\sqrt{\pi a} = \beta\,\sigma\sqrt{\pi a} \\ \beta &= \frac{W}{2\pi a} + \frac{1}{2} \end{aligned}\right\}. \quad (8.24)$$

For larger holes with diameter D and small W, compounding of β must be effected first:

$$\left.\begin{aligned} K^A &= \left(\beta_{DP}\,\beta_{wp}\frac{P}{\sqrt{\pi a}} + \beta_{D\sigma}\,\beta_{W\sigma}\,\sigma\sqrt{\pi a}\right)\Big/2 = \beta\sigma\sqrt{\pi a} \\ \beta &= \frac{1}{2}\beta_{DP}\,\beta_{WP}\frac{W}{\pi a} + \frac{1}{2}\beta_{D\sigma}\,\beta_{W\sigma} \end{aligned}\right\}. \quad (8.25)$$

As the various compounded geometry factors may be based on different reference stresses, the equation should take proper account of, and be modified for, the selected reference stress in the way shown in previous sections.

Geometry factors for almost all asymmetric loading cases can be obtained in this manner. The procedure is to build-up the case from symmetric cases and to subtract superfluous loadings until the original case is obtained with opposite sign. (This is the same procedure as sometimes used in partial integration, where eventually the opposite of the original integral is obtained and the solution follows from $2\,I$ = sum of partials).

A solution for the general case of load bypass is shown in Figure 8.10. In the case of rows of fasteners (rivets, bolts, spotwelds) each fastener transfers part of the total load. The stress intensity is (see also Figure 8.9):

$$K^A = K^{\sigma 2} + \frac{1}{2}\left(K^P + K^{\sigma_1 - \sigma_2}\right) \quad (8.26)$$

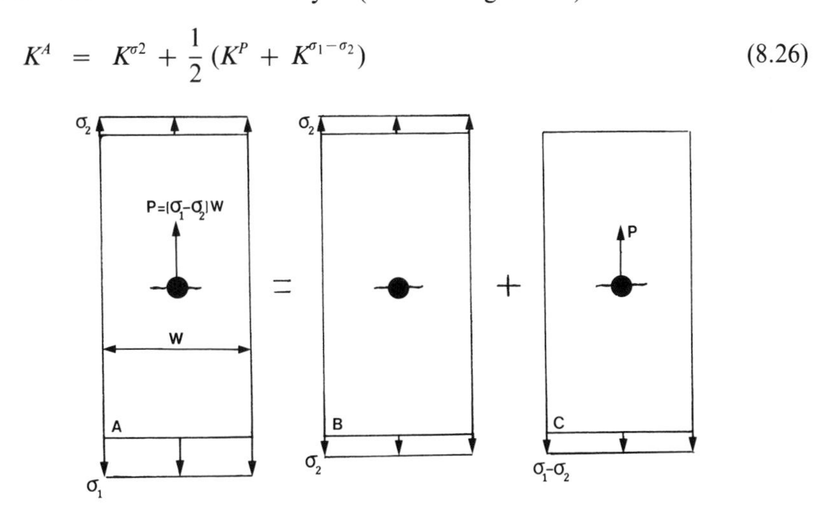

Figure 8.10. General case of asymmetric loading (combine with Figure 8.9).

which can be manipulated in the manner shown in Equations (8.22) through (8.25) to obtain β. Proper accounting for the reference stress is essential.

8.6. Some easy guesses

Geometry factors for complicated configurations often can be obtained in a simple manner, provided one develops some insights. Admittedly, the accuracy may be somewhat limited, but errors are frequently less than a few percent. Since the accuracy of damage tolerance analysis is decided primarily by loads, stresses, material data, and assumptions (Chapters 6, 12), a 5–10% accuracy in β is usually adequate. Several examples of these simple procedures are provided in the following sections, but first some trivial cases are discussed here, upon which the reader can explore other possibilities.

Consider a uniformly loaded plate as in Figure 8.11a. No stresses are acting along the center line. Cutting the plate in two (Figure 8.11b) makes no difference: two half plates can carry the load just as well as one full plate, a fail-safe or multiple-load-path feature often used. Next consider the cracked plate of Figure 8.11c. Provided a/W is small, the geometry factor is approximately $\beta = 1$. Cutting this cracked plate in two is not permissible because there are stresses acting across the center line. If the cut is made these stresses are released (Figure 8.11d, e). It is easy to see that this will open the crack a little

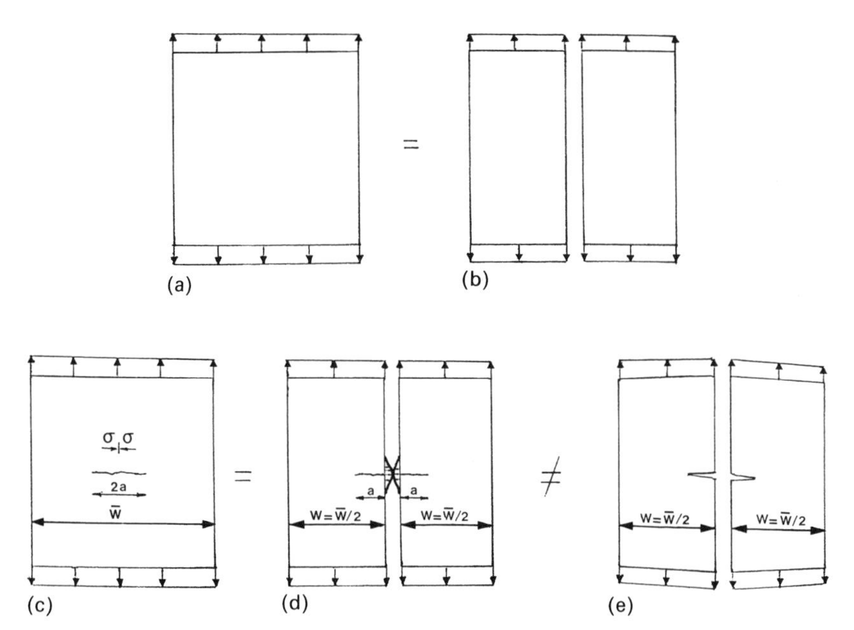

Figure 8.11. Case a = case b; Case c = case d ≠ case e. Cutting in two of uniformly loaded plates.

more: K is higher. Since $K = \beta\sigma\sqrt{\pi a}$ and σ is the same for the two halves (W is large), this implies that $\beta > 1$. The cut stresses existed very close to the crack only (Chapter 2), hence the difference from $\beta = 1$ cannot be large. It is certainly less than a factor of 2. On the other hand one would expect the difference to be more than one percent ($1.01 < \beta < 2$). Intuitively, one would guess the difference to be about 10 percent: $\beta = 1.1$. In actuality the geometry factor for the edge crack is $\beta = 1.12$ for small a/W (Figure 3.3). Hence, if one would guess $\beta = 1.1$, the estimate would have an accuracy of two percent. Better accuracy is certainly not required. Note that a guess of 1.05 would have been within seven percent and a guess of 1.15 within three percent.

For large a/W the differences might be larger. However, let us explore the premise holds for larger a/W. Then β for the single edge crack would be evaluated as $1.12\sqrt{\sec\pi a/2W}$, while the proper solution is the polynomial shown in Figure 3.3 (note that $W = \bar{W}/2$). It is easy to evaluate these two alternatives for various a/W; results are shown in Figure 8.12. The agreement, for the case of $L/W = \infty$, is within a few percent up till $a/W \approx 0.5$. But cracks larger than half the structural size are of no technical interest. Hence, the above 'guesses' are perfectly acceptable from an engineering point of view. (See also Chapter 12 on accuracy.)

Section 8.3 showed the classical solution for a surface flaw. The geometry factor for the back free surface appeared to be $\beta = 1.12$. Clearly, this geometry

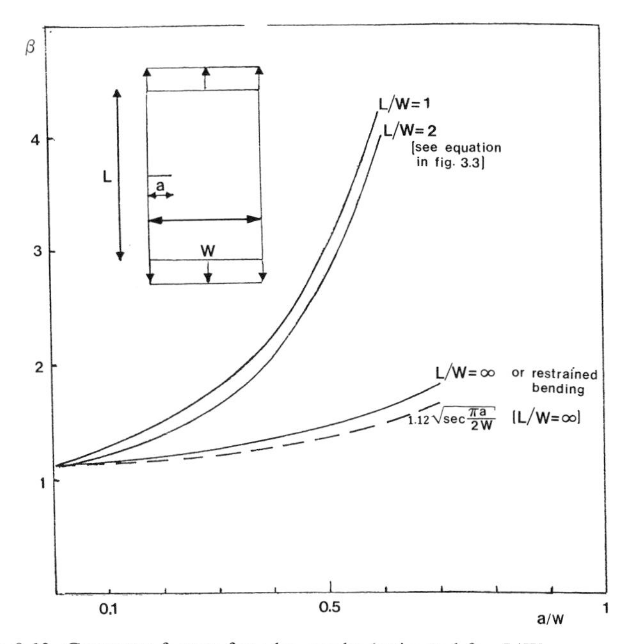

Figure 8.12. Geometry factors for edge cracks (estimated for $L/W = \infty$; see text).

factor has the same origin as the one for the edge crack discussed above. The classical solution is for an embedded elliptical crack (Figure 8.13). Cutting the plate in two to obtain a surface flaw, cuts some stresses acting across the center plane. Again, if one estimated these to be 10%, on would only introduce a two percent error, if any. It is not difficult to figure that the actual factor of 1.12 (Equation 8.5) was simply taken from the edge crack solution above, and as such was an estimate as well. This estimate is not necessarily better than the 10% estimate. Continuing this procedure, the same arguments can be used for a corner crack (Figure 8.13). Again some interface stresses are cut, so that a geometry factor for the side free surface must be introduced. From the above one would estimate $\beta_{SFS}\beta_{BFS} = 1.1^2 = 1.21$. Indeed, most handbooks provide the free surface geometry factor for corner cracks as 1.21. The above 'guess' is 'error free' as compared with the classical solution.

8.7. Simple solutions for holes and stress concentrations

Consider the symmetric case of two very small cracks at a hole in a wide plate (no effect of width). The stress concentration at a hole is $k_t = 3$ (Chapter 2), so that the local stress is 3σ (Figure 8.14a). Hence, the case is equivalent to the one shown in Figure 8.14b (note that the radius of the hole is very large with respect to the crack size if $a \to 0$). The stress intensity is $K = \beta\sigma\sqrt{\pi a}$, where σ is the nominal remote stress. Using the uncracked stress distribution rule discussed in Section 8.4, it follows immediately that $\beta = 3$. One may apply the free surface correction (Section 8.6) and obtain $\beta = 3 \times 1.12 = 3.36$.

Next consider [10] the same situation with long cracks (a/W still small), as depicted in Figure 8.15a. It will not make any difference whether there is a hole or not. If the hole were filled with material (Figure 8.15b) the latter would not carry any load because the crack face is a free surface; the filler material would simply 'go along for the ride' undergoing rigid body motion only. Hence the case of Figure 8.15b is identical to that of Figure 8.15a: the crack behaves as if the

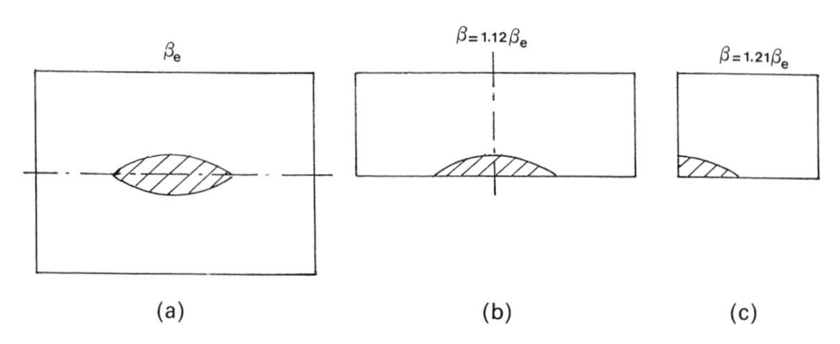

Figure 8.13. Solution for embedded crack applied to part-through cracks. (a) Embedded crack; (b) Surface crack; (c) Corner crack.

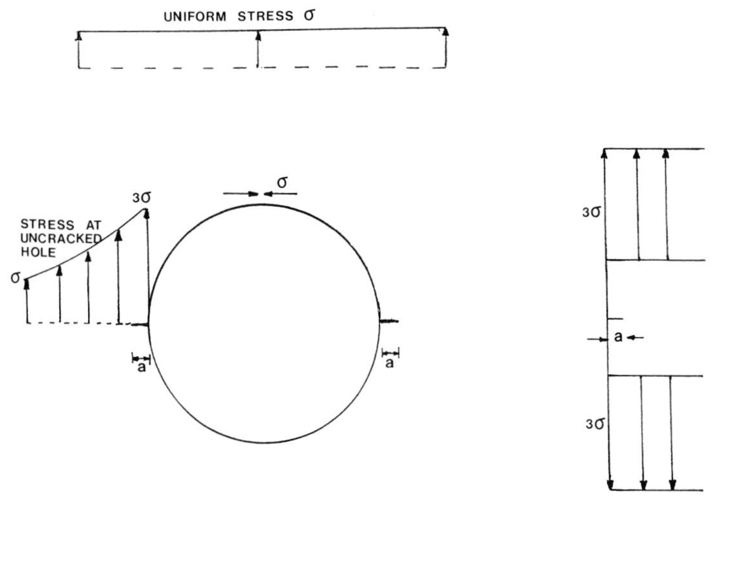

Figure 8.14. Case of two very small cracks at hole. (a) hole; (b) Equivalent case for $a \approx 0$.

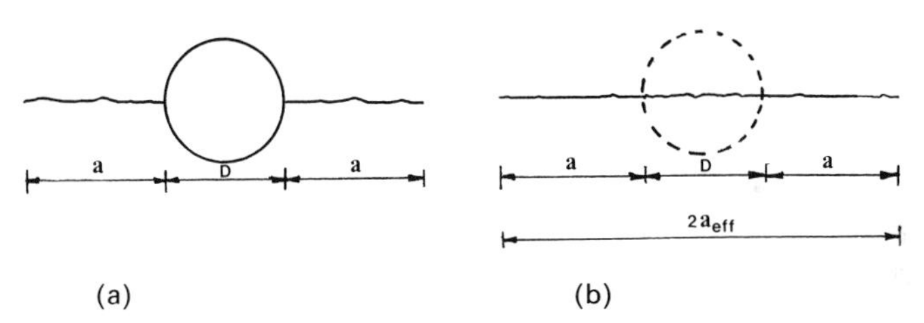

Figure 8.15. Large cracks at hole. (a) Physical crack; (b Effective crack.

hole is part of the crack, and as far as the plate is concerned there is a crack of an effective length $2a_{eff} = 2a + D$. If W is large, the stress intensity is:

$$K = \beta_w \sigma \sqrt{\pi a_{eff}} \quad \text{in which} \quad \beta = 1 \text{ for large } W. \tag{8.27}$$

Expressing K on the basis of the true crack size a, yields (with $\beta_w = 1$):

$$\left.\begin{aligned} K &= \sigma\sqrt{\pi a_{eff}} = \sigma\sqrt{\pi(a + D/2} \\ &= \sqrt{1 + D/2a}\,\sigma\sqrt{\pi a} = \beta\sigma\sqrt{\pi a} \\ \beta &= \sqrt{1 + D/2a} \end{aligned}\right\}. \tag{8.28}$$

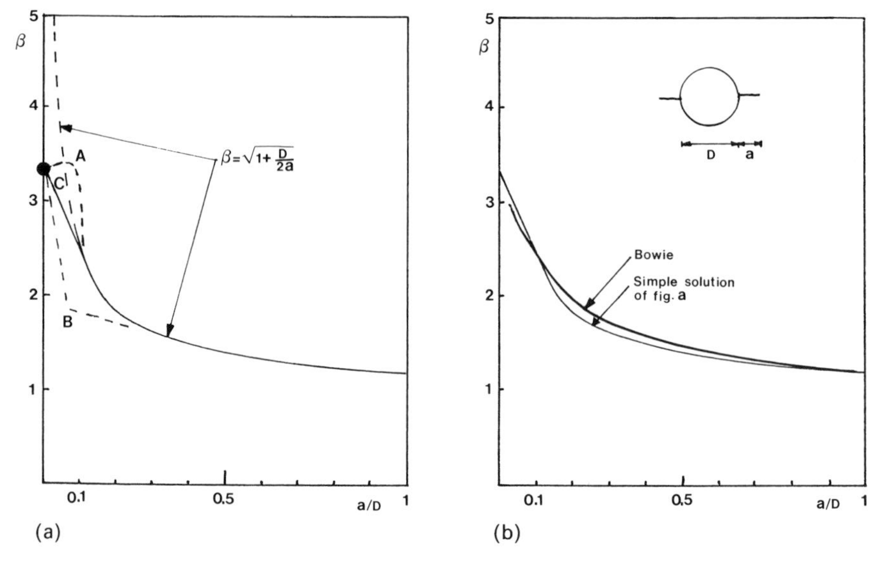

Figure 8.16. β for symmetric cracks at hole. (a) Simple solution; (b) comparison with Bowie's solution.

This β can be calculated for various a/D values and plotted as in Figure 8.16a (curve). Also the point $\beta = 3.36$ for $a = 0$ is plotted. The curve is accurate for large a, but not for small a. The point 3.36 is accurate for small a. The curve between is certainly not like curves A or B in Figure 8.16a. Hence, a line C faired between (0, 3.36) and the curve (accurate for large a) must be very close to the truth, i.e. within a few percent. The result, shown in Figure 8.16b, can be compared with a solution by Bowie [11]. Clearly, the two are very close. The Bowie solution is considered the best available, but it was obtained by numerical methods, and hence, it will have some error. Then it is hard to say whether the Bowie solution or the above simple solution is the better one. But even if one were to accept the Bowie solution as the absolute standard, the simple solution is within a few percent; the former was obtained with great effort and cost, while the latter can be derived literally 'on the back of an envelope', as shown above.

One might argue that the simple solution is hardly worthwhile if a SEMI-rigorous solution is available anyway. However, the above example demonstrates that accurate results can be obtained by simple means. More important, the simple procedure can now be used with confidence to generate solutions to more complicated problems for which no semi-rigorous answers have as yet been obtained, as shown below.

For a single crack at a hole (Figure 8.17) the above procedure leads to $\beta = 3.36$ for small a, and for large a (large W):

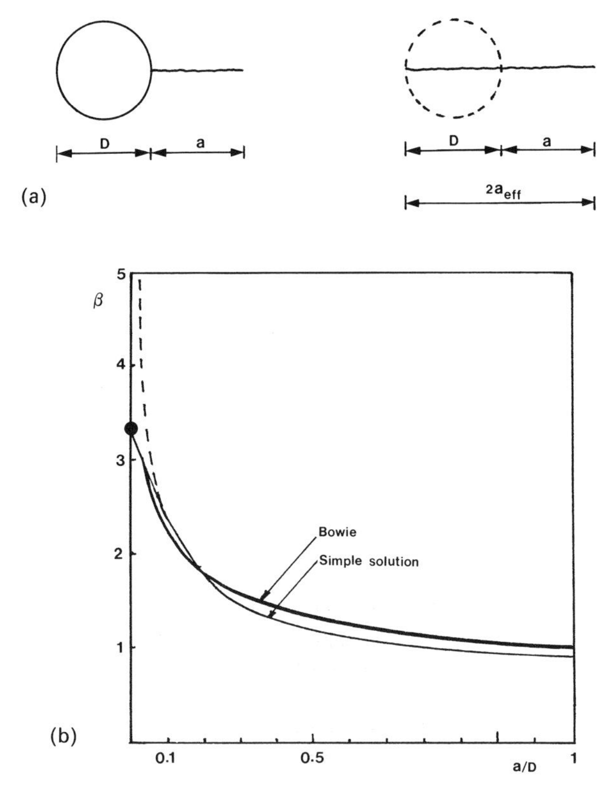

Figure 8.17. One crack at a hole. (a) Single long crack; (b) β for single crack.

$$\left.\begin{aligned} K &= \sigma\sqrt{\pi a_{\text{eff}}} = \sigma\sqrt{\pi(a + D)/2} = \sqrt{1/2 + D/2a}\,\sigma\sqrt{\pi a} \\ &= \beta\sigma\sqrt{\pi a} \\ \beta &= \sqrt{1/2 + D/2a} \end{aligned}\right\} \quad (8.29)$$

which results in Figure 8.17b. For large a this provides $\beta < 1$, which can be understood if it is recognized that the physical crack has only one tip and is defined as a (not $2a$).

Next, consider biaxial loading, as shown in Figure 8.18. Taking one step back, in the uniaxial case, (Figure 8.14) there will be a compressive stress ($-\sigma$; σ being the nominal applied stress) at the poles of the hole. This can easily be demonstrated by pulling a sheet of paper with a circular hole: the paper buckles above and below the hole as a consequence of these compressive stresses. Thus, in a case of biaxial stresses (Figure 8.18a) with $\sigma_l = \sigma_t = \sigma$, there is a tension 3σ due to σ_l at the equator, and a compression $-\sigma$ due to $\sigma_t = \sigma$, resulting in a total stress of $3\sigma - \sigma = 2\sigma$. This leads to $\beta = 1.12 \times 2 = 2.24$ for small

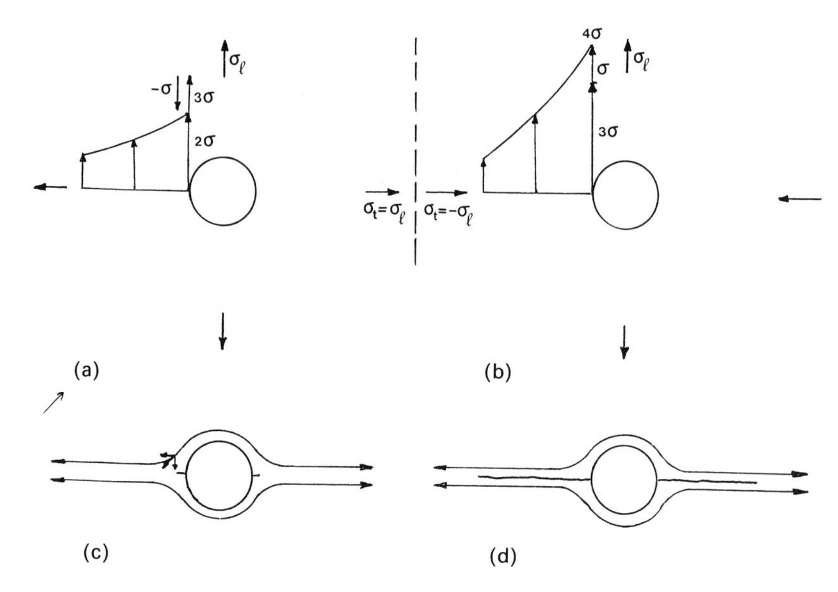

Figure 8.18. Stresses at hole due to biaxial loading. (a) Tension–tension; (b) Tension–compression; (c) Load path deviation at short crack (stress component perpendicular to crack); (d) No load path deviation at tip of long crack.

cracks ($a \approx \partial\, 0$). For long cracks Figure 8.15 and Equation (8.28) apply, because the transverse stress is not deviated at the crack tip: one can cut the plate in two horizontally if only σ_t is present (only when the crack is small will the load be deviated around the hole and have an effect on K causing $\beta = 2.24$ as shown in Figure 8.18c, d). Hence for long cracks Equation (8.28) does apply, and the geometry factor is as shown in Figure 8.19.

This procedure can be extended to any kind of biaxial loading. If $\sigma_t = 0.5\sigma_l$, as in a pressure vessel with end caps, it follows that the equator stress is $3\sigma - 0.5\sigma = 2.5\sigma$. Hence, for small cracks $\beta = 1.12 \times 2.5 = 2.80$. For long cracks again Equation (8.28) applies, and the resulting β is as shown in Figure 8.19. Similarly, geometry factors for other biaxiality ratios can be obtained as shown in Figure 8.19. The same procedure can be followed for single cracks using Equation (8.29).

Should the width W be small, compounding will be necessary, as may be demonstrated by the following example for a symmetric crack (both sides). For small a the geometry factor will still be $\beta = 3.36$ ($a \to 0$; $\sigma_t = 0$). For large cracks:

$$K = \sqrt{\sec \pi a_{\text{eff}}/W}\, \sigma \sqrt{\pi a_{\text{eff}}}. \tag{8.30}$$

Substitution of $a_{\text{eff}} + D/2 = a$ yields:

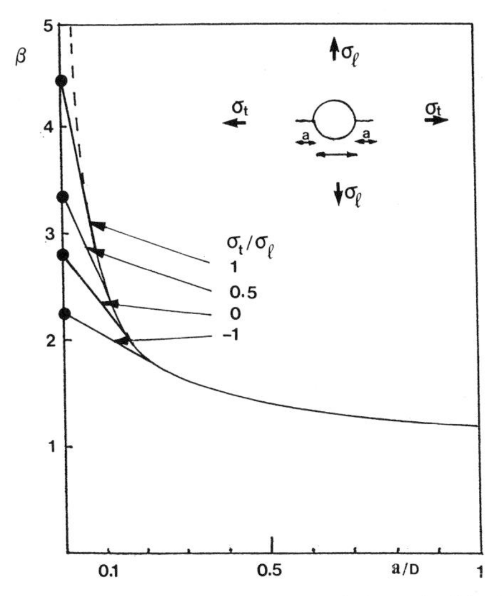

Figure 8.19. Geometry factors for cracks at hole under biaxial loading.

$$\begin{aligned} K &= \beta\sigma\sqrt{\pi a} \\ \beta &= \sqrt{\sec \pi(D/2 + a)/W}\,\sqrt{1 + D/2a} \end{aligned} \qquad (8.31)$$

A curve for the above β can be drawn for various a/D, and the results for small cracks are obtained by fairing between $\beta = 3.36$ for $a = 0$ and the curve in the same manner as described above.

The above method as applied to uniaxial loading with large W was shown to be in excellent agreement with the Bowie solution. Thus it may be expected that the extension of the procedure to more complicated cases will provide geometry factors of good accuracy. The method can be used for cracks at any stress concentration. In the case of Figure 8.20 the stress concentration is $k_t = 1 + 2\sqrt{l/\varrho}$; Chapter 2). If for example $l = 2$ and $\varrho = 0.5$ then $k_t = 1 + 2\sqrt{2/0.5} = 5$. Using the uncracked stress distribution rule (Section 8.4) as above, one obtains $\beta = 5$ for small cracks. For sharp notches it becomes questionable whether one should apply the factor 1.12 as in the case of holes. However, it should be noted that this would introduce at maximum a 12% error. Using engineering judgement and taking the factor as 1.06 would reduce the error to 6 percent. Clearly, the sharper the notch, the smaller the correction should be. A rule of thumb could be to take the factor as $\beta_{FS} = 1 + 0.12/(k_t - 2)$ for $k_t > 3$, and $\beta = 1.12$ for $k_t < 3$. The equation provides $\beta = 1.12$ for $k_t = 3$ (circular hole), but some engineering judgement is required here.

For large cracks, the notch can be considered part of the effective crack ($a_{\text{eff}} = l +$ a; Figure 8.20) as in the case of the holes. Hence:

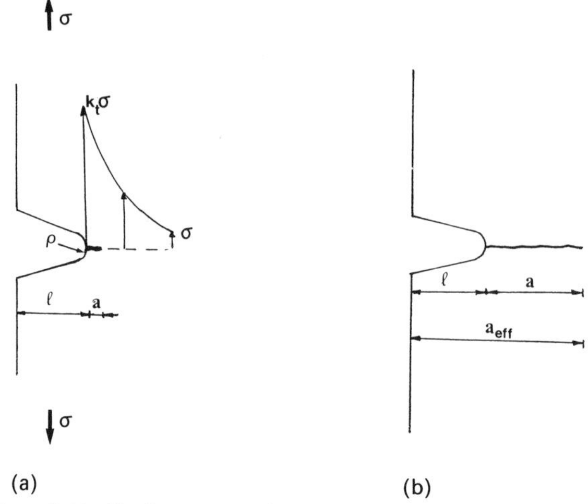

Figure 8.20. Crack at stress raiser. (a) Short crack; (b) Long crack.

$$\left.\begin{aligned} K &= \beta_W \sigma \sqrt{\pi a_{\text{eff}}} = \beta_W (a_{\text{eff}}/W)\, \sigma \sqrt{\pi(a + l)} = \beta\sigma\sqrt{\pi a} \\ \beta &= \beta_W (a_{\text{eff}}/W) \sqrt{1 + l/a} \\ \beta &= \sqrt{1 + l/a}\sqrt{\sec \pi(l + a/W} \qquad \text{for a central notch} \end{aligned}\right\} \quad (8.32)$$

$$\beta = \sqrt{1 + l/a}\left\{1.12 - 0.23\frac{l + a}{W} + 10.56\left(\frac{l + a}{W}\right)^2 - 21.74\left(\frac{l + a}{W}\right)^3 + 30.42\left(\frac{l + a}{W}\right)^4\right.$$

for an edge notch.

By plotting $\beta = \beta_{FS} k_t$ for $a = 0$ and the curve of Equation (8.32), and by fairing for small cracks as in the case of holes a very good approximation of β will be obtained.

A case in point is a crack developing from a hole and causing a cracked ligament as illustrated in Figure 8.21, indeed a very common case. The configuration is evaluated as discussed above. Once the ligament is cracked, a crack will emanate from the other side of the hole. In essence this will be equivalent to a crack at a stress raiser for which $l = d + D/2$ and $\varrho = 1/2D$ (Figure 8.21). Hence, $k_t = 1 + 2\sqrt{(d + D/2)/(D/2)}$, which provides β for small a, while Equations (8.32) provide β for larger a. Using the procedures as described will

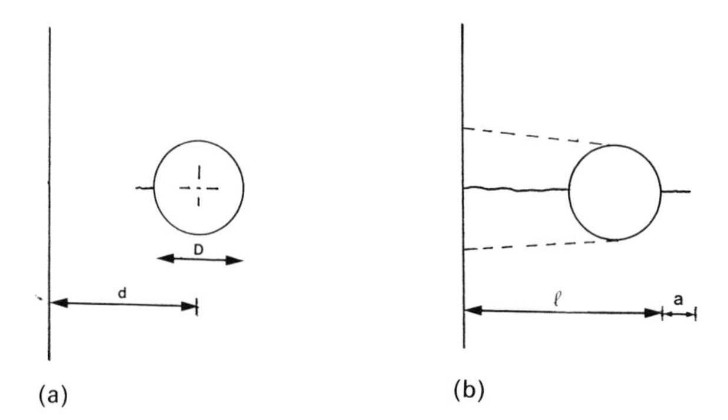

Figure 8.21. Cracking at hole with edge distance d from free surface. (a) First crack; (b) Broken ligament plus crack.

lead to geometry factors with good accuracy (compounding and superposition for various loading cases are to be performed as described).

8.8. Simple solutions for irregular stress distributions

Even if the stress distribution is non-uniform, several simple procedures to obtain geometry factors can be used. The first, and simplest approach would be to approximate the stress distribution by a superposition of uniform tension and a (any) number of bending moments. An example is shown in Figure 8.22. Once this is accomplished, all methods of compounding and superposition as discussed in previous sections can be applied to arrive at the appropriate geometry factors, provided the selected reference stress is adhered to and accounted for conscientiously.

Alternatively, use can be made of the uncracked stress distribution rule, discussed in Section 8.4. The rule states that the stress intensity factor (and thus

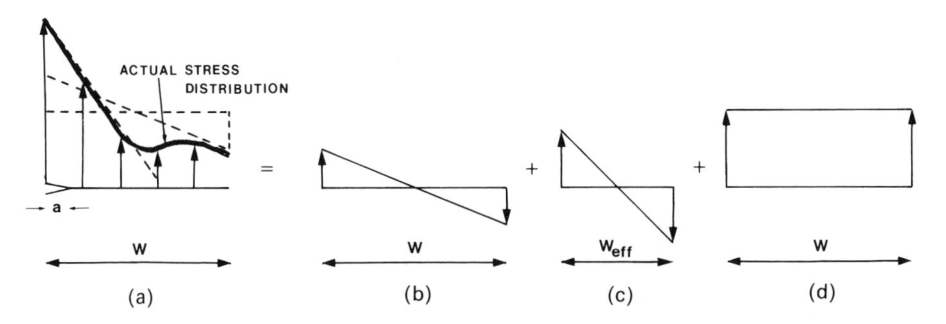

Figure 8.22. Approximation of stress distribution; superposition of b, c, d, should provide good results up to a/W 0.5, which is usually a large enough crack size for practical problems.

β) can be obtained by applying to the faces of the crack stresses that used to be there when there was no crack. Hence, if the stress distribution is known (as it was before cracking), these stresses can be applied to the crack faces to obtain K and β, be it that some work is involved. Naturally, only that part of the stress distribution is used that covers the crack.

Suppose the stress distribution in the uncracked section is as shown in Figure 8.22a. By applying this stress distribution to the crack faces a stress intensity factor can be derived using so-called Green's functions. The latter are shown for central and edge cracks in Figure 8.23. They provide the stress intensity for a point load per unit thickness (stress) at one particular point on the crack face. The total stress intensity can be obtained by (numerical) integration of the Green's functions, using the stresses in the uncracked section as a series of point forces. A certain caution is necessary, because the stresses close to the crack tip are the most influential.

This can be demonstrated by using an example for uniform stress distribution for which β is known. An approximation could be made, using the Green's functions of Figure 8.23 and the resulting rough analysis (performed by hand) would be as in Table 8.2. A more refined approximation would lead to the analysis of Table 8.3. Clearly, the results depend strongly on the approximations close to the crack tip. Solutions of sufficient accuracy can be obtained only if the approximations at the crack tip are adequate, where the adjective 'adequate' is not well defined; engineering judgement is necessary. Computer software [4] using Green's functions will account for this problem.

It should be noted that the above procedure provides the stress intensity factor K. What is needed is the geometry factor β. As $K = \beta\sigma\sqrt{\pi a}$, the geometry

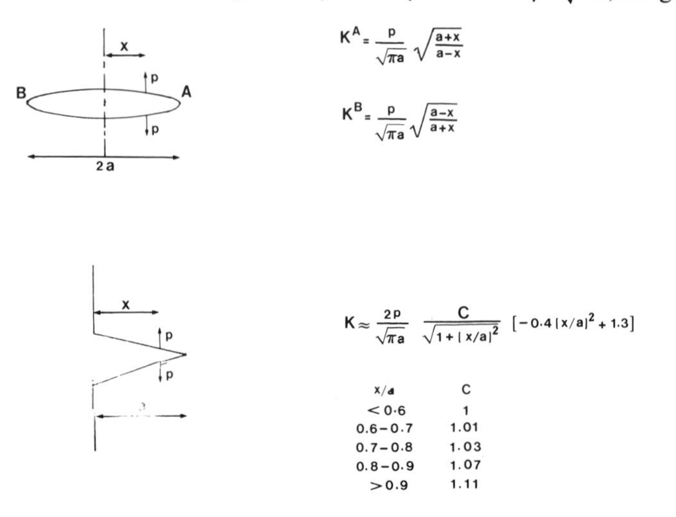

Figure 8.23. Most useful Green's functions.

Table 8.2. Hand analysis using Green's functions with crude steps

Right side (Figure 8.23a)[a]							Left side (Figure 8.23a)[a]				
X	$X_{average}$	ΔX	$P_x = \sigma_{ave}\Delta X$[b]	$\frac{P_x}{\sqrt{\pi a}}$	$\sqrt{\frac{a+x}{a-x}}$	K_1	$P_x = \sigma_{ave}\Delta X$[c]	$\frac{P_x}{\sqrt{\pi a}}$	$\sqrt{\frac{a-x}{a+x}}$	K_2[d]	$K_1 + K_2$
0.0											
	0.1	0.2	0.2	0.113	1.106	0.125	0.2	0.113	0.905	0.102	0.227
0.2											
	0.3	0.2	0.2	0.113	1.363	0.154	0.2	0.113	0.734	0.083	0.237
0.4											
	0.5	0.2	0.2	0.113	1.732	0.196	0.2	0.113	0.577	0.065	0.261
0.6											
	0.7	0.2	0.2	0.113	2.380	0.269	0.2	0.113	0.420	0.047	0.316
0.8											
	0.9	0.2	0.2	0.113	4.359	0.493	0.2	0.113	0.229	0.026	0.519
1.0											
											1.56 +

$$\beta = \frac{K_{tot}}{\sigma\sqrt{\pi a}} = \frac{1.56}{1\sqrt{\pi \times 1}} = 0.88;\ \text{known } \beta = 1;\ \text{error } 12\%,\ \text{but see Table 8.3.}$$

Notes: (1.) Better result is obtained in Table 8.3 with finer steps. (2.) Assumptions: central crack $2a = 2$ inch; uniform stress, $\sigma = 1$. (3.) With above assumptions integration can be done in closed form, but not if stress is non uniform; this is an example for a case with known β so that results can be checked.

[a] There are two sides of the crack ($-x$ and x), both must be accounted for.
[b] In case of non-uniform stress σ_{ave} is average stress over ΔX; in this case $\sigma = 1$ overall.
[c] σ_{ave} at $-X$; in this case $\sigma = 1$ overall.
[d] K at right crack tip due to stresses on left side of crack (Figure 8.23).

Table 8.3. More refined land integration of case of Table 8.2.

Same notes apply as in Table 8.2.

Right side (Figure 8.23a)							Left side (Figure 8.23a)				
X	$X_{average}$	ΔX	$P_x = \sigma_{ave}\Delta X$	$\frac{P_x}{\sqrt{\pi a}}$	$\sqrt{\frac{a+x}{a-x}}$	K_1	$P_x = \sigma_{ave}\Delta X$	$\frac{P_x}{\sqrt{\pi a}}$	$\sqrt{\frac{a-x}{a+x}}$	K_2	$K = K_1 + K_2$
0.0											
	0.1	0.2	0.2	0.113	1.106	0.125	0.2	0.113	0.905	0.102	0.227
0.2											
	0.3	0.2	0.2	0.113	1.363	0.154	0.2	0.113	0.734	0.083	0.237
0.4											
	0.5	0.2	0.2	0.113	1.732	0.196	0.2	0.113	0.577	0.065	0.261
0.6											
	0.65	0.1	0.1	0.056	2.171	0.122	0.1	0.056	0.461	0.026	0.148
0.7											
	0.75	0.1	0.1	0.056	2.646	0.148	0.1	0.056	0.378	0.021	0.169
0.8											
	0.85	0.1	0.1	0.056	3.512	0.197	0.1	0.056	0.284	0.016	0.213
0.9											
	0.925	0.05	0.05	0.028	5.066	0.142	0.05	0.028	0.197	0.006	0.148
0.95											
	0.96	0.02	0.02	0.011	7.000	0.077	0.02	0.011	0.143	0.002	0.079
0.97											
	0.98	0.02	0.02	0.011	9.950	0.109	0.02	0.011	0.101	0.001	0.110
0.99											
	0.9945	0.009	0.009	0.0051	19.043	0.097	0.009	0.0051	0.053	–	0.097
0.999											
	0.9995	0.0001	0.0001	0.0006	63.238	0.038	0.0001	0.0006	0.016	–	0.038 +
1.000										K_{total} =	1.727

$$\beta = \frac{K_{tot}}{\sigma\sqrt{\pi a}} = \frac{1.727}{1\sqrt{\pi 1}} = 0.976\text{; known } \beta = 1\text{; error } 3.5\%.$$

factor is obtained as $\beta = K/\sigma\sqrt{\pi a}$, which means that σ has to be defined. As discussed any reference stress, σ_{ref}, can be used as long as the damage tolerance analysis is based consistently on this same reference stress. The problem of reference stress has been addressed at several places in the preceding sections and specifically in Section 8.2. In the case of a complicated stress distribution, the best option is to select the highest stress in the cracked section as the reference stress. Note: this is not meant to be the highest stress at the crack face, because the latter may vary with crack size. The geometry factor always must be derived for a number of crack sizes. Hence, the highest stress in the section, and not the highest stress on the crack face, is the best reference. Well-designed software will account for this problem as well.

An alternative procedure to obtain geometry factors in the case of complicated stress distributions is to make use of so-called weight functions. Although this is a sound method, it requires a great deal more knowledge of the effect of cracks on strains and displacements, because it is based upon a displacement reference for a known case [12, 13, 14]. It is not particularly suited for hand calculations. Some software [15] includes the use of weight functions and as such it is in the realm of engineering applications. Their use requires more expertise and should not be attempted lightly, unless through a reputed software package.

8.9. Finite element analysis

Finite element analysis can be used in two ways to determine geometry factors, namely indirectly and directly. For the solution of common damage tolerance problems the indirect use of finite element analysis is the most worthwhile. In that case the finite element solution is obtained for the uncracked structure only, and the stress distribution in the section of the future crack is calculated. Subsequently, the uncracked stress distribution rule (Section 8.4) is used to calculate the stress intensity factor with one of the approximate methods described in previous sections, or through the use of Green's functions or weight functions (Section 8.8), especially if software is available to perform either of these tasks.

It is worthwhile pointing out that stress distributions obtained with finite elements are of limited accuracy. Claims that finite element analysis is the most rigorous stress analysis available are exaggarated and naive; a close look at many finite element solutions bears this out immediately. The only 'rigorous' solution is obtained from the differential equations in which the element size ($dx\ dy\ dz$) literally approaches zero. Unfortunately, these differential equations usually cannot be solved. The mere fact that the elements in finite element analysis are of finite size, indicates that the solution is an approximation. Provided the elements are sufficiently small, especially in areas of large stress gradients, very good solutions can be obtained. However, in many practical solutions for complicated structures, model size (degrees of freedom) limitations and coarse modeling in areas of large stress gradients often are cause of limited accuracy. In areas of stress concentrations and of load transfer to other members, accuracies are seldom better than 10 percent for the calculated local stresses. Errors larger than 10 percent are not uncommon.

In the direct use of finite element analysis for the derivations of stress intensity and geometry factors, solutions must be obtained for models with cracks. As there is an extremely large stress gradient at the crack tip, the element sizes around the crack tip must be very small, unless use is made of higher order elements which can model a stress singularity.

The finite element model can provide stresses, strains, displacements, and strain energy only. From these the stress intensity and geometry factor can be obtained by a variety of methods. For example, the universal crack tip stress field solution provides the crack tip stress σ_y for $\theta = 0$ (Chapter 3) as:

$$\sigma_y = \frac{K}{\sqrt{2\pi x}} \; (\theta = 0). \tag{8.33}$$

The finite element solution provides σ_y at various locations (x). By substituting the calculated σ_y, and the distance r for which it applies, in Equation (8.33), the stress intensity is calculated as:

$$K = \sigma_{YFEM} \sqrt{2 \pi x_{FEM}} \, . \tag{8.34}$$

As Equation (8.33) is valid only at very small x (Chapter 3), the stress intensity obtained from Equation (8.34) is in error unless x is extremely small. On the other hand, the calculated value of σ_y contains a larger error the smaller x (unless singular elements are used). In order to circumvent this problem Equation (8.34) can be solved a number of times using the calculated stresses $\sigma_y(r)$ at distances x_1, x_2, x_3, etc. For each combination (σ_y; x) an apparent value of K is obtained from Equation (8.34). None of these is the correct one. One may plot the apparent values as a function of the distance x for which they were calculated. A line drawn through the data can be extrapolated to $x = 0$ – at which point Equation (8.33) is rigorous – , as shown in Figure 8.24. The extrapolated value at $x = 0$ is the sought value of K. For an example see solution to Exercise 14.

Finally, the geometry factor must be obtained. Selecting a reference stress σ_{ref} (this can be the stress as applied to the model or a stress calculated by the model at some convenient location) the geometry factor follows from:

$$\beta = \frac{K}{\sigma_{ref} \sqrt{\pi a}}, \tag{8.35}$$

where a is the crack size in the model. Naturally, the following damage tolerance

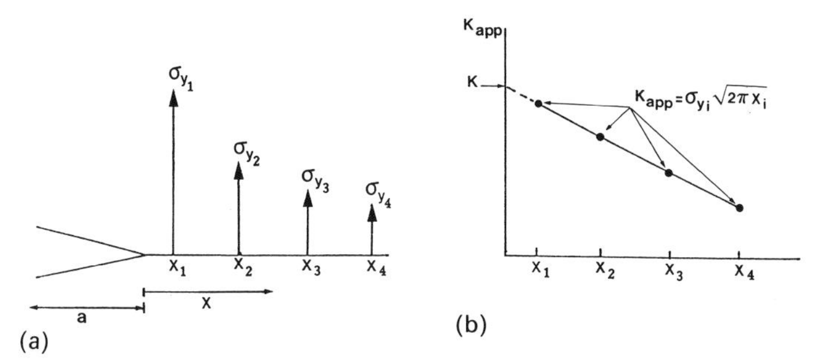

Figure 8.24. Obtaining K from finite element model with crack. (a) Stresses from finite element model; (b) Stress intensity.

analysis should be based upon the same reference stress (Section 8.2). The above provides β for one crack size only, but β must be obtained for a range of crack sizes. Hence, the procedure must be repeated for a number of crack sizes (multiple finite element solutions). It should be noted that in each case the same reference stress must be used – in terms of both location and magnitude – to obtain β from Equation (8.35); any inconsistency may lead to serious errors.

There are a multitude of other methods to obtain β from finite element models with cracks. In analogy with the above approach, use can be made of the displacements. For example, the displacement u at a distance r from the crack tip follows from the general crack tip filed solution as:

$$u = CK\sqrt{r}\, f(\theta), \tag{8.36}$$

which can be used in the same manner as Equation (8.33) to obtain K from the calculated displacements.

Alternatively, β may be derived from the strain energy U. The total strain energy U in the model is calculated for crack size a. Subsequently the crack in the model is extended by one element size (Δa) to $a + \Delta a$, and again the total strain energy calculated. The change in strain energy $\mathrm{d}U/\mathrm{d}a$ is approximated as:

$$\frac{\mathrm{d}U}{\mathrm{d}a} = \frac{\Delta U}{\Delta a} = \frac{U_{a+\Delta a} - U_{\mathrm{a}}}{\Delta a} \tag{8.37}$$

from which K is obtained as (Chapter 3):

$$K = \sqrt{E\,\mathrm{d}U/\mathrm{d}a} \tag{8.38}$$

and finally β in the manner described above.

The above procedure allows larger elements and is acclaimed to have better accuracy than those discussed before. However, some caution is advisable because Equation (8.37) is a differentiation, an inherently inaccurate process. It involves subtraction of two large numbers of equal magnitude. If both $U_{a+\Delta a}$ are accurate within one percent, the accuracy of ΔU is certainly no better than 10%.

If there is a single load on the model the strain energy can be obtained as $U_a = 0.5P\delta_a$, where δ_a is the load-point displacement for crack size a. Similarly, for crack size $a + \Delta a$ the strain energy is $U_{a+\Delta a} = 0.5P\delta_{a+\Delta a}$. Calculation of K and β then proceeds as above from Equations (8.37) and (8.38). For an example see solution to Exercise 15.

Since $J = G = \mathrm{d}U/\mathrm{d}a$ (Chapter 4) the strain energy release rate can be obtained by evaluating the J-integral (Chapter 4) along a convenient path in the finite element model. Subsequently, K and β again are obtained as shown before.

Most finite element codes have post-processors which provide K by several or all of the above methods. It cannot be said *a priori* which method will provide the best results; this will depend upon the configuration, and especially upon the modelling. As in the case of the indirect use of finite element models, the accuracy depends mostly upon modelling of areas with stress concentrations and load-transfer, and upon assumptions made for the boundary conditions.

Damage tolerance analysis always requires knowledge of β (not K) for a range of crack sizes. Hence, several finite element solutions must be obtained for a number of crack sizes, which requires a number of different models; indeed a costly proposition for complex structural configurations. In the literature the accuracy of such analysis is often demonstrated on the basis of simple configurations such as center cracked panels. In the first place those solutions are always known anyway, but more important, structures seldom resemble center cracked panels. Finite element models of cracked real structures are rather more difficult and expensive; besides they require many more assumptions with regard to boundary conditions, load transfer, etc., and small elements in areas of stress concentrations, so that their final accuracy is very limited despite the effort. The simple procedures discussed in this chapter, possibly employing finite element analysis of the uncracked structure, presently are the most viable methods for general engineering applications. In view of the inaccuracies introduced by other assumptions (Chapter 12) they are also the most sensible solutions.

Several other numerical analysis procedures to obtain K are availabe, but they are not within the realm of general engineering applications and as such they are beyond the scope of this book; for a discussion of these and further references, the reader is referred to more basic standard texts.

8.10. Simple solutions for crack arresters and multiple elements

The stress intensity factor is affected by the presence of second elements (stringers, doublers, flanges) or crack arresters to which load can be transferred. If the second element is intact, load transfer from the cracked part to this element will cause a decrease of K and therefore of β.

Consider two wide plates, one without and one with stringers (or doublers), as in Figure 8.25, both subjected to uniform stress σ. For the plate without stringers the geometry factor will be $\beta = 1$ for crack sizes up to $a/b = 1$ ($W \gg b$), as shown in Figure 8.25d. In the plate with stringers, the effect of the latter will be negligible when the crack is small; $\beta = 1$ for $a/b \approx 0$. However, if the crack extends from stringer-to-stringer, the situation is quite different. In an unstiffened plate all load in the section of the crack must bypass the crack inside the plate, which gives rise to $\beta = 1$ in the first place. If the stringers are present a second load path is available: part of the load can now by bypassed outside the plate via the fasteners into the stringers and then back into the plate, again via the fasteners (Figure 8.25c). If part of the load bypasses outside the plate, then the stresses at the crack tip (inside the plate) will be lower. This means that K is lower, which is reflected in a lower β then in the unstiffened plate. Since for the latter $\beta = 1$, the β for $a/b = 1$ in the stiffened plate will be less than 1, as shown in Figure 8.25d. For intermediate crack sizes β will gradually decrease from 1 at $a = 0$ to the lower value at $a/b = 1$.

Thus the stress intensity is lower then in the absence of stringers, which can

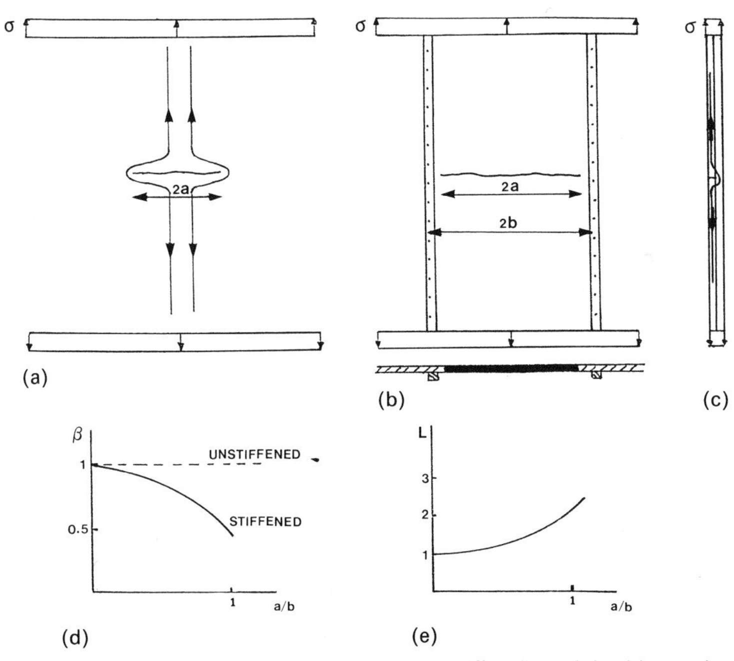

Figure 8.25. Effect of stringers on stress intensity. (a) Unstiffened panel: load bypass in plate; (b) Stiffened panel; (c) Side view; alternative load bypass through stringer; (d) Geometry factor; (e) Stress concentration in stringer.

have a dramatic effect on crack growth and residual strength, as will be discussed in Chapter 9; fracture arrest becomes a distinct possibility. The above example is for mechanically fastened stringers, but similar results are obtained if the second element (doubler or arrester) is welded or an integral part of the structure. For example a transverse web crack in an *I*-beam will show a similar decrease in β when the crack approaches the flanges. The decrease can be significant: $\beta = 0.3–0.5$, depending upon the stiffness of flanges or stringers, the stringer spacing, and the fastener spacing.

The decreases in K (or β) is beneficial for the crack tip stresses (plate), but the side effect is that the stresses in the stringer will be higher and that the fasteners will be subjected to very high shear loads. If the crack is small, the stresses in plate and stringer will be roughly equal; since stringer and plate are attached they must undergo equal strains and equal strains require equal stresses if the moduli of plate and stringer are equal. If the local stress in the stringer at the location of the crack is denoted as σ_l, then $\sigma_l = \sigma$ as long as cracks are small, but $\sigma_l > \sigma$ when $a/b \rightarrow 1$.

A stress concentration factor L will be defined as $L = \sigma_l/\sigma$. From the above arguments it follows that $L = 1$ for $a/b = 0$, and that $L \gg 1$ for $a/b \rightarrow 1$, as shown in Figure 8.25e. Depending upon stringer stiffness and spacing and fastener spacing, L can attain values of 2 to 3 (i.e. the local stress in the stringer will be 2 to 3 times higher than the applied stress). In the same vein the shear on the fasteners (or welds) will increase from essentially zero at $a/b = 0$ to a significant shear load (stress) when a/b reaches 1. Because the load bypass occurs very close to the crack (Chapter 2) only the fasteners close to the cracked section carry the shear. The so called stiffening ratio, μ, which reflects the ratio of cross-sectional area of stringers and skin, and the fastener spacing determine the magnitude of β and L.

For any configuration, β, L, and fastener shear can be readily calculated by means of numerical analysis of closed form solutions [16, 17, 18], or by finite element analysis. The former are preferable because they permit parametric analysis. Due to the high stringer stress and fastener loads plastic deformation may occur, which may affect load transfer and therefore alter β and L. The closed form analysis procedures can properly account for these effects, but then only specific solutions are possible. For cases without plastic effects generic and parametric solutions have been obtained and the results are readily available in handbooks [1, 19], so that they can be used for general damage tolerance analysis (Chapters 9 and 14).

Unfortunately, the handbook solutions [1, 19] only provide β, and neither L nor fastener shear, while all three are necessary for residual strength analysis. However, good estimates of L can be obtained from β as follows. Define the geometry factor for the unstiffened plate as β_u, and for the stiffened panel as β_s.

The reason that $\beta_s < \beta_u$ is the load transfer to the stringer. If there is no stringer, the load carried by the plate in front of the crack tip is:

$$P = \int \sigma_y B \,\mathrm{d}r. \tag{8.39}$$

For $\theta = 0$ the stress field solution provides (Chapter 3):

$$\sigma_y = \frac{K}{\sqrt{2\pi r}} = \frac{\beta\sigma\sqrt{\pi a}}{\sqrt{2\pi r}}. \tag{8.40}$$

Carrying the integration over a distance equal to $r = a$ from the crack tip is sufficient to obtain the bypassed load. Hence Equation (8.39) becomes:

$$P = \int_{r=0}^{r=a} \frac{B\,\beta\sigma\sqrt{\pi a}}{\sqrt{2\pi r}} \,\mathrm{d}r = B\beta\sigma\sqrt{a/2}\int_0^a r^{-1/2}\,\mathrm{d}r \tag{8.41}$$
$$= \sqrt{2}\,\beta\sigma\, aB.$$

Clearly, the additional load carried by the stringer is P_s (stiffened) minus P_u (unstiffened) and the additional stringer stress is $(P_s - P_u)/A_s$ is the stringer cross sectional area. The total stress in the stringer is $\sigma_l = \sigma + (P_s - P_u)/A_s$, so that the stress concentration $L = \sigma_l/\sigma$ becomes, with Equation (8.41):

$$L = 1 + aB\sqrt{2}\,(\beta_u - \beta_s)/A_s. \tag{8.42}$$

The handbook provides β_u and β_s; the value of L can then be obtained from Equation (8.42). By taking β_u and β_s from the handbook for a number of crack sizes between $a/b = 0$ and $a/b = 1$, the stringer stress concentration can be calculated as a function of crack size. An example is shown in the solution to Exercise 15.

Obtaining fastener shear loads is somewhat more precarious, but a sensible estimate can be made as follows. All load is transferred by the fasteners closest to the crack plane: assume that three fasteners above the crack transfer the load into the stringer, three fasteners below the crack transfer the load back into the plate. Together the three fasteners transfer the total load which is $(\sigma_l - \sigma)A_s$. Typically, the fastener closest to the crack transfer most of the load (e.g. 60%), the other two transfer e.g. 30 and 10% respectively. Hence, the highest fastener shear load would be:

$$P_{\text{fastener}} = 0.6\,(L - 1)\,\sigma A_s, \tag{8.43}$$

where L follows from Equation (8.42). Clearly, the fasteners must be made strong enough to carry this shear otherwise the whole scheme will not work (Chapters 9). It turns out that the actual fastener load on the average is only in the order of 60–70% of the value in Equation (8.43) due to fastener hole ovalization (plasticity).

8.11. Geometry factors for elastic-plastic fracture mechanics

The definition of J is (Chapter 4):

$$J = J_{el} + J_{pl} = \beta^2\pi\,\sigma^2\,a/E + H\sigma\varepsilon_{pl}a. \tag{8.44}$$

The geometry factor β is obtained by means of any of the procedures discussed in the foregoing sections. If a stress-strain equation is available, the plastic part of J can be expressed in σ only. The commonly used equation (Chapter 4) is the Ramberg–Osgood equation which provides for the plastic strain: $\varepsilon = \sigma^n/F$. Then the plastic part of J becomes:

$$J_{pl} = H\left(\frac{a}{L}, n\right)\sigma^{n+1}\,a/F. \tag{8.45}$$

By performing non-linear finite element analysis on the cracked structure (direct method in Section 8.9), using the proper n and F values, J_{pl} can be calculated from the integral formulation. With J_{pl} thus known, the stress applied to the model known, and a known, the geometry factor can be extracted as

$$H(\frac{a}{L}, n) = \frac{\mathrm{F}\,J_{pl\mathrm{FEM}}}{\sigma^{n+1}\,a}. \tag{8.46}$$

The procedure must be repeated for various values of a in order to obtain $H(a/L)$ for a certain material with given n and F. If H has to be determined for various materials with different n, the procedure must be repeated for all these different n-values, a costly proposition.

Geometry factors, H, have been determined in this manner for a number of configurations and n-values. These have been compiled in handbooks [20, 21], which provide h_1 instead of H, but H can be derived from h_1 if so desired as discussed in Chapter 4.

Simple methods to obtain H have not yet been devised, but the following procedure might be used. If $n = 1$ then $F = E$ and Equation (8.45) reduces to:

$$J = H\,\sigma^2\,a/E. \tag{8.47}$$

This is obviously the linear elastic case for which (Chapters 3, 4)

$$J = \pi\,\beta^2\,\sigma^2\,a/E. \tag{8.48}$$

Clearly, in this case $H = \pi\,\beta^2$. Since $n + 1 = 2$, the square in σ^2 comes from $n + 1$. Assuming that the square of β also comes from $n + 1$, it follows that either $H = \pi\,\beta^{n+1}$ or $H = (\sqrt{\pi}\beta)^{n+1}$. At present there is no proof whether either of the two expressions is correct. As a matter of fact reasonable agreement is obtained only for small n and small a, so that the approximation may not be very useful. It appears that

$$H\left(\frac{a}{l}, n\right) = \pi\left\{\beta\left(\frac{a}{L}\right)\right\}^{n+1} \tag{8.49}$$

may be used if a is small and n is low, but the solution is up to the user. As forgiving as EPFM analysis is, the results are usually within acceptable engineering accuracy. Hence, H can be obtained from β through Equation (8.49), where β is derived by any of the procedures discussed in the previous sections. The fact that Equation (8.49) does not exactly cover the computed H values does not make the equation invalid, because there is equal reason to suspect the finite element analysis, which will contain increasing errors for larger a/L and larger n.

8.12. Exercises

1. Using the maximum bending stress as the reference stress determine β for a crack with $a/W = 0.5$ from Figure 8.1a. Calculate the residual strength given $K_c = 70\,\text{ksi}\sqrt{\text{in}}$, $W = 10\,\text{inch}$, and $F_{ty} = 60\,\text{ksi}$.

2. Calculate and plot the β-curve for a plate of with an edge crack subjected to combined tension and in-plane bending, using the maximum total stress as a reference; note take six values of a/W at increments of 0.1, starting at 0.1. A remote load P is applied 2 inches from the center of the plate to the side of the crack; $W = 10\,\text{inch}$.

3. Repeat Exercise 2 using the uniform tension stress as a reference.

4. Given that $W = 10\,\text{inches}$, $a = 2\,\text{inches}$, $F_{ty} = 100\,\text{ksi}$ and $K_{tc} = 50\,\text{ksi}\sqrt{\text{in}}$, calculate the residual strength for all cases in Exercises 2 and 3. (Ignore collapse).

5. Calculate and plot the β-curve for a single crack at a hole of diameter D in a plate of width W subjected to uniform tension. Take 10 values of a/D increasing with increments of 0.2.

6. Use the result of Exercise 5 to calculate the rate of fatigue crack growth of a crack of 0.3 inch if $D = 1\,\text{inch}$, $W = 6\,\text{inches}$, $\Delta\sigma = 10\,\text{ksi}$ at $R = 0.2$, $\mathrm{d}a/\mathrm{d}N = 2\text{E-}9\ \Delta K^{2.3}\ K_{\max}^{1.1}$.

7. Modify Exercise 5 for biaxial loading with $\sigma_L/\sigma_T = 3$ and for $\sigma_L/\sigma_T = -0.5$ (σ_T negative). Then repeat exercise 6 for both cases, but with a crack size of $a = 0.1\,\text{inch}$.

8. Determine the β-curve for symmetric cracks at a fastener hole where the fastener takes out 20% of the load for five values of a/D at increments of 0.4 (Note from Exercise 5 that for $a/D > 0.4$ the hole may already be inchconsidered part of the crack. Assume that $W = 8$ inch, $D = 1$ inch, $B = 0.5$ inch and that the stress distribution at the ends is uniform.

9. Calculate the residual strength diagram for the case of Exercise 8 up to $a/D = 2$; $K_c = 50\,\text{ksi}\sqrt{\text{in}}$; $F_{ty} = 80$ ksi.

10. Determine the β-curve for cracks emanating from a semi-elliptical edge notch with a depth d and radius r in a plate of width W subjected to uniform loading. Use 5 a/d values at increments of 0.1; $d/r = 5$.

11. A crack emanated from a hole with diameter D at a distance of e from the edge of the plate. The entire ligament has cracked. Determine the β-curve for the crack emanating at the other side of the hole.

12. A shouldered part with a stress concentration factor of $k_t = 1.5$ at the fillet radius develops a through crack. What is β for small cracks.

13. An elastic finite element analysis with elements of 0.1 inch at the crack tip and a crack of 2 inches produces the following results for the longitudinal stress in the plane of the crack: 48.9 ksi in element 1, 30.0 ksi in element 2, 25.2 ksi in element 3. The applied stress is non-uniform. The highest stress applied in the model is 5 ksi. Calculate β, assuming the given stresses act in the center of the elements, and taking highest stress as a reference.

14. A finite element analysis of a model with a crack is subjected to a point load P. The crack size is 1 inch and the crack tip elements are 0.1 inch. The calculated displacement of the loading point is 0.02 inch. In another run the crack is extended over one element, and the calculated displacement is 0.021 inch. Calculate K if the applied load is 10 000 lbs. Assume unit thickness and $E = 10\,000$ ksi.

15. Assuming $\mu = 0.4$ and a fastener spacing $s/b = 0.1$, calculate the L curve and fastener load curve for 5 values of a/b in increments of 0.1; let Figure 8.25c apply and assume that 60% of the load is transferred by the first fastener. $B = 0.2$ inch, $b = 8$ inch; μ is defined as A_s/bB, where A_s is stringer cross section.

References

[1] D.P. Rooke and D.J. Cartwright, *Compendium of stress intensity factors*, H.M. Stationery Office, London (1976).

[2] G.C. Sih, *Handbook of stress intensity factors*, Lehigh University (1973).

[3] H. Tada et al., *The stress analysis of cracks handbook*, Del Research (1973, 1986).

[4] D. Broek, *GEOFAC, a pre-processor for geometry factor calculation*, Fracturesearch software (1987).

[5] J.C. Newman and I.S. Raju, Stress intensity factors equations for cracks in three-dimensional finite bodies, *ASTM STP* **791** (1983) pp. I-238–I-265.

[6] I.S. Raju and J.C. Newman, Stress intensity factors for circumferential cracks in pipes and rods under tension and bending loads, *ASTM STP* **905** (1986) pp. 789–805.

[7] J.C. Newman and I.S. Raju, *Analysis of surface cracks in finite plates under tension and bending loads*, NASA TP-1578 (1979).

[8] G.G. Trantina et al., Three dimensional finite element analysis of small surface cracks, *Eng. Fract. Mech.* **18** (1983) pp. 925–938.

[9] D. Broek, *Fracture mechanics software*, Fracturesearch (1987).

[10] D. Broek et al., *Applicability of fracture toughness data to surface flaws and corner cracks at hole*. Nat. Airspace Lab (Amsterdam) NLR-TR 71033 (1971).

[11] O.L. Bowie, Analysis of an infinite plate containing radial cracks originating at the boundary of an internal circular hole, *J. Math. and Phys.* **25** (1956), pp. 60–71.

[12] D.P. Rooke et al., *Simple methods of determining stress intensity factors*, AGARDograph 257 (1980) Chapter 10.

[13] M.F. Buckner, A Novel principle for the computation of stress intensity factors, *Z. Angew. Math. Mech.* **50** (1970) pp. 529–546.

[14] P.C. Paris *et al.*, The weight function method for determining stress intensity factors, *ASTM STP* **601** (1976) pp. 471–489.

[15] Anon. *Crack growth analysis software*, Failure Analysis Associates.

[16] H. Vlieger, Residual strength of cracked stiffened panels, *Eng. Fract. Mech.* **5** (1973) pp. 447–478.

[17] T. Swift, Development of the fail safe design features of the DC – 10, *ASTM STP* **486** (1974) pp. 164–214.

[18] T. Swift, Design of redundant structures, *AGARD LSP* **97** (1978), Chapter 9.

[19] C.C. Poe, *The effect of riveted and uniformly spaced stringers on the stress intensity factor of a cracked sheet*; AFFDL-TR-79-144 (1970) pp. 207–216.

[20] V. Kumar et al., *An engineering approach for elastic-plastic fracture analysis*, Electric Power Res. Inst. NP-1931 (1981).

[21] V. Kumar et al., *Advanced in elastic-plastic fracture analysis*, Electric Power Res. Inst. NP-3607 (1984).

CHAPTER 9

Special subjects

9.1 Scope

This chapter covers a number of special subjects. Although the procedures discussed in Chapters 3 and 4 for residual strength analysis, and those for crack growth analysis discussed in Chapters 5 and 7 remain unaffected in principle, slight complications arise e.g. in the analysis of surface flaws, corner cracks and multiple cracks, or in the case that residual stresses are present intentionally or inadvertently. In other cases, such as leak-break analysis or in a situation where load transfer to other members may set up conditions for fracture arrest, the interpretation of the analysis results may be somewhat different than usual. Engineering procedures to deal with such problems are discussed in this chapter. Examples are presented

The final sections provide a brief review of engineering solutions to mixed-mode loading cases and a short discussion on damage tolerance of composites.

9.2. Behavior of surface flaws and corner cracks

The classical solutions for stress intensity and geometry factors of elliptical surface flaws and corner cracks was discussed in Chapter 8, Section 3. More recently solutions for such part-through cracks of all kinds (surface flaws and corner cracks in tension and bending, surface flaws in circular bars, corner cracks at holes, etc) were obtained by Newman et al. [1, 2, 3, 4]. The latter solutions are generally acclaimed to be of better accuracy, and may be preferable above the classical solution.

For the following discussion, it does not matter which solution is used. The only issue of importance is that the geometry factor (and therefore the stress intensity) varies along the crack front. In contrast to the case of a through-the-thickness crack with an essentially straight front where K and β are the same everywhere along the crack front and where one can speak of THE stress intensity, in the case of part-through cracks the stress intensity and β-depend

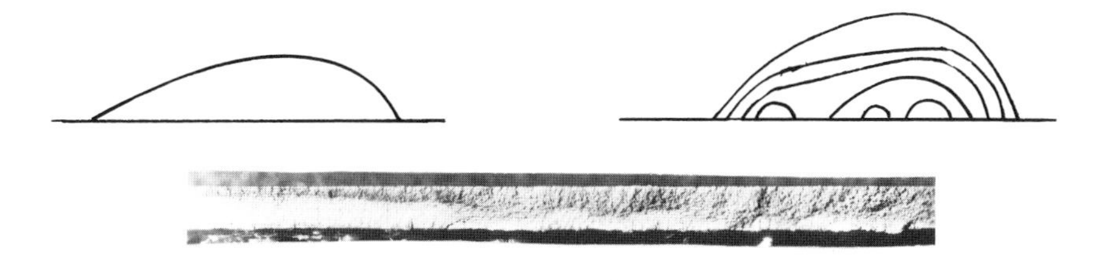

Figure 9.1. Non-elliptical surface flaws and corner cracks.

upon location. To illustrate the behavior of part-through cracks, a surface flaw under uniform tension will be used as an example and use will be made of the classical solution. In principle these issues are the same for all other part-through cracks regardless of the loading and the geometry factor solution used.

Surface flaws and corner cracks are not necessarily elliptical. In many practical cases (Figure 9.1) the shape of a surface flaw is irregular; especially when there are multiple crack initiation points the linking up of the various small cracks often leads to a non-elliptical crack (Figure 9.1). Provided geometry factors for these irregular cracks are available, the damage tolerance analysis can proceed in a similar manner as for elliptical flaws. However, ready-made geometry factors are available only for elliptical cracks, reason why in practical analysis all part-through cracks are assumed to be elliptical. If the flaw had indeed an irregular front, this assumption may cause considerable error in the analysis.

The geometry factors for a surface flaw in tension were discussed in Section 8.3. For the following illustration we will consider a case where a/B is small so that the front free surface factor, $\beta_{\mathrm{FFS}} \approx 1$. The stress intensities for the crack extremities (Section 8.3) are then

$$\left.\begin{aligned} K^A &= \frac{1.12}{\sqrt{Q}}\,\sigma\,\sqrt{\pi a} \\ K^c &= \frac{1.12}{\sqrt{Q}}\sqrt{\frac{a}{c}}\,\sigma\,\sqrt{\pi a} \end{aligned}\right\}. \qquad (9.1)$$

In this equation, $Q = \phi^2$, is a function of crack aspect ratio a/c. Its value can be obtained from Figure 8.3. Some literature provides a set of curves for Q where the lines are labled for various ratios of applied stress/yield strength (σ/F_{ty}). The curves stem from the time that attempts were made to compensate for plasticity by means of a so-called 'plastic zone correction' to the crack size [5]. This practice has been long abandoned as it is cumbersome and inadequate. Nevertheless some analysts and computer codes still use these curves. However, if the practice is not followed for other cracks, there is no reason to use it for surface flaws. Moreover, if a plastic zone correction is made for the surface flaw, it should also be made in the evaluation of K_{Ic} and of ΔK in the rate diagram, which it is not. Thus, only the line for $\sigma/F_{ty} = 0$ (elastic solution) is applicable, which is the one shown in Figure 8.3. (Modern solutions [1-4] do not use a plastic zone correction either).

For residual strength analysis the fracture criterion (Chapter 3) is:

$$\text{Fracture if } K = K_{Ic} \tag{9.2}$$

Indiscriminate use of this criterion would lead to:

$$\frac{1.12}{\sqrt{Q}}\sigma\sqrt{\pi a} = K_{Ic}. \tag{9.3}$$

The assumption is implied that fracture indeed occurs when the highest stress intensity anywhere equals the toughness. As the highest stress intensity occurs at the deepest point, it is K^A that would be used in Equation (9.3). If the flaw is circular ($\sqrt{a/c} = 1$) the stress intensity is the same everywhere ($K^A = K^C$), and the use of Equation (9.3) is certainly justified. But, in the case of a long elongated flaw ($\sqrt{a/c}$ small), the stress intensity at C, K^C, is still be considerably less than the toughness when Equation (9.3) is satisfied. One could then argue that fracture probably will be postponed, as is indeed borne out by some test data [5, 6, 7]. Thus the use of Equation (9.3) may be somewhat conservative. As it cannot be assessed theoretically how much fracture would be postponed, it is safe engineering practice to use the conservative Equation (9.3)

It is emphasized that the toughness used must be (the plane strain fracture toughness), K_{Ic}. At through-the-thickness cracks constraint is dictated by thickness because the length of the roll of highly stressed material wanting to undergo (Poisson) contraction is equal to the thickness (Chapters 2 and 3). In the case of part-through cracks, the length of this role, as determined by the length of the crack front, bears no relation to thickness. The contraction of the roll is fully constrained by surrounding elastic material (Figure 2.8), so that plane strain prevails at least in the interior at the deepest point of the flaw. Hence, the use of the plane strain fracture toughness is indicated for ALL part-through cracks, regardless of thickness.

Fatigue crack growth (and stress corrosion cracking) is also dictated by the

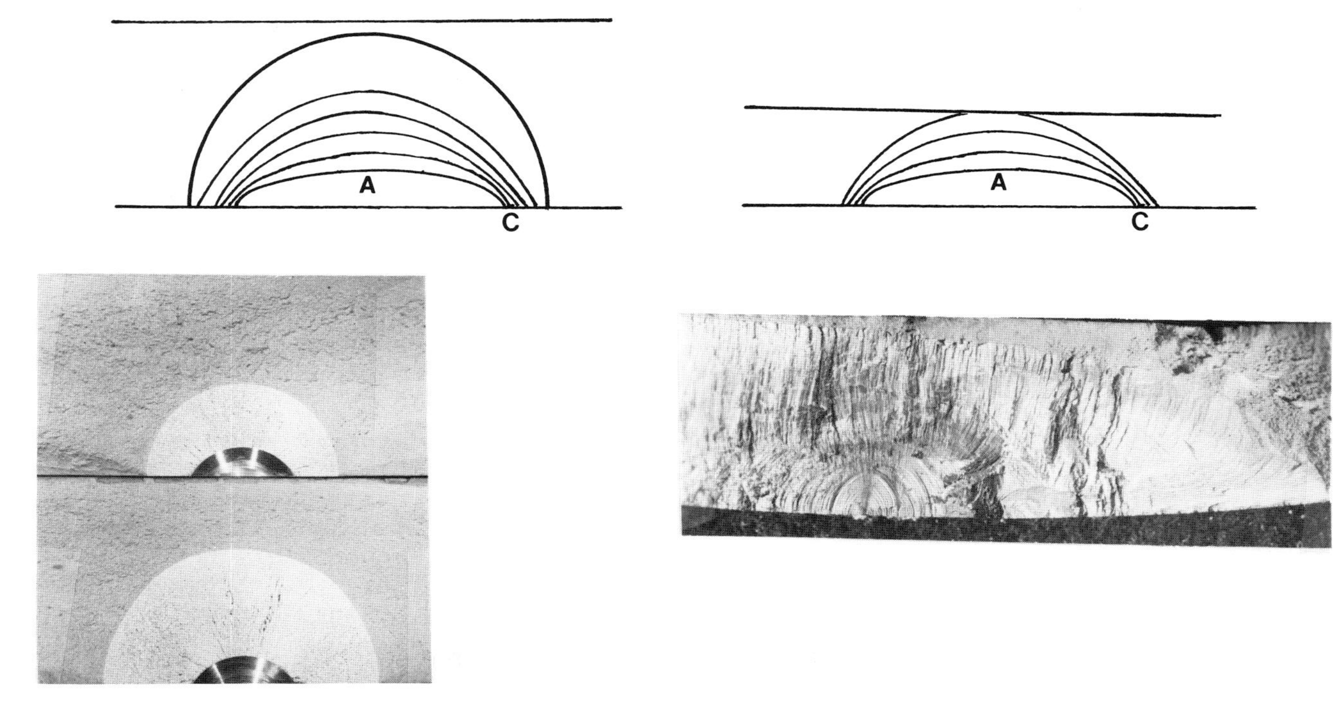

Figure 9.2. Development of surface flaws by fatigue. Lower right: originally circular flaw becomes elliptical due to bending (stress gradient).

stress intensity. As K^A is larger than K^C, more growth will occur in a certain cycle at A than at C (Figure 9.2). Hence, the flaw shape, a/c, will change and therefore Q will change accordingly. In the next cycle K^A is still less than K^C and again a will grow more than c. As a consequence, the flaw aspect ratio decreases and the shape begins to approach a circle. Once the flaw has become circular, the stress intensity will be the same everwhere, K^A being equal to K^C because $\sqrt{a/c} = 1$. Then also, the growth at a will be the same as at c and the flaw will remain circular. This development from elliptical to circular can be commonly observed in cases of uniform stress (Figure 9.2), provided the thickness is large enough for a circular shape to be reached. If the latter is not the case, there will not be enough material space and the crack may reach the front free surface before it has attained the circular shape. In the case of out-of-plane bending, the stress gradient through the thickness may cause K^A to decrease; in such a case a circular shape may not be reached either (Figure 9.2). Similarly, for corner cracks at holes, where K^A is affected by k_t more than is K^C, the end situation may not be a circular shape. (Figure 9.1 lower right)

As an example consider a flaw under uniform stress with an aspect ratio of $a/c = 0.25$. In that case $K^C = \sqrt{a/c}\ K^A = 0.5\ K^A$ by Equation (9.1), i.e. the stress intensity at C is only half that at A. Assuming that e.g. a Paris equation (Chapter 5) applies with an exponent of 4, ($\mathrm{d}a/\mathrm{d}N = C\,\Delta K^4$), the growth at C will be only $(0.5)^4 = 0.063$ times the growth at A: the growth in depth will be 16 times faster than the growth in length. This is demonstrated by the results of an actual crack growth analysis displayed in Figure 9.3a.

This presents a complication for the crack growth analysis. In order to obtain the growth of a, the flaw shape must be known because Q must be evaluated first. But the new flaw shape cannot be known unless the growth of c is known and vice versa. This problem can be solved by modifying the crack growth analysis to account for the growth of a and c simultaneously. In every cycle K^A and K^C are calculated for the current flaw shape. Then growth da at a and dc at c are assessed; the new flaw shape $a + \mathrm{d}a/(c + \mathrm{d}c)$ permits evaluation of K^A and K^C for the next cycle. The analysis then automatically provides the changing flaw shape as in Figure 9.3. A good computer code for crack growth analysis will include this feature. It should be noted that growth of a and c cannot be computed independently as the flaw shape must be known in each stage of the analysis; simultaneous analysis is a pre-requisite. Also, the analysis may have to make use of different rate data for a and$c$ if the crack growth properties in depth direction (LS) are markedly different from those in width direction (LT).

The above problem can be avoided if the flaw is assumed to be circular to begin with. Note however (Figures 9.3a and b) that this causes dramatic difference in projected crack growth life. This is another demonstration of the fact that assumptions have much more effect on the results of an analysis than small errors in geometry factors for example. When assuming a circular flaw it

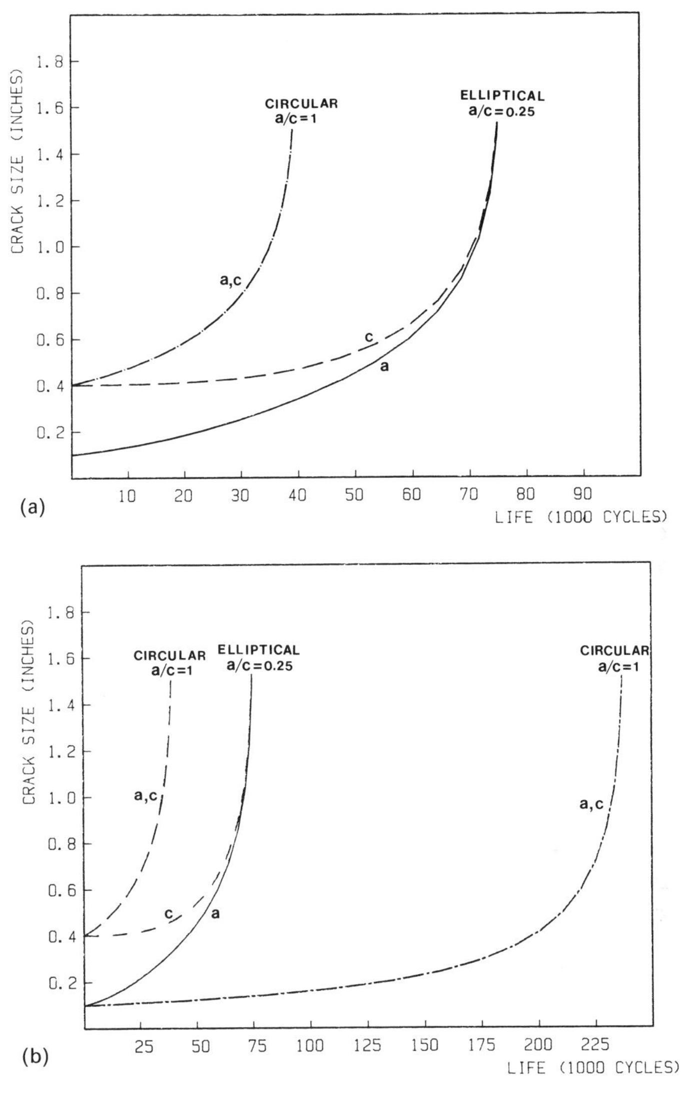

Figure 9.3. Effect of flaw shape assumptions. (a) Elliptical versus circular flaw assumption; (b) Alternative assumptions.

is hardly worthwhile worrying about the accuracy of β. The cicular flaw assumption may lead to a conservative answer, it does not provide a realistic answer. Besides, if there is a stress gradient through the thickness then even for a circular flaw $K^A \neq K^C$, and growth of a and c still would have to be assessed simultaneously. Whether the assumption of a circular flaw is conservative in that case will depend upon structure and loading.

Table 9.1. Hand calculation for change of shape of surface flaw.

1	2	3	4	5	6	7	8	9	10	11
a	c	$a/2c$	a/B	ϕ Figure 8.3	β_{FFS} Figure 8.3	$\beta^A = 1.12\,\beta_{FFS}/\phi$	Δa	$\Delta c = (\sqrt{a/c})^m\,\Delta a$	New a $(a + \Delta a)$	New c $(c + \Delta c)$
0.1	0.300	0.17	0.10	1.08	1.07	1.11	0.05	0.010	0.15	0.310
0.15	0.310	0.24	0.15	1.23	1.07	0.97	0.05	0.017	0.20	0.327
0.20	0.327	0.31	0.20	1.32	1.05	0.89	0.05	0.024	0.25	0.351
0.25	0.351	0.36	0.25	1.42	1.06	0.84	0.05	0.030	0.30	0.381
0.30	0.381	0.39	0.30	1.48	1.05	0.80	0.10	0.070	0.40	0.451
0.40	0.451	0.44	0.40	1.55	1.06	0.77	0.10	0.083	0.50	0.534
0.50	0.534	0.47	0.50	1.58	1.05	0.72	0.10	0.091	0.60	0.625
										Almost circular

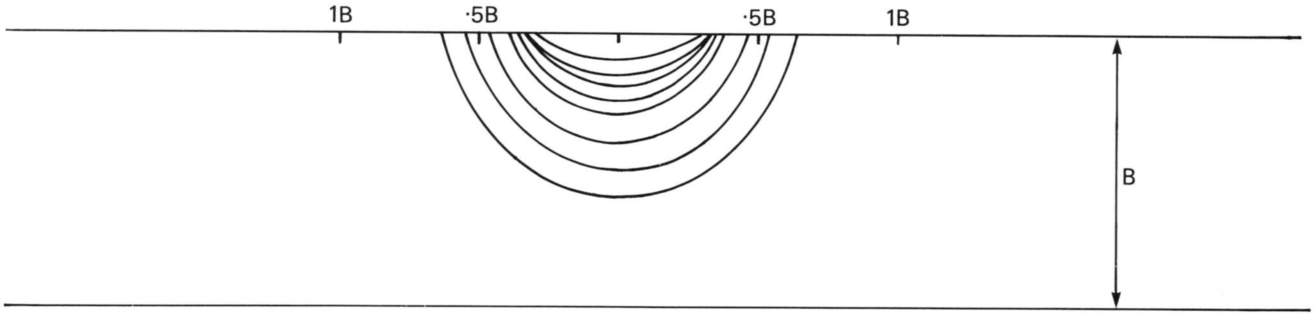

1B ·5B ·5B 1B

B

Notes:

1. Find power m in rate equation (in this case $m = 3$).
2. In this example thickness $B = 1$ inch; start is $a/c = 0.1/0.3 = 0.333$.
3. Work table horizontally; c in first line is starter value.
4. Take small increments Δa.
5. Calculate new a and c and use in next line.
6. Submit columns 1 and 7 as β-table to computer program for crack growth.

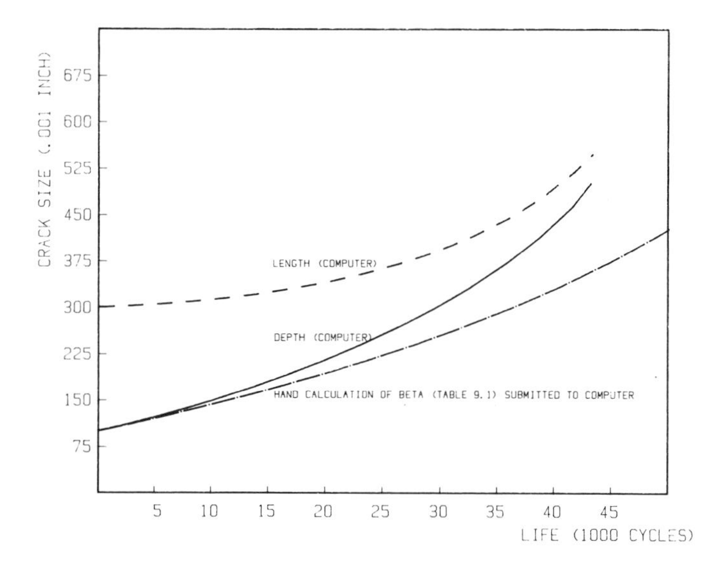

Figure 9.4. Comparison of computer calculation operating on a and c simultaneously, and result for a, using β of Table 9.1 in computer analysis.

In many analyses and in some damage tolerance requirements (e.g. military airplanes) part-through flaws indeed are assumed to be circular. The reason for the assumption may be either the desire for conservatism, or simply the lack of a computer capability to assess growth of a and c simultaneously. If the lack of a computer capability is the problem one is not necessarily forced to assume a circular flaw, because a quick engineering judgement of flaw shape development can be made as follows.

One approximates the rate data by a simple equation, preferably a Paris equation. Then a hand-calculation is performed as in Table 9.1. The depth a is increased by small increments Δa and the associated growth of c is assessed, going through the table horizontally. In the next line Δa is assessed for the new flaw shape, etc. The end result is a table of $\beta(a)$ which now accounts for the changing flaw shape. This table of $\beta(a)$ is used as input in the crack growth analysis, which the provides the 'proper' growth of a. The associated c values follow from the table. Of course the proper Paris exponent should be used (the value of 4 in the table is an example only). The growth curve for a was recalculated in this manner and the result is compared with a similtaneous computer analysis of a and c as shown in Figure 9.4. Clearly there is reasonable agreement. A few more steps as in Table 9.1 will further improve the result. The procedure still works when there is retardation because retardation is mostly determined by the ratio of stress intensities in successive cycles, and of course these ratios are dictated by the stressess only, and not by β. Similarly, if there are stress gradients through the thickness, the procedure of Table 9.1 can be used by including the effect in β^A (columns 5-7 in the table),

Circular flaw assumptions may lead to results far different from crack growth in service. On the other hand if that assumption is not made, another one may be necessary. Unless the initial flaw shape is known from tests or service cracks, one must make an assumption for the initial value of a/c. It needs no further explanation that the analysis results will depend strongly upon the assumption made, and there is a risk that it leads to unconservative results. In this respect an assumption of $a/c = 1$ (circular), may well be preferable. Unrealistic results are not due to short comings of the procedure. With the proper input the analysis result is reliable. It is the assumption that determines the magitude of the error.

9.3. Break-through; leak-before-break.

Whether leak-before-break is at issue or not, a clear distinction must be made between break through of a part-through crack due to crack growth and break through due to fracture. Crack growth of a surface flaw by fatigue or stress corrosion may be terminated by fracture before break-through, upon which the flaw extends by the fracture processes such as rupture or cleavage. The latter happens when K reaches K_{Ic} due to the gradually increasing stress intensity or because one higher load occurs in a sequence of smaller ones. If fracture does not occur, the flaw will extend gradually by crack growth until the front reaches the front free surface (Figure 9.5). When this situation is reached, the stress intensity at D and E will be very high (protruding wedge of material) and crack growth will occur rapidly from D to F and from E to G. The latter takes so few cycles that it is safe to assume that F and G are reached immediately upon break-through. The crack has then become a through crack with a length $2a$ (equal to the original $2c$ at break-through). Conitnued growth is dictated by the stress intensity of the through-crack (different β). It is possible that the new stress intensity of the through-crack is higher than the toughness. In that case fracture ensues immediately and subsequent flaw extension occurs rapidly, the results being 2 half structures.

If the flaw is in the wall of a pressure vessel, its break-through causes a leak, but when fracture immediately follows, this leak actually constitutes a break. If on the other hand the stress intensity of the through crack is still less than the toughness, further crack extension will occur by crack growth, the structure remains intact for the time being, and a simple leak will occur. This is a case of leak-before-break, which is desirable because a leak usually can be readily retected. Provided leaks would not cause fires or other dangers, inspections for cracks would not be necessary; one could simply wait for a leak and then repair. Indeed leak-before-break is a desirable damage tolerance property.

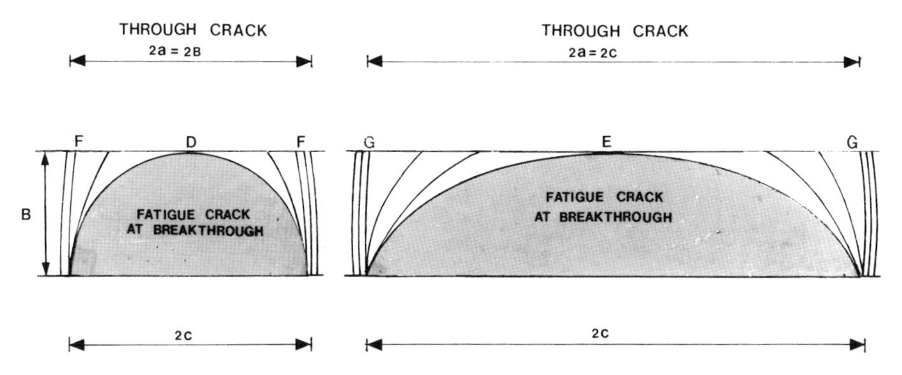

Figure 9.5. Break-through of fatigue crack. Left: circular; right: elliptical.

The smallest through-the-thickness crack that can be formed is of a length equal to twice the thickness ($2a = 2B$); this will happen when the crack was already circular ($a = c = B$) at the time of break-through (Figure 9.5 left). If the initial flaw is very slender, there may not be enough room for the crack to become circular. In that case the through crack that develops will have a length equal to the major axis of the ellipse at break-through, $2a = 2c$. Should the stress intensity of the through crack still be below the toughness, the through crack will continue growth by one of the cracking mechanisms. As the crack now has another configuration, the appropriate β must be used. Various possible configuration changes of different part-through cracks are shown in Figure 9.6. Most computer codes for crack growth analysis have a facility that effects this geometry change automatically.

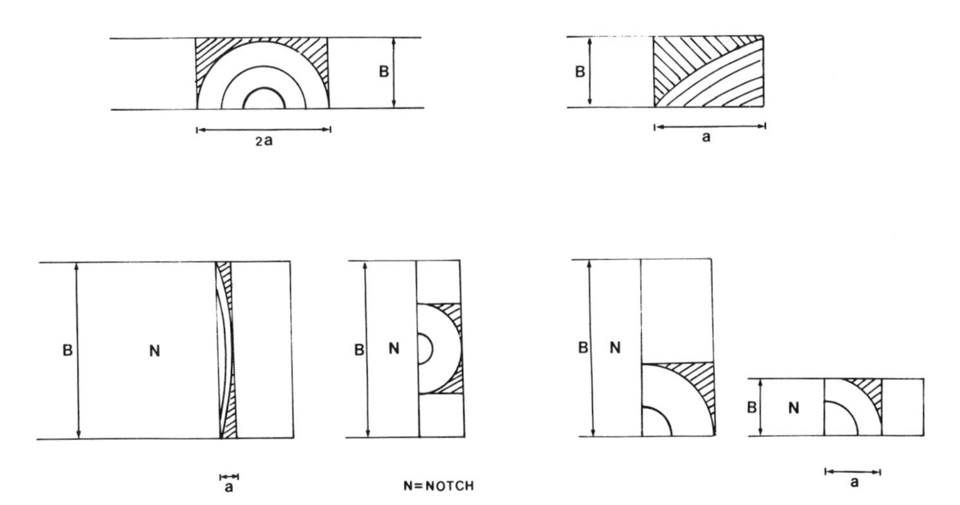

Figure 9.6. configuration changes due to break-through.

Should the stress intensity reach the toughness before the crack has grown through the thickness then a (fast) fracture will ensue. The flaw development is essentially the same as in the case of crack growth, except that the break through now occurs suddenly from a crack smaller than break-through size (Figure 9.7). Upon break-through a leak occurs in the same manner as before and a configuration change takes place. Again the question is whether the stress intensity for the new configuration is above or below the toughness. If $K > K_c$ (or K_{Ic}) fracture will continue (break), but if $K < K_c$ (or K_{Ic}) the through crack will be sub-critical, fracture will stop and only a leak is formed: leak-before-break, and further growth would be by crack growth instead of fracture.

Fracture of the part-through crack is governed by plane strain. However, the constraint for the through crack is governed by thickness, and there is a distinct possibility that plane stress or a transitional condition prevails for the through crack. In that case the toughness is higher (Figure 3.6) as well, (leak only). Clearly then, the chances for leak-before-break are better when the wall is thinner. The longer the through crack, the smaller the chances that $K < K_c$, because K depends upon the through-crack size a. Thus, a circular flaw has a better chance of leading to a leak than an elongated crack, the former causing a through crack of size $2a = 2B$, the latter a through crack of larger size.

Suppose that break through occurs by fracture of a small crack. Also suppose that K of the ensuing through crack is less than the toughness, K_C or K_{Ic}, whichever is applicable. Fracture being a fast event, the question arises whether it is indeed arrested or whether it will continue even though K is less than the toughness. Although dynamic fracture analysis [5, 8] in principle can answer this question, the analysis is rather involved and beyond the realm of day-to-day engineering fracture mechanics. Besides, the solution depends upon the assumptions. Pragmatically, a simple assumption can be made. If K is equal to the toughness or slightly below, the dynamic effects will probably prevail and a break occur. Accounting for a reasonable dynamic effect of 15%, a leak-before-break will occur when K is less than 85% of the toughness K_c or K_{Ic}, whichever is applicable. Instead of using 0.85 K_c (or 0.85 K_{Ic}) sometimes the so-called arrest toughness, K_a, is used (ASME requirements). Even with the use of the arrest toughness [5] the question of state of stress of the through crack remains, as the through crack could be in plane stress. In view of the fact that measurement of the arrest toughness is difficult and data are scant, the pragmatic approach of using a 15% dynamic effect may be the only possible avenue. An example of analysis is given in the Solutions to Exercise 1-4.

It should be noted here that leak-before-break assessment requires establishment of complete residual strength diagrams (Chapter 3) for the part-through crack as well as the through crack. Various a/c ratios must be considered. Considering only one crack size or stress can easily lead to erroneous

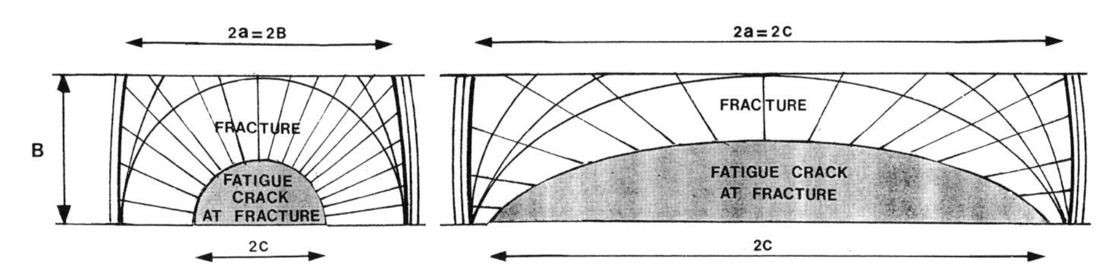

Figure 9.7. Break-through by fracture. Left: circular; right: elliptical.

conclusions, especially in this case, where geometry factors depend upon crack size as well as shape.

9.4. Fracture arrest

In all cases of load transfer to second elements, fracture arrest is possible; arrest occuring when the fracture reaches the second element. The resulting large damage is more easily detectable which is desirable for fracture control. If a structure has crack arrest capabilities, superficial inspections for large damage might be adequate. In the literature both the terms 'crack arrest' and 'fracture arrest' are used. As the condition relates to fracture and not to crack growth only the term 'fracture arrest' is used here.

Fracture arrest principles will be illustrated first on the basis of stiffened panels. Subsequently, the idea will be generalized to multiple element structures and 'crack' arresters. Arrest analysis procedures were developed mainly by the aircraft industry [5, 9-11]. At present, most large commercial aircraft are designed with fracture arrest capability. As a matter of fact, damage tolerance requirements for commercial aircraft (Chapter 12) tend to make this almost inescapable. Thus, it is justified to start off with an illustration of aircraft practice. An application to ship hulls is shown in Chapter 14.

It was shown in Section 8.10 (Figure 8.25) that load transfer to stiffeners causes a decrease of β, as shown in Figure 9.8. In turn, this load transfer causes a stress concentration L in the stringer. As in all fracture analysis a complete residual strength diagram must be determined to avoid misinterpretation. The residual strength diagram is calculated (Chapter 3,) on the basis of the criterion

$$\text{Fracture if } K = K_c \quad \text{or} \quad \beta\sigma\sqrt{\pi a} = K_c. \tag{9.4}$$

Note that the use of the plane stress or transitional toughness is appropriate here, because the plate will usually be of too small a thickness for plane strain, while all cracks considered are through the thickness. For large W, the geometry factor will be essentially equal to unity for unstiffened panels ($\beta = 1$). Using Equations (9.4) the fracture stress for the unstiffened panel is calculated for a number of crack sizes and plotted as the dashed line in Figure 9.8c. Appropriate

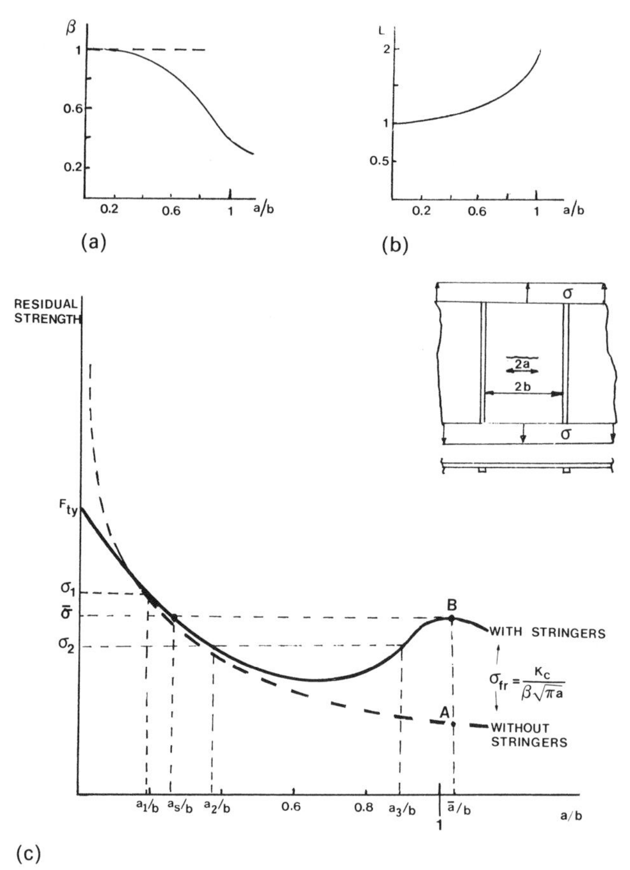

Figure 9.8. Residual strength of stiffened panel. (a) Geometry factor; (b) Stringer stress concentration; (c) Residual strength diagram.

corrections should be made for small crack sizes where σ is close to F_{ty}, as discussed in Chapter 3.

Subsequently, the residual strength diagram for the stiffened panel is established, also using Equation (9.4). For small cracks ($\beta \approx 1$) the residual strength curve is the same as for the unstiffened panel (Figure 9.8c), but for the case of $a/b = 1$, the β for the unstiffened panel is much less than 1 (Chapter 8). If for example $\beta = 0.4$ then the residual strength of the stiffened panel will be $1/0.4 = 2.5$ times as high as that of the stiffened panel. This defines point *B* in Figure 9.8c to be 2.5 times as high as point *A*. For intermediate crack sizes the residual strength curve for the stiffened panel will be as shown (see β in Figure 9.8a).

Apparently the residual strength diagram of the stiffened panel exhibits a relative minimum and a relative maximum at $a/b = 1$. Paradoxically, the residual strength under the presence of a long crack of $a = b$ is higher than for certain smaller crack sizes; the strength increases beyond the relative minimum when the crack is longer.

Consider a case where a (fatigue) crack of size a_1 has developed (Figure 9.8c). Should a high stress (load) occur of magnitude σ_1 a complete fracture will ensue. Next consider a case where the crack grows by fatigue until size a_2. At a stress of magnitude σ_2 again a fracture will ensue, but it will only run to a_3 and be arrested at a_3. Note that the curve identifies all points where $K = K_c$. The fracture will proceed to a_3, but then again $(K < K_c)$; although the crack size increases, the decrease in β is such that $\beta\sqrt{\pi a}$ decreases. In order for fracture to continue from a_3 the stress must be increased to $\bar{\sigma}$. The damage will then increase by fracturing to $\bar{a}$, but this fracturing will be controlled tearing. At the point ($\bar{\sigma}$, $\bar{a}$) fracture will again be fast and unstable.

For ALL cracks larger than a_s, fracture will be arrested, and the residual strength will still be $\bar{\sigma}$, as illustrated in Figure 9.8c. Naturally, for all cracks smaller than a_s an immediate total fracture would result, but this would require a very high stress. If the minimum permissible residual strength, σ_p (Chapters 1, 11, 12) is $\sigma_p < \bar{\sigma}$ the structure can always sustain a damage of $\bar{a}$ as shown by the horizontial line in Figure 9.8c. Of course, a fracture can occur at stresses lower than $\bar{\sigma}$ $(a > a_s)$, but such a fracture would be partial only: it would be arrested at or close to the stringer upon which the residual strength still would be $\bar{\sigma}$.

Indeed, if the structure is designed such that $\bar{\sigma} > \sigma_p$, large damage can be sustained under all circumstances where $\sigma < \sigma_p$, so that fracture control is facilitated by inspection for large damage. For reasons of stiffness airframes are built of skin-stiffened structure anyway; it then is a small step to provide the above fail-safety by apropriate selection of stringer size and spacing. (As discussed in Chapter 8, the reduction of β by load transfer depends upon stringer stiffness and spacing, i.e. upon the stiffening ratio μ.)

Up till this point the stiffeners and the fasteners have been ignored. When the fracture approaches the stringer the stresses in the stringer will increase significantly because of load transfer from plate to stringer, the very reason for which the plate experiences lower crack tip stresses and lower β. This was demonstrated in Figure 8.25. The stringer stress concentration, L, is depicted in Figure 9.8b. Load transfer to the stringer has to be accomodated by the fasteners causing high fastener shear. As the stringer is uncracked it will fracture when its local stress is equal to its tensile strength $\sigma_l = L\sigma = F_{tu}$. Thus stringer fracture occurs when the remote stress is $\sigma_{fs} = F_{tu}/L$. Using L from Figure 9.8b, this criterion provides the stringer fracture line as in Figure 9.9a.

A fatigue crack a_1 in Figure 9.9a will cause a fracture at a stress σ_1, and arrest at a_2. The stress can then be increased to $\bar{\sigma}$ (fracture proceding to $\bar{a}$) upon which the stringer will break. Then load transfer to the stringer is no longer possible

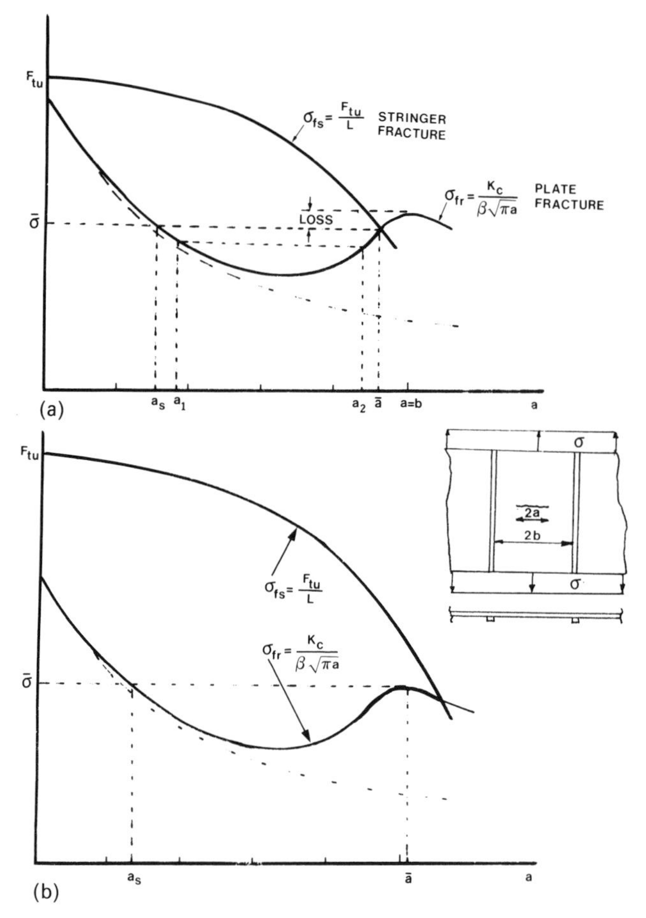

Figure 9.9. Effect of stringer on residual strength diagram. (a) Stringer critical; (b) Plate critical (stringer material with higher F_{tu} than in Figure a).

and that total fracture ensues. This means that $\bar{\sigma}$ is reduced to the level of the intersection between stringer fracture line and plate fracture line. (Compare Figures 9.8c and 9.9a). This is an undesirable situation. It can be improved by selecting a stringer material with a higher F_{tw}. The diagram of Figure 9.9a will then change into that of Figure 9.9b. Again, proper design will affect the crack arrest capabilities to the degree required; other ways to change a stringer critical case into a skin critical case will be shown later in this section (Figure 9.17)

It is important to note here that the stringer fracture is detemined by F_{tu} because the stringer is uncracked. Should the stringer be cracked already (which is unlikely), its strength would be much less and be determined again by the toughness, so that it would fail at lower stress. In such a case arrest may not be possible.

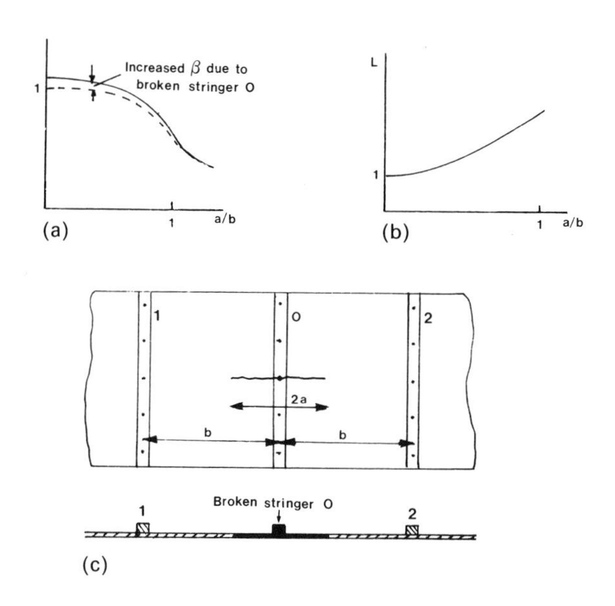

Figure 9.10. Crack strting at fastener; with broken stringer. (a) Geometry factor; (b) Stress concentration in stringers 1 and 2; (c) Configuration.

Fastener failer due to high shear must be considered as well. The fastener forcers can be calculated accurately or estimated as discussed in Chapter 8. In an uncracked structure the fasteners carry hardly and shear, so that they are often light, their primary function being to hold plate any stringers together. However, in the case of a crack the fasteners must transfer load to the stringer so that they are subjected to high shear. Thus, in order to make fracture arrest possible, heavier fasteners may be necessary.

It may seem that the case discussed above is not a relevant one, because cracks will not occur between stringers. If cracks develop they will do so at a fastener hole at a stringer (Figure 9.10). The stringer crossing the crack will be highly stressed, and develop a crack as well. It will not take long before this stringer breaks. The plate then will have to carry the load of the broken central stringer, so that K is increased; $\beta > 1$. However, from then on the problem is as before: there is a crack between the two adjacent stringers. By simple redefining the stringer spacing as b, instead of $2b$ as before, all previous arguments will hold (Figure 9.10b), with slight modifications to β and L; these parameters can be obtained from handbooks as discussed in Chapter 8. Typically, the stringer or frame spacing in a commercial jet is from 8–12 inches. These structures are designed to sustain a two-bay crack (16–24 inches) with the central frame or stringer broken.

If stringers have higher stiffness, or are more closely spaced, they will transfer more load and be more effective in reducing β. The smaller the fastener spacing,

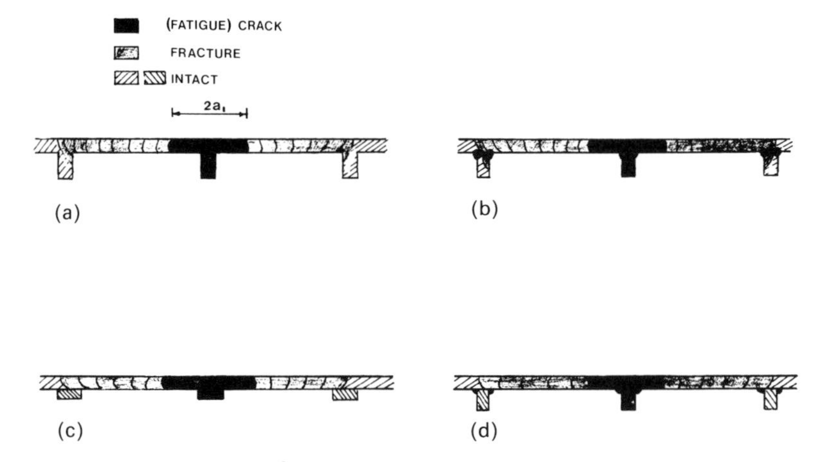

Figure 9.11. Behavior of stiffened plates with 'zero' fastener spacing. (a) Integral; fracture severs stringer; (b) Welded; fracture severs stringer; (c) Adhesively bonded (physical separation between plate and stringer; (d) fillet weld (physical separation).

the more effective load transfer to the stringers. This can be understood intuitively, because the closer the fasteners at either side of the crack tip the more the stringer will prevent the crack from opening (lower K and lower β). Hence, small fastener spacing improves arrest capabilities, as long as the fasteners are of sufficient size to carry the shear.

The smallest 'fastener' spacing is obtained if the stringer is integral or continuously welded (Figure 9.11a,b). Unfortunately, this does not permit fracture proceed into the stringer, severing the latter in the process. There being no physical separation between plate and stringer the fracture will include the stringer, upon which total fracture ensues. An alternative way to obtain 'zero'

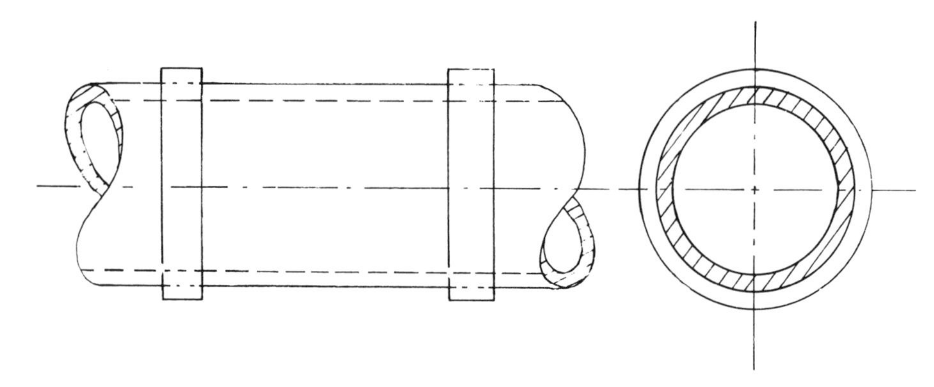

Figure 9.12. Possible arresters for pipeline.

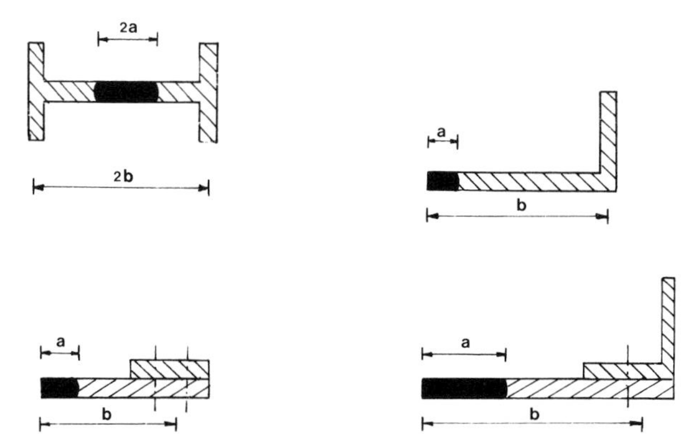

Figure 9.13. Various kinds of second elements.

fastener spacing is adhesive bonding or fillet welding (Figure 9.11c, d). Because of the physical separation the running fracture cannot penetrate the stringer. The bonds or welds must be capable of transferring the load to the stringer by shear.

Although in the above discussions the word stringer was used, one need only replace the work 'stringer' by 'arrester' or 'second element', and the arguments will be equally applicable. For example, a crack arrester for a pipeline could be designed as in Figure 9.12. and the above discussions would apply. Similarly the effects of second elements such as shown in Figure 9.13, whether integral or not

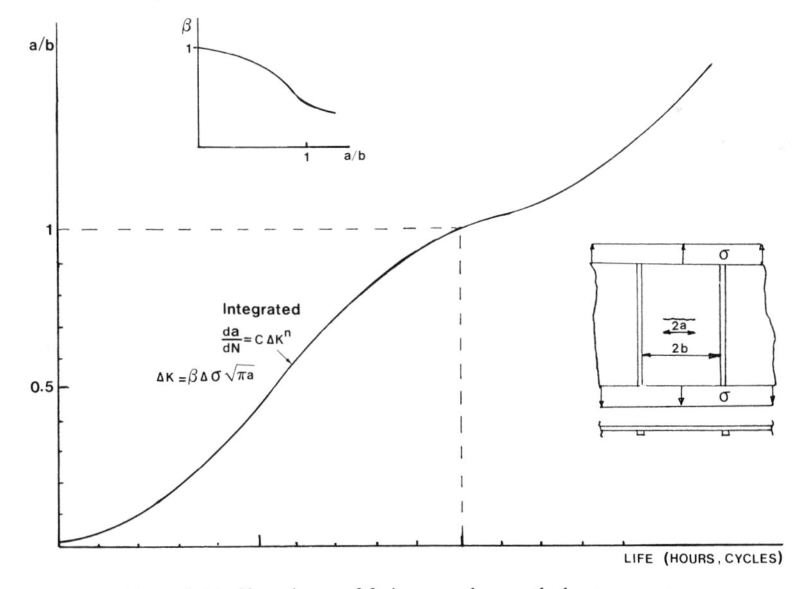

Figure 9.14. Slow down of fatigue crack growth due to arrester.

can be assessed in the manner described. Analysis procedures are already in place and β for many configurations can be obtained from handbooks, while L can be obtained from β (Chapter 8).

In crack growth (as opposed to fracture), the β-reductions discussed are effective as well. If K remains less than K_c the crack will continue to grow by fatigue or stress corrosion. Crack growth is also dictated by K; it will be slower if K decreases due to decreasing β. If fracture does not occur crack growth will be as shown in Figure 9.14.

Note that the decrease in β will slow down crack growth even when the stringers are integral or welded (Figure 9.11), although these configurations may not be effective for fracture arrest. Thus the effect of second elements on fatigue crack growth will be significant also in configurations such as in Figure 9.13, provided no fracture occurs. In this respect integral or welded stringers will be more effective than mechanically fastened stringers; only in the case of fracture will their effectiveness be less to non-existent.

The possibility that crack growth may be intermitted by fracture must be counted on. If the loading is of variable amplitude a high load in the sequence may cause fracture and arrest (Figure 9.15). Subsequent growth again occurs by fatigue, be it that all of a sudden the crack is much longer due to the fracture. An intermediate fracture will reduce the 'life' considerably. The long cracks between arresters can be sustained (residual strength) so that inspections might focus on the detection of obvious damage, but the latter can be sustained only for a short time, because crack growth continues. Hence, the inspections (though more easy) must be repeated frequently.

Fracture is a fast process. The dynamics of the problem might suggest that fracture would continue even if $K < K_c$. However, the only way the fracture could be driven past the stringer (or arrester) would be by the kinetic energy of the fracturing structure. This kinetic energy is relatively small and the question really is whether this energy can be absorbed by the arrester. If the arrester is uncracked, which is normally the case, it easily can absorb this energy because the plastic deformation energy of an uncracked part is very large. This is demonstrated by a numerical example in Chapter 14. It is also borne out by many experiments on stiffened panels [9-11] of which some results are shown in Figure 9.16. Design for specific arrest requirements is possible, because–as discussed in Chapter 8–the effect of the various parameters in the problem can be readily assessed. Essentially all possibilities for design and design improvement are shown in Figure 9.17. A detailed numerical example of arrester design appears in Chapter 14.

9.5. Multiple elements, multiple cracks, changing geometry

Analysis of structures with multiple elements proceeds as discussed in the previous section. The second element may be mechanically fastened, welded or integral (flanges of L, I or U sections, as in Figures 9.11 and 9.13). For many of these cases β can be found in a handbook [12], although superposition and compounding are usually necessary (Chapter 8). Note: the case of an I or L section may not appear specifically in the handbook; instead there will be a case of a crack in a thin member approaching an area of increased thickness; the flange width should then be interpreted as the increased thickness; alternatively the flange can be treated as a stringer; if load transfer does occur its effect on β should be included in the crack growth analysis.

Figure 9.18 illustrates cases where the presence of one crack influences the other, i.e. extension of crack 1 increases/decreases β of the other crack. Therefore, these cracks must be analysed simultaneously. The procedure for simultaneous growth is the same as for the simultaneous growth of a and c of a surface flaw as discussed on Section 9.2. If the computer code does not have this capability the following shortcut can be made in a manner similar as in Table 9.1. Both cracks are subject to the same stress history (relatively, as there only will be a proportional difference in stresses). As a consequence the growth of crack 2 during a certain small interval or cycle can be prorated to the growth of crack 1. At a certain stage with a_1 and a_2, the geometry factors are β_1 and β_2, both of which now depend upon a_1 as well as upon a_2. A simple rate equation (e.g. Paris) is used to determine *a priori* how Δa_1 of a_1 changes due to the growth of a_2, and a table is made just as in the case of a surface flaw (Table 9.1). The end result is a table of β for a_1 in which the growth of a_2 is accounted for. This

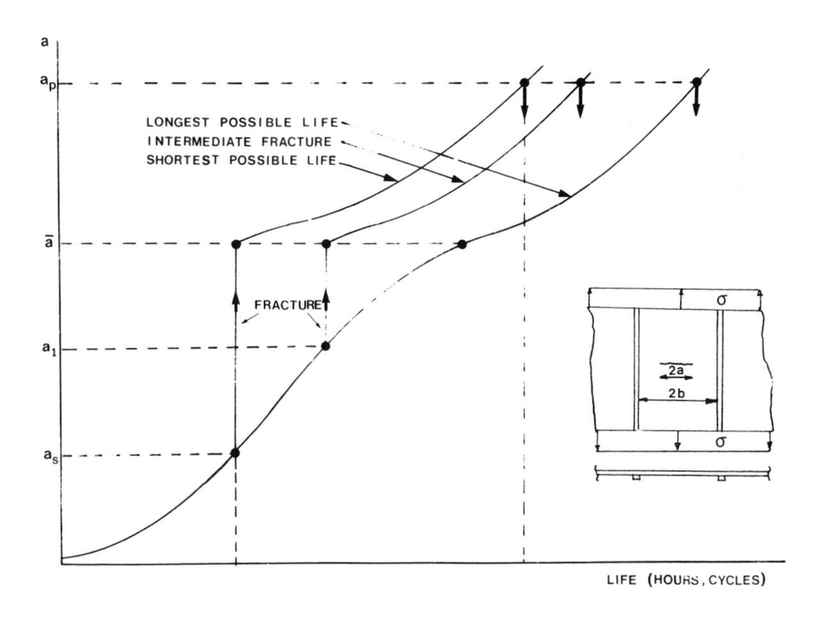

Figure 9.15a. Life as affected by intermediate fracture and arrest. Intermediate fracture and arrest.

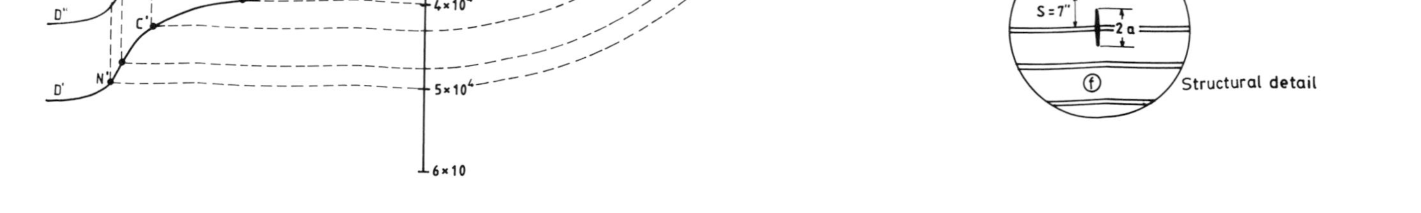

Figure 9.15b. Life as affected by fracture and arrest. Actual case in aircraft structure.

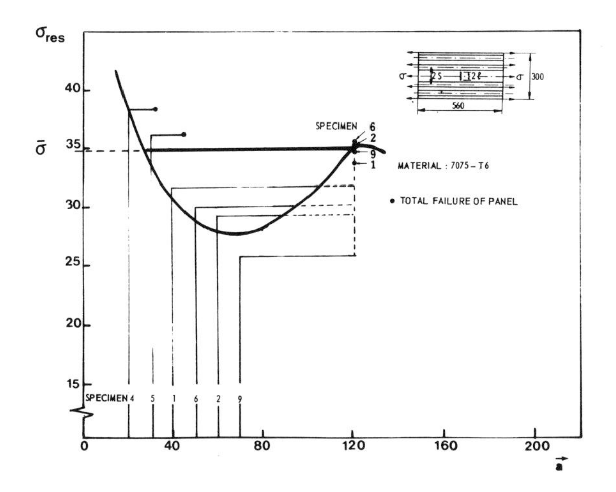

Figure 9.16. Test data for stiffened panels [10].

table is submitted to the crack growth program, which now calculates the growth curve for a_1. The associated a_2-sizes follow from the pre-established table, as does c in Table 9.1

A changing geometry is also a common problem. For example, (consider the case of Figure 9.19) with a crack starting in the thin flange (plane stress; high K_c). Although for such cases β sometimes can be found in a handbook, an

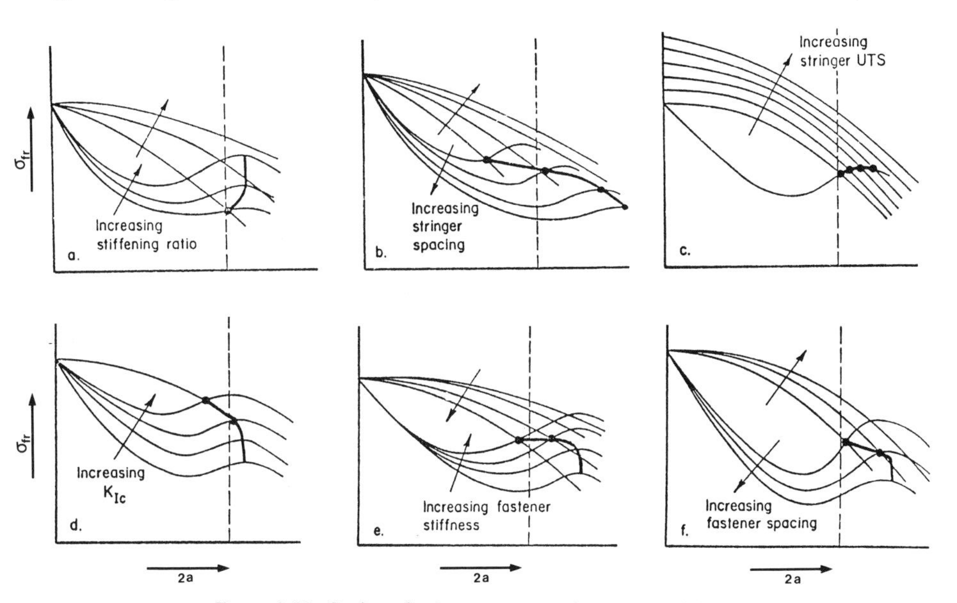

Figure 9.17. Options for improvement of arrest capability.

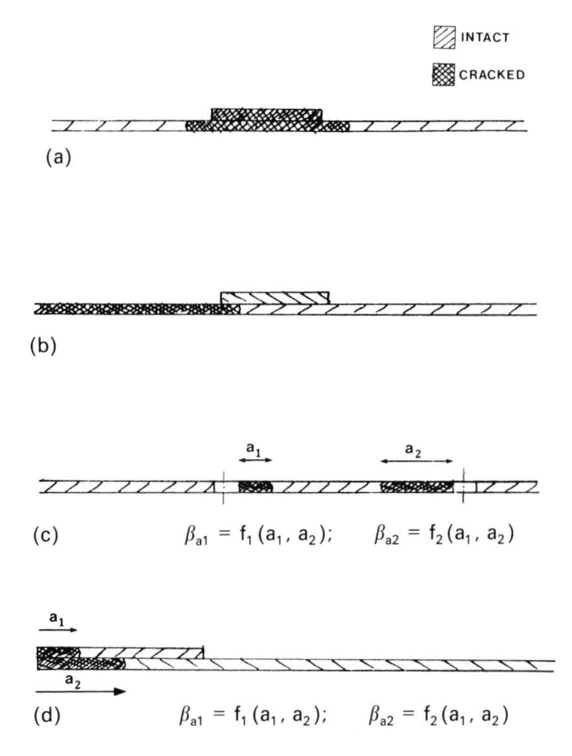

Figure 9.18. Cases of multiple cracks. (a) Increased β due to load transfer of broken 2nd element; (b) Decreased β due to intact 2nd element; (c) Interacting cracks in same element; (d) Interacting cracks in different elements.

engineering procedure is shown below; similar approaches for other configurations are possible. For an inspector the crack size would be a in both cases, but it would be too conservative to use the same definition of a in the residual strength analysis, as this would be implicitly assuming that the load path represented by area A is cut and that this load has to bypass the crack. But area A does not carry load because it does not exist and therefore only the cut load of area B (the flange) has to bypass and crack. Thus it would be more reasonable to assume the configuration of Figure 9.19c where only the real cut load path is added to a fictitious crack size, a_{eff}, rather than to the real crack size. When executed carefully, the procedure presents no difficulty as shown below.

As long as the crack is in the flange (high K_c), the problem is essentially the one of Figure 9.19b for which the residual strength diagram can be calculated (Figure 9.19d) in accordance with the standard procedure. The left part of the curve for large a would be incorrect but that is irrelevant because it is not used anyway. Subsequently, the residual strength diagram is calculated for the body (Figure 9.19c) with an effective crack size, a_{eff}, the flanges accounted for in a_{eff}.

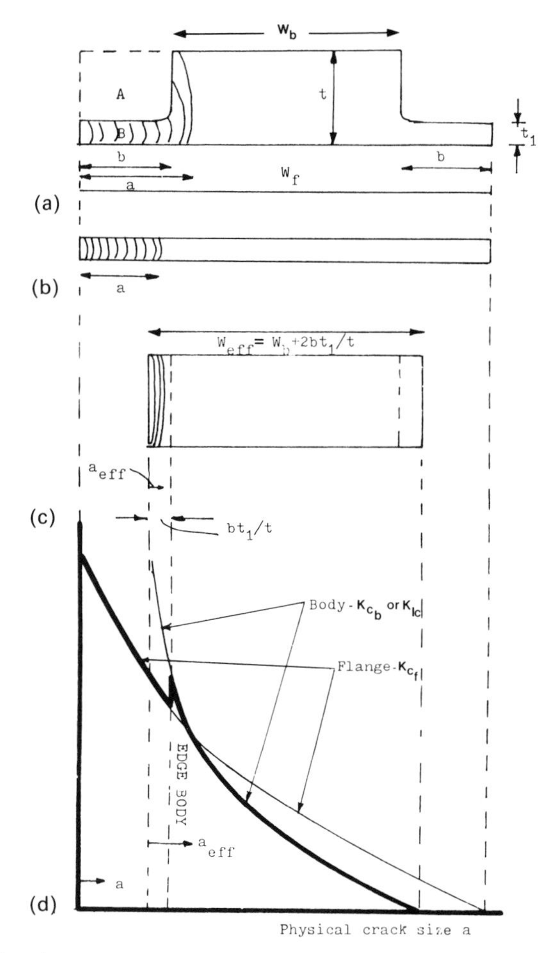

Figure 9.19. Changing geometry (approximation). (a) Configuration; (b) Flange; (c) Body; (d) Residual strength.

The residual strength diagram is plotted on the same scale as that of the flange (Figure 9.19d) but in such a manner that $a_{eff} = 0$ coincides with the location of the fictitious edge of the body. Note that K_c for the body is lower (larger thickness) which must be accounted for in the analysis; the body may be so thick that plane strain prevails (K_{Ic}).

The two residual strength curves are now combined; it is of course the lowest residual strength curve that is of interest. The discontinuity is artificial and the change-over can be considered as the average of the two curves. In this particular case, a more rigorous analysis can be made, because a handbook solution for a thickness change is available [12].

This example demonstrates that the complete residual strength diagram is required to determine the permissible crack size. Also note that the abscissa of the residual strength diagram is indeed on the scale of the physical crack size, the only one of interest for inspection. Naturally, the problem can be solved by using handbook solutions, but the above example demonstrates that where such solutions are not available good engineering approximations can be made.

A changing geometry may also be encountered when multiple elements are cracking simultaneously, as in the case of Figure 9.20. It may not be realistic to consider only the case that the entire reinforcement is cracked. The engineering approach illustrated below can provide a reasonable and slightly conservative result. First consider the two members more or less as independent parts and then the combination.

When the cracks are small, there hardly will be any load transfer between the members. Thus a residual strength curve can be calculated for each. However, for longer cracks there will be load transfer and therefore the net section yield (collapse) condition is meaningless for the reinforcement; and can be ignored for the reinforcement. The independent residual strength curves are shown in Figure 9.20b. Finally, the residual strength curve is obtained for the case that

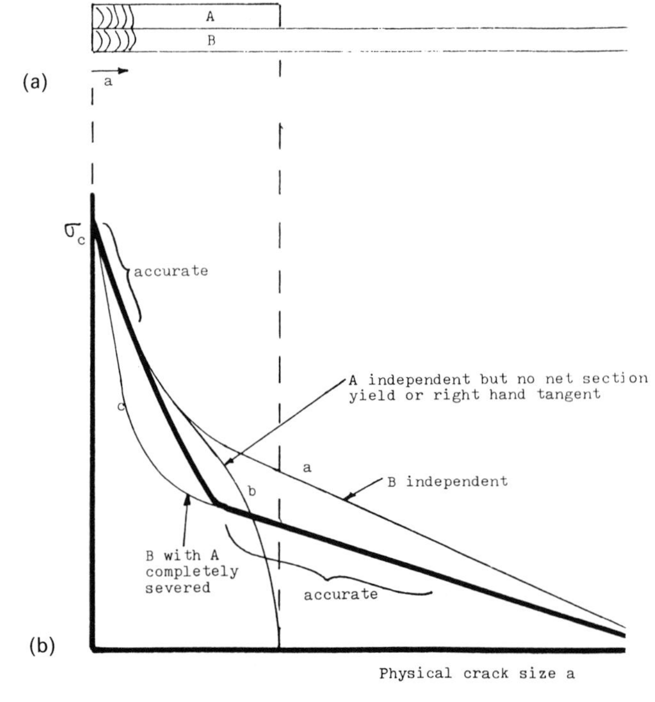

Figure 9.20. Changing geometry with two members (approximation). (a) configuration; (b) Residual strength.

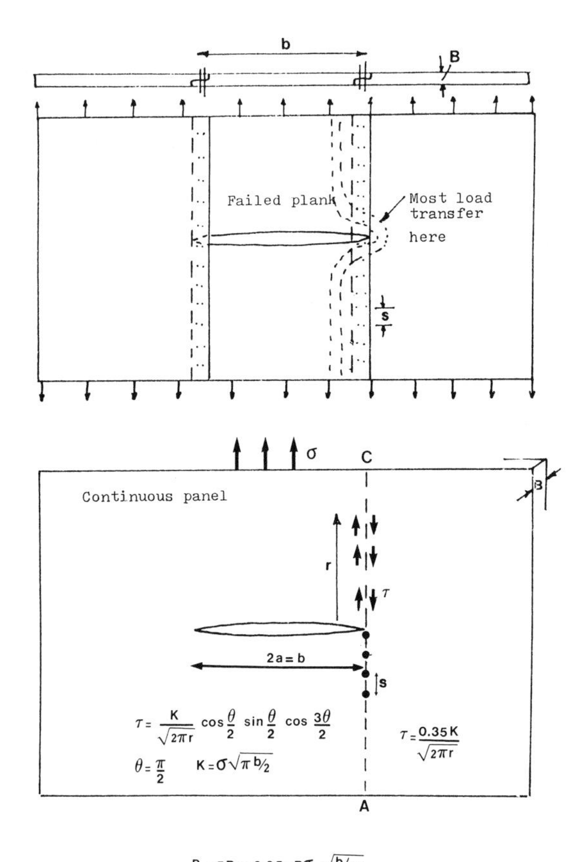

Figure 9.21. Estimate of fastener forces in case of failed plank (top: planks; bottom: full plate equivalent).

A is completely severed; β for this can be found in handbooks, the resulting curve is shown in Figure 9.20b as well.

When cracks are small member *A* will be intact and curve *c* is not applicable. Thus the curve starts out as the lowest of curves *a* and *b* (in this case *a* and *b* coincide). Where curves *b* and *c* intersect, member *A* will be failed and curve *c* applies. On the basis of these two criteria, the solid line can be drawn as a good approximation of the residual strength curve. The result is certainly approximative, but the top of *b* and the tail of *c* are accurate. Not too many different curves can be drawn between the top part of *b* and the tail of *c*. Therefore the interpolation must be reasonable.

In the example of Figure 9.21 the centre plank may experience fracture instability. The structure may qualify as fail safe if the remaining planks can still carry σ_p. In the example of three planks this would cause an average stress of $1.5 \times \sigma_p$ in the two remaining planks which will be less than the actual strength of these (intact) panels, if they were initially designed with a safety factor of 1.5 or more. The question is, however, whether the load of the failed panel can be transferred to the intact planks. Load transfer has to occur through the fasteners. Displacement compatibility requires (see discussion Figure 2.2) that the load in the center plank remains there until close to the break and that almost all its load will be transferred by shear by a few fasteners close to the break. If these fasteners would fail under this shear the longitudinal splice would zip open and the structure would still fail. Thus fastener strength is the crucial point, and two rows of fasteners are usually required.

A good way to find the fastener forces is through finite element analysis, but this will require assumptions with regard to the fastener stiffness; the fasteners will yield and must therefore be represented as non-linear springs. If they are simply assumed elastic the result of a finite element analysis will be no better than of the following engineering approach to obtain a first estimate.

Compare the structure with the continuous panel in Figure 9.21b. There will be shear stresses along line A-C. This is essentially the same as the case of three planks, but the shear in the vertical plane would have to be transmitted by the fasteners. The shear stresses in the continuous panel are known from Equations (3.1) as shown in Figure 9.21b. For $\theta = \pi/2$ they are: $\tau = 0.35\,K/\sqrt{2\pi r}$. As a first approximation it may be assumed that the total shear force over the fastener distance, τBs, should be transmitted by the adjacent fastener or, if there are two rows of fasteners, shared equally by two fasteners. Then the forces on the fasteners nearest the fracture can be estimated as $P = 0.35\,B\sigma\sqrt{b/2s}$, since $a = b$ and σ is the applied stress. This is probably a conservative estimate, because fastener and hole deformation will tend to spread the shear somewhat to more remote fasteners. The estimate can be improved by integrating over s instead of taking $r = s/2$.

In cases of rapidly decreasing K fracture arrest conditions may be set up. The most prominent examples are stringer-stiffened structures as discussed in Section 9.4, in general, it applies to all built-up structures where load transfer to uncracked members or flanges is possible. Other cases of decreasing K may be encountered if cracks are in fields with negative stress gradients, or in cases of displacement control, when stresses decrease owing to reduced stiffness due to the presence of the crack (thermal stresses are a prominent example). Finally, fracture arrest conditions may occur due to changing constraint (plane strain to plane stress), as discussed in Section 9.3, or if a crack in a thick part starts penetrating a thinner part. In such cases arrest is not due to decreasing K but due to increasing toughness, K_c.

As in the discussions of stiffened panels, the term fracture arrest is used. Crack arrest, (arrest of crack growth) will seldom occur: a decreasing K will simply

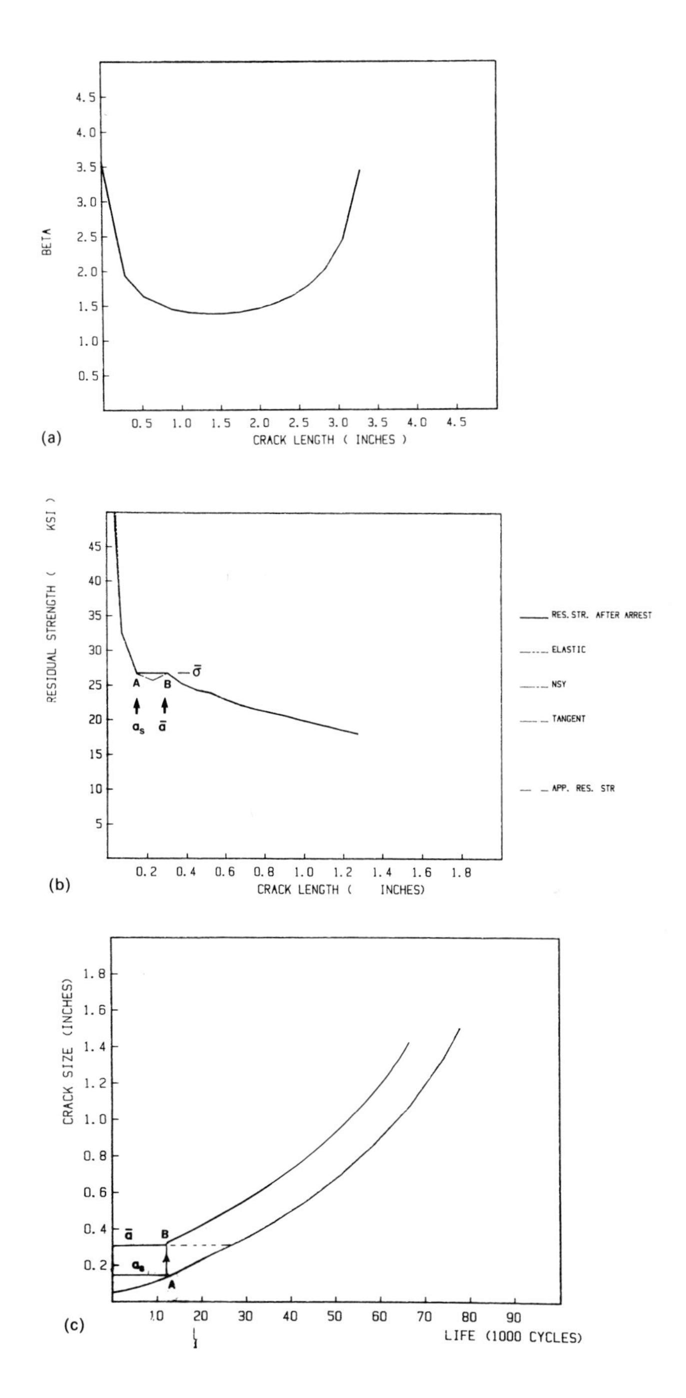

Figure 9.22. Case of decreasing β. Lug with bad bypass. (a) Geometer factor; (b) Residual strength; arrest condition between a_s and $\bar{a}$; (c) Crack growth; lower line, but upper line if fracture burst from a_s to $\bar{a}$ (smaller jump possible).

slow down crack growth by fatigue (Figure 9.14). In the case of stress corriosion cracking crack arrest may occur if K drops below K_{Iscc}.

Cases of decreasing K are quite common, but fracture arrest is not always of significance. For the case of a lug the geometry factor was derived in Equations (8.25). This geometry factor decreases rapidly because of the term $W/\pi a$. When the crack is long, the effect of the finite Width, W, will act to increase β again. Hence, for a particular case as in Figure 9.22, β may be as shown. The residual strength diagram follows from $\sigma = K_c/\beta\sqrt{\pi a}$ with appropriate corrections for small a (Chapter 3). Thus the residual strength diagram is as shown in Figure 9.22b.

Clearly, there is a dip in the residual strength diagram, so that fracture arrest will occur for all cracks between a_s and $\bar{a}$, the residual strength over this range of crack sizes being essentially equal to $\bar{\sigma}$ as in the case of stiffened panels discussed in Section 8.4. This arrest is not of much consequence for fracture and residual strength other than that the residual strength does not drop below $\bar{\sigma}$ until the crack exceeds $\bar{a}$. However, in the interpretation of the fatigue crack growth curve this effect should not be ignored.

The fatigue crack growth curve is shown in Figure 9.22c. Indeed a slow down between a_s and $\bar{a}$ is discernible because of the decreasing K. It is quite possible, however, in the case of variable amplitude loading that a high load occurs between a_s and $\bar{a}$. Should this load have sufficient magnitude to cause fracture, the damage will progress by fracture from e.g. a_s to $\bar{a}$ (Figure 9.22b). Further growth at lower stresses now occurs by fatigue as shown in Figure 9.22c, but the life between $\bar{a}_s$ and $\bar{a}$ is lost. Such 'fracture bursts' are commonly observed in service failures of structures with decreasing K. Thus, it may be prudent in the analysis of such a case to ignore the crack growth life between a_s and $\bar{a}$ (Figure 9.22c) for a case of variable amplitude loading. Whether or not such prudence is taken must remain an engineering decision. Fracture mechanics can analyse the problem, but it cannot predict which loads will occur in the future.

9.6. Stop holes, cold worked holes and interference fit fasteners.

If a service crack is detected an immediate repair is not always opportune. In such a case drilling so called stop holes may be a good temporary repair. A stop hole eliminates the sharp crack tip, but leaves a notch. It will take some time before the crack reinitiates, as shown schematically in Figure 9.23a. This reinitiation time is the 'gain' effected by the stop hole.

The sharper the notch the higher is the stress concentration, and the sooner the crack reinitiates. The stophole configuration can be considered an elliptical noth (Figure 9.24) for which;

$$k_t = 1 + \sqrt{\frac{l}{r}}, \tag{9.5}$$

Where $r = D/2$ is the radius, of the stop hole (Chapter 2), $l = a$ for the configuration of Figure 9.23b. Hence, larger diameter stop holes are more effective. As a matter of fact, the configuration of Figure 9.23c would be the most effective, as can be seen from Equation (9.5), but it is not a desirable one for other reasons. Also, stop holes will have more effect for smaller cracks as borne out by Equation (9.5).

Once the crack reinitiates the effective length of the new crack will be $a_n = a + D/2$ for the configuration of Figure 9.23d. See also Chapter 8 on cracks emanating from notches. This longer crack will grow faster than the original crack. From this, as well as from Equation (9.5) and Figures 9.23 a through d, it can be concluded that the best position for the stop hole would be the one of Figure 9.23b. However, in that situation one would never be sure that the crack tip is eliminated; if it is not, the 'stop hole' becomes just a hole and will have no effect at all. Hence, the best compromise is a stop hole centered at the crack tip (Figure 9.23d).

The beneficial effect of a stop hole can be improved considerably by cold work. Plastic deformation of the rim of the hole will stretch the material permanently. However, since this material must fit in the surrounding elastic material, it must be compressed to its original size. Thus the cold work (plastic deformation) will create a residual compressive stress around the rim of the hole in much the same manner as in the case of an overload (Chapter 5). Its workings are discussed later in this section.

There are many ways to effect the cold work (plasticity). Methods as crude as hard blows with a blunt instrument are effective and may be useful in tests, but somewhat more sophisticated procedures are indicated for practical applications. The procedure most uniformly used is hole expansion by pulling a tapered pin through the hole. Figure 9.24 shows how this expansion (cold work) improves the beneficial effect of stop holes [13]. The figure shows cases of very large hole expansions; normally one would use 3-5% in order to avoid introducing new cracks.

The above arguments are not limited to stop holes. Clearly, if structural holes are cold worked to begin with, the residual compressive stresses around the rim would postpone crack initiation as well as reduce crack growth rates once the crack has started. It has become common practice in aircraft production to cold work all sigificant holes. Obviously, the procedure is not limited to aircraft structures, but can be used just as effectively in any other structure, because it is based on a principle, not a specific material or structure.

Devices for hole expansion are commercially available, but simple tools can

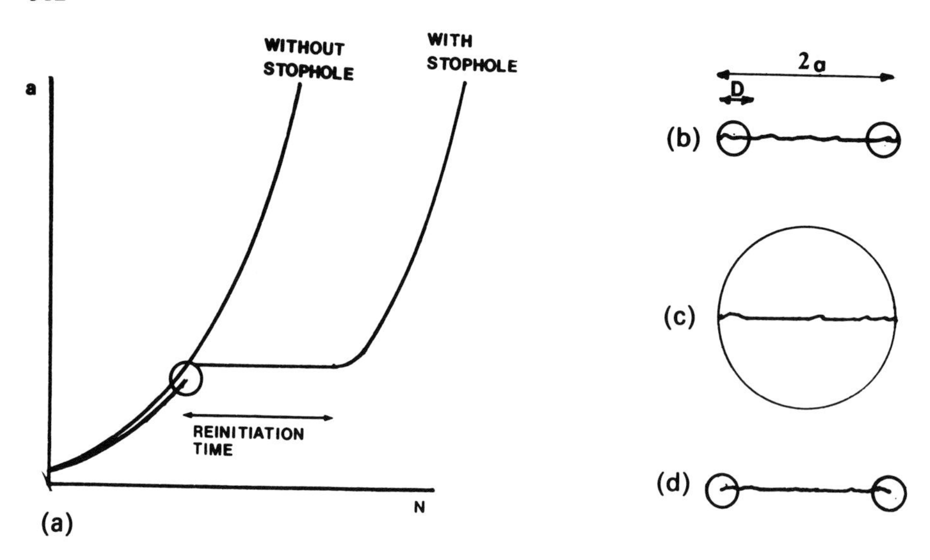

Figure 9.23. Temporary repair with stopholes. (a) Principle, (b), (c), (d) Various configurations.

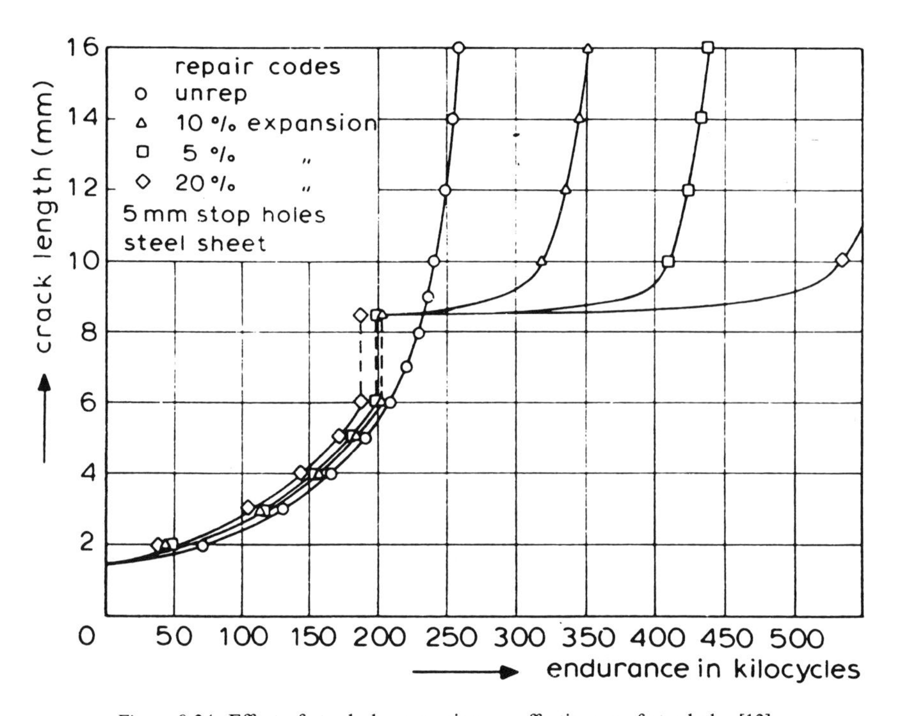

Figure 9.24. Effect of stophole expansion on effectivenes of stopholes [13].

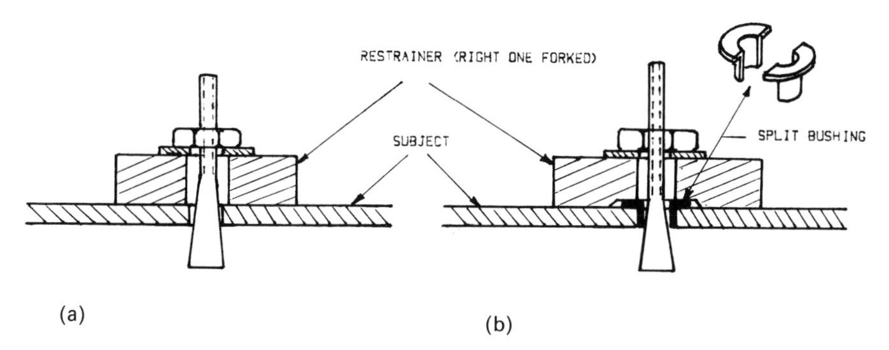

Figure 9.25. Simple devices for expanding holes (more sophisticated devices commercially available). (a) Two side access; (b) One side access from top.

be made without great expenditure. If the hole is accessible from both sides, the device can indeed be very simple as shown in Figure 9.25a. A tapered pin for e.g. three or five percent hole expansion is inserted from the bottom. A common wrench is used to pull the pin through. As the pin will cause some damage (scratches) to the hole subsequent slight reaming is advisable, but the detrimental effect of such damage usually will be outweighed by the beneficial effect of the residual stresses.

If the hole is accessible from one side only, a device with a split bushing must be used. This is illustrated in Figure 9.25b. The tapered pin in this case has a maximum diameter equal to or less than the hole diameter. It is inserted from the top. Subsequently, the two halves of the split bushing are dropped into the hole from above. The remaining hole is now too small for the pin. If the pin is pulled through, the hole will be expanded because the bushing is split and can expand. The bushing is removed after the pin is through. Commonly, the bushing is not re-usable. Again, more refined tools are commercially available.

The principle underlying the effect of cold work will be discussed later, because it is more illustrative to consider interference fits first. Oversize fasteners, such as e.g. taper locks (trade name) improve life to crack initiation and decrease the rate of growth of small cracks. In such a case the hole is also stretched and it is often argued that the effect is the same as of cold work, which is not the case. In cold work, the oversized pin is removed, allowing the material to relax and causing the build-up of residual compressive stresses. In the case of an interference fit the oversized fastener is left in, causing residual tensile stresses.

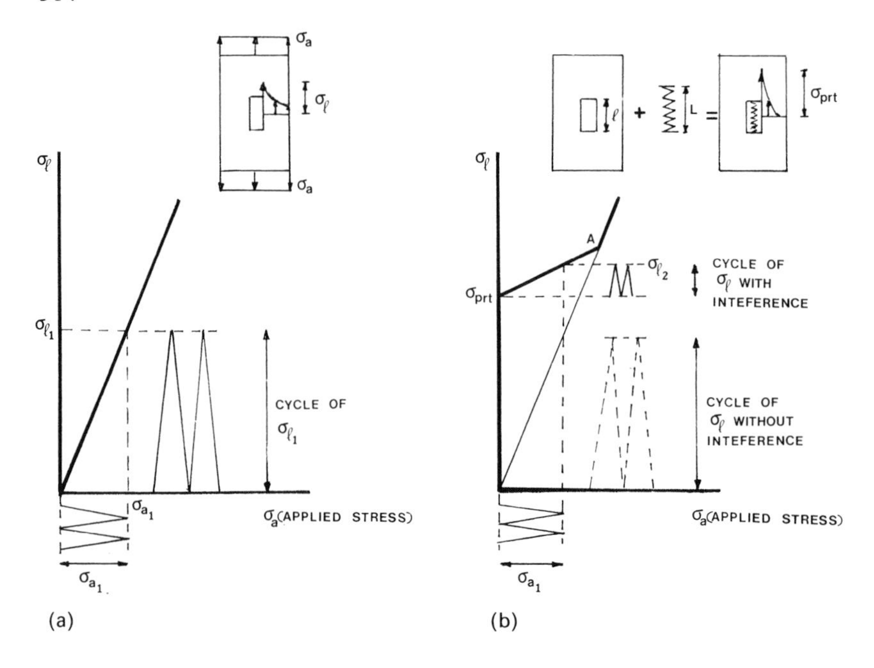

Figure 9.26. Effect of interference fit fastener on cycle of local stress. (a) Open hole; (b) Interference fit fastener.

In order to simplify the explanation, consider a rectangular hole, as in Figure 9.26a. If the applied stress varies between 0 and σ_{a1} then the local stress cycles between 0 and σ_{l1} (stress concentration $k_t = \sigma_1/\sigma_a$). Now put an oversize spring L in the hole l (Figure 9.26b insert). As the spring is too large (interference), there will be a pre-tension, σ_{prt} at the edge of the hole as shown in Figure 9.26b.

Assuming elastic behavior it can be established how the local stress will vary during cycling of the applied stress. If the plate is stretched, the hole will stretch. Eventually the size of the hole will be $l + \Delta l = L$. Then the spring will be loose and it can be taken out. From then on the behavior will be as that of an open hole. Hence, eventually point A will be reached while the stress starts in σ_{prt}. Whether the line σ_{prt}–A is straight or curved does not matter for the explanation. If the applied stress varies from 0–σ_{a1}, then the local cycles from σ_{prt}–σ_{l2}, instead of from 0–σ_{l1}. Hence, the stress range is reduced, R is increased. Because $\Delta\sigma$ has more effect than R, the life is increased and crack growth is slower.

Cold work does exactly the opposite, as shown in Figure 9.27. Consider a rim of material around the hole. This ring is expanded and is then too large so that it is squeezed by the surrounding elastic material after the pin is through. The edge of the hole is then under a compressive stress. Because the rim was yielded, forcing it back to its original size requires yielding in opposite direction. Hence,

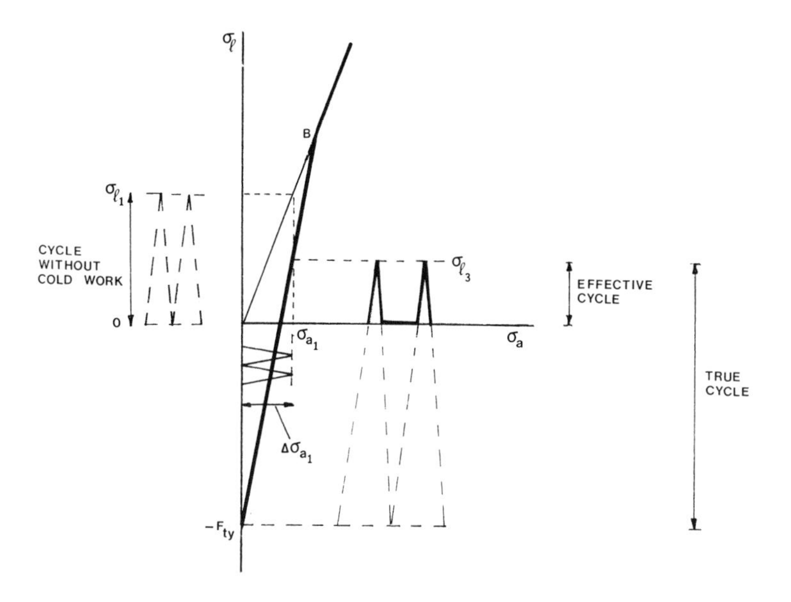

Figure 9.27. Effect of cold-worked hole on local stress cycle.

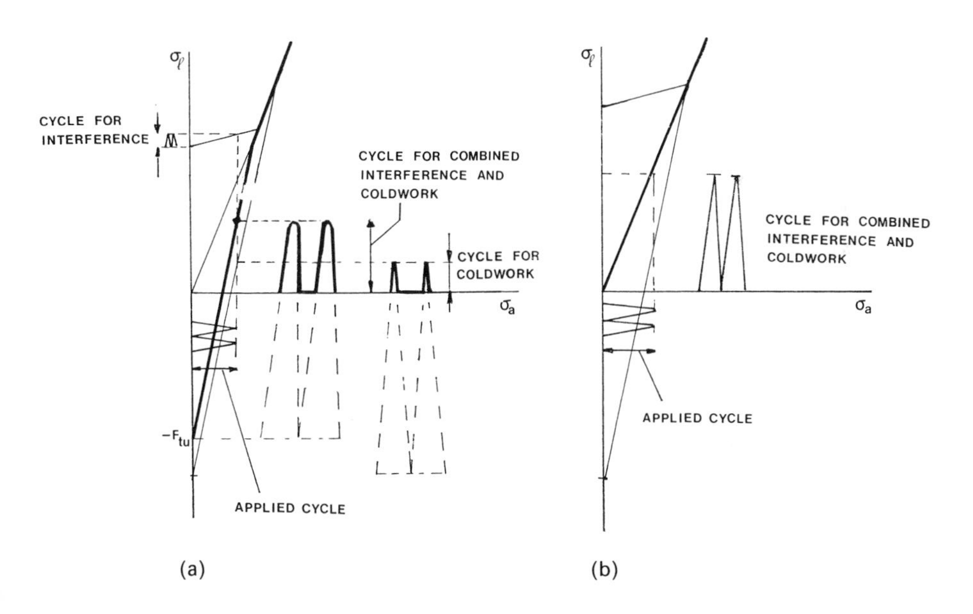

Figure 9.28. Combination of cold work and interference. (a) Net positive effect; (b) Cancellation of effects.

the residual compressive stress will be equal to at least F_{ty} (Chapter 5). If the applied stress is high, the compressed ring will eventually be relieved (point B in Figure 9.27) and from there on the line for an open hole is applicable. If the applied stress varies from 0 to σ_{a1} the local stress varies from $-F_{ty}$ to σ_{l3}. Hence R is reduced, but $\Delta\sigma$ is much larger than before (it was 0–σ_{l1}). However, most of this cycle ($-F_{ty}$ to 0) is in compression. Only the tension part is effective, so that the local stress is from 0–σ_{l3} which is less than from 0–σ_{l1}. Hence, the life is longer; crack growth is slower.

It is sometimes argued that the combination of cold work and interference gives an even larger improvement. However, a simple assessment of the situation shows that in general this cannot be so, as shown in Figure 9.28. The local stress range is effectively larger. Indeed if $\sigma_{prt} = F_{ty}$ the two effects cancel completely (Figure 9.28b). Hence, the combination is worse than either interference or cold work alone. Claims that combinations are better are fortuitous, for which there can be two reasons (Figure 9.28):

(a). The range from zero in cold work is somewhat larger than with the combination ($R = 0$), but the difference may be small in certain cases. If the interference reduces fretting (which it will) the combination may give longer life.

(b). Figures 9.26–9.28 are not entirely true to nature because they are for elastic behavior but plastic deformation occurs locally.

Yet, it is fortuitous if the combination is better in a certain test. Whether it is or not depends upon spectrum (plastic deformation at high loads), the material and the hardness of the interference pin or bushing (fretting). It will be easy to show cases where the opposite is true. The foregoing explanations, showing that the effects are due to opposite causes bears this out. Claims that cold work and interference are basically the same, clearly are unfounded.

9.7. Residual stresses in general

The previous section shows how residual stresses affect crack growth. Residual stresses are static. Thus, in general their only effect is a change in R, not in $\Delta\sigma$, unless they are at a stress concentration as in the case of holes discussed in the Section 9.6. In the case of uniform residual stresses, the situation is in accordance with one of the cases shown in Figure 9.29.

If the residual stress is negative R will be lower. Since only the positive part of a cycle is effective for crack growth (Chapter 7) also the effective $\Delta\sigma$ (with $R = 0$) may be less as is shown in figure 9.29. It should be noted here that residual stresses may gradually disappear as a consequence of the cycle stresses, a phenomenon called 'shake down'. If crack growth is by stress corrosion only the sustained stress is of importance. The sustained stress is equal to the residual stress plus the applied stress. Hence, tensile residual stresses will always promote

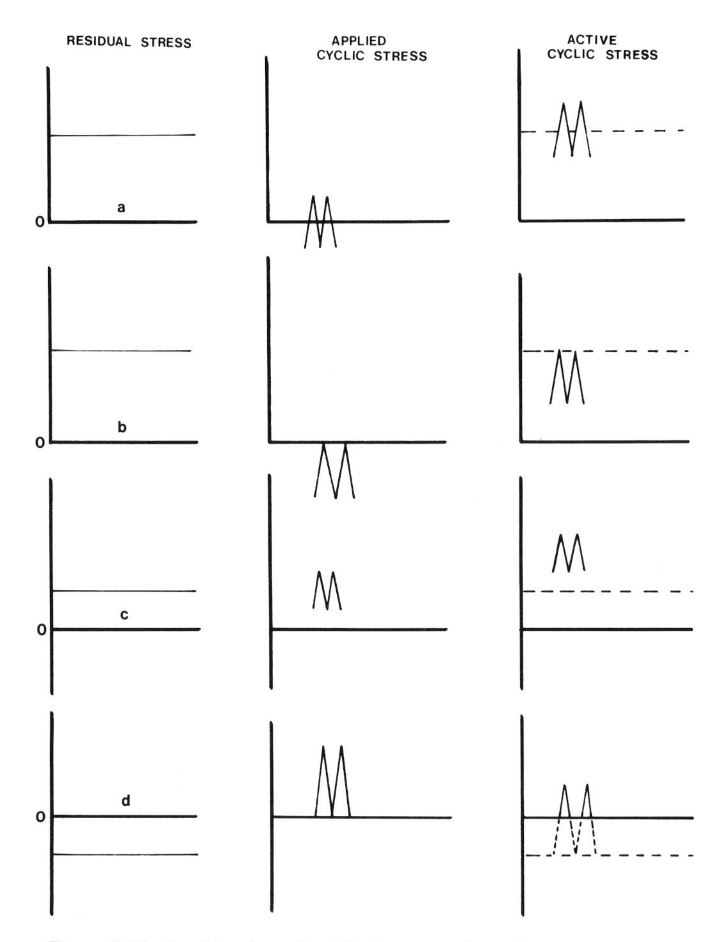

Figure 9.29. Combination of residual stress and applied cyclic stress.

stress corrosion crack growth and compressive residual stresses will supress it. This implies that all measures taken to introduce compressive residual stresses (shot peening, hole expansion, etc.) will be beneficial. Interference fit fasteners (tension) are detrimental for stress corrosion. It is stressed once more that in the case of stress concentrations residual stresses may affect $\Delta\sigma$ as well as R (Figures 9.26 and 9.27).

For residual strength (fracture) the applied stress is always additive to the residual stress (superposition). Tensile residual stress is adverse, compressive residual stress beneficial. These are qualitative statements. It is quite another matter to quantify the residual stress effect. This requires determing β and K for the residual stresses. In principle, this can be done using Green's functions or weight functions (chapter 8), but principle is not practice. The stress intensity (or

β) can be determined easily enough when the residual stresses are known, but usually, they can be estimated only.

The above arguments apply to welded structures as well. Residual stresses are always present especially in the heat affected zone (HAZ). If the user knows the residual stresses at a weld, the procedures discussed in this book are applicable. The work by Rybicki *et al.* [14, 15] may be helpful in determining residual stresses at welds. Thus fracture mechanics is applicable to welded structures provided the input is correct. Admittedly, there are some complications, but there are complications for other structures as well as was clearly shown in foregoing sections. A crack (or fracture) propagating in a weld or a HAZ must be analyzed using the properties (K_{Ic}, da/dN) of the weld metal or of the HAZ. Again, this is a matter of input, not a shortcoming of fracture mechanics. It may be difficult to obtain the properties of welds and HAZ, but it is not impossible; it may be costly and objectionable, it does not constitute a shortcoming of fracture mechanics. Naturally, some questions will always remain, but many of these can be answered by engineering procedures (see examples in Chapter 14). The problem of crack and fracture in welds is no more complicated than those in many other structures. Each have their specific problems. A proper

Table 9.2. Crack tip stresses for Modes *II* and *III*

Mode *II*

$$\sigma_x = \frac{-K_{II}}{\sqrt{2\pi r}} \sin\frac{\theta}{2}\left(2 + \cos\frac{\theta}{2}\cos\frac{3\theta}{2}\right)$$

$$\sigma_y = \frac{K_{II}}{\sqrt{2\pi r}} \sin\frac{\theta}{2}\cos\frac{\theta}{2}\cos\frac{3\theta}{2}$$

$$\tau_{xy} = \frac{K_{II}}{\sqrt{2\pi r}} \cos\frac{\theta}{2}\left(1 - \sin\frac{\theta}{2}\sin\frac{3\theta}{2}\right)$$

$$\sigma_z = \nu(\sigma_x + \sigma_y)$$

$$K_{II} = \beta_{II}\tau\sqrt{\pi a}$$

Mode *III*

$$\tau_{xz} = -\frac{K_{III}}{\sqrt{2\pi r}} \sin\frac{\theta}{2}$$

$$\tau_{yz} = \frac{K_{III}}{\sqrt{2\pi r}} \cos\frac{\theta}{2}$$

$$\sigma_x = \sigma_y = \sigma_z = \tau_{xy}{}' = 0$$

$$K_{III}{}' = \beta_{III}\tau\sqrt{\pi a}$$

solution may require some expenditures, but service failures are usually more expensive.

9.8. Other loading modes; mixed mode loading

The crack tip stress fields for modes *II*and *III* can be derived in the same way as for mode *I*. The solutions even have the same format as for mode *I*, as can be seen in Table 9.2; they contain a stress intensity factor K. Throughout this book the denotation K has been used for stress intensity, and since all discussions concerned the same loading mode, there could be no confusion. However, when considering different modes a distinction must be made between the stress intensity factors of different modes, K_I, K_{II} and K_{III}.

Dimensional analysis in the manner used in Chapter 3, readily shows what the format of K_{II} and K_{III} must be. All crack tip stresses must be proportional to the applied stress (elastic loading), hence, e.g. $K \div \tau$. Certainly, the crack tip stresses must be larger if the crack size a is larger, and since there is a root of a length (r) in the denominator, the only possibility is that $K \div \tau\sqrt{a}$, otherwise the expression for stress would not have the proper dimensions (note that all functions of θ are dimensionless).

Dimensional analysis does not provide any constants that might appear. These would have to be derived from a formal analysis of the problem. Again, it turns out that for infinite panels the proportionality factor is $\sqrt{\pi}$, so that $K_{II} = \tau\sqrt{\pi a}$. The dimensional argument remains the same when the body is of finite size, so that as in the case of mode *I* (see Chapter 3) the general expression for K becomes $K_{II} = \beta_{II}\tau\sqrt{\pi a}$, and similarly $K_{III} = \beta_{III}\tau\sqrt{\pi a}$. Again β must be dimensionless and still depend upon geometry, such dependence can be only in the form $\beta(a/L)$, where L is a generalized length parameter. All geometrical effects must be reflected in β, so that the structural complexity will be accounted for in the geometry factors. They are obtained by means of the same procedures as discussed for mode *I* (Chapter 3). Although handbooks contain β_{II} and β_{III} for some geometries, the mode *II* and *III* geometry factor are not available in such abundance as β_I

Fracture will occur when the crack tip stresses become too high, which leads to the fracture conditions: Fracture if $K_{II} = K_{IIc}$, or if $K_{III} = K_{IIIc}$. The toughness K_{IIc} or K_{IIIc} in principle could be measured in a test on any specimen, provided β_{II} or β_{III} is known for the configuration. Analysis of the residual strength for any other structure would then proceed on the basis of the above fracture criterion in the same manner as for mode *I*. Hence, for the analysis of the SEPARATE modes no new analysis techniques would be needed.

In practice modes *II* and *III* do not occur separately, but always in combination with mode *I*, e.g. *I–II*, or *I–III* or *I–II–III*. The treatment of combined modes is somewhat more complicated. Before considering these combinations

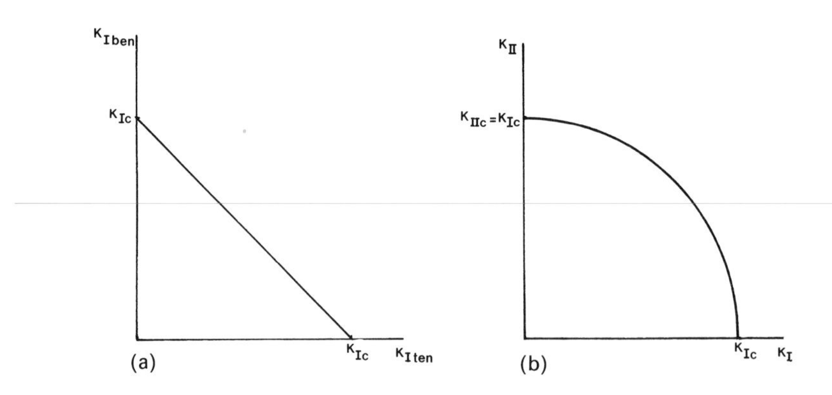

Figure 9.30. Fracture locus for combined loading. (a) Combined modes I; (b) Combined modes I and II.

of different modes, a brief review of combined mode *I* loadings (e.g. tension plus bending), is in place; it was discussed in detail in Chapter 8.

In combined mode *I* loadings, addition of crack tip stresses leads to:

$$\left.\begin{aligned} \sigma &= \frac{K_{I\,\mathrm{ten}}}{\sqrt{2\pi r}} f(\theta) + \frac{K_{I\mathrm{ben}}}{\sqrt{2\pi r}} f(\theta) \\ &= \frac{K_{I\,\mathrm{tan}} + K_{I\,\mathrm{ben}}}{\sqrt{2\pi r}} f(\theta) = \frac{K_{\mathrm{tot}}}{\sqrt{2\pi r}} f(\theta) \\ \text{so that} \\ K_{\mathrm{tot}} &= K_{\mathrm{ten}} + K_{\mathrm{ben}}\,. \end{aligned}\right\} \tag{9.6}$$

This is the superposition principle (Chapter 8): stress intensity factors of the SAME mode are additive, because the crack tip stress field solution is universal. This means that the functions of θ in the terms of Equation (9.6) are identical, so that the terms can be combined by making $f(\theta)$ explicit. Fracture occurs when $K_{\mathrm{tot}} = K_C$ (or K_{Ic}). In combined bending and tension for example, this leads to $K_{\mathrm{ten}} + K_{\mathrm{ben}} = K_c$, which is represented by a straight line in Figure 9.30a. Each combination (K_{ten}, K_{ben}) falling on this straight line represents a fracture case.

Now compare this with a combination of different modes, e.g. *I–II*. Addition of the crack tip stresses leads to:

$$\sigma = \frac{K_I}{\sqrt{2\pi r}} f(\theta) + \frac{K_{II}}{\sqrt{2\pi r}} g(\theta) \tag{9.7}$$

Because the function *F* and *g* of θ are different, the equation cannot be manipulated like equation (9.6) to provide an explicit stress intensity. Hence, in

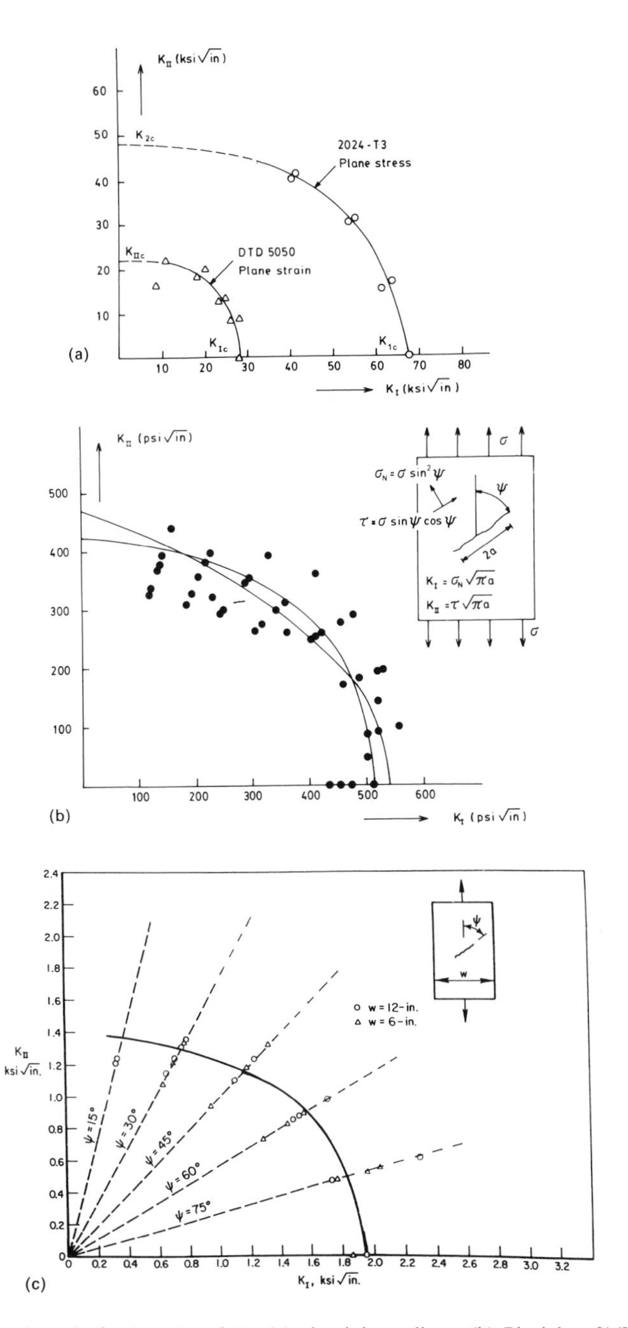

Figure 9.31. Mixed mode fracture test data. (a) aluminium alloys; (b) Plexiglass [16] (Courtesy ASME); (c) Paper. Courtesy Mead Paper Co.

combined mode loading stress intensities cannot be superposed; another route must be followed to solve the fracture problem.

The energy conservation criterion, being a general law of physics, should apply to the problem. It was shown in Chapter 3 that the energy release rate is equal to the derivative of the strain energy, $\mathrm{d}U/\mathrm{d}a$. Fracture will occur when the available (release) energy is sufficient to deliver the fracture energy, $\mathrm{d}W/\mathrm{d}a$. Hence, the fracture criterion is:

$$\left(\frac{\mathrm{d}U}{\mathrm{d}a}\right)_{\mathrm{tot}} = \frac{\mathrm{d}W}{\mathrm{d}a}. \tag{9.8}$$

Since energy is a scalar, superposition is freely permitted, so that the equation becomes:

$$\left(\frac{\mathrm{d}U}{\mathrm{d}a}\right)_{I} + \left(\frac{\mathrm{d}U}{\mathrm{d}a}\right)_{II} = \frac{\mathrm{d}W}{\mathrm{d}a}. \tag{9.9}$$

As shown in Chapter 3, the value of $(\mathrm{d}U/\mathrm{d}a)_I$ is K_I^2/E. Obviously, $(\mathrm{d}U/\mathrm{d}a)_{II}$ must have the same dimension (units), which means that it must be proportional to K_{II}^2/G, where G is the shear modulus. But since $G = E/2(1 = \nu)$, it follows that $(\mathrm{d}U/\mathrm{d}a) = CK_{II}^2/E$. The proportionality factor C can be obtained only from a formal analysis of the problem; it turns out that $C = 1$. Thus Equation (9.9) can be written as (plane stress).

$$\frac{K_I^2}{E} + \frac{K_{II}^2}{E} = \frac{\mathrm{d}W}{\mathrm{d}a}. \tag{9.10}$$

This would be the fracture criterion for combined mode I and II. For it to be valid in general, it must hold for the case that $K_{II} = 0$. Then it follows readily (see also Chapter 3) that $\mathrm{d}W/\mathrm{d}a = K_{1c}^2/E$. The equation must also hold when $K_I = 0$, in which case $\mathrm{d}W/\mathrm{d}a = K_{2c}^2/E$. This would lead to the conclusion that $K_{2c} = K_{1c}$. (Since apart from a factor $(1-\nu)$ the above applies to plane stress as well as plane strain, the toughness is denoted here as K_{1c} and K_{2c}, where the Arabic 1 and 2 signify mode I and mode II respectively, independent of the state of stress.)

Equation (9.10) then reduces to:

$$K_I^2 + K_{II}^2 = K_{1c}^2. \tag{9.11}$$

Equation (9.11) represents a circle with radius K_{1c} as shown in Figure 9.30b. Every combination of K_{II} and K_I falling on this circle represents a fracture case. (Compare with Figure 9.30a.)

Application of two modes of loading in a test generally requires two loading axes. Since most testing machines have only one loading axis, most tests to validate the criterion of Equation (9.11) have been performed on plates with oblique cracks (Insert Figure 9.31b), where both modes I and II occur due to one loading axis. The stress intensities are $K_I = \beta_I \sigma_N \sqrt{\pi a}$, and $K_{II} = \beta_{II} \tau \sqrt{\pi a}$, where both β_I and β_{II} are aproximately equal to 1 when the plate is large. In a fracture test one measures the value of σ at fracture. From this σ_N and τ at fracture can be determined as in Figure 9.31b, and subsequently the values of K_I and K_{II} at fracture calculated. The data point can be plotted in K_I–K_{II}-space. For the fracture criterion of Equation (9.11) to be true, data points of tests with different K_I/K_{II} (different angles) must fall on a circle with radius K_{1c} (or K_{2c}).

Three such data sets are shown in Figures 9.31. A test in which the crack angle = 90 degrees is not a viable possibility. Thus the test data for large crack angle (large K_{II}) are dubious. But even if such data points are ignored, the hypothesis does not seem to be substantiated: the data do not fall exactly on a circle.

The question is now where the hypothesis could be wrong. Equation (9.8) is certainly true, and therefore Equation (9.9) must be correct. Hence, the error must have been introduced between Equations (9.9) and (9.10). It was shown in Chapter 3 that in mode I, indeed $\mathrm{d}U/\mathrm{d}a = K_I^2/E$. In the derivation of this equation an implicit assumption was made however, namely that fracture occurs in a self-similar manner: da in the same direction as a. This seemed trivial at the time: it is everbody's experience that the fracture will proceed in the manner indicated (Figure 9.32a). Pulling a piece of paper with a tear will confirm this. In the same vein, self-similar fracture was assumed when writing $(\mathrm{d}U/\mathrm{d}a)_{II} = K_{II}^2/E$. However, it is not obvious that the fracture proceeds self-similarly in a combined mode I–II case. One feels intuitively that it does not;

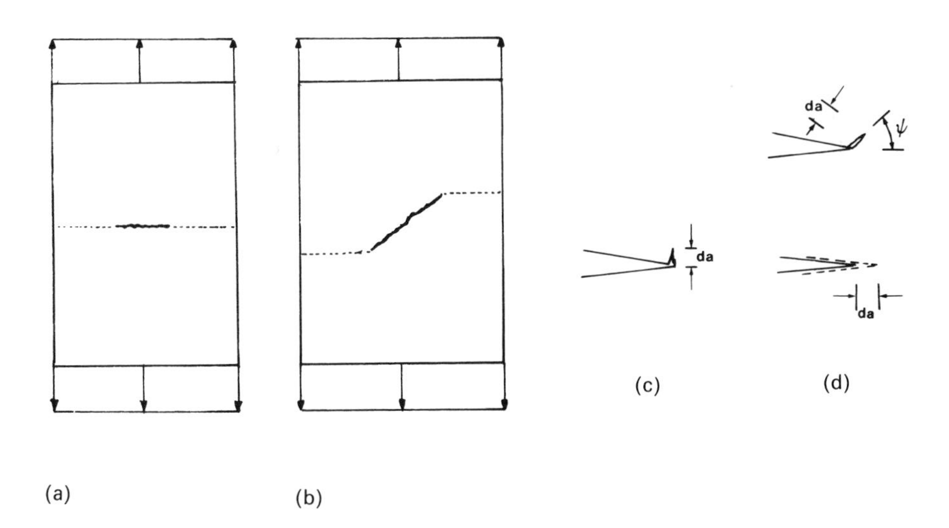

Figure 9.32. Fracture direction. (a) Anticipated fracture path in mode I; (b) Anticipated fracture path for mode I-II; (c) vertical da; (d) Direction of da.

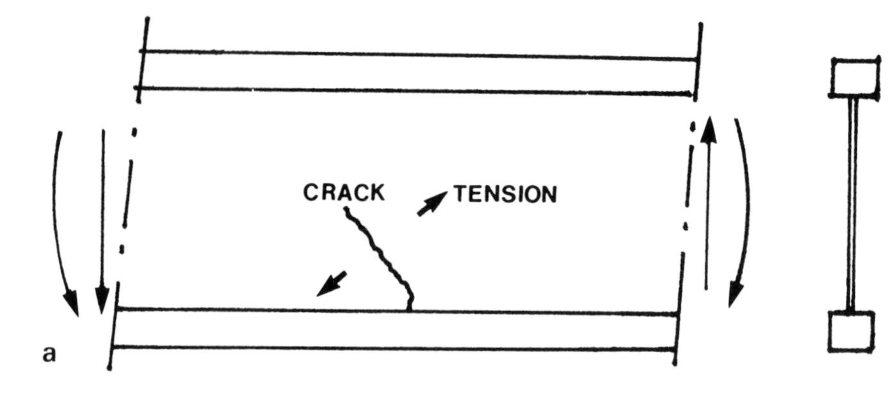

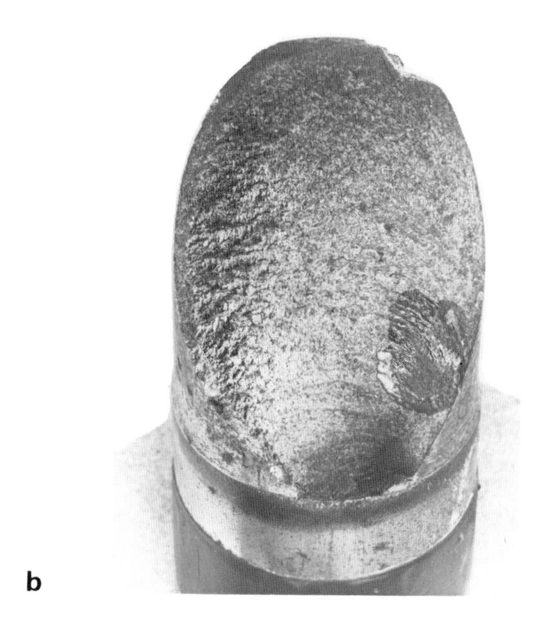

Figure 9.33. Crack perpendicular to principal stress (tension) (a) Shear web. (b) Pure torsion.

it will proceed as shown in Figure 9.32b. This can be readily demonstrated by pulling a sheet of notepad paper with an oblique (e.g. 45 degrees) slit.

The energy release rate $(\mathrm{d}U/\mathrm{d}a)_{\mathrm{tot}}$ depends upon the angle at which fracture proceeds. It is obvious that a vertical da (figure 9.32c) will cause much less energy release than a horizontal da. Indeed a horizontal da gives the highest energy release in mode I, reasons why fracture proceeds as in Figure 9.32a. In combined mode loading fracture will proceed in that direction for which the total energy release is the largest, because Equation (9.8) is first satisfied for this direction. Hence, in developing Equation (9.9), one should write:

$$\alpha_I(\psi)\,\frac{K_I^2}{E} + \alpha_{II}(\psi)\,\frac{K_{II}^2}{E} = \frac{\mathrm{d}W}{\mathrm{d}a}, \tag{9.12}$$

where α_I and α_{II} are both functions of ψ (Figure 9.32) which can be formally evaluated by calculating $\mathrm{d}U/\mathrm{d}a$ as a function of the angle ψ of da. It can then be determined for which ψ the left hand side of Equation (9.12) becomes a maximum. This provides not only the angle ψ at which the fracture proceeds, but also the values of K_I and K_{II} necessary for fracture. The question is whether this is still necessary.

There is no doubt that a fracture in mode I will proceed in a self-similar manner. Neither is there any doubt that the fracture as in Figure 9.32b will proceed as shown (simple tests on paper bear this out). In both cases this is the direction perpendicular to the maximum principal stress. This is immediately clear for pure mode I, but it can be demonstrated for the combination as well [5, 16].

This being the case, the question must be asked: how the oblique crack (mode I/II) in the tests represented in Figure 9.31 could develop in the first place. The answer is simple: the cracks did not develop; they were cut intentionally for the purpose of the tests. Had the crack developed by fatigue, they would have grown horizontally and not under an angle. As a matter of fact this is a common observation. Cracks in shear webs and pure-torsion bars develop under 45° as shown in Figures 9.33. Even if such cracks initially are forced in another direction, they will soon turn and find a path in which they experience mode I only. Cracks tend to turn away from the shear loading so that the mode I loading is the only relevant one. Hence, cracks such as in Figure 9.31 would never develop.

The above behavior of service cracks is also observed in tests. A fatigue crack in a combined shear-tension field [17] followed a curved path. Finite element analysis of a model in which the crack was represented in the way it actually developed in the test showed that K_{II} dropped immediately to zero. Its initial value was high because it was forced by the notch, but the crack turned readily away into a direction of mode I only. Thus all tests data in Figures 9.31 and behavior of forced slant cracks or forced combined mode tests in general, are of academic interest but of no practical value if natural cracks develop in mode I.

Essentially then the entire problem of combined mode loading seems of little practical interest. This categorical statement must be blunted somewhat, however. There are cases where the crack may indeed be forced into a combined mode situation. A good example would be a torque tube with a circumferential weld. Because the weld HAZ is often the weakest path, the crack will stay in the HAZ where it experiences both mode I and II. Other examples of this forced behavior can be thought of. In such cases combined mode crack growth and fracture criteria may have to be used. One may then turn to formal analysis [5] which leads to considerable complication. Instead, engineering pragmatism would dictate simpler and more cost-effective ways. The simplest approach would be to use the circle of Equation (9.11). In most engineering cases K_{II} will be small with respect to K_I and for small ratios K_{II}/K_I the circle is a very good approximation as can be readily seem from Figures 9.31. It could be improved somewhat by recognizing that data suggest that $K_{2c} \approx 0.8\, K_{1c}$ and that the data fall more closely on an ellipse:

$$\left(\frac{K_I}{K_{1c}}\right)^2 + \left(\frac{K_{II}}{0.8\, K_{1c}}\right)^2 = 1. \tag{9.13}$$

The use of the criterion is simple. For example, if $K_{1c} = 50$ ksi$\sqrt{\text{in}}$, it follows that $K_{2c} = 40$ ksi$\sqrt{\text{in}}$. When considering a case where $\beta_I = \beta_{II} \approx 1$ (other cases can be treated in the same manner after determination of β_I and β_{II}), it can be calculated how much shear can be applied if e.g. $\sigma = 16$ ksi and $a = 2$ inches. It follows that $K_1 = 16\sqrt{\pi \cdot 2} = 40$ ksi$\sqrt{\text{in}}$. Fracture occurs in accordance with Equation (9.13):

$$\left(\frac{40}{50}\right)^2 + \left(\frac{K_{II}}{40}\right)^2 = 1$$

so that $K_{II} = 24$ ksi$\sqrt{\text{in}}$ and $\tau = 24/\sqrt{\pi \cdot 2} = 9.6$ ksi. Thus if the tension is 16 ksi, an additional shear of 9.6 ksi will cause fracture. Similarly, if the shear is e.g. 16 ksi, and $a = 1$ inch, the value of K_{II} is $16\sqrt{\pi} = 28.4$ ksi$\sqrt{\text{in}}$, and fracture occurs when $K_I = 50\sqrt{1 - (28.4/40)^2} = 35.2$ ksi$\sqrt{\text{in}}$, i.e. at $\sigma = 35.2/\sqrt{\pi} = 19.9$ ksi.

The treatment of fatigue crack growth could follow similar lines the 'effective' ΔK_I being determined by:

$$\Delta K_{I\text{eff}} = \sqrt{\Delta K_I^2 + (0.8\, \Delta K_{II})^2} \tag{9.14}$$

and by using the mode I rate data $(\mathrm{d}a/\mathrm{d}N)$ with $\Delta K = \Delta K_{I\text{eff}}$.

In the above it was assumed that K_I and K_{II} are in phase, as would occur in the shear web of a beam or a torque tube with bending and torsion. If they are not in phase, the problem is different. For example consider the span of a bridge.

Tension, K_I, is at a maximum when a truck passes over the crack location. The shear, K_{II}, on the other hand changes sign when the truck passes the crack location. Thus K_I and K_{II} would be out of phase. Similar conditions occur when torsion and bending vary independently. An example would be an aircraft wing. Normally torsion and bending are in phase, but a new situation develops with 'flaps down'. Again torison and bending are in phase, but their ratio is different than with 'flaps up'.

Although the above cases are realistic engineering problems, none of the mixed mode hypotheses have even begun to address this situation; up till now, virtually all research has addressed either academic problems of forced mode *I*/*II* or *I*/*III*, or concentrated on in phase loading for crack growth. Thus the problems discussed in the previous paragraph must be addressed by engineering methods. The only sensible procedure would seem to be to treat the problem as mode *I*, with the crack growing perpendicular to the highest principle stress for the worst mode *I* case, and to assume the stress range is the largest difference in principal stresses.

In conclusion, it can be stated that the combined mode problem is largely academic. In most cases of combined mode loading the crack develops into a direction of only mode *I*. In practice, probably 90% of the engineering problems are mode *I* to begin with. Of the 10% combined mode loading cases, probably 80% of the cracks will turn immediately upon initiation and experience mode *I* only. This brings the total of engineering problems to 98% mode *I*. Of the remaining few per cent, most will be in phase, so that the engineering analysis of Equations (9.19), (9.13) and (9.14) would apply. In those few cases where out of phase loading is significant no hypothesis is available at all. The pragmatic approach discussed above is the only way out.

9.9. Composites

Fracture mechanics methods to deal with crack growth and fracture in composites are practically non-existent. Summaries of many ongoing research programs [18] reveal that the 'development' is still in the stage of explaining and interpreting experimental observations. Research programs involving 'finite element analysis' were not very successful either. Crack growth and fracture analysis require a crack growth and fracture criterion based on the physics of the problems. Finite element analysis, by itself does not provide this; it provides stresses and strains but these are useless without a fracture criterion.

There is a risk that tests performed to investigate fracture criteria use specimens of inadequate size, incorporating similar problems of interpretation as in the case of small metal specimens. (Chapter 3 and 10). As composites are expensive, the quest for small specimens is understandable.

It is not surprising that the damage tolerance assessment of composites aircraft structures [18] is entirely based on full-scale tests. The component is subjected to a simulated lifetime of cyclic stresses and environmental changes. At the end of this 'life' it must still be able to sustain the design load. Subsequently, damage is inflicted by a weight of prescribed size dropped from a prescribed height. The testing is continued, the damage increasing in size during load and environmental cycling for another lifetime. At the end of the next 'life' the structure must still have the minimum permissible residual strength, σ_p.

In the case of composites various failure mechanisms can be operative, namely: matrix cracking, fiber fracture, fiber decohesion and finally laminate decohesion. Compare this with a stringer stiffened structure as discussed extensively in Section 9.4. There, various failure mechanisms are possible as well, namely: skin cracking (matrix cracking), stringer failure (fiber fracture) and fastener failure or adhesive debonding (fiber dechohesion). Thus, the composite problem is similar in many ways to the stringer-plate problem for which solutions are readily available. Composite research might make use of these established procedures. Composite fibers are essentially finely distributed stringers. Admittedly, multiple laminates with different fiber orientations will complicate the problem. But building upon the skin-stringer solution would seem more promising than finite element analysis.

Paper and cardboard are composites, be it that they have short randomly oriented fibers. Clusters of fibers and sometimes individual fibers can be seen in a sheet of bond held against the light. No fracture tests are cheaper than tests

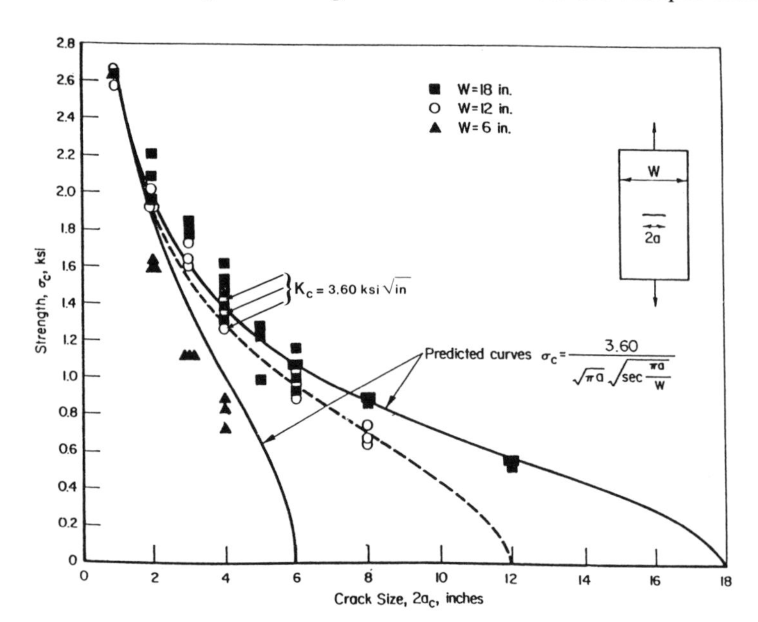

Figure 9.34. Residual strength prediction and data for random fiber composite (paper board). Courtesy Mead Paper Co.

on paper and cardboard to check analysis procedures. Figure 9.34 shows test data for center-cracked card-board panels. A 'toughness' $K_c = 1.34\sqrt{\sec 2\pi/12}\sqrt{2\pi} = 3.6\,\text{ksi}\sqrt{\text{in}}$ follows from the average fracture stress for the three data points indicated by arrows. Using common procedures of residual strength analysis as discussed in Chapters 3 and 10, the residual strength curves for other panel sizes were predicted. Figure 9.34 compares predictions (curves) with tests (data points). Clearly, the results are in accord with predictions. There are discrepancies, but these are well-explained by common material scatter, and no larger than in metals. In the case of card board, one must account for anisotrophy and inhomogeneity of the material and the problem seems unmanageable from the point of view of analysis. However, if one can make predictions (curves in Figure 9.34) as well as shown using a very simple concept, the engineering problem is solved. Similarly the problem of crack growth and fracture in composites can be solved. However, solutions as e.g. the one discussed above must probably come from engineering-oriented (development) research. Academic fracture mechanics research may help to improve understanding, but its track record for solving engineering problems is not encouraging.

9.10. Exercises

1. Using the values for a, determine the residual strength diagrams for longitudinal suface flaws (internal as well as external) with $a/c = 1$ and $a/c = 0.3$ in a pressurized cylinder of 10 inch diameter with a wall thickness of 0.5 inch; assume that the hoop stress is pR/B (thin wall solution. $K_{Ic} = 35\,\text{ksi}\sqrt{\text{in}}$; $F_{ty} = 70\,\text{ksi}$. Assume β of figure 8.3 is applicable. Note: use a for residual strength analysis, and $c = a/(a/c)$.

2. Given that the pressure in the cylinder of Exercise 1 cycles between 0 and 3000 psi, will the cylinder leak or break? Assume $\beta = 1$ for the through cracks.

3. Assuming that $F_{ty} = 50\,\text{ksi}$ instead of 70 ksi in Exercises 1 and 2, and $K_c = 80\,\text{Ksi}\sqrt{\text{in}}$, will the cylinder leak or break?

4. If $\text{d}a/\text{d}N = C\,\Delta K^3$, determine the changing shape of the crack starting at $a/c = 0.3$ and $a/c = 1$ starting at $a = 0.1$ inch. Assume same conditions as in previous exercises.

5. Assume Figure 8.25 applies. Given is: $B = 0.2$ inch, $b = 8$ inches, W large, $\mu = 0.4$, $F_{ty} = 60\,\text{ksi}$, $F_{tu} = 75\,\text{ksi}$, $K_c = 70\,\text{ksi}\sqrt{\text{in}}$. Calculate the residual strength diagram (including stringer failure line). Is this case stringer critical or skin critical? What would you change to make this a skin critical case? If

the allowable shear stress of the fasteners is 100 ksi what size of fasteners do you need? What is the bearing stress?

6. Suppose that in the case of Exercise 5 the normal fatigue stresses have a range of 15 ksi at $R = 0.2$. Occasional higher stresses occur. One of these occurs when the crack size is $a = 2$ inches, causing a fracture. Assume that $da/dN = 3E{-}9\,\Delta K^{2.1} K_{max}^{0.9}$.
   (a) At which stress did the fracture occur?
   (b) At which crack size does arrest occur?
   (c) What was the rate of fatigue crack growth before arrest?
   (d) What is the rate of fatigue crack growth after arrest?
   (e) Why could you have obtained the result of d without any calculation, and what is the general consequence of this observation?

7. A stop-hole repair is made of an edge crack of 0.75 inch length for which $\beta = 1.12$. Calculate β for the crack emanating from the stop hole. Diameter stop hole is 0.25 inch, its center at the crack tip.

8. Given that $K_c = 50\,\text{ksi}\sqrt{\text{in}}$, what was the residual strength of the crack in Exercise 7 before stop hole drilling, and what is the residual strength immediately after reinitiation of the crack from the hole? $F_{ty} = 100$ ksi.

9. A hole with a diameter of 0.25 inch and an edge distance of 0.625 inch develops a crack that grows through the entire ligament. Calculate β for the crack emanating on the other side.

10. Supposing a residual tensile stress of 20 ksi exists at a potential crack location. The applied stress cycles between 0 and 10 ksi. Calculate the rate of growth with the Forman equation if $C_F = 2 \times 10\,E{-}7$, $m_F = 2$ and $K_c = 80\,\text{ksi}\sqrt{\text{in}}$; Assume $\beta = 1$; $a = 1$ inch.

11. Would the crack of Exercise 10 grow by stress corrosion if $K_{Iscc} = 20\,\text{ksi}\sqrt{\text{in}}$?

12. To pre-existing crack a mode II shear stress of $\tau = 20$ ksi, as well as a tension are applied. The toughness of the material is $80\,\text{ksi}\sqrt{\text{in}}$. Assuming $\beta = 1$ calculate allowable tension if the crack is of 2 inch length.

References

[1] J.C. Newman and I.S. Raju. Stress intensity factors equations for cracks in three-dimensional finite bodies. *ASTM ATP 791* (1983) pp. I-238-I-265.

[2] I.S. Raju and J.C. Newman, Stress intensity factors for circumferential cracks in pipes and rods under tension and bending loads *ASTM STP 905* (1986) pp. 189-805.

[3] J.C. Newman and I.S. Raju. *Analysis of surface cracks in finite plates under tension and bending loads*, NASA TP-1578 (1978)

[4] G.G. Trantina *et al.*, Three dimensional finite element analysis of small surface cracks, *Eng. Fract. Mech. 18* (1983).

[5] D. Broek, *Elementary engineering fracture mechanics,* Nijhoff (1986) 4th edition.

[6] D. Broek *et al., Applicability of toughness data to surface flaws and corner cracks at holes.* Nat. Aerospace Lab. NLR TR 71033 (1971)

[7] L.R. Hall and R.W. Finger. *Fracture and fatigue crack growth of partially embedded flaws*, AFFDL TR 70-144 (1970) pp. 235-2626.

[8] M.F. Kanninen and C.H. popelar. *Advanced fracture mechanics,* Oxford University Press (1985).

[9] T. Swift. Development of fail safe design features of the DC-10, *ASTM STP 486* (1974) pp. 164-214.

[10] H. Vlieger and D. Broek. *Residual strength of cracked stiffened panels.* AGARDograph 176 (1974) Chapter V.

[11] H. Vlieger. Residual strength of cracked stiffened panels. *Eng. Fract. Mech. 5* (1973) pp. 447-478.

[12] D.P. Rooke and D.J. Cartwright, *Compendium of stress intensity factors* HM Stationery Office, London (1976)

[13] H.P Van Leeuwen *et al., The repair of fatigue cracks in low alloy steel sheet,* National Aerospace Institute, Amsterdam, Rep NLR TR-70029 (1970)

[14] E.F. Rybicki and R.B. Stonesifer, Computation of residual stresses due to multi-pass welds in piping systems, *J. of Press. Vessel Techn. 101* (1979) pp. 149-154.

[15] E.F. Rybicki *et al.*, A finite element model for residual stresses and deflections in girth-butt welded pipes, *J. of Press Vessel Techn 100* (1978) pp. 256-262.

[16] F. Erdogan and G.S. Sih. On the crack extension in plates under plane loading and transverse shear. *J. Basic Eng. 85* (1963) pp. 519-527.

[17] D. Broek and R.C. Rice. *Prediction of fatigue crack growth in rails.* Battelle rept to US DOT TSC (1977)

[18] Various Authors, *Tough composite materials,* Noyes (1985).

[19] J.R. Soderquist *Certification of civil composit aircraft structure* SAE paper 811061 (1981).

CHAPTER 10

Analysis procedures

10.1. Scope

This chapter is essentially a summary of all foregoing chapters and serves as an introduction to those following on fracture control and damage tolerance requirements. As such it is repetitive in many ways and contains many cross-references to other chapters. Nevertheless, despite its repitition, the reader may find it useful as a general tie-in of the general procedures, and as a reference to other parts of this book where the issues are discussed in more detial.

10.2. Ingredients and critical locations

The block diagram in Figure 10.1 shows the general outline of the damage tolerance analysis. After selection of the critical location to be analysed, the ingredients (input) for the analysis must be obtained. Apart from the general stress distribution the load spectrum (and environmental spectrum) must be known. These lead to the basic input for the analysis, namely

(a) Stress history, Chapter 6.

(b) Materials data ($\mathrm{d}a/\mathrm{d}N$, K_{Ic}, K_c, J_R, F_{ty}, F_{col} etc), Chapter 7.

(c) Geometry factor(s), Chapter 8.

Material data are generally obtained from data handbooks. Care must be taken that they are for the proper material direction. Assumptions are often necessary, as discussed extensively in Chapter 7.

The stress history is obtained on the basis of the load spectrum and the stress analysis. The input may be in tabular form. If an exceedance diagram is available the stress history can be generated using procedures as discussed in Chapter 6. Decisions with regard to sequencing, clipping and truncation must be made. Some computer codes perform the stress history generation automatically from the exceedance diagram [4] input and provide options for truncation and clipping. The many issues involved were discussed in Chapter 6.

Geometry factors must be obtained using one of the methods reviewed in

Table 10.1. Analysis (and fracture control) complications for various types of structures (subjective rating)

Item	Complexity due to:	Ships	Offshore	Pipelines	Nuclear	Airplanes	Chemical processing	Heavy machinery
Load and spectrum	Spectrum	8	8	2–7[a]	2	9[b]	2	7–8
	Load response	7–8	6–7	1–4[a]	3–4	8–9[c]	2	6–7
	Load and weight configurations	6–9	2	2	2	8	2	3
	Environment	6	8	3–8[a]	9	4	9	7–9
Stresses	Structural geometry	7	6	3	5	9	3–5	8
	Stress analysis	7	6	4	6	8–9	5	8
Cracks and fracture	Detail design	8	6	4	5	9	4–5	6
	Residual stress	6	8	5	4	2	7	7–8
	Welded joints	8	8	7	6	1	8	8
	Fastened joints	1	1	1	3	8	2	1
Active fracture control	Inspection	6	8	4–7[a]	9	6	6–7	5
	Total complexity	70–74	67–68	36–52	54–55	74–76	50–54	66–71

[a] High numbers for submarine pipeline.
[b] Maneuvers, gusts, and taxi-loads.
[c] Including control surfaces.

Chapter 8. Libraries of geometry factors are sometimes included in computer codes. This is useful mostly for standard geometries. When compounding, superposition and other methods are necessary, a pre-processor performing all of the procedures discussed in Chapter 8 is more versatile [4].

After the input is complete residual strength analysis (Chapters 3, 4, 9) is performed to determine the permissible crack size, a_p. Crack growth analysis (Chapters 5, 9) then provides the life to a_p (Chapter 1), upon which fracture control decisions can be made (Chapter 11, 12). The latter computations are usually performed by employing proven computer software. As such they are essentially the easiest part of the analysis. Nevertheless, the analyst should be aware of the procedures used in order to appreciate the possible consequences of approximations and judgements, which largely determine the accuracy of the analysis. Preparation of the input is the most difficult and most crucial task. Every type of structure presents its own special difficulties (Chapter 9); as a consequence the complexity of the analysis and the difficulties are more or less of the same level regardless of the application. This can be appreciated from Table 10.1, where the various complicating factors are subjectively rated on a scale from 1 to 10. With few exceptions the total rating is of the same order of magnitude. Naturally, these ratings are somewhat arguable, but they were based mostly upon actual experience with damage tolerance analysis for these structures, so that they certainly provide a good indication of complexities.

10.3. Critical locations and flaw assumptions

Perhaps the most important step in the analysis is the very first (Figure 10.1), namely the selection and identification of the critical location. It is also the most difficult to discuss. Analysis must be performed for those structural elements and details liable to develop cracks. The assessment of these critical locations requires sound judgement and should take into account the following factors:

(a) Areas of high stress (stress calculations and strain measurements).

(b) Stress concentrations due to discontinuities and eccentricities, particularly in the areas mentioned sub a.

(c) Location where stresses would be high if adjacent elements failed (load transfer, Chapter 9)

(d) Eccentricities and stress concentrations resulting from fracture of one member of a back-to-back or multiple load path structure.

(e) Details prone to cracking according to service experience or tests.

(f) Areas where fretting may occur.

(g) Areas of secondary stress due to displacements of other parts.

Rather than relying on computer stress analysis the damage tolerance analyst should develop a second nature for identifying stress concentrations. Especially in complex structures the load path should be 'envisioned'; where the load must

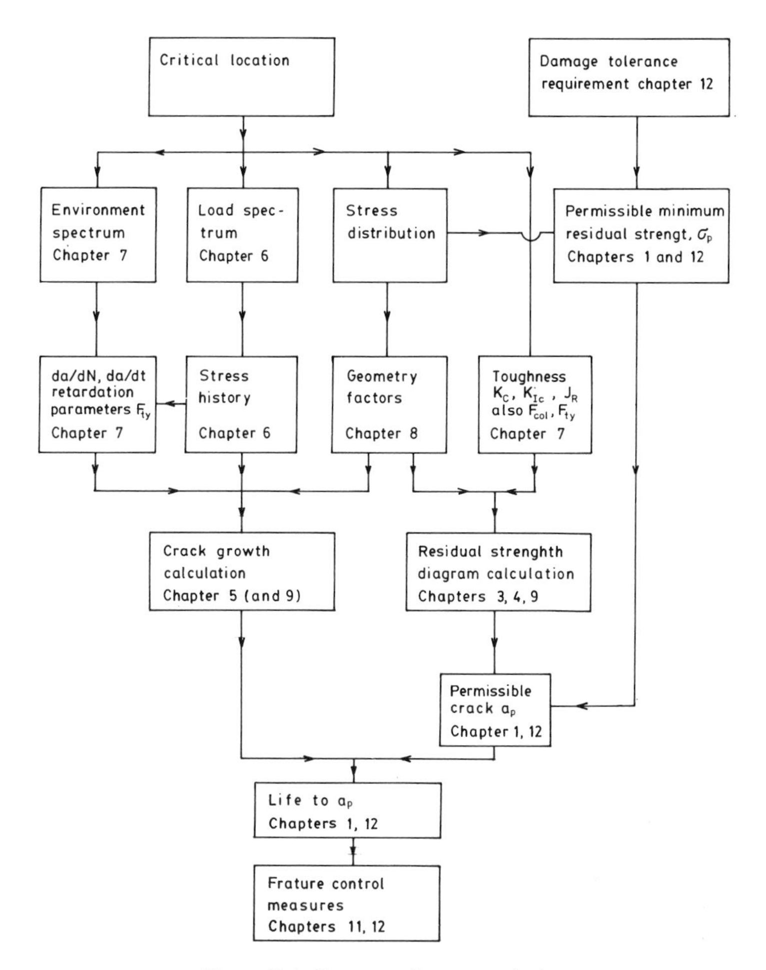

Figure 10.1. Damage tolerance analysis.

go and where it cannot go depends upon stiffness distributions and location of reaction forces. For design details the load-flow lines discussed in Chapter 2 are useful.

'Hidden' stress concentrations should be recognized on the basis of stiffness distributions and eccentricities. For example the bolted joint in Figure 10.2a may be perfect from a static design point of view. At the design load ample plastic deformation occurs at the bolt holes for the load to be evenly distributed over the bolts, each bolt then carrying 1/4P. However, the design is poor from the point of view of fatigue. This can be readily seen from a comparison of stiffness as shown below.

First note that during normal service loading the stresses are elastic . Now

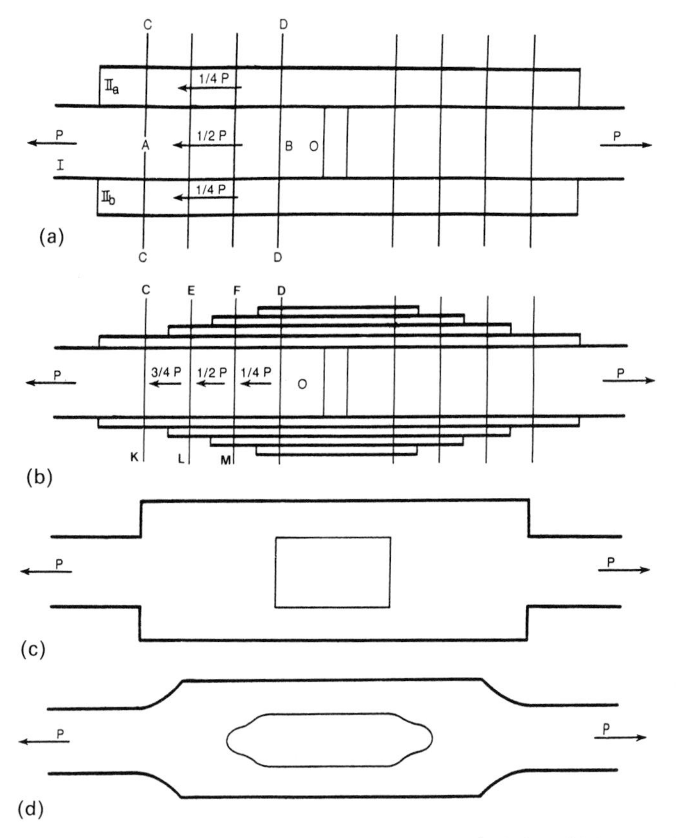

Figure 10.2. Hidden stress concentration in bolted joint. (a) Poor design; (b) Improved design; (c) Full body equivalent of *a*; (d) Full body equivalent of *b*.

consider the area of the joint between bolts C and D. Parts IIa and IIb together have the same stiffness (equal thickness) as part I. Since the parts are attached at C and D they must stretch equally: they undergo equal strain. But for equal strain they must have equal stress. Therefore IIa and IIb together carry half the load, and part I carries half the load. This means that half of the load is transferred at bolt C, the other half at bolt D; the two other bolts carry no load at all. Equal strain in the parts between C and D, causes the holes of the centre bolts in the three parts to remain lined up: hence the bolts cannot transfer load.

A much better design is shown in Figure 10.2b. If bolt C is to transfer 1/4P then there will be 3/4P between GH, while there will be 1/8P between CE and KL each. The parts being attached, they must have equal strain (stress). Therefore the thickness of CE and KL each must be $1/8 \times 3/4 = 3/32$ of the center part, and so on, in order to carry loads of $\frac{1}{4}$P, $\frac{1}{2}$P and $\frac{3}{4}$P.

The design of Figure 10.2b provides a gradual change in stiffness. The equivalent in full body design is shown is Figure 10.2d. No designer would conceive the configuration of Figure 10.2c, but the design of Figure 10.2a is quite common. Apart from the stress concentration of the bolt holes per se, there is a 'hidden' stress concentration of a factor of 2 (the bolt carrying 1/2P instead of 1/4P). Cracking will occur at C and D, and the analysis must account for a bolt load of $\frac{1}{2}$P.

Once the critical areas have been established, it is still not trivial how the analysis should proceed. One must also know (or assume) how cracks will develop. This assumption is very influential to the results of the life calculation and may overshadow other errors (Chapter 12).

In the case of surface cracks or corner cracks there is usually little choice but to assume that the flaw is elliptical, otherwise geometry factors are not available. Yet many surface flaws are not elliptical. A circular flaw is often assumed. As discussed in Chapter 9 this assumption may have more effect on the results of the analysis than all acclaimed analysis shortcomings together (see also Figure 7.8). One might select to go through the costly exercise of generating geometry factors by means of 3-D finite element analysis, but if the wrong flaw shape assumption is made also this may be in vain.

For example, nuclear pressure vessels tend to develop cracks at the nozzle (Figure 10.3) where cyclic (thermal) stresses occur due to the 'cold' water returning from heat exchanger or turbine. The stress distribution is as shown in Figure 10.4a. Stress intensity factors were obtained [1] by 3-D finite element analysis (Figure 10.4b) on the basis of the flaw assumptions of Figure 10.4c. Consider the case of crack front 3. The stress intensity is higher at the free surface. This means that crack growth will be faster there, so that crack front 4 will no longer be a circle (as was assumed). Indeed, model tests confirmed this faster growth at the surface, as shown in Figure 10.5b. Subsequent fracture tests overpredicted (unconservative) the fracture stress as shown in Figure 10.5.

It is strongly emphasized that this is not intended to suggest that the descrepancies in Figure 10.5 are due to the flaw assumptions. On the other hand, if the tests had been done first, proper flaw assumptions could have been made, and at least one area of uncertainty could have been eliminated. The example shows that answers do not lie in costly analysis. Unfortunately, engineering insight and foresight tend to be lost in the computer world. Any of the methods discussed in Chapter 8 (possibly combined with the FEM stress distribution for the uncracked model) would have given equally useful results, the (flaw) assumptions being the crucial point.

Similarly, flaw development presents a problem, in particular when the structure consists of multiple elements. It would be hard to decide which assumptions should be made with regard to the sequencing and interaction of cracks 1 through 8 in a case such as in Figure 10.6: Of the many possible assumptions none will occur in reality, and no useful answer will be obtained

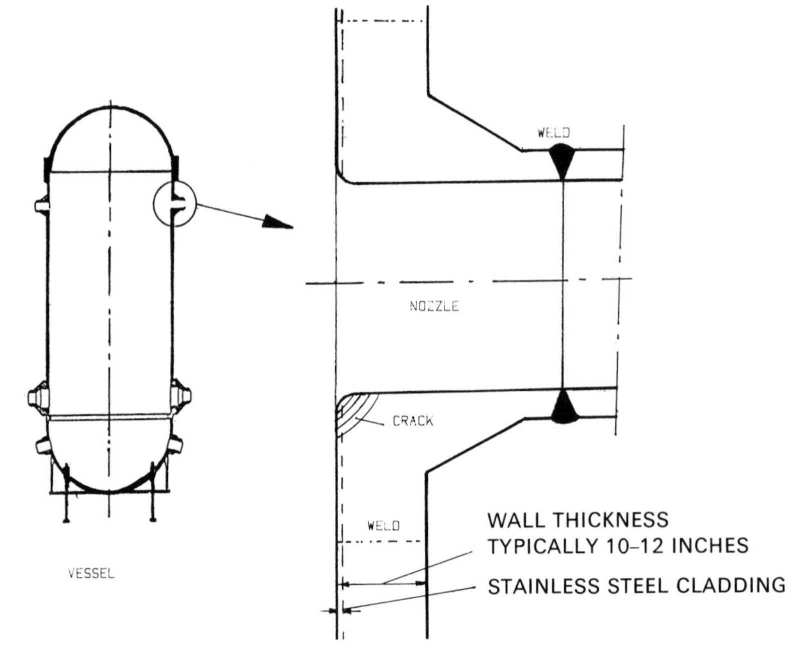

Figure 10.3. Nozzle crack in nuclear pressure vessel..

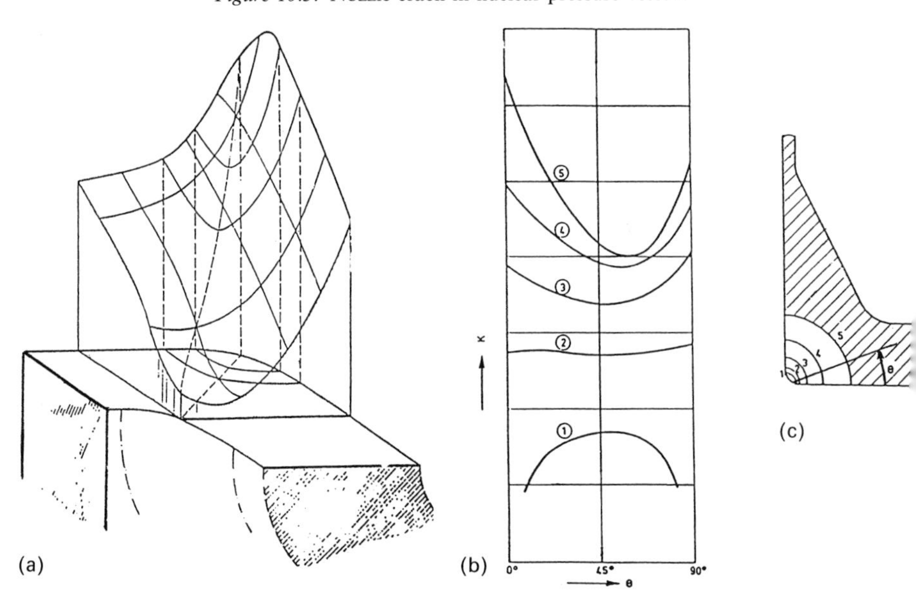

Figure 10.4. Analysis of nozzle corner cracks [1]. (a) Stress distribution at nozzle corner; (b) Calculated stress intensity on basis of a and c. (c) Flaw assumption.

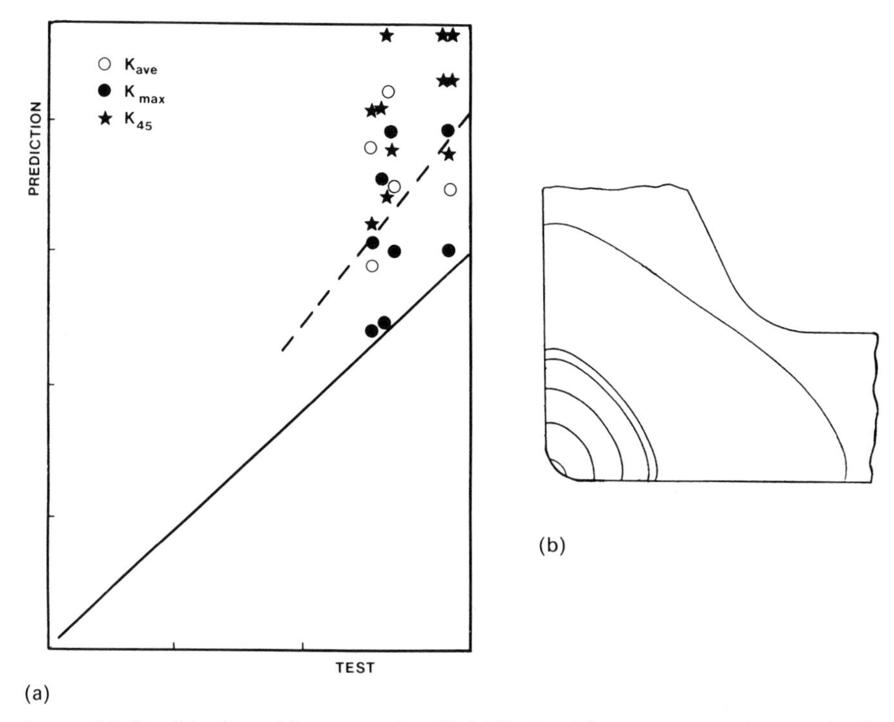

Figure 10.5. Possible effect of flaw assumption [1]. (a) Predicted fracture stress and test results; (b) Actual crack shapes in test.

unless the effect of different assumptions is analysed and the worst scenario adopted. The latter must follow from multiple analyses; it cannot be foreseen. Naturally, if the structure is really damage tolerant (can sustain large damage) no assumption may be necessary. The analysis could start by noting that eventually all of the area in Figure 10.6b will be cracked, and the inspection interval could be based upon the time for this damage to grow to a_p (Chapter 11, 12).

10.4. LEFM versus EPFM

The concepts of residual strength analysis were discussed in Chapters 3 and 4; special situations were reviewed in Chapter 9. Collapse must always be considered as a competing condition; the lower result of fracture and collapse analysis is the residual strength. In principle, the fracture analysis is not essentially different from conventional design analysis as demonstrated in Figure 10.7.

A question may arise as to whether to use LEFM or EPFM. In this respect Figure 10.8 provides some guidelines. Most cases can be treated with LEFM

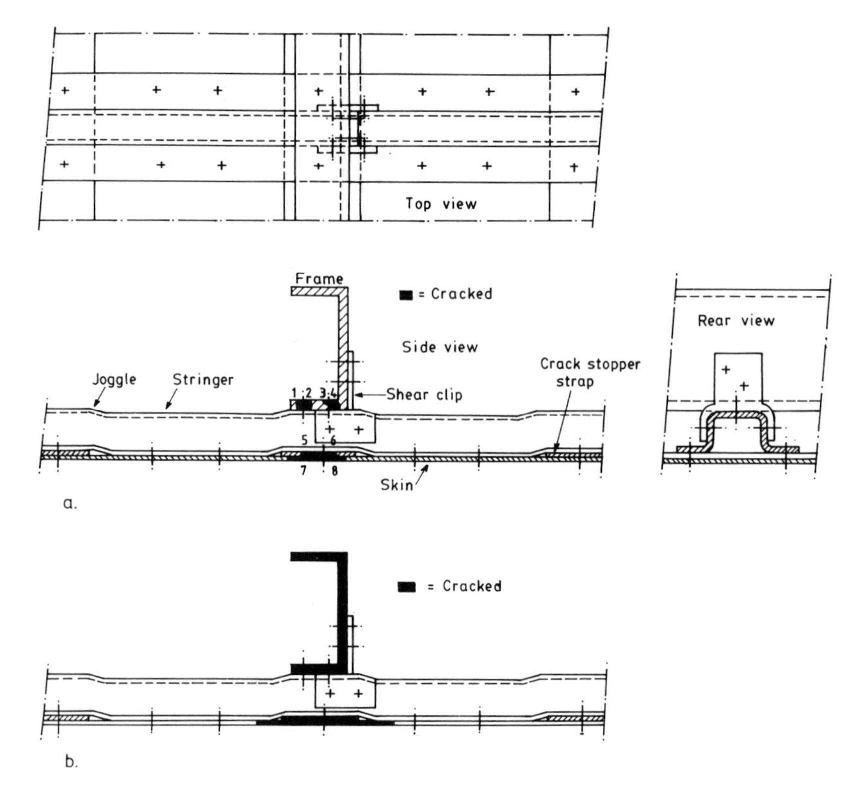

Figure 10.6. Crack development and sequencing in multiple element structure (example: aircraft fuselage) (a) Eight different cracks; (b) total assumed damage.

and collapse, be it that some approximations may have to be made for small cracks (fracture at high stress), but such approximations are still necessary with EPFM, as shown later in this chapter. The significance of EPFM is exaggarated in the literature. Judicious use of LEFM in conjuction with collapse analysis will provide the residual strength with good engineering accuracy in most cases. The intrinsic inaccuracy of the toughness measured in terms of J_R (Chapter 7) usually causes EPFM analysis to be no more accurate than LEFM even for very ductile materials, in particular because collapse must still be separately evaluated. Besides, fracture control usually depends only slightly on permissible crack size. (Chapter 11, 12) and is more influenced by crack growth.

In this respect the so-called 'failure analysis diagram' developed [2] in Great Britain is of great interest, because it provides much insight. The whole gamut of fractures from brittle to fully plastic can be represented in the failure analysis diagram as shown in Figure 10.9a. The stress intensity is plotted along the

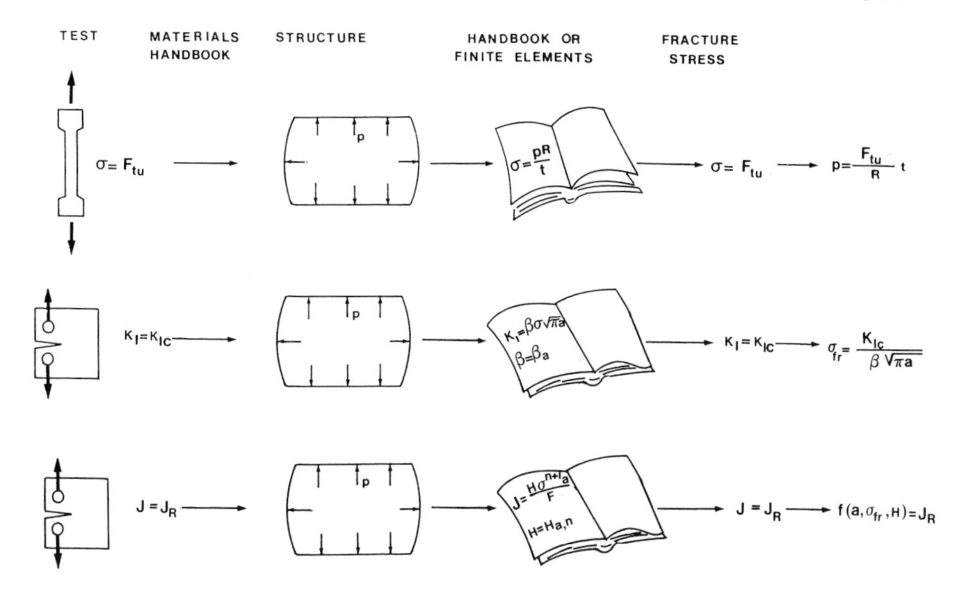

Figure 10.7. Fracture analysis (bottom lines) as compared to conventional design analysis (to line).

ordinate and stress along the abscissa. The stress is limited by collapse and the stress intensity by the toughness.

The limits of K_c at the one end and collapse at the other end, require that there is a limiting line going from K_c to σ_{fc}. The end portions of this contour must be straight so that there is only a relatively small curved part. The top horizontal part is governed by LEFM and the vertical portion by collapse. The curved part is the regime of EPFM. For some applications it may be permissible to approximate the diagram by two straight lines (Figure 10.9b). The failure analysis diagram is usually presented in normalized form by plotting K/K_c and σ/σ_{fc} so that the intercepts with the axes are at 1 (Figure 10.9c). This normalization makes the diagram universal.

The use of the failure analysis diagram can best be demonstrated by an example. Consider a material with $K_{Ic} = 50\,\text{ksi}\sqrt{\text{in}}$ and $F_{\text{col}} = 60\,\text{ksi}$. Assume that the 'structure' is a center cracked panel, 12 in wide with a crack $2a = 2$ inch subjected to a stress of 10 ksi. The nominal stress at collapse would be:

$$\sigma_{fc} = \frac{W - 2a}{W} F_{\text{col}}, \tag{10.1}$$

which for the given case becomes: $\sigma_{fc} = 60(12 - 2) = 50\,\text{ksi}$. Thus, $\sigma/\sigma_{fc} = 10/50 = 0.20$. The stress intensity is $K = \beta\sigma\sqrt{\pi a}$, so that with $\beta \approx 1$ its value is $K = 10\sqrt{\pi \times 1} = 17.7\,\text{ksi}\sqrt{\text{in}}$. Thus, $K/K_c = 0.35$. Now the point

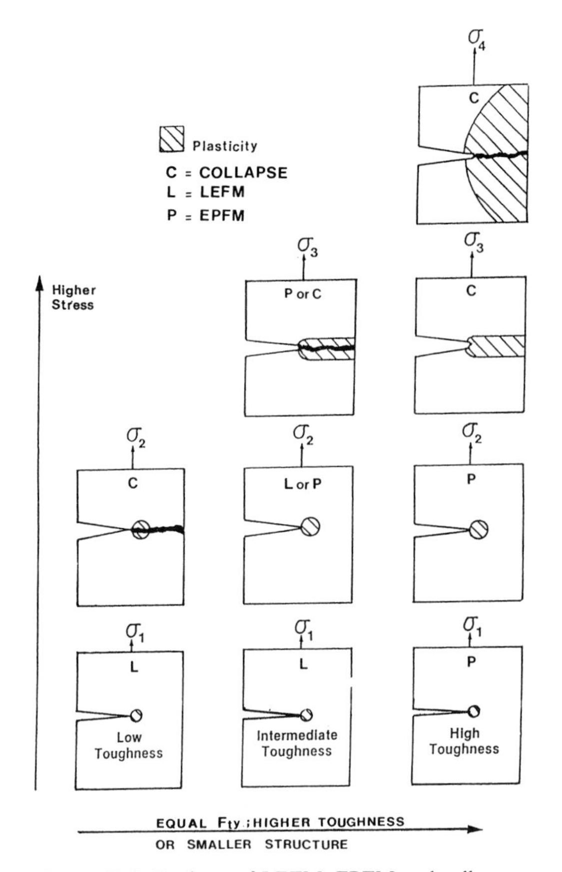

Figure 10.8. Regimes of LEFM, EPFM and collapse.

$\sigma/\sigma_{fc} = 0.20$ with $K/K_c = 0.35$ can be plotted in the diagram as point A in Figure 10.10. If the stress is raised to 20 ksi, the stress intensity becomes $K = 20\sqrt{\pi \times 1} = 35.5\,\text{ksi}\sqrt{\text{in}}$ and $K/K_c = 35.5/50 = 0.71$. Further, $\sigma/\sigma_{fc} = 20/50 = 0.40$. This produces point B in the diagram

Clearly, K/K_c and σ/σ_{fc} increase proportionally. Therefore, a straight line through the origin provides all combinations of K and σ. Fracture occurs where this line intersects the fracture locus. Judgement of the proximity of fracture can be made from the distance between a point and the contour. Extension of the line will also show whether fracture occurs by LEFM or collapse. In the present example LEFM applies (point C).

Consider another case with a crack of $2a = 0.6$ in. The stress at collapse is $\sigma_{fc} = 60(12 - 0.6)/12 = 57$ ksi. For example, with $\sigma = 20$ ksi one obtains $\sigma/\sigma_{fc} = 20/57 = 0.35$, and $K = 20\sqrt{\pi \times 0.3} = 19.4\,\text{ksi}\sqrt{\text{in}}$ or $K/K_c = 19.4/50 = 0.39$, plotted as point D. The extension of the line OD predicts that

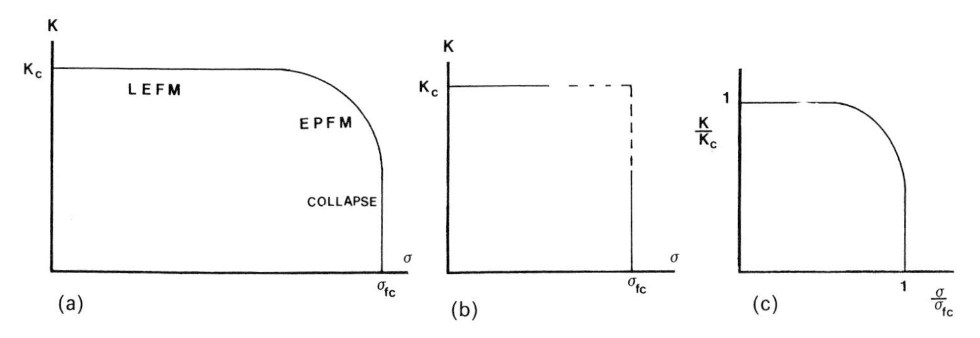

Figure 10.9. Principle of failure analysis diagram[2]. (a) Regimes of fracture failure; (b) Questionable area in transition region; (c) Normalized diagram.

fracture will occur at $\sigma/\sigma_{fc} = 0.84$ so that the fracture stress is $0.84 \times 57 = 48$ ksi. This point is not on the LEFM part in accordance with the fact that the fracture stress is close to σ_{fc}. The tangent to the LEFM curve discussed in Chapter 3 predicts that for a crack of $2a = 0.6$ the residual strength is 47 ksi, as shown in Figure 10.10b, which is almost exactly the same number as obtained above. Hence, the failure analysis diagram can be constructed on the basis of the tangent. The residual strength diagram for the above case was calculated on the basis of $\sigma_{fr} = K_c/\beta\sqrt{\pi a}$, and the tangent constructed. For crack sizes of e.g. $2a = 0.25$, 0.5 and 1 inch the fracture stresses following from the tangent are 55, 50 and 40 ksi respectively. With these stresses, the following values for K/K_c and σ/σ_{fc} are obtained.

Point	E	F	G	On tangent
$2a =$	0.25	0.5	1	in
$\sigma_{res} =$	55	48	40	ksi (from tangent)
$K = \beta\sigma\sqrt{\pi a} =$	34.5	42.5	50	ksi$\sqrt{\text{in}}$
$K/K_{Ic} =$	0.69	0.85	1	
$\sigma_{fc} = (W - 2a)\mathrm{F}_{col}/W$	58.8	57.5	55	ksi
$\sigma_{res}/\sigma_{fc} =$	0.94	0.80	0.73	

Then the three points F, and G and E in Figures 10.10 a can be constructed so that the curved part of in Figure 10.10a can be drawn. Clearly in this case the curved part can still be treated with the LEFM tangent approximation.

In its normalized form the failure analysis diagram is the same for all materials and structures, regardless of K_c and F_{col}. Given the scatter in material behavior, an approximation of the curved part will suffice for many purposes. A more precise diagram (curved part) can be drawn based on J, but obviously this cannot be too different.

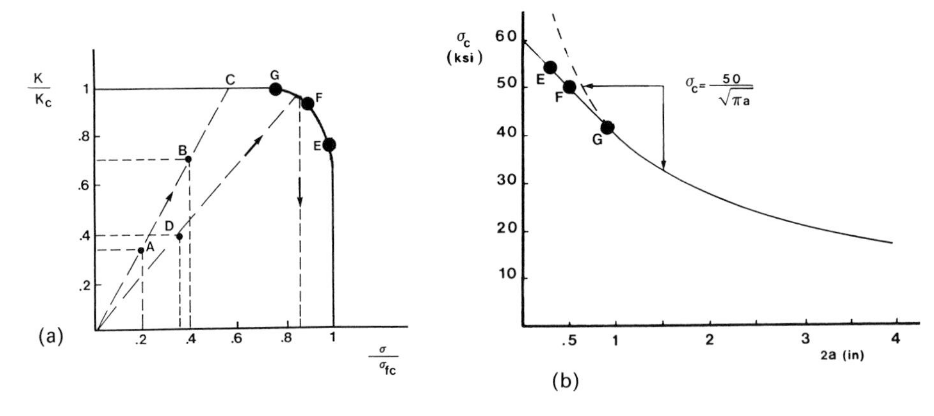

Figure 10.10. Approximation of curved part by LEFM. (a) Failure analysis diagram; (b) Tangent.

The use of the diagram for a particular application requires the calculation of the stress at collapse and the stress intensity at a given stress and crack length. This permits calculation of and K/K_c at that stress, which can be plotted on the diagram.

The diagram presents a means for a judgement of the proximity of fracture and it shows what kind of fracture to expect, putting the three areas of fracture analysis, LEFM, EPFM and collapse in perspective. It is useful in conjunction with (not instead of) the residual strength diagram, as it can be derived from the latter as shown above. Its significance is possibly that it illustrates that from a technical point of view the EPFM fracture criterion is not very sensitive, and that also in EPFM collapse must treated separately as a competing condition.

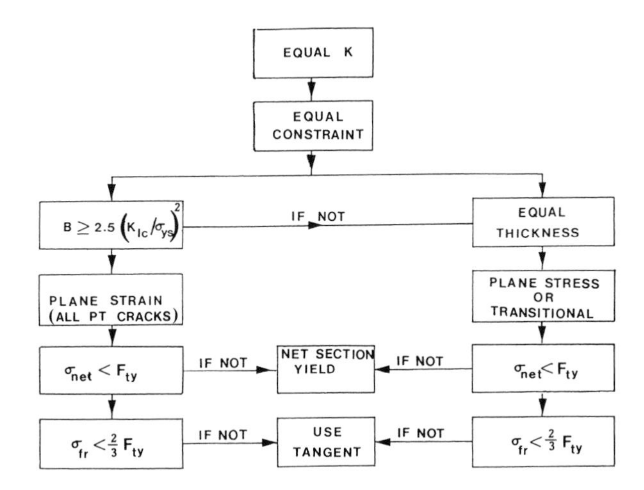

Figure 10.11. Conditions to be considered for LEFM residual strength analysis.

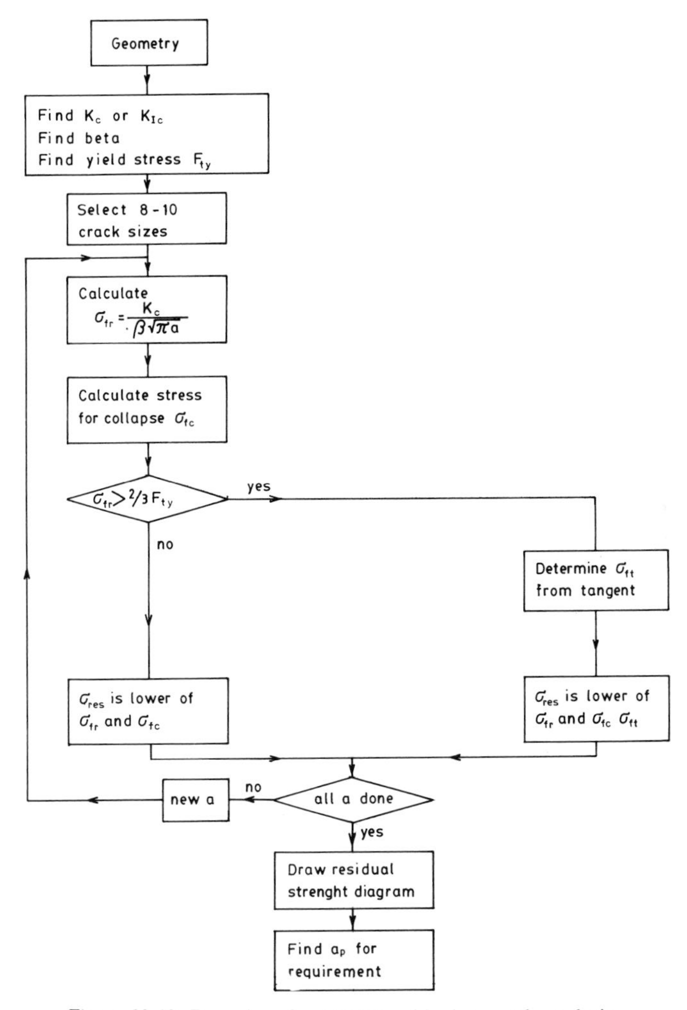

Figure 10.12. Procedure for LEFM residual strength analysis.

This discussion would not be complete without mentioning of the so-called R-6 diagram [3]. The latter is an extension of the failure analysis diagram to account for stable fracture, but it opens no new avenues, as the residual strength whether in LEFM or EPFM can be readily calculated as shown in the following sections.

10.5. Residual strength analysis

The concepts and conditions for residual strength analysis with LEFM were discussed in Chapter 3. Figure 10.11 summarizes the conditions, while Figure

10.12 shows the procedure. Note again that for small cracks unconservative results would be obtained, so that a tangent approximation to the calculated curve must be made. Similarly, for high toughness, or in the general case for smaller structures, collapse may prevail.

Generally the residual strength analysis (LEFM, EPFM and Collapse), will be performed using some sort of computer software but hand calculations are well possible. It should be noted once again, that the analysis will be the same for any kind of structure, provided one finds β (or H) for the structural geometry as described in Chapter 8. Since the center crack is easy to envision, examples here will be based on the center crack.

The stress intensity is $K = \beta\sigma\sqrt{\pi a}$. Fracture occurs when the stress intensity is equal to the toughness, K_{Ic} or K_c whichever is applicable. Fracture occurs if: $\beta\sigma\sqrt{\pi a} = K_{Ic}$ (or K_c). Thus for given crack size the fracture stress is found as:

$$\sigma_{fr} = \frac{K_{Ic}(\text{or } K_c)}{\beta\sqrt{\pi a}}. \tag{10.2}$$

Table 10.2. Critical crack size calculation (not advisable; see test)

$K_{Ic} = 60\,\text{ksi}\sqrt{\text{in}}, \quad F_{ty} = 120\,\text{ksi}.$

What is the critical crack size for an applied stress of 30 ksi in a center cracked panel of 30 inch width?

Solution

According to Figure 3.3, the stress intensity is:

$$K_I = \sigma\sqrt{\pi a}\sqrt{\sec\frac{\pi a}{W}}; \quad \beta = \sqrt{\sec\frac{\pi a}{W}}$$

$\pi a/W$ in radians!

$$a_c = \frac{1}{\pi}\frac{K_{Ic}^2}{\sigma^2\sec\dfrac{\pi a}{W}}; \quad \text{First assume } \beta = 1, \text{ so that } a = \frac{1}{\pi}\left(\frac{K_{Ic}}{\sigma}\right)^2;$$

$$a_c = \frac{1}{\pi}\left(\frac{60}{30}\right)^2 = 1.27\,\text{in} \quad \text{This means } \beta = \sqrt{\sec\frac{\pi\times 1.27}{12}} = 1.209;$$

$$\text{Then } a_c = \frac{1}{\pi}\left(\frac{60}{1.029\times 30}\right)^2 = 1.20\,\text{in} \quad \text{This means } \beta = \sqrt{\sec\frac{\pi\times 1.2}{12}} = 1.025;$$

$$\text{Then } a_c = \frac{1}{\pi}\left(\frac{60}{1.025\times 30}\right)^2 = 1.21\,\text{in}.$$

The last a_c differs only slightly from the previous one so that iteration complete: $a_c = 1.21$ in, and $2a_c = 2.42$ in.

Table 10.3. Calculation of residual strength center crack; $F_{ty} = 120\,\text{ksi}$, $K_{Ic} = 60\,\text{ksi}\sqrt{\text{in}}$ and $W = 12$.

$2a$	a	$\beta = \sqrt{\sec\dfrac{\pi a}{W}}$	$\sqrt{\pi a}$	$\sigma_{fr} = K_{Ic}/\beta\sqrt{\pi a}$	$\sigma_{fc} = \dfrac{W-2a}{W} \times F_{ty}$	σ_{res}
0.5	0.25	1	0.89	67	115	67
1	0.50	1	1.25	48	110	48
2	1	1.02	1.77	33	100	33
4	2	1.07	2.51	22	80	22
6	3	1.19	3.07	16	60	16
8	4	1.41	3.54	12	40	12
10	5	1.97	3.96	7.7	24	7.7
11	5.5	2.77	4.16	5.2	10	5.2

This fracture stress is the residual strength unless collapse occurs at a lower stress, or unless σ_{fr} is close to F_{col} so that the tangent must be used (Figures 10.11 and 10.12.

It is possible to calculate the crack size at fracture directly, or rather, the permissible crack size a_p for the minimum permissible residual strength, σ_p. It followed from Equation (10.1) that

$$a_p = \frac{1}{\pi}\left(\frac{K_{Ic}}{\beta\sigma}\right)^2. \tag{10.3}$$

Unfortunately, this equation cannot be solved directly, because β depends upon a; hence the value of β is not known before a is known and vice versa. Often β is given in graphical form, or at best in terms of a complicated polynomial. Therefore Equation (10.3) can be solved only by iteration. A numerical example is shown in Table 10.2. Note that it would still be necessary to check for collapse.

Similar calculations would show that for a stress of 30 ksi the permissible crack size would be 1.27 in, and for a stress of 33 ksi, it would be 1.05 inch. The chance of knowing the acting stress to an accuracy of 10% is small, but 33 ksi as opposed to 30 ksi makes a difference of almost 20% in the critical crack size. Thus, one calculation may be inadequate. If the complete residual strength diagram is calculated instead, one would immediately see from the slope of the curve how inaccuracies in stress would affect the permissible crack size. Calculating the entire residual strength diagram is no more work than calculating crack size with Equation (10.3) because no iteration is necessary, as shown in the example in Table 10.3, which is for the same case as the previous. Besides the calculation of Table 10.3 provides much more information.

Consider a situation where β is a long polynomial in a/w (as e.g. for an edge crack, see Figure 3.3). In that case the iteration procedure would be more cumbersome. Last but not least, the iteration does not always converge.

Improper choice of the starting value of β may cause diversion instead of conversion. Then the procedure must be started all over again. Finally, if the actual result would fall on the tangent, the whole procedure would be erroneous, as fracture would occur below K_c.

These examples illustrate that calculating a permissible crack size directly by iteration is inefficient. It is advisable under all circumstances to calculate a complete residual strength diagram. The calculation is no more work, while the complete diagram can provides much more information. The results of Table 10.3 were plotted in Figure 10.13. It is now immediately obvious in which regime the fastest drop in strength occurs. Similarly, it can be seen in which regime errors in stress have the largest effect on permissible crack size. It is a simple matter to sketch operating stress levels in the diagram which facilitates a judgement of the criticality of a certain crack or loading situation.

The variability of toughness within an apparently homogeneous material can be significant, and the heat-to-heat variation of the toughness may even be larger. A scatter in fracture toughness on the order of 15% is not unusal. One can account for the effect of scatter in the residual strength diagram (Figure 10.13). The figure demonstrates the usefulness of constructing complete residual strength diagram when evaluating the criticality of a structure with respect to cracks.

A problem occurs in the calculation of the residual strength for small cracks because Equation (10.2) predicts that σ_{fr} approaches infinity if a approaches zero. It was shown in Chapter 3 that a tangent to the curve is a good approximation. Also collapse must be considered. Assuming (Chapter 3) that in LEFM the collapse strength, F_{col}, is equal to the yield strength, F_{ty}, the stress for collapse of a center cracked panel follows from Equation (10.1) Collapse conditions for other cases were shown in Chapter 2 in Table 2.1.

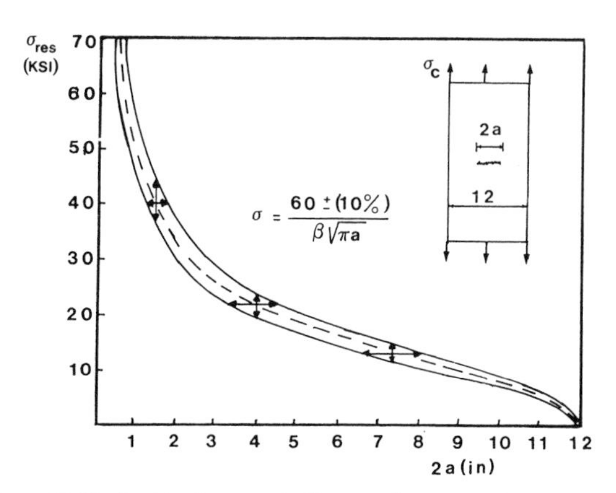

Figure 10.13. Residual strength diagram for c ase of Table 10.3.

Table 10.4. Residual strength calculation for 3 materials with different toughness and same yield. $F_{col} = F_{ty}$.

(a) 12-inch wide center cracked panel.

$2a$	a	$\beta = \sqrt{\sec(\pi a/W}$	$\sqrt{\pi a}$	$\sigma_{fr} = K_{Ic}/\beta\sqrt{\pi a}$			$\sigma_{fc} = (1 - 2a/W)F_{ty}$	σ_{res} lower of σ_{fr} and σ_{fc}		
				$K_{Ic} = 40$ A	$K_{Ic} = 80$ B	$K_{Ic} = 160\,\text{ksi}\sqrt{\text{in}}$ C	$F_{ty} = 60\,\text{ksi}$; all cases	A	B	C
0.5	0.25	1	0.89	44	88	176	57.5	*	*	57.5
1	0.50	1	1.25	32	64	128	55	32	*	55
2	1	1.02	1.77	22	44	88	50	22	*	50
4	2	1.07	2.51	14	28	56	40	14	28	40
6	3	1.19	3.07	11	22	44	30	11	22	30
8	4	1.41	3.54	8	16	32	20	8	16	20
10	5	1.97	3.96	5.2	10.4	20.8	10	5	10	10
11	5.5	2.77	4.16	3.4	6.8	13.6	5	3	5	5

(b) 60 inch-wide center cracked panel.

$2a$	a	β	$\sqrt{\pi a}$	σ_{fr}			σ_{fc}	σ_{res}		
				$K_{Ic} = 40$	$K_{Ic} = 80$	$K_{Ic} = 160\,\text{ksi}\sqrt{\text{in}}$	$F_{ty} = 60\,\text{ksi}$	A	B	C
0.5	0.25	1	0.89	44	88	176	59.5	*	*	*
1	0.50	1	1.25	32	64	128	59	32	*	*
2	1	1	1.77	22	44	88	58	22	*	*
5	2.50	1	1.80	14	29	57	55	14	29	*
10	5	1.02	3.96	10	20	40	50	10	20	*
20	10	1.07	5.60	7	13	27	40	7	13	27
30	15	1.19	6.86	5	10	20	30	5	10	20
40	20	1.41	7.93	4	7	14	20	4	7	14
50	25	1.97	8.86	2	5	9	10	2	5	9

* On tangent from F_{ty} (see figure 10.14).

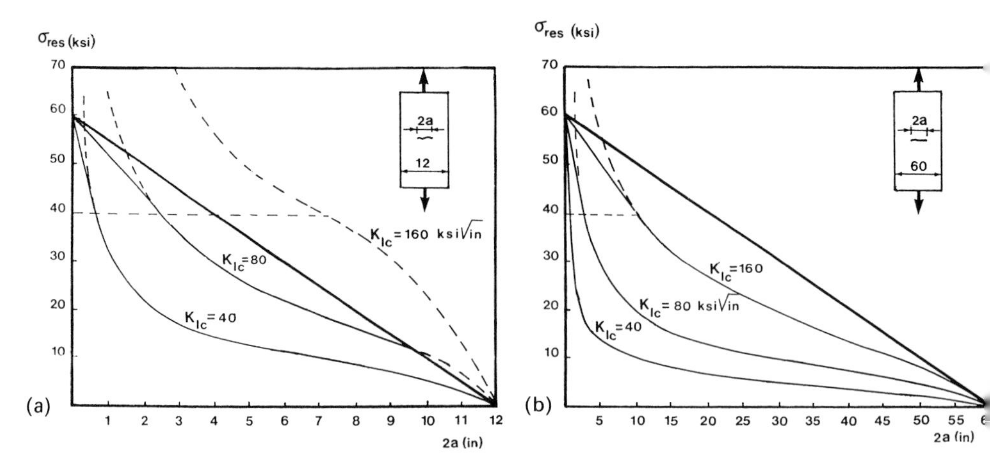

Figure 10.14. Residual strength diagrm for cases of Table 10.4. (a) 12 inch panel; (b) 60 inch panel.

Accounting for all of the above, residual strength analysis can be performed as in Table 10.4. In order to show that collapse may occur even if the toughness is low, two panel sizes were considered. Also some of the results may fall on the tangent regardless of how low the toughness. This becomes clear from the residual strength diagrams obtained by plotting the data of table 10.4 as in Figure 10.14 (Note: in order to avoid clutter and to better demonstrate the effect, F_{ty} was assumed the same for all three materials; naturally, the principle would not change if the materials had different F_{ty}.

Figure 10.14 illustrates one more reason for which it is important to construct a complete residual strength diagram. The line for collapse can be constructed and it can be seen immediately whether or not the residual strength should be found from collapse, LEFM or from the tangent to the curve. Also, note that the whole curve has to be determined before the tangent can be constructed.

In the case of plane stress or transitional fracture the question arises as to whether one should use K_{eff} instead of K_c (Chapter 3). From a technical point of view K_{eff} is the relevant toughness. Naturally, the calculation procedure remains the same regardless of the toughness used.

The limitations of LEFM are by no means as severe as sometimes suggested, if sensible approximations are made where necessary, as shown. Besides EPFM is no panacea either.Calculations are more involved and often not more accurate than the approximations used in LEFM border cases.

The concept and equations for EPFM residual strength analysis were discussed in Chapter 4. The calculation is simplest when stable fracture is ignored (as it is in LEFM). In many cases this will lead to slightly conservative results. The calculation procedure for that case is shown in Figure 10.15. The question of constraint (Chapters 4, 7) has not been resolved for EPFM, so that

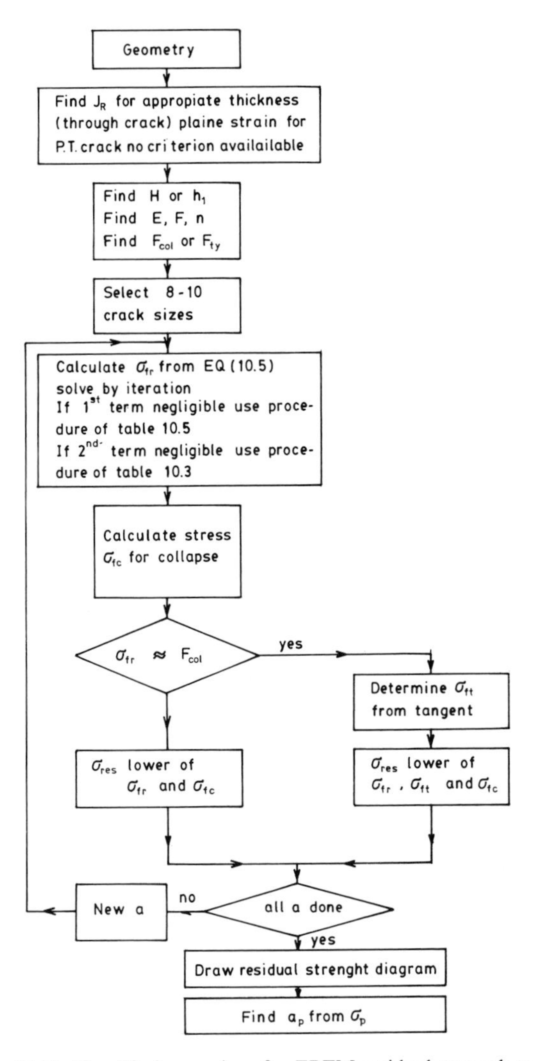

Figure 10.15. Simplified procedure for EPFM residual strength analysis.

'toughness' values must be used for the appropriate thickness in the case of through cracks. For part-through cracks (plane strain) there is no obvious answer; the use of conservative values is recommendable.

Generally the fracture equation (Chapter 4):

$$\frac{\pi\beta^2\sigma a^F}{E} + \frac{H\sigma^{n+1}a}{F} = J_R \tag{10.4}$$

must be solved for σ_{fr}, the fracture stress. Solution is possible only by iteration which is best done by means of appropriate computer software [4]. If either of

Table 10.5. Residual strength analysis using EPFM for four cases, ignoring stable fracture.

a (in)	a/W	H	$\sigma_{fr} = (FJ/Ha)^{1/(n+1)}$		$\sigma_{fc} = (1 - 2a/W)\ F_{cal}$
			$J_R = 0.83$ kips/in	$J_R = 5.3$ kip/in	$F_{col} = 56$ ksi
$W = 8$ inch					
0.5	0.063	8.72	38.9	53.3	49
1.0	0.125	13.5	32.4	44.1	42
1.5	0.188	24.6	27.4	37.3	35
2.0	0.250	57.9	22.6	30.8	28
2.5	0.313	190	17.9	24.4	21
3.0	0.375	1239	12.7	17.3	14
3.5	0.438	35389	7.1	9.6	7
$W = 24$ inch					
1.5	0.063	8.72	32.6	44.4	49
3.0	0.125	13.5	27.1	86.9	42
4.5	0.188	24.6	22.8	29.6	35
6.0	0.250	57.9	18.8	25.7	28
7.5	0.313	190	14.9	20.3	21
9.0	0.375	1239	10.5	14.4	14
10.5	0.438	35389	5.9	8.0	7

$n = 5$; $F = 1.9\,\mathrm{E}10\,\mathrm{ksi}^5$.

the terms in Equation (10.4) is small with respect to the other, hand calculations are a simple matter. The case of the second term being negligible is a trivial one. because then the case reverts to LEFM and it is solved as discussed above (toughness $\sqrt{EJ_R}$). If the first term is negligible solution proceeds as in the numerical example in Table 10.5.

The results of this example are plotted in Figure 10.16. Note again that the same problems exist as in LEFM. An approximation must still be made for small crack sizes (because for a approaching zero Equation (10.4) leads to infinite stress); also collapse must be evaluated separately. A complete residual strength diagram is again necessary for a complete evaluation. Almost all remarks in this section made regarding LEFM apply to EPFM residual strength analysis as well, as can be judged easily from a comparison of Figures 10.16 and 10.14.

There is an additional nuisance of EPFM. The criterion is expressed in terms of strain energy and therefore the 'toughness' or fracture resistance (J_R) is a meaningless number for direct comparison of alloys. This can best be demonstrated on the basis of the LEFM expressions. For example, consider a steel and an aluminum alloy both with the same toughness (e.g. $K_c = 50\,\mathrm{ksi}\sqrt{\mathrm{in}}$). This means that the fracture resistance in terms of G (which may be called G_{ic} or R) irrespective of its denotation is $G_c = R\ (= J_R) = K_{Ic}^2/E$. For the aluminum with $E = 10\,000$ ksi one obtains $G_c = 50^2/10.000 = 0.25$ kips/in, and for the steel with E 30 000 ksi; $G_c = 50^2/30\,000 = 0.083$ kips/in.

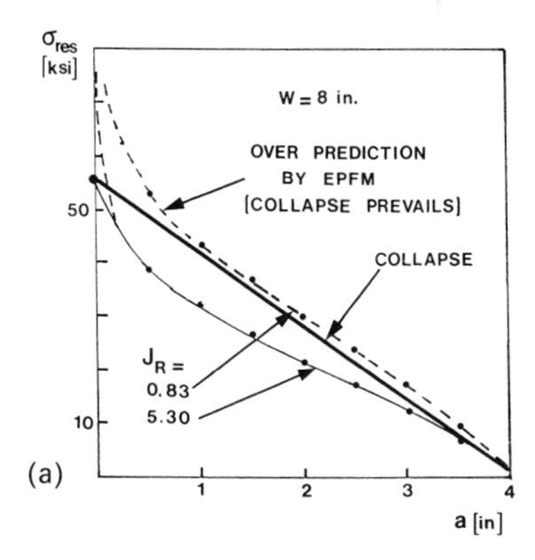

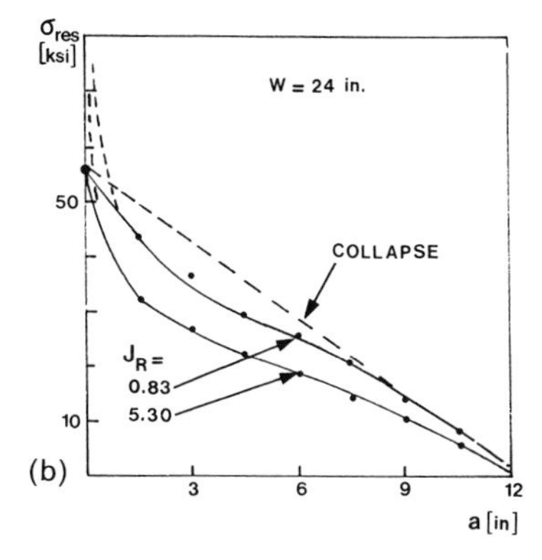

Figure 10.16. Residual strength analysis with EPFM (Tble 10.5). ● Calculated points from EPFM (Table 10.5). (a) Panel of 8 inch width; (b) Panel of 24 inch width.

If the fracture resistance of these two materials is quoted as 0.25 and 0.083 kips/in respectively, one is inclined to believe that the aluminum alloy is a better material. Yet, as both materials have same toughness of 50 ksi $\sqrt{\text{in}}$ their residual strength curves are identical. From the engineering point view, the important issue is the stress that can be sustained at a certain crack size. Clearly, this is the same for both materials as can be readily seen because both materials have the same toughness. The above numbers of 0.25 and 0.083 are meaningless without knowledge of E.

In the case of EPFM the situation is worse. Not only is the 'stiffness' F involved, but also the strain hardening exponent n. Thus comparitive J_R values give no indication at all which is the better material from the point of view of residual strength. The material with the lower J_R may well have the higher residual strength. This is demonstrated in Table 10.6.

10.6. Use of *R*-curve and J_R-curve

Whether the R-curve (LEFM) or the J_R-curve (EPFM) is used, the procedure is essentially the same. The solution is the easiest when done graphically, using

Table 10.6. Examples of lower fracture stress for higher J_R.

J_R (kips/in)	n	F ksi^n	a[a] in	$H((a/W), n)$ $(a/w) = 0.125$	$\sigma_{fr} = (FJ_R/Ha)^{1/(h+1)}$ (ksi)
3	5	1E11	2	13.5	47.2
5	7	3E13	2	23.3	36.6
5	10	4E14	2	50.8	22.4
1	5	1E11	2	13.5	39.3
3	7	3E13	2	23.3	34.3

[a] Same configuration and same a in all case.

Table 10.7. Calculation of G lines for graphical solution.

a_{total} [a] $(a + \Delta a)$	β [b]	$G = \pi\beta^2\sigma^2 a/E$ $E = 10\,000$			
		$\sigma = 25$	$\sigma = 30$	$\sigma = 33$	$\sigma = 35$
0	1	0	0	0	0
0.5	1	0.031	0.045	0.171	0.192
1	1	0.062	0.090	0.342	0.384
1.5	1	0.093	0.135	0.256	0.577
2	1	0.124	0.180	0.342	0.780

Note: Plot calculated G-lines and R-curve as in Figure 10.17. The G-lines for $\sigma = 35$ ksi appear about tangent; hence $\sigma_{fr} \approx 35$ ksi.

[a] More intermediate crack sizes should be taken if $\beta \neq 1$, because then lines will be curved.

[b] β should be calculated for the cae at hand; $\beta = 1$ assumed here.

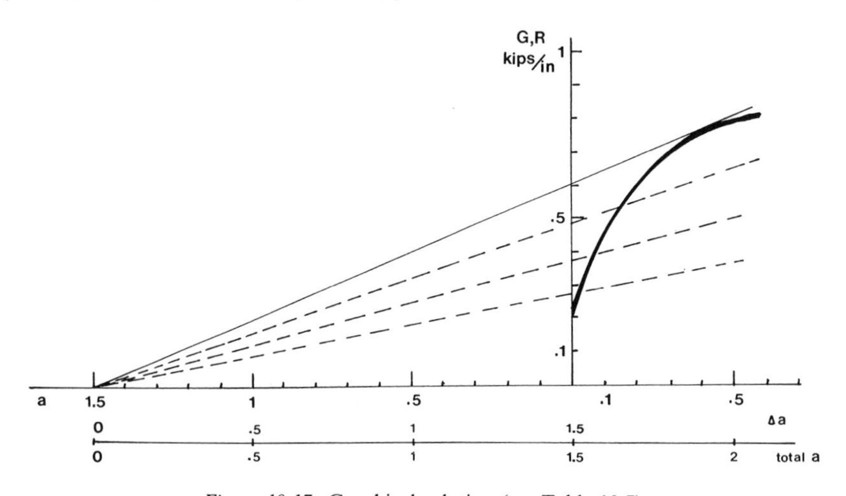

Figure 10.17. Graphical solution (see Table 10.7).

plots such as in Figure 3.17 and 4.2. The G or J lines are represented by the equations:

$$G = \frac{\pi\beta^2\sigma^2 a}{E}$$

$$J = \frac{\pi\beta^2\sigma^2 a}{E} + \frac{H\sigma^{h+1}a}{F}. \tag{10.5}$$

First the R or J_R curve (material data) is plotted. Then G(or J) are calculated in tabular form for a number of crack sizes and five or six values of the stress. The fracture stress is the one associated with that curve which is tangent to the R or J_R curve. An example is shown in Table 10.7 and Figure 10.17. Clearly, the procedure works the same for EPFM using the second of Equations (10.5).

Table 10.8. Logic for iterative solution

$$\frac{\pi\beta\sigma^2 a}{E} + \frac{H\sigma^{n+1} a}{F} = J_R \text{ (or } R)$$

if either term is negligible solve directly, otherwise solve by iteration (software).

(a) Solve equation for commencement of fracture to obtain σ_i (J_R or R for $\Delta a = 0$).
(b) Increment by small δa.
(c) $a_{\text{new}} = a_{\text{old}} + \delta a$; $\Delta a_{\text{new}} = \Delta a_{\text{old}} + \delta a$.
(d) Calculate dJ/da or dG/da as $\dfrac{J_{a+\delta a} - J_a}{\delta a}$ or $\dfrac{G_{a+\delta a} - G_a}{\delta a}$.
(e) Obtain J_R or R for Δa_{new} from J_R-curve or R-curve.
(f) Calculate dJ_R/da or dR/da as $\dfrac{J_{R\Delta a_{\text{new}}} - J_{R\Delta a}}{\Delta a}$ or $\dfrac{R_{\Delta a\text{new}} - R_{\Delta a}}{\Delta a}$.
(g) if dJ/da $\geqslant$ dJ_R/da or dG/da $\geqslant$ dR/da then done (tangency).
(h) Otherwide go back to b.

Table 10.9. LEFM; *R*-curve analysis.

Crack. L *a* inches	Δa inches	Stable growth and instability			dG/da	dR/da
		Stress ksi	*G* kips/in	*R* kips/in		
1.500	0.000	14.002	0.100	0.100	0.000	0.000
1.530	0.030	14.115	0.100	0.104	0.078	0.133
1.561	0.061	14.223	0.104	0.108	0.080	0.133
1.592	0.092	14.326	0.108	0.112	0.082	0.133
1.624	0.124	14.424	0.112	0.116	0.084	0.133
1.656	0.156	14.516	0.116	0.121	0.086	0.133
1.689	0.189	14.604	0.121	0.125	0.088	0.133
1.723	0.223	14.687	0.125	0.130	0.090	0.133
1.757	0.257	14.765	0.130	0.134	0.092	0.133
1.793	0.293	14.838	0.134	0.139	0.094	0.133
1.828	0.328	14.857	0.139	0.143	0.097	0.107
1.865	0.365	14.860	0.143	0.147	0.098	0.100
1.902	0.402	14.861	0.147	0.150	0.100	0.100
Fracture instability, dG/da > dR/da, max load						
1.940	0.440	14.859	0.150	0.154	0.101	0.100
1.979	0.479	14.854	0.154	0.158	0.000	0.100
2.058	0.558	14.836	0.158	0.166	0.00	0.100

Center cracked panel; plane stress.
Calculation of initiation and instability in load control
Yield strength = 50 ksi
Ult. tens. st. = 70 ksi
Collapse str. = 50 ksi
Modulus = 10000 ksi
Crack length = 1.5 inches
$2a$ = 3 inches
Width = 12 inches
Thickness = 0.2 inches

Note:
in each step G is
increased to match R of
previous step

case numerical examples are shown in Table 10.9 for LEFM (R), and in Table 10.10 for EPFM (J_R). So many iterations are necessary both in LEFM and EPFM that the procedure is best done by computer [4]. Tables 10.9 and 10.10 are result of computer analysis.

10.7. Crack growth analysis

The crack growth analysis procedure is shown in Figure 10.18. Extensive details, algorithms and examples were shown in Chapters 5 and 6, specific problems illustrated with examples in Chapter 9. Crack growth analysis, it was pointed out, requires appropriate computer software [e.g. 4]; in few cases an analysis by hand is possible (constant amplitude). The following example may serve as a general illustration.

It is anticipated that cracks might occur at the edge of the reinforcement plate of a nozzle in a pressure vessel Figure 10.19). The pressure fluctuates approxi-

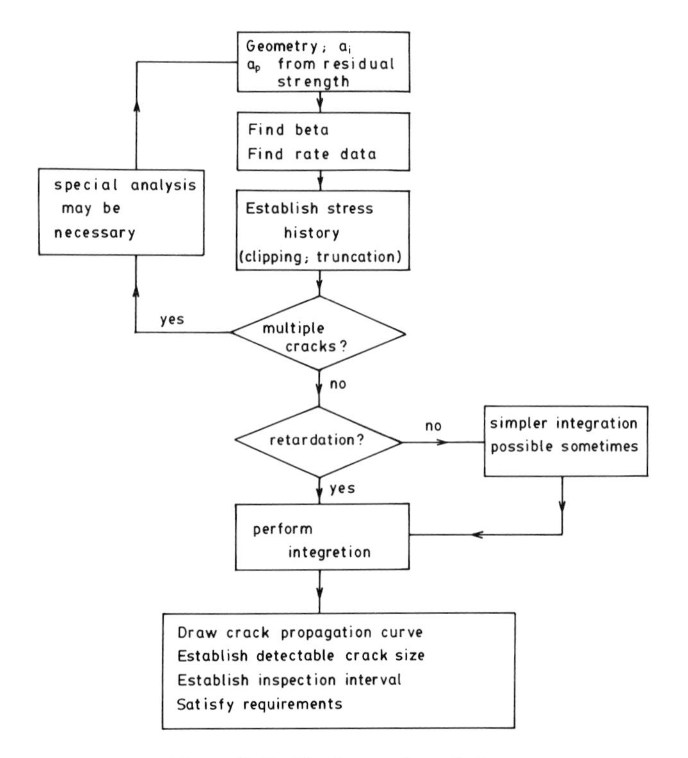

Figure 10.18. Crack growth analysis.

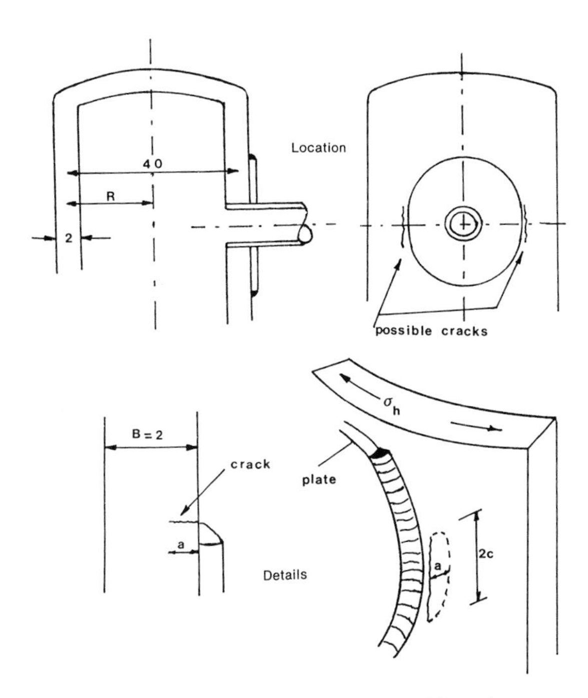

Figure 10.19. Examples of nozzle weld crack.

Computational solutions are more involved because they often require (many) iterations. The logic for the procedure is shown in Table 10.8. For this mately once per two hours between $p = 1500$ and 3000 psi. Under certain conditions if the process stagnates the pressure may reach 3600 psi, but this is an unlikely event. Failure of the vessel would cause explosion damage and involve loss of life. The fracture control objective is to reduce the probability of failure to essentially zero even if this would involve high fracture control costs. A conservative analysis is indicated.

Cracks are expected to occur in a plane perpendicular to the hoop stress. As a first approximation, the vessel is treated as a thin-wall cylinder. The hoop stress (Figure 10.19) is $\sigma_h = pR/B = p \times 20/2 = 10p$. Hence, $\sigma_h = 30.000$ psi (30 ksi) at maximum operating pressure (and 36 ksi at overpressure) with fluctuations of 15 ksi with a stress ratio of $R = 15/30 = 0.5$. The yield strength of the steel is $F_{ty} = 60$ ksi. The toughness is 80 ksi $\sqrt{\text{in}}$ at operating temperature. Fatigue crack growth data appear to be reasonably well described by the Walker equation: $\mathrm{d}a/\mathrm{d}N = 10^{-10}\,\Delta K^{3.8}/(1\text{-}R)^2$.

Cracks will initate as surface flaws. Several of these may initiate simultaneous-

Table 10.10. Computational solution in EPFM

Load/mom inkips	Crack. Length inches	J-appl kips/in	Stress ksi
0.000	1.000	0.000	0.000
124.386	1.000	0.035	11.330
248.772	1.000	0.358	22.659
373.159	1.000	1.796	33.989
497.545	1.000	6.009	45.318
621.931	1.000	15.641	56.648
621.933	1.002	15.690	56.648
621.933	1.004	15.739	56.648
Max load; instability $dJ/da > dJ_R/da$			
621.933	1.006	15.789	56.648
621.932	1.008	15.838	56.648
621.930	1.010	15.888	56.648

Crack length inches	Δa inches	Stress ksi	J-appl kips/in	J_R kips/in	dJ/dA	dJ_R/da
1.0000	0.0000	56.65	15.64	15.63	24.49	24.5
1.0020	0.0020	56.65	15.69	15.68	24.53	24.5
1.0040	0.0040	56.65	15.74	15.73	24.58	24.5
Max load; instability $dJ/da > dJ_R/da$						
1.0060	0.0060	56.65	15.79	15.78	0.00	24.5
1.0080	0.0080	56.65	15.84	15.83	0.00	24.5
1.0100	0.0100	56.65	15.89	15.88	0.00	24.5

Cylindrical container
Bending; through crack; circumferential; bending or tension
Outside diameter = 6 inches
Wall thickness = 0.5 inches
Crack size $2a$ = 2 inches
Yield strength F_{ty} = 30 ksi
Ultimate tensile strength F_{tu} = 60 ksi
Collapse stress F_{col} = 40 ksi
Reference stress σ_0 = 40 ksi
Reference strain ε_0 = 1.73913E-03
α = 15.99831
Strain hardening exponent, n = 3.4
Young's modulus = 23000 ksi
Plastic modulus = 1.006 E + 07 ksi ∧ 3.4
Nominal stress at collapse = 44.39756 ksi

Table 10.11. Approximate analysis of case of Figure 10.19.

(a) *Residual strength.*
$B = 2$ inch; $K_{Ic} = 80\,\text{ksi}\sqrt{\text{in}}$; $a/2c = 0.1$; $F_{ty} = 60$ ksi $\beta = 1.12\,\beta_{FFS} k_t/\phi$; $\phi \approx 1$ (Figure 8.3); $k_t = 1.3$; $\beta \approx 1.46\,\beta_{FFS}$

Crack depth a (in)	$\sqrt{\pi a}$	a/B	β_{FFS} from Figure 8.3	β	$\sigma_c = (K_{Ic}/\beta\sqrt{\pi a})$
0.1	0.56	0.05	1.02	1.49	96
0.3	0.97	0.15	1.07	1.56	53
0.5	1.25	0.25	1.15	1.68	38
0.7	1.48	0.35	1.28	1.87	29
1	1.77	0.50	1.60	2.34	19
1.2	1.94	0.60	2.00	2.86	14

(b) *Crack growth.*

$$(\mathrm{d}a/\mathrm{d}N) = \frac{10^{-10}}{(1-R)^2}\Delta K^{3.8};\ \beta \text{ as above}$$

Assumed: constant flaw shape; $R = 0.1$; 1 cycle/2 hours $\Delta\sigma = 15$ ksi.

a	Δa	$\sqrt{\pi a}$	a/B	β	ΔK	$(\mathrm{d}a/\mathrm{d}N) = (10^{-10}\,\Delta K^{3.8}/(1-R)^2$	$\Delta N = \Delta a/(\mathrm{d}a/\mathrm{d}N)$	N	Hours
0.05	–	–	–	–	–	–	–	0	0
0.075	0.025	0.49	0.025	1.47	10.8	1.13×10^{-6}	22181	22181	44362
0.1	0.025	0.56	0.05	1.49	12.5	1.96×10^{-6}	12727	34908	69816
0.15	0.05	0.69	0.075	1.50	15.6	4.55×10^{-6}	10968	45876	91752
0.20	0.05	0.79	0.1	1.53	18.2	8.19×10^{-6}	6106	51982	103964
0.25	0.05	0.89	0.125	1.55	20.7	1.34×10^{-5}	3744	55726	111452
0.30	0.05	0.97	0.15	1.56	22.7	1.89×10^{-5}	2637	58363	116726
0.40	0.10	1.12	0.20	1.61	27.0	3.67×10^{-5}	2728	61091	122182
0.50	0.10	1.23	0.25	1.66	31.2	6.34×10^{-5}	1574	62665	125330
0.60	0.10	1.37	0.30	1.75	36.0	1.09×10^{-4}	914	63579	127158

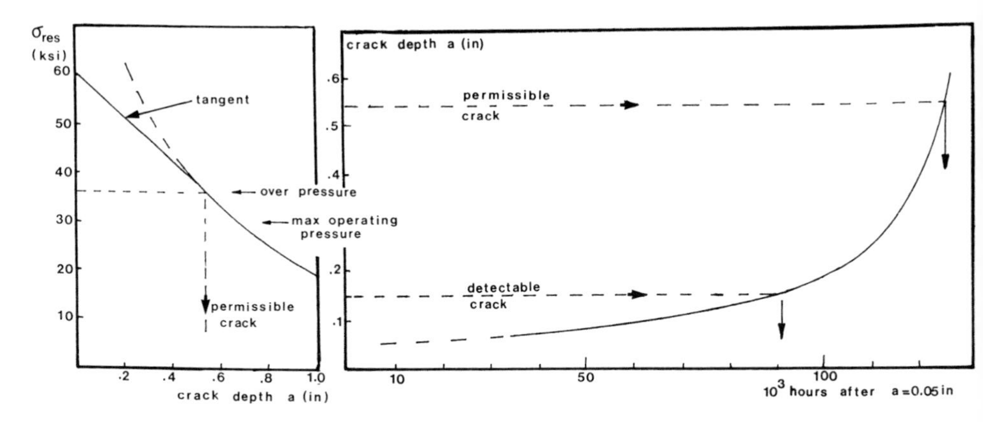

Figure 10.20. Results of Table 10.11 for case of Figure 10.19.

ly and join up to form a long slender flaw, or a slender flaw may form immediately. For flaws with large aspect ratios, ϕ will approach 1 (Figure 8.3). A stress concentration factor, $k_t = 1.3$ is included to account for the stress concentration due to the weld. A hand analysis is shown in Table 10.11; the results are plotted in Figure 10.20.

Cracks may or may not occur, but if they do, they will cause catastrophic failure. Therefore, the fracture control is to be based upon inspection (Chapter 11). Since the maximum permissible crack size (for the case of overpressure) is 0.54 inch, much smaller cracks must be detected. Ultrasonic inspection (Chapter 11) is probably best suited for the purpose. Considering the circumstances (height, surrounding piping) the probability of detecting cracks with a depth less than 0.15 in is considered low regardless of their length. In that case, the remaining life to failure would be 126 000-92 000 = 34 000 hours or 3.9 years (Figure 10.20). The calculations were generally conservative, but nevertheless a factor of 4 on life is considered advisable. Thus, the inspection interval is taken as 3.9/4 = 1 year. Further analysis should be performed to evaluate cracks of various depth-length ratios. Clearly, more refined analysis is possible with sophisticated computer software [4]. It should be realized however, that this would not necessary lead to better answers. (Chapter 12; Sources of error).

Most practical problems are more complicated than the above example. Questions of stress history, sequencing, clipping and truncation, retardation parameters and so on, must be considered, as discussed in Chapters 5 and 6. The above merely shows the principles. Analysis is best performed using reliable software, but the reliability of software is largely determined by the user (input); with a sound knowledge of the principles behind the software the effects of assumptions can be assessed (Chapter 12).

10.8. Exercises

1. Given that $K_{Ic} = 55\,\text{ksi}\sqrt{\text{in}}$ and that $\sigma_p = 16\,\text{ksi}$ calculate a_p using iteration, and Equation (10.3). The configuration is a single edge crack (Figure 3.3) with $W = 10$ inches; $F_{ty} = 100\,\text{ksi}$; $B = 0.8$ inch

2. Calculate the complete residual strength diagram for the case of Exercise 1 and determine a_p. How much time did you save and how much more information did you get, comparing Exercises 1 and 2?

3. Calculate and plot the residual strength diagrams for three panels with center cracks; $W = 4$, 10 and 30 inches respectively. $F_{col} = F_{ty} = 60\,\text{ksi}$, $K_c = 65\,\text{ksi}\sqrt{\text{in}}$. In each case determine the permissible crack size if the permissible minimum residual strength is 20 ksi, and if it is 35 ksi, and if it is 51 ksi, use 10 crack sizes in each case.

4. Repeat Exercise 3 for the case that $J_c = 0.75$ kips/in; $F_{col} = 60\,\text{ksi}$. Use H as in Table 10.5. Assume $F = 1.2\ E\ 11$, $n = 5$, and neglect elastic term of J. Take crack sizes so that a/W in each case comes out at values for which H is known, so that you do not need to interpolate.

5. Calculate graphically the fracture stress for a fatigue crack of $a = 2$ inch, using the information of Table 10.7 and the R curve of Figure 10.18; How much stable fracture will occur?

6. Repeat Exercise 5 using the procedure of Table 10.9, assuming $\beta = 1$. Compare with results of Exercise 4.

7. Repeat Exercises 5 and 6 using F and n of Exercise 4. Assume that J_R curve is same as R-curve in Exercise 5, and assuming $W = 16$ inch.

8. On the basis of Exercise 7 estimate K_c from J_R ($E = 10\,000$ ksi) Then repeat Exercise 4 using LEFM procedures.

9. Compare results of Exercise 3–8 above and draw your own conclusions.

References

[1] M.J. Broekhoven and M.G. Ruytenbeek, *Fatigue crack extension in nozzle junctions.* 3rd SMIRT Conf., paper G4.7 (1975)

[2] G.G. Chell, A procedure for incorporating thermal and residual stresses into the concept of a failure analysis diagram. *ASTM STP 668* (1979)

[3] J.M. Bloom, Prediction of ductile tearing using a proposed strain hardening failure assessment diagram, *Int. J. Fract. 16* (1980) pp. R 163-167.

[4] D. Broek, *Fracture mechanics software,* FractuREsearch (1987).

CHAPTER 11

Fracture control

11.1. Scope

Chapters 2 through 5 provided the concepts of damage tolerance analysis, while Chapter 6 through 10 were concerned with input and analysis practice. This chapter considers the use of the analysis for fracture control. Crack growth and fracture analysis is not an end by itself. Its sole purpose is to provide a basis for fracture control.

Fracture control can be exercised in many different ways. Apart from a review of fracture control options, this chapter provides procedures for the use of analysis results in scheduling inspections, repair and replacements, proof tests and so on. In view of the nature of the problem, the discussions do not provide clear-cut recipes. Even more so than the analysis, fracture control measures require engineering judgement and pragmatism. The considerations upon which such judgement may be based are reviewed. Some damage tolerance requirements already specify the fracture control procedure, as discussed in Chapter 12. These can be understood in the light of the possible fracture control measures presented here.

After a summary of fracture control options, the selection of inspection intervals on the basis of analysis results is disussed. Fracture control by inspection is probably the most universal; safety depends upon the timely detection and repair of cracks. The sole purpose of the damage tolerance analysis is then to establish the inspection procedure and the inspection interval. In view of the cost of analysis, it is important that this be done rationally. The analysis efforts are futile if the inspection interval is still determined haphazardly. The chapter is concluded with a survey of itemized fracture control plans, but in the light of the cost of fracture and fracture control.

11.2. Fracture control options

Structural strength is affected by cracks. The residual strength as a function of crack size can be calculated, using fracture mechanics concepts. A condition has

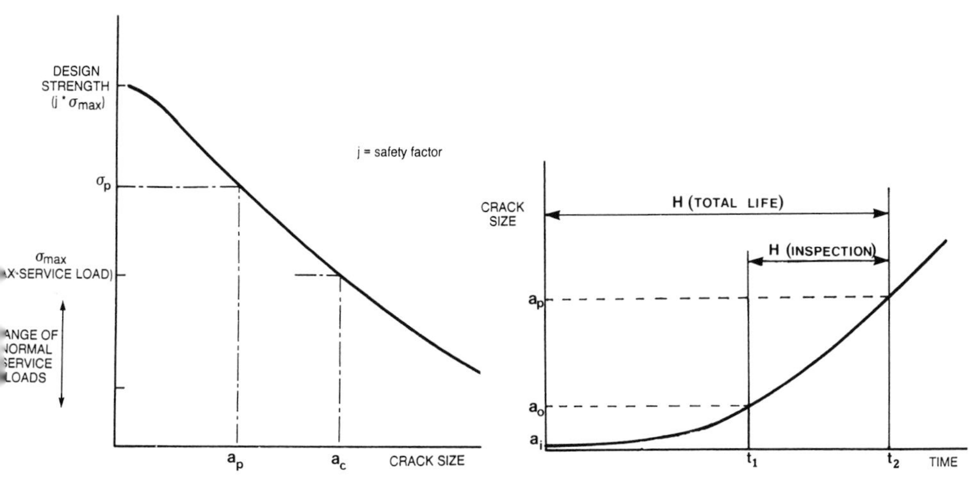

Figure 11.1. Time available for fracture control. (a) Residual strength diagram providing a_p; (b) Crack growth curve providing H.

to be set (generally by offical rules, regulations or requirements) as to the lowest acceptable strength in the case of cracks, i.e. the minimum permissible residual strength, σ_p. When the residual strength diagram has been calculated (Figure 11.1a) the maximum permissible crack size, a_p, follows from the minimum permissible residual strength.

The other information from analysis is the crack propagation curve. It shows how a crack develops by fatigue or stress corrosion as a function of time. The maximum permissible crack, a_p, following from the residual strength analysis of Figure 11.1a, can be plotted on the calculated crack growth curve as in Figure 11.1b.

There are several ways in which this information can be used to exercise fracture control. In all cases, the time period, H, to reach a_p (Figure 11.1b) is the essential information needed. As no crack is allowed to grow beyond a_p, repair or replacement is dictated by H. The following options are available for the implementation of fracture control.

(a) Periodic inspection; repair upon crack detection.
(b) Fail safe design; repair upon occurrence of partial failure.
(c) Durability design; replacement or retirement after time H.
(d) Periodic proof testing; repair after failure in proof test.
(e) Stripping; periodic removal of crack.

Damage tolerance requirements sometimes prescribe the fracture control procedure. For example military airplane requirements prescribe methods (a) and (c), commercial airplane requirements prescribe methods (a) and by their intent promote (b). Requirements will be discussed in Chapter 12.

The above fracture control options are discussed below.

Table 11.1. Inspection methods

Method	Principle	Comments
Visual	Naked eye, assisted by magnifying glass, lamps and mirrors.	Only at places easily accessible.
Penetrant	Coloured liquid is brushed on to penetrate into crack, then washed off. Quickly-drying suspension of chalk is applied (Developer). Penetrant in crack is extracted by developer to give coloured line.	Only at places easily accessible.
Magnetic particles	Liquid containing iron powder. Part placed in magnetic field and observed under ultraviolet light. Magnetic field lines indicate cracks.	Only for magnetic materials. Parts must be dismounted and inspected in special cabin.
X-ray	X-rays pass through structure and are caught on film. Cracks are delineated by black line on film.	Method with versatility and sensitivity. Small surface flaws difficult.
Ultrasonic	Probe (piezo-electric crystal) transmits high frequency wave into material. The wave is reflected by crack. Time between pulse and reflection indicates position of crack.	Universal method; variety of probes and input pulses.
Eddy current	Coil induces eddy current in the metal. In turn this induces a current in the coil. Under the presence of a crack the induction changes.	Cheap method (no expensive equipment) and easy to apply. Coils can be made small enough to fit into holes. Sensitive.
Acoustic emission	Measurement of the intensity of waves emitted in the material due to plastic deformation at crack tip.	Inspection while structure is under load. Continuous surveillance possible. Interpretation difficult.

(a) *Periodic inspection*
Safety is insured when cracks are eliminated before they impair the strength more than acceptable: they must be repaired before reaching a_p. Therefore, any cracks must be discovered before that point by means of periodic inspection. Various inspection methods are possible, viz. visual inspection (including loupes and magnifying glasses), penetrant, eddy current, ultrasonic, X-ray, and acoustic emission. These methods are summarized in Table 11.1; for details on these procedures the reader is referred to the extensive literature on the subject.

Whatever the inspection method, there is a certain crack of size, a_0, detection of which is questionable; during inspections before a_0, the crack is unlikely to be discovered. This implies that discovery and repair must occur in the time interval H between a_0 and a_p as shown in Figure 11.1. Should an inspection take place at time t_1, the crack will be missed, and should the next inspection be at t_2, after an interval H, the crack would already be too long (having reached a_p). Hence, the inspection interval, I, must be shorter than H. It is often taken as $I = H/2$, but a more rational procedure to determine inspection intervals can be employed, as discussed later in this chapter.

Damage tolerance analysis, to obtain residual strength and crack growth curves, is performed solely to determine H and from this the inspection interval. Safety is maintained by providing a sufficient number of inspections (at least 2) during H, to ensure crack discovery before a_p. Naturally, a crack once discovered must be repaired at the operator's earliest convenience. Since a_p is a permissible and not a critical crack, and since detection will commonly occur at sizes less than a_p, immediate repair may not be necessary, but any complacency will defy all analysis and inspection efforts.

Regardless of how long or short H (the inspection interval) or the inspection procedures used, safety is maintained, with some reservations as discussed later in this chapter. Whether inspections must be performed every day (e.g. $H = 2$ days) or every year ($H = 2$ years), there will always be two inspections between a_0 and a_p. Although a daily inspection might be cumbersome, the achieved safety is not really different in the cases of daily or yearly inspections. If a crack is missed in daily inspections a potential fracture will occur sooner, but if a crack is missed in yearly inspections fracture will occur nonetheless before the year is over. If short inspection intervals are undesirable, one has the option of selecting a more difficult but more refined inspection procedure with a smaller a_0. Then H, and hence the inspection intervals, will be longer. Fewer inspections are necessary, but the cost of individual inspections may be higher.

It does not matter either at which time the crack initiates, as illustrated in Figure 11.2. Inspections scheduled at e.g. $H/2$ interval will always give two opportunities for detection regardless of when crack growth begins, provided that inspections are scheduled at $H/2$ interval starting from hour zero (even if initially the chance of a crack is small). Similarly, if the interval is chosen as $H/3$,

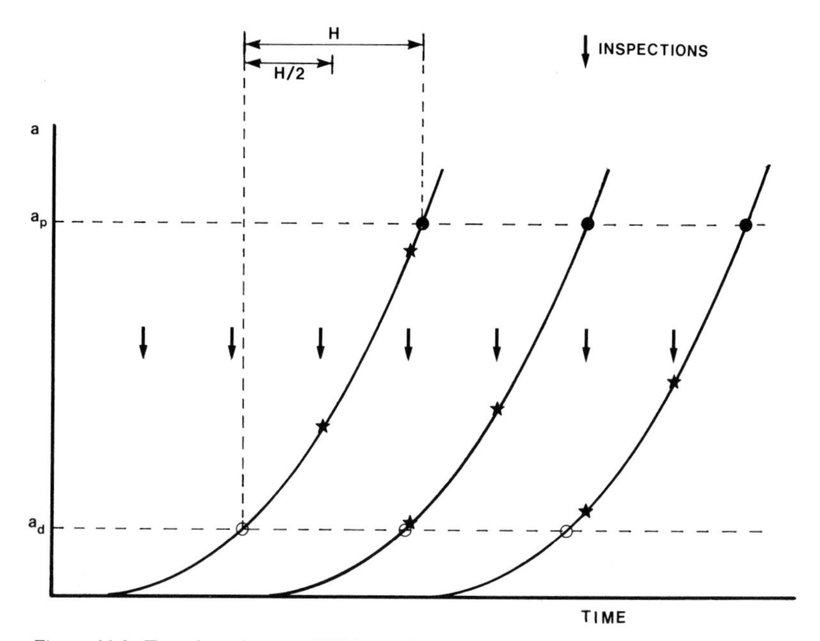

Figure 11.2. Two detection possibilities with interval $H/2$, no matter when crack starts.

there will always be three inspections between a_0 and a_p, whether the crack occurs early or late, or whether simple visual or other inspections are used (different a_0 and H).

(b) *Fail safety*

Providing fail safety by means of crack arresters or multiple load paths, is essentially a variation of method (a): cracks or failed members must be detected and repaired. The only difference is that the structure is designed for tolerance of large damage which is more readily apparent.

For example, crack arresters can be designed specifically for such a large permissible crack size (Chapter 9) that cracks will be obvious in superficial (but frequent) inspections. Alternatively, a pipeline or pressure vessel can be designed so that cracks will cause leaks rather than breaks. As a leak is presumably obvious, no special inspections are necessary other than frequent checks for leaks (leak-before-break is also discussed in Chapter 9). In either case, the structure must be designed to provide for a large and obvious permissible crack or leak, otherwise the effort is in vain and one of the other fracture control methods must be employed.

The same can be accomplished by providing multiple load paths (parallel members), as discussed in Chapter 9. When properly designed, the structure can still sustain σ_p when one member fails. Inspection for cracks would not be required, but regular checks for failed members would be. Of course member failure must be obvious, otherwise there is no advantage; a second member will soon develop cracks when it must carry additional load.

(c) *Durability*

If no inspections can or will be done, a small crack, a_i, could be assumed to exist initially in the new structure (Figure 11.1). The time H, for the crack to grow from a_i to a_p is then the available safe life. In that case the structure or component must be retired or replaced after e.g. $H/2$ hours. This is called the durability approach.

This may prove a wasteful method. Since H must be very long, heavier and more costly design may be necessary. If no inspections are performed, there is no other choice than replacement after $H/2$ hours. In the case of inspections, the structure can essentially be operated forever (Figure 11.2). If no cracks should occur, this would be evident; if they do occur they are detected and repaired and so eliminated. In the durability approach without any inspection, cracks of size a_p could be present after H, but there would be no way of knowing their presence.

A major problem with the durability approach is the necessary assumption for the size of the initial crack. This is an odd problem for structures and components that are essentially defect free. In that case the initial flaw may just represent an equivalent crack; at best it is the size of a flaw that can pass initial quality control. In welded structures the assumption of an initial flaw is more realistic. Welds often contain defects such as porosity or lack of fusion. In particular the latter is a sharp defect equivalent to a crack of about equal size.

(d) *Proof testing*

If the toughness is very low the maximum permissible flaw, a_p, may be smaller than the detectable crack, a_d (Figure 11.3). This can also be the case when the structure is so large that inspections for cracks are impractical (e.g. a 1000 miles pipeline), so that the 'detectable' crack size is effectively infinite. In such situations proof testing is another fracture control option.

Let at a certain time a component be subjected to a proof stress, σ_{proof}. Fracture would occur if a crack a_{proof} were present, as shown in Figure 11.3b. Conversely, if no fracture occurs, the maximum possible crack is a_{proof}, so that a safe operational period H (for growth from a_{proof} to a_p) is ensured (Figure 11.3a). If the proof test is repeated every H hours, a period of at least H hours of safe operation is available after each successful proof test. Should failure occur during the proof test (a_{proof} present) then a repair or replacement is made. The life can be extended forever if no proof test failures occur, provided proof tests are always conducted at the proper interval, H.

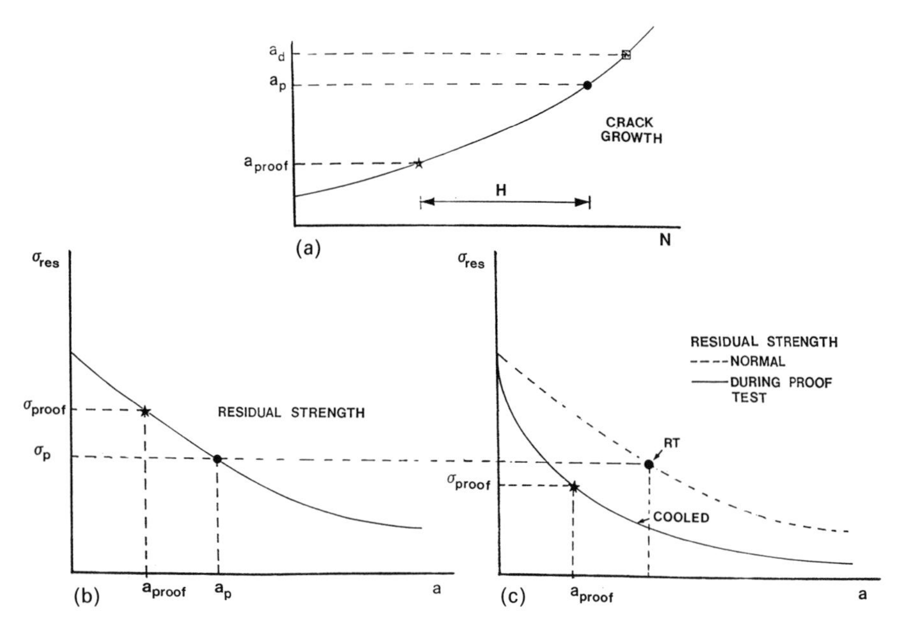

Figure 11.3. Proof testing. (a) Crack growth; (b) Proof test; no crack larger than a_{proof}; e.g. hydrostatic test of pipeline or pressure vessel; (c) Lower proof test load after cooling; example USAF F-111.

Pipelines and pressure vessels are eminently suitable for the proof test approach. A line or vessel normally filled with gas or dangerous chemicals can be proof tested (hydro-tested) with water. A failure during the proof-test would happen under controlled circumstances, causing a water leak only. In many cases hydro-tests are already performed anyway. Selecting the proof stress level and interval on the basis of fracture mechanics analysis, H, would give these a rational foundation.

Proof tests on structures other than pressurized containers are often hard to perform. However, if the component can be removed and easily loaded, the option is available and has been exercised (wing hinges of F-111 aircraft). Cooling the structure or component during the proof test causes a drop in toughness. This permits the use of lower proof stresses to 'detect' the same a_{proof}, as shown in Figure 11.3c. After the test and warm up, the original toughness and residual strength are automatically restored.

(e) *Stripping*

Stripping is another option for fracture control in components with permissible crack sizes so small that they defy detection. If at a certain time the crack has reached the permissible size a_p, machining (stripping) away a surface layer δ would reduce the crack size to $a_s = a_p - \delta$ (Figure 11.4b). As it would take H

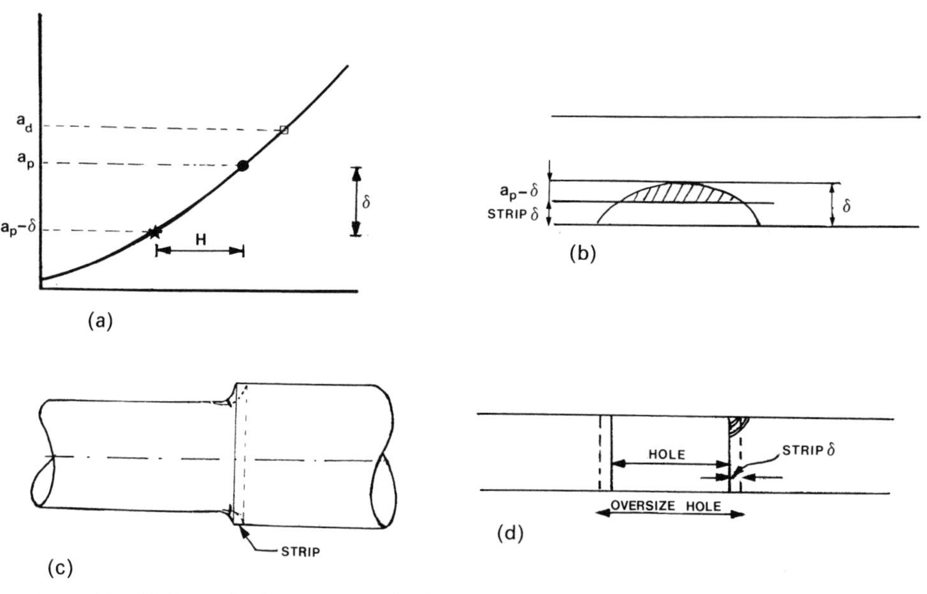

Figure 11.4. Stripping for fracture control. (a) Crack growth; (B) Stripping; (c) Practice at fillet; (d) Practice in oversizing holes.

hours for the crack to grow from a_s to a_p (Figure 11.4a), H hours of safe operation would be available after stripping. After these H hours the crack could be of size a_p again (note that cracks are too small for detection), hence the stripping of δ would have to be repeated every H hours.

It would seem that stripping cannot be repeated too many times, but it should be realized that the stripping layer δ is very small indeed, the cracks not being detectable. Furthermore, it should be known where the cracks occur, as for example in a fillet radius (Figure 11.4c, d). Machining the fillet radius by a small amount is repeatable many times without affecting the general stress level. Shot peening after the operation, to introduce residual compressive stresses, would be beneficial. Another case where stripping is feasible and often performed is a fastener hole. Oversizing the hole and using an oversize fastener can be repeated several times. A safe period H is available after each such operation.

11.3. The probability of missing the crack

All information needed to determine H and the inspection interval can be calculated, except for the detectable crack size. The latter has to be obained from inspection experience, which is not very well documented in the open literature. Yet, H, or the length of the inspection interval is very sensitive to the (choice of) detectable crack size, because the slope of the crack growth curve is small for

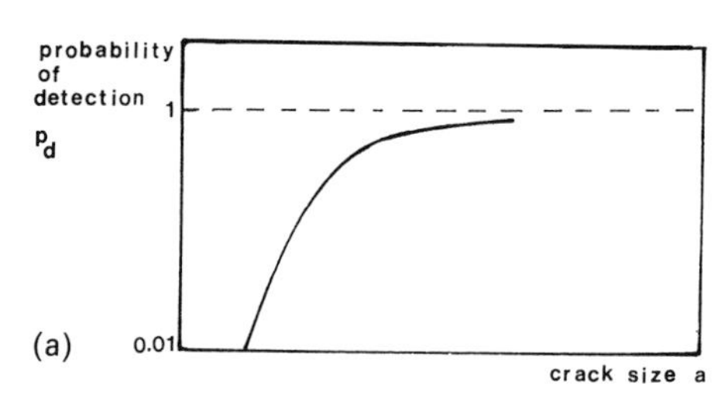

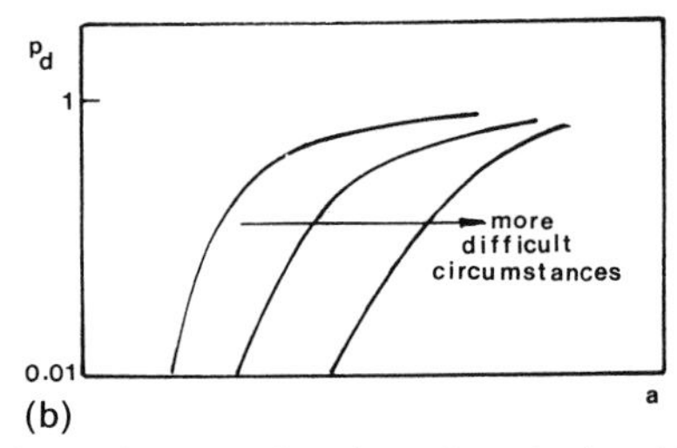

Figure 11.5. Probability of crack detection in one inspection as a function of crack size. (a) Probability of detection in one inspection; (b) Effect of circumstances on detectability.

small cracks. Consequently, an ASSUMPTION with regard to the detectable crack size, may have more weight in the determination of the inspection interval than the painstaking and costly damage tolerance analysis. This is unsatisfactory. A more rational procedure for establishing inspection intervals is desirable, as discussed in this and the following two sections.

Detection of cracks larger than the 'detectable size' is not a certainty. It is affected by many factors: (a) the skill of the inspector; (b) the specificity of the assignment: e.g. one specific location, as opposed to a whole wing or bridge; (c) the accessibility and viewing angles; (d) exposure; part of a crack may be hidden behind other structural elements; (e) possible corrosion products inside the crack, and so on.

Typically, the probability of crack detection depends upon crack size in the manner shown in Figure 11.5. There is a certain crack size, a_0, below which detection is physically impossible. For example, for visual inspection this would be determined by the resolution of the eye, for ultrasonic inspection by the wave length, and so on. In reality, a_0 is larger than these physical limits.

The probability of detection is never equal to 1 even for large cracks; any crack may be missed. It follows that the probability curve must have the general form as shown in Figure 11.5, which can be described by the equation:

$$p = 1 - e^{-\{(a-a_0)/(\lambda-a_0)\}^{\alpha}}, \tag{11.1}$$

where a_0 is the crack size for which detection is absolutely impossible (zero probability of detection), α and λ are parameters determining the shape of the curve. The equation gives the probability, p, that a crack of size a will be detected in one inspection by one inspector. This inspector may not detect the crack. The probability of non-detection is $1-p$. Note that Equation (11.1) is a curve-fitting equation; it does not make any statement about the statistics involved, despite the fact that a similar equation is used for certain statistical procedures. Equation (11.1) is merely a convenient format for describing the general shape of the curve. Any other equation providing a similar curve shape would be equally useful.

A crack is subject to inspection several times before it reaches the permissible size. At each inspection there is a chance that it will be missed. At successive inspections, the crack will be longer, and the probability of detection higher, but there is still a chance that it goes undetected.

Consider 100 cracks growing at equal rates, (same population), all in the same stage of growth (same size). Let the probability of detection at a certain inspection be $p = 0.2$. The probability that a crack will be missed is then $q = 1-p = 0.8$. That means that 80 cracks will go undetected. At the next inspection the cracks are longer; let the probabilty of detection then be $p = 0.6$, so that $q = 0.4$. Thus of the remaining 80 cracks $0.4 \times 80 = 32$ cracks will go undetected, etc. Apparently the cumulative probability that a crack will be missed in successive inspections is $Q = q_1 \times q_2 \times \ldots \times q_n$. In the above example $Q = 0.8 \times 0.4 = 0.32$: of the 100 cracks 32 cracks remain undetected after two inspections. The cumulative probability of detection is $P = 1-Q$. In the above example $P = 0.68$: of the 100 cracks 68 were detected after two inspections, but 32 were missed.

The cumulative probability of detection is then:

$$P = 1 - \prod_{i=1}^{n} (1 - p_i) \tag{11.2}$$

where p at each crack size follows from a curve such as in Figure 11.5 or from an equation such as Equation (11.1).

Figure 11.6 shows what happens if inspection intervals are determined as $I = H/2$, where H is the time required for crack growth from a_d to a_p. The

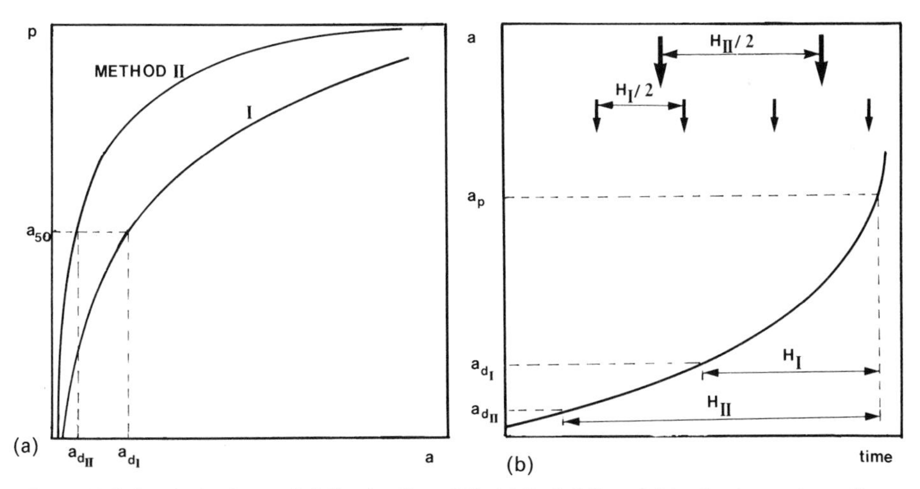

Figure 11.6. Inspection intervals following from $H/2$. (a) Probability of detection in one inspection of crack size a, with two methods; (b) Crack growth curve and inspection intervals for two inspection methods.

detectable crack size, a_d, is often taken more or less arbitrarily, but it might be selected as a crack with a certain probability of detection. For example, the detectable crack size could be defined as a_{50}, a crack with 50% probabilty of detection. Such a criterion certainly has appeal, because it seems consistent; yet it still leads to inconsistencies.

For the case of Figure 11.6 either method *I* or method *II* could be prescribed. The detectable crack sizes, a_{50}, lead to different inspection intervals, $H_I/2$ and $H_{II}/2$. Inspections would take place as indicated by arrows in Figure 11.6b. The cracks would be inspected for the first time when they still have a size smaller than a_{50}. At this first inspection, the probability of detection is not zero (unless $a < a_0$). There a distinct probability, p_1, that the crack is already detected during that first inspection; the probability that it is missed is $q_1 = 1 - p_1$. At the next inspection the crack is larger, and the probabilty of detection is p_2, etc. By the time the crack reaches a_p, it has been inspected n times. The cumulative probability of the crack having been detected at any one of these inspections follows from Equation (11.2).

The probability-of-detection curves are different for the two procedures in Figure 11.6, but either method would be satisfactory on the basis of the criterion $I = H/2$, where I is the inspection interval. The more involved method *II* (for example ultrasonic) inspection with a conveniently long interval, and the easier method *I* (for example visual) inspection with the more cumbersome shorter

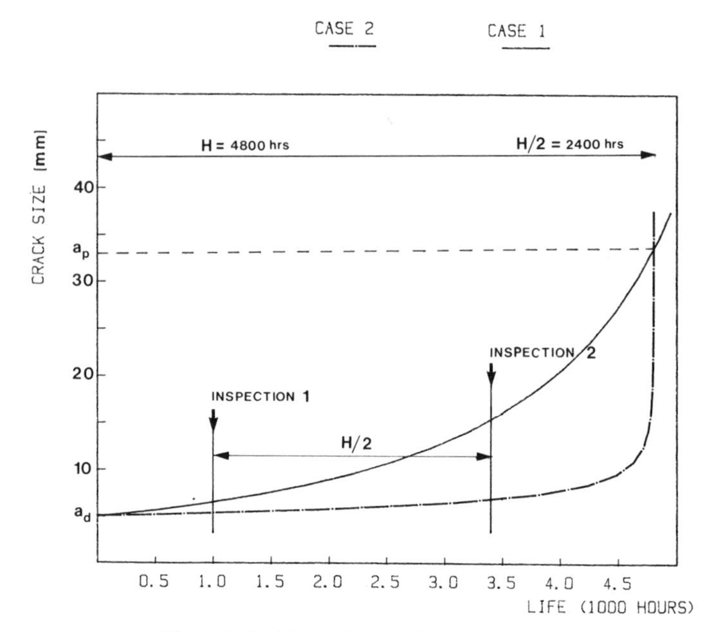

Figure 11.7. Alternative crack growth curves.

interval are equally acceptable under the criterion $I = H/2$. However, the safety level certainly would not be the same in both cases: the cumulative probability of detection certainly would be different, as can be readily appreciated from Figure 11.6 and Equation (11.2).

The effect is even more striking for the case of different crack growth curves, as illustrated in Figure 11.7. Such different growth curves are e.g. the result of different β (different geometries). Let the method of inspection envisaged be one with a 'detectable crack size a_{50}' of 5 mm, and let in both cases the maximum permissible crack size be $a_p = 33$ mm. Then H would be 4800 hours for both cracks. This would lead to an inspection interval of $I = H/2 = 2400$ hours in both cases. Consequently, both cracks would be inspected equally often. But crack 1 would have a much better chance of being detected than crack 2, the latter being much shorter at any of the inspections. Although the same criterion was applied for both cracks, not the same level of safety is achieved. A better rationale for determining inspection intervals [1] is discussed in the following sections.

11.4. The physics and statistics of crack detection

A sizeable body of inspection data was generated by Lewis et al. [2] for several large and small structural components inspected by many inspectors. An example of the results is shown in Figure 11.8. The figure shows the detection ratio, defined as the number of detections of a crack of a certain size divided by

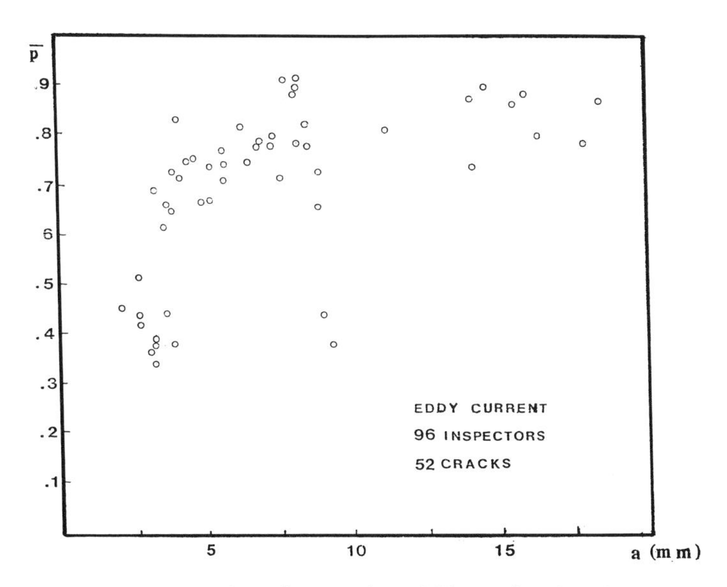

Figure 11.8. Detection ratio, $\bar{p}$; from Lewis et al [2], as a function of crack size a.

the number of inspectors. If the number of inspectors is large, as it was, the detection ratio will approach the true probability of detection.

The number of inspectors covered in Figrue 11.8 is 96. For convenience, consider a situation where 100 inspectors would look at the same crack for which the probability of detection is known to be p. The results of these 100 inspectors are known beforehand for any probability p, because the distribution is binomial, as if these 100 inspectors were picking marbles out of a vase with white (detect) and red (miss) marbles. They either pick a white or a red one. For a certain probability, p, the band in which e.g. 90% of the observations must fall can be constructed *a priori*.

For example, for $p = 0.8$ and $q = 0.2$, the standard deviation of this binomial distribution is:

$$s = \sqrt{(pqn)} = \sqrt{(0.8 \times 0.2 \times 100)} = 4, \qquad (11.3)$$

where p is the probability of detection, $q = 1-p$ the probability to miss, and n the number of inspectors. Since the 90% band is given by $1.64\ s = 1.64 \times 4 = 6.4$, the results are pre-determined. The expectation is for 80 inspectors to detect and for 20 to miss. There is a five percent chance that only 8–6.4 = 73 inspectors would detect, and a five percent chance that 80 + 6.4 = 86 inspectors would detect the crack. Conversely, if 73 inspectors out of the 100 detect a certain crack, the probability of detection could still be as high as 0.8; if 86 inspectors would detect, the probability of detection could still be as low as 0.8. The 90% band for the detection ratio of 100 inspectors MUST be between 0.73 and 0.86 if the actual probability of detection is $p = 0.8$. Hence, if only 20 out of 100 inspectors detect a crack, the probability of detection is definitely less than 0.8.

It follows that if 100 inspectors 'look' at two different cracks of the same size, and report 75 detects (25 misses) for one crack, and 30 detects (70 misses) for the other, then the probability of detection of these two cracks is different: although they may have the same size, they belong to different categories with different probability-of-detection curves. For any such curve, the 90% band for 100 inspectors is pre-established by Equation (11.3) as shown in Figure 11.9. If some data obtained by 100 inspectors fall outside this band, the cracks involved belonged to different categories (populations).

For example, the data in Figure 11.8 show that certain cracks of approximately 9 mm size were detected with a ratio of 0.4, while others of the same size were detected with a ratio of 0.9. The above discussion shows that the spread for 96 inspectors cannot be that large, so that these cracks of approximately 9 mm were not of the same population. Ergo, the data set in Figure 11.8 contains more than one population and more than one probability curve.

The 'probability-of-detection' curve for cracks of any type in any structure is

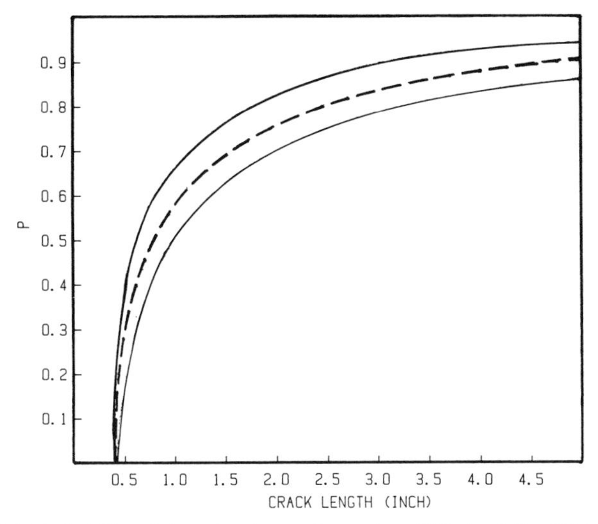

Figure 11.9. Probability of detection in one inspection; 90% band for 100 inspectors.

low and follows the lower bound of the data in Figure 11.8. Indeed, this low curve should be used if inspectors were assigned just to look for cracks in a large structure somewhere. That is not the way inspections are specified in practice.

The data of Figure 11.8 are repeated in Figure 11.10 to show categories of difficulty of inspection. Figure 11.10 also contains data [3] for a case where

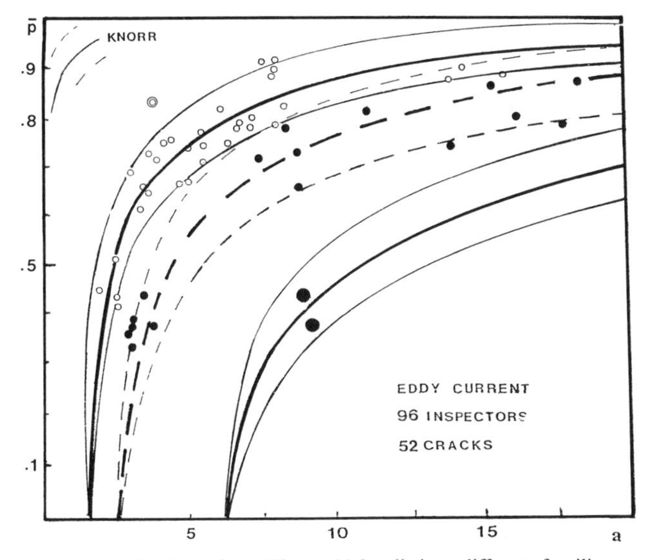

Figure 11.10. Data from Figure 11.8 split into different families.

inspectors were assigned one particular location in different aircraft: they knew where to inspect and what type of crack to look for (high specificity), as is normally the case; their results are much better. Had all inspectors represented in Figure 11.10 been given specific assignments their results would have been better as well. As the assignment was for any cracks in large areas the results are varied, namely the way they came out.

The 90 percent bands, known *a priori* as discussed on the basis of Equation (11.3) and Figure 11.9, can be drawn for the 96 inspectors involved for certain probability curves. Three such bands are shown in Figure 11.10. They indicate that the data cover at least three different populations, determined by specificity (some components were small, others very large) and accessibility (some cracks were easily accessible, others were not).

A category or population can be defined qualitatively only. An 'ideal' population is formed by the increasing crack sizes during successive stages of growth of a single crack. These cracks of different length are indeed of the same class of difficulty (access). Similarly, cracks of different sizes at the same location in a number of identical components would be in the same category. But two such crack types would be in different populations if one were located in a niche with a difficult viewing angle, the other in a smooth flat surface. The data reported in [3], shown in Figure 11.10, belong to more than one population.

A comprehensive damage tolerance assessment will identify types of cracks and their locations. Inspections will be prescribed for specific locations with known access. The probability-of-detection curve can be defined for the specific circumstances of the inspection envisaged.

Clearly, probability-of-detection curves obtained in the laboratory are not very relevant. In the first place the inspectors know that there are cracks, otherwise the experiment would not be conducted; this introduces bias. In the second place laboratory specimens are ideally accessible under comfortable circumstances. Third, the assignment is very specific: small specimens and usually one specific location. Results of such investigations can provide data for the most ideal circumstances only. Cracks in the tension bars of a suspension bridge will be in a different category. In each case a different probability of detection curve applies.

Available data obtained under realistic circumstances for structures [2] are useful, provided it is realized that they cover more than one population. For inspections of high specificity and/or easy access their upper bound applies, while for general inspections and/or poor access their lower bound applies. As long as inspection assignments are accounted for, the relevant probability curve can be determined from those data. Thus a re-evaluation of the data would be highly worthwhile. In the mean time, the data set is still useful, because of its extent. Not only eddy current inspection was covered, but also X-ray, penetrant and ultrasonic inspection.

11.5. Determining the inspection interval

Inspection must be prescribed with due account of accessibility and specificity. This can be done if the critical locations are properly identified. Specificity and accessibility determine the applicable probability-of-detection curve. Categories of accessibility and specificity must be established first [1], and for each of these the probability curve established on the basis of available data [1, 4].

The length of the inspection interval should be established such as to provide a consistent safety level (cumulative probability of detection), independent of the shape of the crack growth curve, the accessibility, and the specificity of the inspection. The aimed for cumulative probability of detection could be set for example at 95 or 98%, and be specified in damage tolerance requirements. Given the calculated crack growth curve and permissible crack size, and the probability-of-detection for the relevant specificity and accessibility, the cumulative probability of detection can be calculated for different lengths of the inspection interval by means of Equation (11.2). When the results are plotted, the interval for the desirable probabilty of detection can be obtained from the curve [1, 5]. The interval will be different for different inspection methods, different crack growth curves, accessibility and specificity, but the cumulative probability of detection is always the same (equal safety). The problems discussed in Section 11.3 are then eliminated automatically.

A computer [5] can perform the calculation for different interval lengths, provided the crack growth curve calculated in the damage tolerance analysis and the applicable parameters to Equation (11.1) are provided as input. For a certain inspection interval it finds the crack sizes, at which the inspection will take place, from the crack propagation curve. At each inspection (crack size) the probability of detection follows from the probability curve with the parameters appropriate for the inspection method and category. Equation (11.2) is then applied to obtain the cumulative probability. An example of a hand calculation for 2 inspection interval-lengths is shown in Table 11.2. For a complete analysis more inspection intervals-lengths must be considered. In a computer program the procedure can be further refined by accounting for the fact that the crack size at the first inspection may vary in accordance with the time of crack initiation (Figure 11.2).

Typical computer results [1, 5] are shown in Figures 11.11 and 11.12 for the two crack propagation curves in Figure 11.7. As discussed in Section 11.3 the criterion $I = H/2$ would assign the same inspection interval to both cracks. According to Figures 11.11 and 11.12 this would lead to different cumulative probabilities of detection. In order to ensure the same probability in both cases the intervals must be shorter in case 1. Figure 11.13 shows how the probability of detection of the case 1 crack would be affected by accessibility and specificity.

Although refinements can be made, the above provides a rational procedure

Table 11.2. Hand calculation of cumulative probability of detection for two inspection intervals.

Inspection interval = 500 hours							Inspection interval = 1000 hours						
1	2	3	4	5	6	7	8	9	10	11	12	13	14
Inspection	Hours	a from Fig. 11.7	p_d from Equation (or figure)	p_{miss} $(1 - p_d)$	p_{miss} cumulative $\Pi \times p_{miss}$	p_{detect} cumulative $(1 - 6) \times 100\%$	Inspection	Hours	a from Fig. 11.7	p_d from Equation	p_{miss} $(1 - p_d)$	p_{miss} cumulative $\Pi \times p_{miss}$	p_{detect} cumulative $(1 - 13) \times 100\%$
	0	5						0	5				
1	500	5.5	0	1	1		1	1000	6.5	0.27	0.73	0.7300	27
2	1000	6.5	0.27	0.73	0.7300	27	2	2000	9	0.54	0.46	0.3358	66
3	1500	7.5	0.42	0.58	0.4234	58	3	3000	13	0.69	0.31	0.1041	90
4	2000	9	0.54	0.46	0.1948	80	4	4000	19	0.80	0.20	0.0208	98
5	2500	11	0.63	0.37	0.0721	93			Total at a_p:				98
6	6000	13	0.69	0.31	0.0223	97							
7	3500	15.5	0.75	0.25	0.0056	99.4							
8	4000	19	0.80	0.20	0.0011	99.9							
9	4500	28	0.88	0.12	0.0001	99.99							
			Total at a_p:			99.99							

Notes: (1.) Crack growth curve (calculated) must be known (Case 1 of Figure 11.7 assumed). (2.) Columns 3 and 10 follow from crack growth curve. (3.) p_d in columns 4 and 11 follow from curve as in Figure 11.10 or from Equation (11.1). In this case Equation (11.1) assumed with $a_0 = 6$, $\lambda = 11$, $\alpha = 0.5$. (4.) Columns 6 and 13 from Equation (11.2): Column 5 or 12 times previous number in same column.

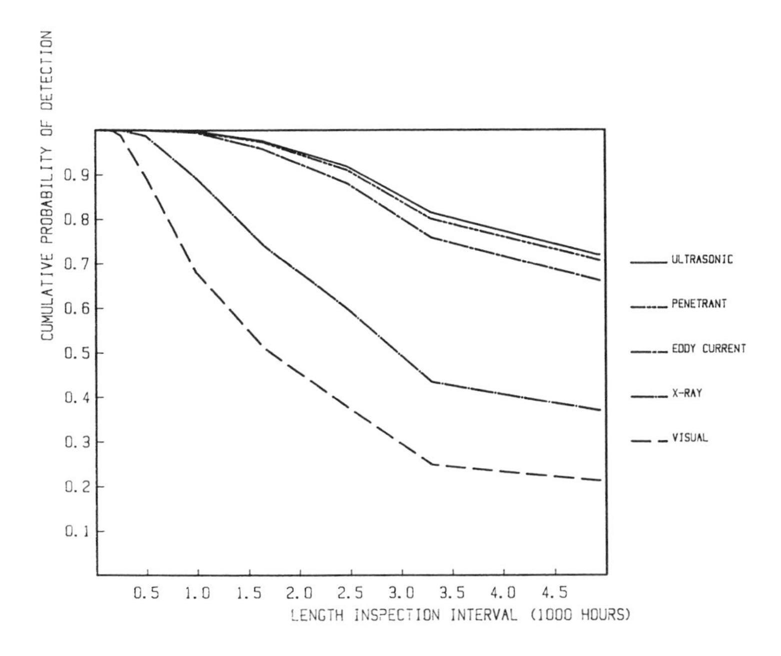

Figure 11.11. Cumulative probability of detection; Case 1 crack growth curve of Figure 11.7 (small area; good access).

to establish inspection intervals for which the probability of detection is independent of the inspection technique, the crack growth curve, and the assignment. The procedure is finding acceptance in the aircraft industry. As calculation of the crack growth curve and a_p requires expensive damage tolerance analysis, using the results for determination of inspection intervals as $I = H/2$ is unsatisfactory indeed. The above procedure is a much better approach.

11.6. Fracture control plans

The optimum fracture control plan depends upon the consequences of a fracture. If the number of fractures experienced is considerd to be at an acceptable level with a certain fracture control plan at acceptable costs, the plan is close to optimum. Before implementation of a fracture control program the objectives must be identified. If a structure can sustain assumed damage under an assumed loading condition, it is not necessarily safe despite all analysis. Before defining the permissible residual strength or permissible crack size, the desired level of safety should be established, even if only qualitatively. It will appear that every component and structure sets different fracture control requirements.

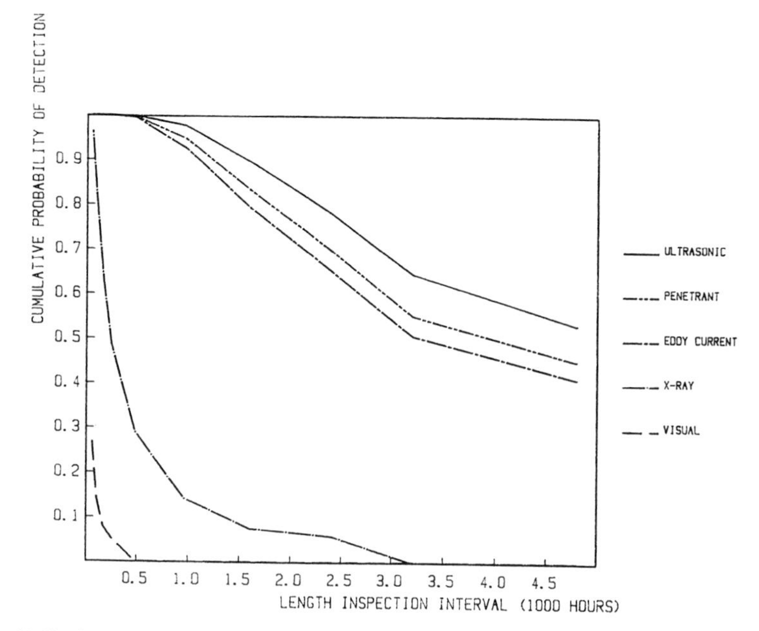

Figure 11.12. Cumulative probability of detection; Case 2 crack growth curve of Figure 11.7 (small area; good access).

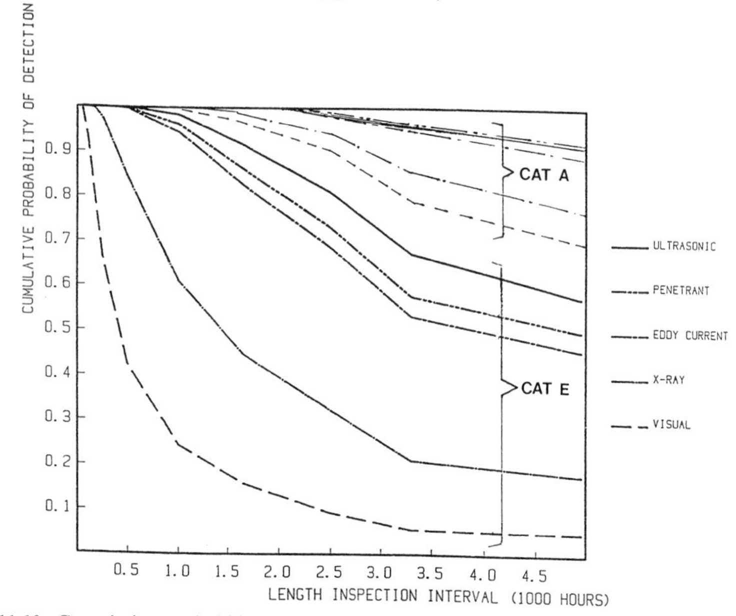

Figure 11.13. Cumulative probability of detection Case 1 crack growth curve of Figure 11.7. Cat. A; One location: Good accessibility Cat. E: Large area; access not easy.

The consequences of a fracture must be acceptable, the fracture control measures in accordance with the acceptable risk. The structure must have adequate damage tolerance to meet this risk. Designer or manufacturer prescribe the details of the fracture control plan, the operator implements this plan through maintenance, inspection, repair, replacement, proof testing, and possibly load monitoring. The plan must be suitable for a particular structure, component or part; it also must be suitable for the potential operators. Professional operators of pressure vessels, airplanes and the like, can implement more complex fracture control measures than the general public operating automobiles. When fractures can be adequately controlled by selecting materials of sufficient toughness, the fracture control plan is indeed simple. But here the concern is with those cases where fractures can have serious consequences and where material selection alone does not provide adequate safeguards against such fractures.

(a) *Detectable cracks*
Table 11.3 shows the ingredients of fracture control plans for structures in which cracks are detectable by inspection. If initial defects will not grow during service, Plan I is applicable. If defects, whether initial or developing later, may grow under service loading, a crack eventually will become critical, unless it is timely discovered and repaired. If the permissible crack is large and readily apparent Plan IIa is applicable, otherwise inspections should be scheduled in accordance with Plan IIb.

(b) *Cracks not detectable by inspection*
Cracks may not be detectable, either because their permissible size is so small that it defies inspection, the location not accessible, or the structure so large that inspections are not feasible. Plans IIIa and IIIb are applicable in such cases (Table 11.4). If stripping (Section 11.2) is possible, Plan IIIc may present an alternative. This could be for cracks at fastener holes (oversizing of holes and use of oversize fasteners), or for cracks at fillets if the component can be easily removed. Plan IV in Table 11.5 involves proof testing to show that no cracks larger than a_{proof} are present (Section 11.2). If larger cracks are present, a failure will occur during the proof test, but this failure must not be catastrophic in its consequences. The latter should be ensured by the use of water instead of gas for proof testing pressurized containers, testing small sections at a time, evacuation of surroundings, and if possible, cooling so that low pressures are sufficient.

(c) *After crack detection*
All fracture control calls for immediate repair or replacement when a crack is discovered. This is not always convenient. Large savings may be realized if

Table 11.3. Fracture control plans for anticipated cracks that are detectable by inspection

Plan I For initial defect not expected to grow by fatigue.
- Calculate permissible size of defect.
- If stress corrosion can occur, calculate which size of defect can be sustained indefinitely given the K_{Iscc} of the material.
- Inspect once using a technique that can reliably detect defects of above sizes.
- Eliminate all detected defects larger than above.

Plan II
For all defects (initial or initiating later) that will grow during service – these will reach critical size.

Alternative IIa
- Show by analysis (or tests) that the structure can sustain without failure such large defects that the damage will be obvious (e.g., readily apparent leak or failed component; fail safety).
- Repair when damage is discovered.

Alternative IIb
- If above cannot be shown, calculate permissible crack size.
- Establish crack size that can be detected reliably with inspection technique envisaged.
- Calculate time for crack growth.
- Implement periodic inspection based on crack growth calculation, using adequate factor or procedure of Section 11.5.
- Start inspection immediately as time of crack initiation is not known.
- Repair or replace when crack is detected.

Table 11.4. Fracture control plans for anticipated cracks not detectable by inspection because they are too small.

Plan III For parts where a is so small that it defies inspection.

Alternative IIIa
- Calculate life.
- Replace/retire after calculated life expires (using adequate factor).

Alternative IIIb
- Make best estimate of possible initial defects.
- Calculate permissible crack size a_p.
- Calculate crack growth life from initial defect size to a_p.
- Replace/retire after calculated life expires (using adequate factor).

Alternative IIIc If stripping is possible.
- Calculate permissible crack a_p.
- Establish feasible stripping depth, δ (see Section 11.2).
- Calculate crack growth life, H, from $(a_p - \delta)$ to a_p.
- Repeat stripping of δ at intervals ($H/2$.)

Table 11.5. Fracture control plan for anticipated cracks not detectable by inspection because inspection is not feasible, but proof testing is possible.

Plan IV For components or structures that can be proof tested and where failure during proof testing is not a catastrophy.

- Determine feasible proof test pressure or load.
- Calculate maximum crack size a_{proof} that could be present after proof test (see Section 11.2).
- Calculate maximum permissible crack a_p.
- Calculate crack growth time, H, from a_{proof} to a_p.
- Repeat proof test before H has expired (using adequate factor).

Table 11.6. Fracture control plans for cracks discovered in service

Plan V For detected cracks for which no analysis is done.
- Repair or replace unconditionally.

Plan VI
For detected cracks for which analysis is done (if immediate replacement is impractical).

Alternative VIa
Show that larger defect can be sustained.
- Check growth daily; drill stop hole if possible.
- Prepare for repair or replacement at earliest convenience.

Alternative VIb
Determine exact size and shape.
- Find materials data; if possible cut test specimens from structure.
- Obtain reliable load and stress information.
- Calculate a_p.
- Calculate time, H, for growth to a_p.
- Prepare for repair or replacement before H (with adequate factor) expires.
- Check growth daily; drill stop hole if possible.
- If crack grows faster than calculated, update prognosis and speed up replacement or repair actions.
- If possible reduce operational loads.
- Repair or replace as soon as possible.

Plan VII
For structures identical to those in which a crack was detected.
- Use parts of cracked or failed structure to obtain material properties.
- Implement one of Plans II a–b, III a–b, IV.

remedial action can be scheduled for the next major overhaul or shut down, or when at least operations can continue until a new part or component has been manufactured and received. Whether or not this is possible depends upon the fracture control plan in force. A well-conceived Plan IIa already contains information on crack growth and residual strength. Using this information as an initial safeguard, operation can be continued but the analysis should be updated and Plans VIa or VIb (Table 11.6) be put into action. As it is often difficult to measure the exact size and shape of the crack, the more stringent plan VIb may be indicated.

A crack may be discovered accidentally in a structure not subject to a fracture control plan. When no analysis is to be done, Plan V is the only possible course. Otherwise, Plans VIa or b can be used. Recurrence of the incident can be prevented using Plan VII.

11.7. Repairs

The sole objective of damage tolerance analysis is to establish fracture control measures so that cracks can be eliminated before they become dangerous by either repair or replacement of the component. The objective is not to determine whether a crack appearing in service can be sustained. If a crack appears it must be repaired; there is no excuse for a fracture resulting from known cracks regardless of what analysis predicts. Uncertainties in analysis (Chapter 12) are such that cracks once discovered must be eliminated or repaired. Naturally, replacement or repair is not always convenient immediately upon crack discovery, but damage tolerance analysis is not intended to show how long 'one can live with' cracks. Rather it provides the information to enable timely discovery for repair, or replacement. The above are repititious statements of the same issue, but repetition is justified as the objective of damage tolerance analysis is too often misinterpreted: it is to prevent fractures, not to evaluate how long discovered cracks can be sustained.

If crack discovery demands repair a new damage tolerance analysis problem arises. Not only must the repair be adequate to restore strength, it must be analyzed for damage tolerance again. Unfortunately, repairs are often treated too casually. A simple cover plate usually does not suffice (Figure 11.14). In view of Figure 10.2 such a repair may rather aggravate the situation and cause new cracks in due time. The stiffness of the cover plates may introduce a more severe situation than existed before; the solution may be 'worse than the disease'. (Consider the fact that the increased stiffness will attract loads to the bolt holes; attached parts must undergo the same displacement – strain – so that the stiffer part will take most of the load/stress.)

Repairs must be designed to cause gradual transfer of loads and stresses as discussed on the basis of Figure 10.2. A new damage tolerance analysis must be performed for the repair. It should not be assumed that the repair is a permanent solution. Fracture control measures must be reinstated for the repair.

The above may seem trivial at first sight. However, severe accidents have occurred as a result of inappropriate repairs; in a recent case more than 400 people lost their lives owing to inadequacy of a 'so-called repair'. Because it is so important, the above will be reiterated again. Damage tolerance analysis is

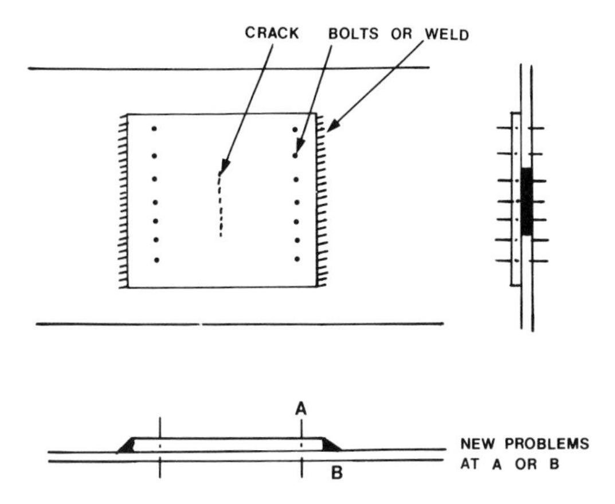

Figure 11.14. Unsatisfactory repair (for better solution see Figure 10.2b).

meant to prevent fractures and not to assess how long discovered cracks can be sustained. Upon discovery cracks must be repaired at the earliest possible convenience. Repairs are not necessarily a 'final solution'. It is too easy to perform an inadequate repair. Damage tolerance analysis of the repair must be performed, and fracture control measures must be taken for the repaired structure; all issues discussed in Chapters 11 and 12 must be accounted for. Efforts to reduce stresses by including a (stiff) load bypass or second elements often make the situation worse than it was, because the load will be attracted to the stiff element (see Figure 10.2).

11.8. Statistical Aspects

Many of the parameters and variables playing a role in fracture control vary beyond control of human beings. Usually, the statistical variabilty is dealt with in a deterministic way by assuming that estimates of the average values provide adequate answers to engineering problems. The answers are factored to account for variability. Sometimes variability is accounted for by taking 90 or 95 percent exceedance values.

All material properties, including ultimate tensile strength and yield strength, show variability (scatter). Fracture toughness and crack growth properties do too. A scatter in fracture toughness of 10 to 15% is not unusual; variability by a factor of about 2 of fatigue crack growth rates is normal. In most cases the structural loads are statistical variables. The pressure in a vessel may be well controlled, but random fluctuations may occur. The loads on bridges vary widely depending upon traffic; they can be estimated but cannot be known until after the fact. Finally, crack detection is governed by statistical variables. There is a non-zero probability that a crack will be missed. In spite of sophisticated fracture control, the probability of fracture will never be zero. Ideally the fracture control plan should be based upon the acceptable probability of failure. Because of the variability described, a safety factor is necessary if a deterministic analysis is performed.

In addition, there are errors due to shortcomings and limitations of the analysis, due to the limited accuracy of loads and stress history, and due to simplifying assumptions (Chapter 12). The magnitude of the necessary safety factor then depends upon the 'total uncertainty'. There is a natural tendency to cover every uncertainty when it appears by taking conservative numbers: highest estimates for loads and stresses, low estimates for toughness, upper bound growth rates, worst crack configurations, and so on. This amounts to a compounding of 'safety factors' of unknown magnitude which may lead to conservative answers, but the final conservatism is unknown. For the effects of all these assumptions see the discussion on accuracy in Chapter 12. It is preferable to use best estimates and average data and to apply a factor of known magnitude at the end of the calculation. Ideally, regulating societies or authories should establish rules and recommendations for safety factors, as they do for general design. Otherwise safety factors must be decided upon on a case-by-case basis.

As an example, consider a crack growth analysis to determine an inspection interval. The rate of crack propagation may depend upon ΔK to the 4th power. If there is a possible uncertainty of 10% in the loads, 10% in the stresses following from these loads, and 10% in β, the potential error in ΔK may approach 30%. The effect on da/dN will be a factor of $(1.3)^4 = 2.86$. If growth rates can vary by a factor of 2, the calculated life might be off by a factor of 5.72. One could then apply a factor of 5.72 on the calculated life, by scheduling 6 inspections: $I = H/6$.

Statistical fracture mechanics have been developed, in which all variables are accounted for by the rules of statistics. Such procedures are of great interest, provided the statistical distribution parameters could be known. If these have to be estimated the more complicated technique may not lead to more reliable answers. The simple way of applying statistical fracture mechanics is to determine the statistical distributions of all input variables. By employing a Monte Carlo technique a value for each input parameter can be selected and a deterministic analysis performed. Subsequently, new input values are selected, again with the Monte Carlo technique, and another deterministic analysis performed. This process is repeated many times, so that eventually a distribution of answers is obtained. The latter can then be analyzed statistically to determine the probability of failure, given that certain fracture control measures are implemented.

The problem with the latter procedure is to establish the statistical distribution of the input. This can be done only if assumptions are made with regard to interdependence. For example, the statistical distributions of such input as F_{ty}, K_{Ic}, and da/dN are dependent, because these properties are intrinsically related to the material. Both the toughness and F_{ty} for a certain alloy may show variations by 15%. However, if F_{ty} falls at the low end of its range, it is more than likely that K_{Ic} will fall at the high end of its range, and vice versa. By assuming that these properties are independently variable, the physics of the problem are violated no matter how elegant the subsequent statistics and mathematics. The problem could be analyzed if the physics of the dependence were known, but they are not. Determining this dependence on the basis of data would require many more test data than are usually available. Although statistical fracture mechanics are of interest, it would seem that much more development is needed for general engineering applications.

11.9. The cost of fracture and fracture control

The acceptable consequences of failure form the basis for the fracture control philosophy. These consequences must be weighed against the probability of failures other than by fracture. Establishing the acceptable consequences of fracture is an economic as well as an ethical problem; they must be considered in the light of other circumstances endangering life. From a technical point of view, the problem can be dealt with only if the consequences of a failure in terms of economic, ecological, and human loss can be quantified (expressed in cost) and compared to the cost of fracture control. Then the cost effectiveness of fracture control measures can be compared with their effect. If the probability of fracture is low and the consequential cost of fracture manageable, costly analysis and a costly fracture control plan cannot be justified. It is morally difficult to assign a cost to a human life, but practically it is not. An individual buying life insurance, in principle assigns a value to life, although courts of law may ignore this personal assessment and appropriate higher values. Be that as it may, a monetary value is assigned.

Let the total cost of a single fracture be S and probability of fracture P, then the expected cost of fracture is PS. Obviously, if P were equal to e.g. 10^{-5}, it would not be wise to use a fracture control measure costing $10S$. This would be insurance against a loss of $10^{-5}S$ at a premium of $10S$.

The potential costs of fracture include;

(a) Loss of human lives.
(b) Impact on environment, including natural habitat.
(c) Litigation expenses.
(d) Replacement of structure.
(e) Damage to buildings and surrounding structures.
(f) Down time (loss of production).
(g) Goodwill loss of sales and contracts.

The total potential cost of fracture is the sum, S, of the above. The anticipated cost is $P*S$.

The costs of fracture control include:

(a) Damage tolerance analysis (20 000–50 000 man hours for an airplane).
(b) Coupon tests and verification tests.
(c) Inspections (or stripping or proof tests).
(d) Repairs or periodic replacements.

Some of these are incurred by the manufacturer, some by the operator, but the manfuacturer's cost (including those of fracture) are obviously calculated in the price, so that eventually all costs are incurred by the operator.

The costs of fracture control as listed above, can be easily assessed, but determining those of fracture is more difficult. Some items can be estimated, others can be 'guessed' only. Besides, the anticipated cost of fracture depends upon the probability of fracture, which is the most difficult to estimate. Nevertheless, the princicple applies, whether the numbers are 'hard' or 'soft'.

To facilitate the discussion, consider a qualitative Fracture Control Index (FCI), a higher FCI signifying more extensive fracture control measures. The probability of failure decreases within increasing FCI (Figure 11.15a). The

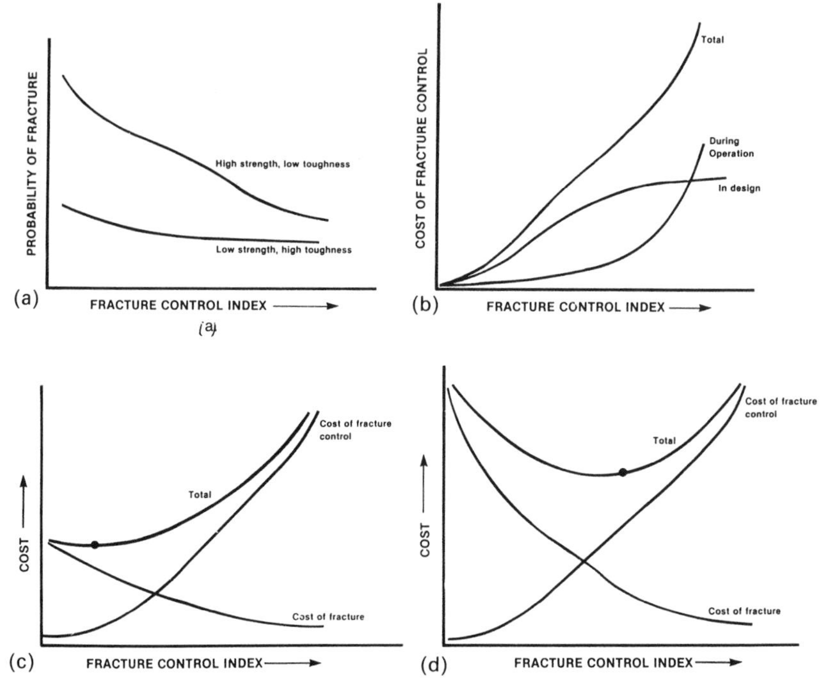

Figure 11.15. Cost of fracture and fracture control. (a) Probability of fracture; (b) Cost of fracture control; (c, d) Total cost.

decrease is faster for high strength materials, because low strength (high toughness) materials have a lower probability of failure in general. The cost of fracture control increases with FCI, both for design and operation (Figure 11.15b; the curves show trends only).

The probability of failure can be translated into an expenditure, and the costs plotted versus the FCI (Figure 11.15c). The minimum of the total-cost curve indicates the most economic fracture control. For high-cost structures and high-strength materials, the minimum shifts to the right so that more extensive fracture control is warranted (Figure 11.15d). If only parts of the structure are fracture critical, the cost of fracture control would pertain mostly to those parts, and fracture-control costs could be much lower. (If fracture of a given component would cause loss of structure a higher FCI is warranted only for that specific part). Should the cost lines be different than assumed, a minimum may not be achievable. The probability of fracture of various components of a system may be different; then the probability of fracture can be made the lowest for those components for which fracture control is the easiest. This permits acceptance of a somewhat higher probability for components for which fracture control is more difficult, while the total probability could remain the same.

The above is but a qualitative assessment of the problem. Nevertheless, it touches upon the relevant issues. When the logical process of decision-making leads to a fracture control plan involving analysis, information on loads must be available. The cost of obtaining load data must be expended. Any analysis without detailed information on loads, load history and stresses is wasteful. The decision maker should be aware of the obtainable accuracy in analysis (Chapter 12) and of the statistical aspects of fracture control as discussed. The decision maker, if aware of the above considerations and of the sources of inaccuracy, will not embark on finite element analysis to obtain geometry factors when loads and load history are not known accurately. Cheaper, approximate analysis will suffice in such cases; uncertainties should be covered by safety factors. Where fracture control calls for inspection, the decision maker should appreciate that even detectable cracks may be missed. Inspection intervals should be determined rationally as discussed in Section 11.5, otherwise all analysis, regardless of accuracy is futile. If the cost of fracture control (including analysis) far exceeds the cost of fracture, a simple fracture control plan should be selected. Analysis then may serve as a guideline; it may bound the problem. But in such cases rough assessments should suffice.

11.10. Exercises

1. Determine the inspection interval on the basis of the criterion $I = H/6$ for structures with crack growth curves as in Figure 11.7, assuming the 'detectable crack size' is 5 mm, and the permissible crack size 33 mm.

2. Using the results of Exercise 1, determine the cumulative probability of detection for the two cases, assuming that the middle probability of detection curve in Figure 11.10 is applicable.

3. Repeat Exercise 2 for the case that $\alpha = 0.5$, $a_0 = 5\,\text{mm}$, $\lambda = 8\,\text{mm}$.

4. Select three inspection intervals, 500, 1000, and 1500 hours. Determine the cumulative probability of detection for each of the cases of Exercise 2 using the upper curve of Figure 11.10; then estimate the required inspection interval for a cumulative probability of detection of 95%. Compare the results with those obtained in Exercise 2.

5. Assuming the crack propagation curves of Figure 11.7, determine a proof test interval. Assume that $K_c = 50\,\text{MPa}\sqrt{\text{m}}$, $\beta = 1$, and select proof test conditions that would eliminate cracks larger than 15 mm. What is the required proof stress?

6. A large component made of a material with $K_{Ic} = 30\,\text{ksi}\sqrt{\text{in}}$ and $F_{ty} = 200\,\text{ksi}$ is subjected to service stresses of 100 ksi. Crack growth from $a = 0.01$ inch to a_p has been calculated to cover two years of operation given that $\sigma_p = 150\,\text{ksi}$. The crack occurs at a fillet. Determine a stripping depth; assume $\beta = 1$.

7. Fracture of a certain structure is assessed at S_h. Replacement of the critical part would cost S. No other costs are anticipated. Analysis costs are S_a per hour. Load data are available. Extensive analysis including finite element evaluation would require h hours. A total T of these structures are anticipated to be in operation. The probability of a fracture is estimated to be P. Which course of action would you recommend?

References

[1] D. Broek, Fracture control by periodic inspection with fixed cumulative probabilty of crack detections, *Structural failure, product liability and technical insurance*, Rossmanith Ed. pp. 238–358, Interscience Enterprises, Ltd (1987).

[2] W.H. Lewis et al., *Reliability of non-destructive inspections*, SA-ALC/MME 76-6-38-1 (1978).

[3] E. Knorr, Reliability of the detection of flaws and of the determination of flaw size, *AGARDograph* **176**, pp. 396–412 (1974).

[4] U. Gorenson, Paper presented at ICAF meeting, Toulouse (1983).

[5] D. Broek, *IPOCRE*, Software FractuREsearch Inc. (1985).

CHAPTER 12

Damage tolerance substantiation

12.1. Scope

Previous chapters dealt with analysis procedures, (Chapters 2–5), the ingredients needed for the analysis (Chapters 7–10) and with the use of the results for fracture control (Chapter 11). This Chapter concentrates on the general scope of the analysis, its relationship to tests (verification and substantiation), the assumptions and sources of error, and the design options for improvement of damage tolerance. In short, it considers the analysis in the framework of damage tolerance provisions.

Damage tolerance analysis substantiation is governed by damage tolerance requirements if any are in effect. Indubitably, the damage tolerance requirements for commercial and military aircraft are the most widely enforced, and presently there is considerable experience with their use. Thus a discussion in some detail of the aircraft requirements is certainly worthwhile, even for readers not concerned with aircraft, as it will bring out the good and bad aspects of requirements in general. Although incidental rules may be in effect here and there, the only other requirements addressing damage tolerance directly are embedded in the ASME boiler and pressure vessel code. These will be reviewed as well. Requirements for the use of arrester strakes in ships are discussed in one of the examples in Chapter 14.

Compliance with requirements is an issue in this chapter, but the discussions concentrate on the damage tolerance substantiation in general. In particular the effects of the assumptions on the accuracy of the analysis – and thus on fracture control and safety – will be discussed. Damage tolerance requirements may enforce certain assumptions, and so be of greater effect on accuracy than the analysis itself (Section 12.8).

12.2. Objectives

The objectives of damage tolerance provisions have been discussed at various

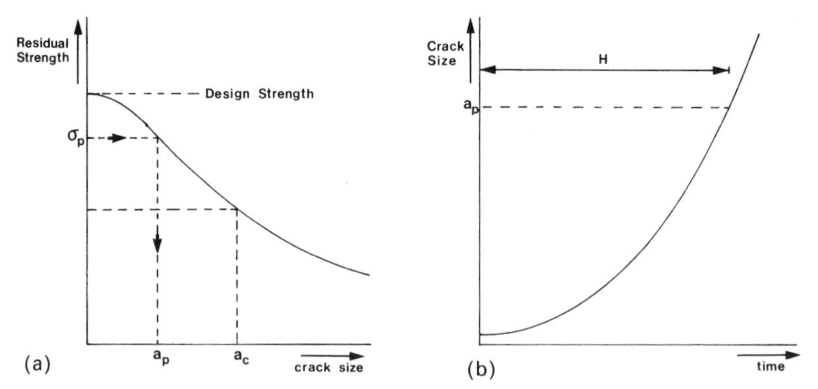

Figure 12.1. The engineering problem. (a) Residual strength curve; (b) Crack growth curve.

places in this book. They are briefly summarized here, in the context of damage tolerance requirements.

A new structure can sustain the design load, which is higher than the maximum expected service load, because of safety factors on loads or allowable stresses. The probability of occurrence of the design load is small, but finite for many structures, so that the probability of failure is not zero. When cracks are present, the strength is less than the design strength so that fracture may occur during extreme, or even normal operation. A fracture control plan is established to prevent such fractures. The residual strength diagram as produced by the fracture analysis is depicted in Figure 12.1a. Determination of this diagram is the first order of business in damage tolerance analysis.

Although certainly repetitive, the basis of fracture control is stated again. Fracture control is not always completely effective. The adoption of fracture control measures can be rationalized only if they achieve the goals of the fracture control philosophy. The latter forms the basis of rules and regulations, which may also prescribe the specific fracture control measures to be used. For example, inspection is essentially the only fracture control option for commercial airplanes. The fracture control philosophy will also lead to a decision with respect to the lowest strength that will ever be permitted. As the strength decreases during crack growth, the safety factor against fracture is reduced. A damage tolerance requirement sets a limit to the remaining safety factor, by specifying the minimum permissible residual strength, σ_p (Figure 12.1a). After the residual strength curve has been calculated, the specified minimum residual strength, will enable determination of permissible crack size, a_p. Larger cracks will cause the strength to be less than σ_p. By implication cracks larger than a_p are not permitted.

Next, analysis must provide the time of crack growth by fatigue or stress corrosion to a size a_p (Figure 12.1b). When this curve and the largest permissible crack size are known, the time, H, for a crack to develop to size a_p is obtained;

it is the time available for fracture control. As the crack may not grow beyond a_p, the structure or component must either be replaced or the crack must be discovered (by inspection or proof testing) and repaired before H expires. In either case, fracture control is based upon H (Chapter 11).

The damage tolerance analysis must provide:

(a) The residual strength as a function of crack size.

(b) The permissible crack size.

(c) The crack growth time H.

(d) The size of a pre-existing flaw that can be permitted in a new structure (in some cases).

(e) The interval for inspection, proof testing, or stripping, or the replacement time.

Many low-stress fractures occurred during the early years of the industrial era. Better materials and detail design reduced their number to acceptable proportions. With the introduction of all-welded structures, the number increased again and once more was controlled by better materials (higher transition temperature) and detail design.

The modern era brought about a new generation of fracture-prone structures, often operating in hostile environments and at extreme temperatures where material behavior is less predictable. Among these are offshore platforms, certain chemical plants, nuclear plants, aircraft, etc. These can be realized only if weight and costs are controllable. This drives the design to high quality materials and high operating stress. Refined stress analysis has improved confidence so that the high quality materials are operating closer to their limits than the materials of the past, which increases the risk of crack formation.

The oldest remedy, material improvement, may still have potential. But material improvements often are immediately exploited by increasing stresses to reduce weight and costs. Another conventional remedy, improved detail design, can be exploited by means of modern stress analysis, again to increase allowable stresses, rather than to reduce fracture risks. This creates a vicious circle and the danger of malperformance remains.

It is a compelling necessity to exploit new developments for further progress and higher performance. Thus, low-stress fractures must be prevented by fracture control. But fracture control technology (e.g. inspection techniques) is also subject to further development. Improvements making it more efficient will be exploited by designing closer to the limits. Fracture control is also part of the vicious circle, and hence must be based on a more or less time-independent philosophy. Damage tolerance requirements should reflect this philosophy.

12.3. Analysis and damage tolerance substantiation

Damage tolerance analysis provides the information needed to exercise fracture control; it is only one link in the chain of fracture prevention. The manner in which the analysis results are used to implement fracture control was discussed extensively in Chapter 11. The damage tolerance substantiation consists of the proof that the damage tolerance requirements can be met (these may be self-imposed). Although this proof is provided by analysis, uncertainties and engineering judgements often require tests to verify that the analysis is adequate (a full-scale test will often be part of the damage tolerance substantiation of airplanes).

Figure 12.2 shows the elements of the damage tolerance substantiation program. The small center box pertains to the fracture mechanics analysis. Material data handbooks may be useful, but some tests are often necessary to

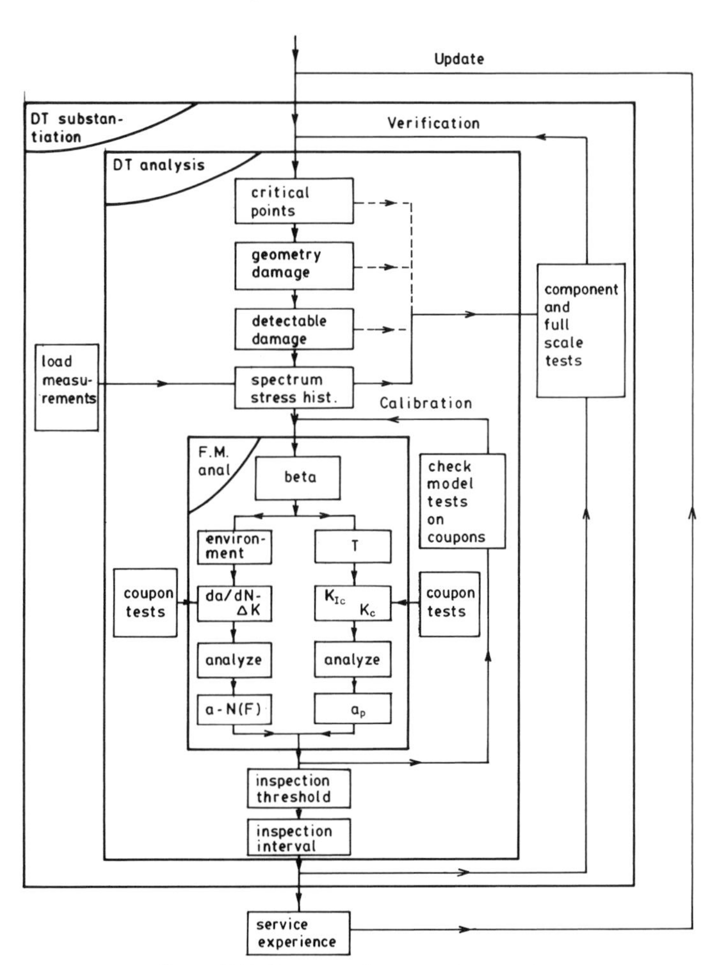

Figure 12.2. Damage tolerance substantiation.

substantiate and adjust the analysis. These may include (semi-random loading) crack growth tests, arrest tests, and tests for calibration of retardation models, effects of clipping and truncation. Such tests establish confidence that the analysis procedure is adequate for the substantiation of all other critical locations, loading cases and crack cases. Component tests (and/or full-scale tests) are used to substantiate the analysis, at least for some of the most critical locations. They provide guidelines for the engineering judgements involved. This phase is not a check of the models per-se, but of the analysis capability in general, including the basic stress analysis.

The ultimate check of damage tolerance is in the service experience. Feed backs on crack detection and (perhaps most important) measured load/stress histories are extremely useful for analysis updates and refinement of inspections, inspection intervals, and replacement schedules. In view of the many assumptions and judgements, expenditures can be saved (and safety improved) by monitoring service loads, and by updating analysis and fracture control plans during operation.

12.4. Options to improve damage tolerance

The time available for fracture control is H, which is governed by the residual strength (a_p) and the crack growth curve. In essence, fracture safety is not affected by the length of H (Chapter 11). If H is short, frequent inspections must be scheduled, or the component replaced soon. As long as all fracture control decisions are indeed based on H, safety will be maintained. But long inspection intervals or replacement times are desirable from an economic point of view.

The question then is which measures can be taken to improve the situation when H is too small to be economically acceptable. The following avenues are open (Figure 12.3).

(a) *Use of a material with better properties (Figure 12.3b).*
A higher toughness will provide a somewhat larger a_p, but generally speaking, is not of great influence on H; most of the life is in the early phase of crack growth. Increasing toughness (a_p) only affects the steeper part of the curve which has only a small effect in general. Should a_p be non-detectable then the effect of increased toughness is more significant. An average reduction in rates by a factor of 2, immediately increases H by a factor of 2. Protective coatings may also help, but it should be noted that surface layers only protect the free surface while the crack surface is still unprotected.

(b) *Selection of a better inspection procedure (Figure 12.3c)*
Improving the inspection technique, i.e. by selecting a more sophisticated inspection procedure, reduces detectable crack sizes. This usually has a very significant effect on H, because of the small slope of the initial part of the crack

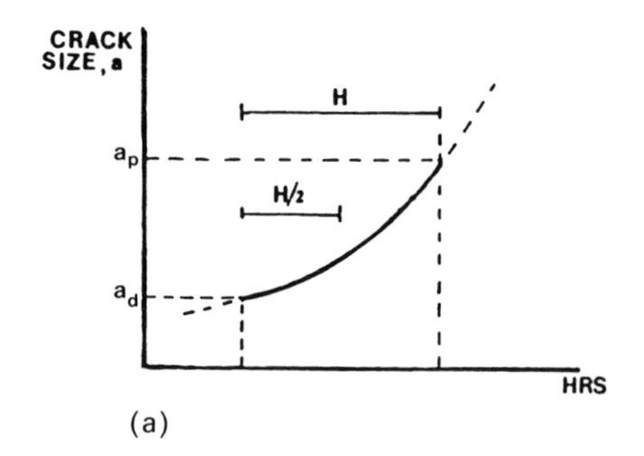

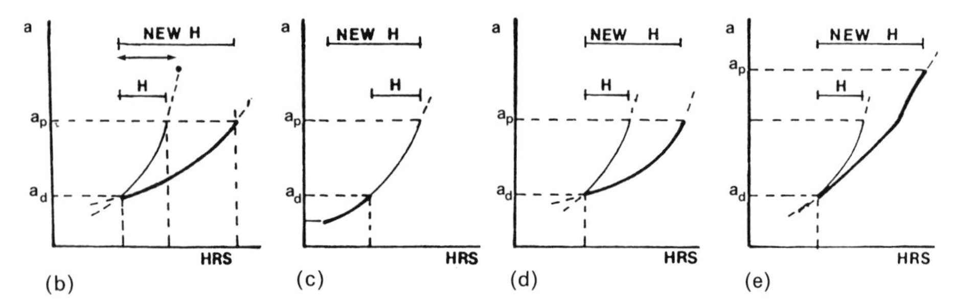

Figure 12.3. Options to increase inspection interval. (a) Crack growth time from a_d to a_p; (b) Better material; (c) Other inspection method; (d) Redesign or lower stress; (e) Redundancy or arresters.

growth curve. Note that inspection intervals (Chapter 11) on the basis of probability of detection are still governed by *H*. The penalty will be a more difficult inspection, but the inspection interval is longer; fewer inspections are needed.

If the structure is not inspected and the component replaced after *H* hours better quality control can be used to reduce a_0 and so increase the replacement life *H*. In the case of proof testing, higher proof loads or lower temperatures would reduce a_{proof} and hence increase the proof test interval (*H*).

(c) *Redesign or lower stress (Figure 12.3d)*

The crack growth curve is governed by the stress intensity. Reducing the stress by e.g. 15% will reduce *K* by 15%. As crack growth rates are roughly proportional to K^4 a 15% reduction in stress will increase *H* by a factor $1.15^4 = 1.75$. Such a stress reduction seldom requires a general 'beef up' of the structure; cracks occur where the local stresses are high, and the stress reductions are needed only locally: reduction of stress concentrations, larger fillet radii, less eccentricity will hardly add material, cost, or weight.

Redesign may also affect *K*; a reduction in β is just as effective as a reduction in σ. In the above example of the reduced stress concentration, the effect is actually in $\beta(k_t)$ instead of in σ (the nominal stress does not change). Last but not least, the redesign may be in the production procedure so that cracks occur in cross-grain direction instead of along exposed grain boundaries (Figure 7.8).

(d) *Providing redundance and arresters (Figure 12.3e)*
Building the structure out of more than one element provides multiple load paths (Chapter 9). In a well-designed multiple load path structure, only inspections for a failed member might be necessary, provided the fasteners can transfer the load of the failed member by shear. (Chapter 9). Similarly, stringers and arresters (Chapter 9) can improve the design and increase H.

All of the above options can be exercised during design. It is crucial, therefore that damage tolerance assessments commence in the early design phase when modifications are still possible. Once the design is finalized, the options for improvement are drastically reduced. Essentially for finalized designs and existing structures, only option *b* remains (Figure 12.3c), although doublers or arresters sometimes can be added later.

12.5. Aircraft damage tolerance requirements

Damage tolerance is the ability of the structure to sustain damage in the form of cracks, without catastrophic consequences, until such time that the damaged component can be repaired (Commercial Aircraft Requirement), or (Military Aircraft Requirements) until the economic service life is expired and the airplane or component retired. Damage tolerance can be achieved more easily by incorporating fail-safety features, such as redundancy, multiple load paths and crack arresters. Fail-safe structures can sustain larger damage, but if unattended this damage will eventually still cause a catastrophic failure. Hence, fail-safety features by themselves do not prevent fracture: the partial failure (e.g. the failed load path) still must be detected and repaired; even if the structure is fail-safe, inspection is essential to achieve safety. Without fail-safety features the structure can still be damage tolerant, provided cracks are detected and repaired before they impair the safety. Fail-safety features merely alleviate the inspection problem. The Military Requirements also use the damage tolerance analysis for a durability requirement (Chapter 11). Cracks growing from a presumed initial size must be sustainable throughout the economic service life.

The requirement that damage can be safely sustained should be interpreted to mean that the probability of failure must remain acceptably low. If cracks are left unattended, the probability of fracture will eventually become equal to 1 (fracture will occur). Thus, a criterion for lowest strength permitted must be based upon the acceptable probability of fracture.

(a) *Requirements for commercial airplanes* [1]
The U.S. Federal Aviation Requirements [FAR. 25b], enforced in a similar way in other countries, stipulate that the residual strength shall not fall below limit load P_L, so that $P_p = P_L$. The so called limit load is, generally speaking, the load anticipated to occur once in the aircraft life. Given P_p, the residual strength diagram provides the maximum permissible crack size, a_p. It should be noted again that a_p is not a critical crack, but is the maximum permissible size under the regulations. It would be critical only if the load P_p would occur; the probability of P_p coinciding with the occurrence of a_p is extremely small, so that an acceptably low probability of fracture is indeed achieved.

In essence the above is the complete requirement; indeed, little more is necessary. To satisfy the requirement, the manufacturer is obliged to design in such a manner that cracks can be detected before they reach a_p and to prescribe to the operator where and how often to inspect. Similarly, the operator is obliged to follow the manufacturer's inspection instructions. Fracture control by far rules must be exercised by inspection. The excuse that some cracks are non-inspectable is not maintainable. Every crack will become detectable if large enough. Thus, the requirement forces tolerance of damage large enough for detection (Figures 10.6 and 12.9), which promotes fail safe design with multiple load path and crack arrest features.

In a competitive field, it is in the manufacturer's best interest to ensure easy inspection; designs with high residual strength and large a_p (Figure 9.9) will ensure long inspection intervals. If too heavy a burden is put on the operator, the latter will prefer a competitive airplane requiring less and easier inspections. Hence, the requirements do not have to prescribe the inspection intervals and detectable crack sizes. If the design requires an unacceptably small inspection interval, options for improvement as discussed in the previous section can be exercised. These drive the design to fail-safety, and ease of fracture control for operators. The requirement, as simple as it is, accomplishes its objectives, a safe highly damage tolerant structure, at the lowest cost.

Although there is a problem in the definition of detectable cracks, an (arbitrary) specification of detectable size would not improve the requirement, because detectability depends upon the type of structure, its location and accessibility. The best way to determine inspection intervals is based on the cumulative probability of detecton discussed in Chapter 11. A useful improvement of the requirements would specify the desirable cumulative probability of inspection.

(b) *Military aircraft requirements* [2]
The U.S. Air Force requirements (adopted by some other forces as well) distinguish three types of structures, namely Slow Crack Growth (SCG), Multiple Load Path (MLP) and the Crack Arrest Fail Safe (CAFS) structure

(Chapter 9). When a crack in SCG structure would cause fracture instability the airplane would be lost. In both MLP and CAFS structure large damage is permissible (Chapter 9). Therefore, the requirements for MLP and CAFS structures are less stringent than for SCG structures. The term SCG is a misnomer; slow crack growth is always desirable. As opposed to MLP and CAFS, SCG is rather a Non-Fail-Safe structure (NFS). The commercial requirements do not need to make these distinctions because they automatically promote fail-safety features (see above). Since the primary Military Requirements are less stringent for MLP and CAFS than for NFS structures, they also should promote fail-safe structures.

The minimum permissible residual strength is somewhat higher than in the commercial requirements. This is certainly necessary for fighter airplanes and trainers which experience the 'limit' load more often than once in their life, but since the requirements cover all military airplanes, one would have expected a different residual strength requirement for transport airplanes.

Up to this point, the military requirements differ only slightly from the commercial requirements. As argued, no further rules are needed for commercial aircraft, because it is in the manufacturer's best interest to build an easily inspectable airplane. This does not quite hold for U.S.A.F. procurements: in that case, the regulator is also the operator. Accordingly, the inspection interval is specified as 1/4 life, for which there is a compelling reason. Although conceptual designs are competitive, the final design and production usually reside at a sole contractor, under which monopoly the operator must be protected. Yet, a fixed inspection interval, regardless of location and accessibility, may promote convenience more than safety. For NFS structure the inspection interval (I = 1/4 life), requires a factor of two, so that H must be equal to 1/2 life. For MLP and CAFS on the other hand, no factor is required so that the H need be only 1/4 life.

The requirements also consider durability. A small initial crack must be assumed present in the new structure; it must then be shown that this initial crack will not grow to a_p within the economic service life for CAFS and MLP structure or within twice the life for NFS structure. Formerly, these intial crack sizes were prescribed to be 0.02 inch for MLP and CAFS and 0.05 inch for NFS; they have since become negotiable. Note that these flaw sizes are not based upon quality control (the latter is no more difficult for NFS than for CAFS, but the initial flaws are different). These primary requirements are contrasted with the commercial requirements in Table 12.1 and Figure 12.4. Apart from these, there are several secondary requirements.

A fracture instability occurring in CAFS and MLP structure results in a large (arrested) crack. While this large damage is more easily detectable, it is not necessarily obvious. A large crack in a stringer-stiffened wing will cause fuel leakage and might be detected during a cursory visual inspection. However, a

Table 12.1. Comparison of Federal and Military Requirements for aircraft.

Damage tolerance	Federal	Military	
		MLP and CAFS	NFS
Minimum residual strength P_p	P_{LL}	$[1 + \alpha]P_{LL}$	$[1 + \alpha]P_{LL}$
Detectable crack size	a_d	a_d or fixed	a_d or fixed
Growth period	H	$H = \frac{1}{4}$ life[a]	$H = \frac{1}{2}$ life[a]
Inspection interval	$\leqslant = \frac{H}{2}$[b]	$H = \frac{1}{4}$ life[a]	$\frac{H}{2} = \frac{1}{4}$ life[a]
Post arrest instability damage	2 bays or[b] failed member	2 bays or failed member	
Post instability growth: if			
Ground evident		1 flight	
Walk around visual		50 flights	
Special visu		2 years	
Depot level		$\frac{1}{2}$ life[a]	
Safe Life			
Non-attended hypothetical crack			
P_p		$[1 + \alpha + \beta)P_{LL}$	$[1 + \alpha + \beta]P_{LL}$
Initial crack size		0.02 inch	0.05 inch
Growth period		1 life[a]	2 lives[a]

[a] Life is economical service life (or design goal)
[b] By implication

similar crack in a pressure bulkhead may still not be discovered until the next major overhaul, while further growth by fatigue of the large crack will be fast. To cover this there are secondary requirements for post-arrest, the severity of which depends upon the detectability of the post-arrest damage. It can be so-called ground evident, walk around visual, or detectable only at the next major overhaul. These secondary requirements are also shown in Table 12.1. In the commercial requirements this post-instability is not covered explicitly.

Two other secondary requirements are of interest. These concern so-called 'continuing damage' and 'dependent damage'. Continuing damage is a conservative 'invention' to facilitate crack growth calculations. Its effect is illustrated in Figure 12.5. When a crack grows into a hole it is effectively terminated; a certain time is required for reinitiation of a crack at the other side of the hole (Chapter 9). Crack growth analysis is powerless in calculating the reinitiation time. Continuing damage conveniently provides for immediate reinitiation through a mandatory assumption of a pre-existing crack of 0.005 inch at every hole in the structure. Crack growth thus calculated will follow curve C in

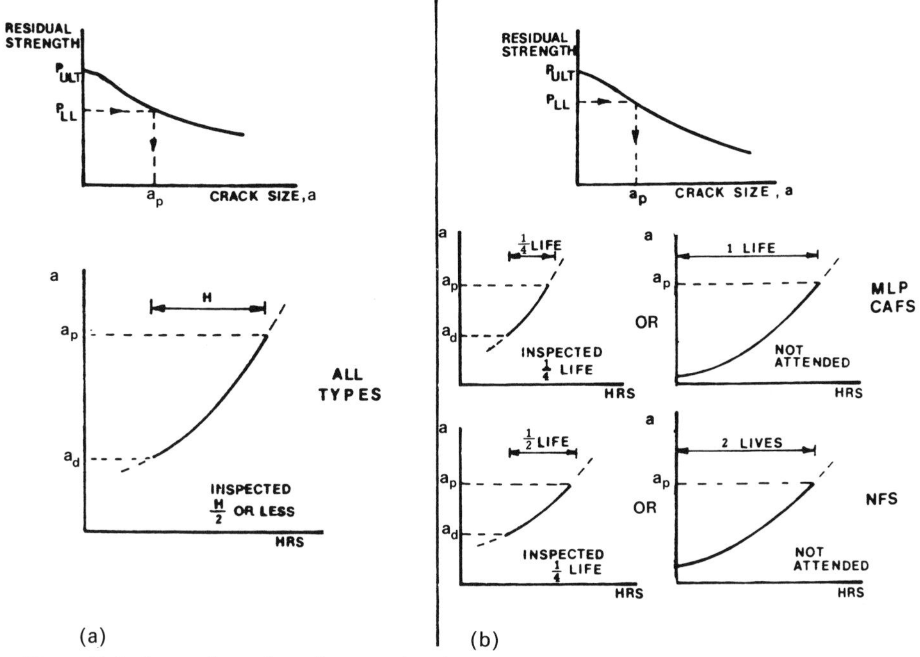

Figure 12.4. Comparison of requirements for aircraft damage tolerance. (a) Federal; (b) Military.

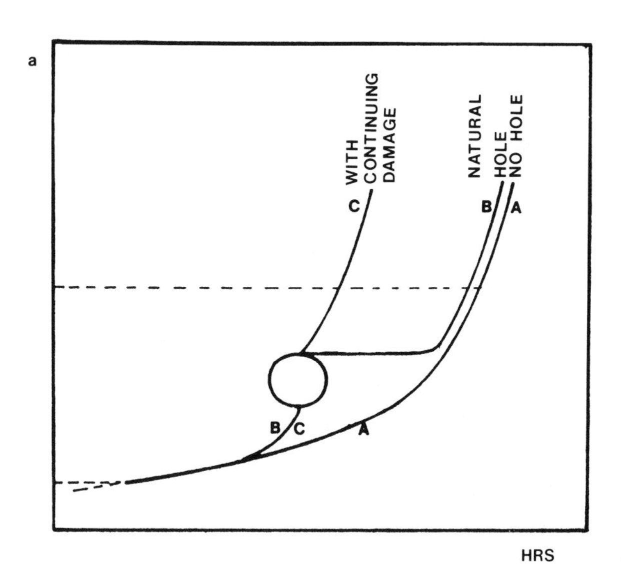

Figure 12.5. Crack running into hole.

Figure 12.5. Further artificiality is introduced because the 0.005 inch continuing damage cracks in some cases must be assumed to grow simultaneously with the main crack, in other cases they are assumed dormant until reached by the main crack. It is questionable whether natural cracks will comply with these PRESCRIPTIONS (assumptions).

When the crack approaches the hole it accelerates. After reinitiation it is longer by the diameter of the hole (curve B in Figure 12.5). It has been shown experimentally for through-the thickness cracks [3] that regardless of distance and diameter of the hole these two effects approximately cancel the gain due to reinitiation. A crack growth calculation ignoring the hole (curve A) would provide a 'good' answer within the accuracy of analysis. Admittedly, examples can be given where this simple solution is not so obvious. Besides, in a row of fasteners several holes may crack simultaneouly, so that continuing damage is indeed present. Some assumption may be necessary, but whether that assumption should be prescribed quantitatively in an official requirement is questionable.

Another secondary requirement concerns so-called dependent damage. This is based on the notion that if a crack initiates in one of multiple parts, joined together by one fastener, the holes for which are drilled in a 'stack', cracks in the other parts will follow soon. The cracked part loses stiffness and, therefore, sheds its load to the other parts which then will crack as well. However, by the nature of the load shedding through adjacent fasteners, cracking of the other parts should typically occur at the next fastener. Because of the assumption of pre-existing cracks – supposedly due to manufacturing – parts joined by one fastener must be assumed to have the same initial damage if the fastener hole in all parts is drilled in one operation in a stack.

Non-fail-safe structure (NFS) is penalized because an initial crack of 0.05 inch has to be assumed instead of 0.02 inch for fail safe structure, and because growth period H must cover two lives (durability) or two-inspection intervals (inspectable), as opposed to one life or one inspection interval for fail-safe structures (Figure 12.4). Accepting the assumption of initially cracked structure, an initial 0.05 inch crack is more conservative than an 0.02 inch crack. But, whether or not the longer crack has a much shorter life depends upon its location. For example, five types of cracks are compared in Figure 12.6. Crack type B, starting at 0.02 inch, would have a life of 34 900 flights; starting at 0.05 inch the life would be (34 000 − 19 000) = 15 000 flights. Thus the assumption of the larger initial crack implies an additional safety factor of more than two.

In the case of crack type D on the other hand, the life of an 0.02 inch crack would be 9000 flights, whereas an 0.05 inch crack would have a life of (9000 − 2000) = 7000 flights. The difference is much smaller here, so that for cracks of this type the additional safety factor is much less. For many types of cracks, it can be readily foreseen how much NSF structure is penalized, and as Figure 12.6 shows, the extra conservatism is small for some types of cracks. A numerical example is shown in the solution to Exercises 1–3. Besides, the

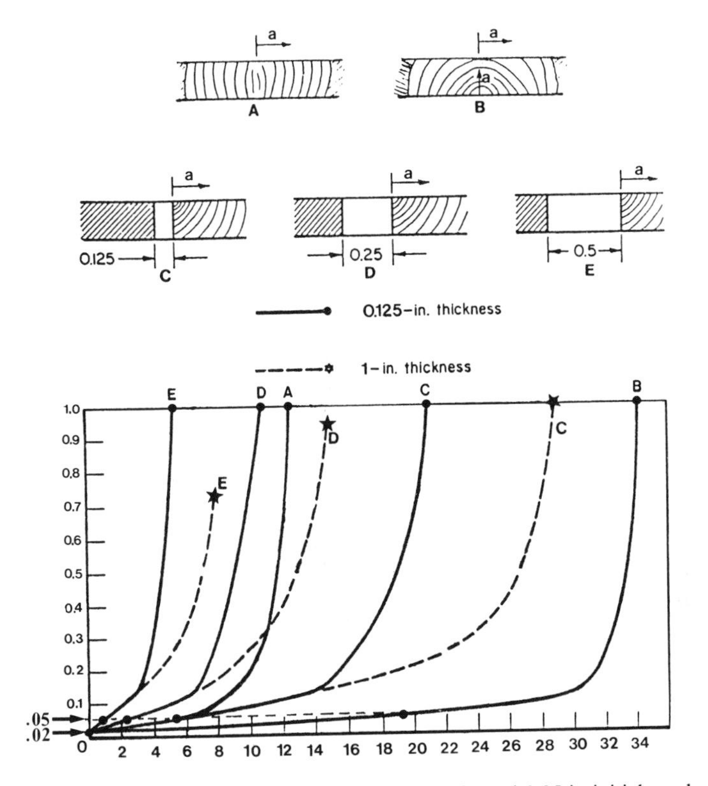

Figure 12.6. Differences in crack growth life for 0.02-in and 0.05-in initial crack.

question arises whether NSF structures are built so differently that they have larger initial damage than fail-safe structures.

Although the requirements for non-inspectable structure (durability) may be defendable, commercial requirements are based on the fact that all cracks are detectable sooner or later (Figures 10.6 and 12.9). If a hidden crack becomes long enough it will eventually run into inspectable area. For such large damage to be sustainable, only very damage tolerant designs can satisfy the requirements, so that the design must be made more damage tolerant.

The assumed initial cracks are small. This means that crack growth is largely influenced by adjoining structural elements. Due to the dependent damage assumptions, neighboring holes are cracked as well. Finally all cracks are prescribed to be circular. But natural cracks have no obligation to the requirements; natural crack growth will be different from the calculated growth. Thus, the calculation becomes hypothetical due to too many assumptions.

A lug such as in Figure 12.7 would be considered well designed, because it provides multiple load paths. However, because the three holes in the fork will be commonly drilled they must be assumed to have the same initial crack. Since the three prongs are equally stressed, the three cracks will grow at the same rate and will reach a_p at the same time. Hence, the design has (artificially) lost its fail-safety (all three prongs will break at the same time). A one-prong lug (worse design) could more easily pass the requirements (one crack only). The latter would be cheaper to make. Thus the requirements may accomplish the opposite of what was intended due to the assumptions.

If it is demonstrated on paper that a structure can sustain postulated damage, there is still no proof of damage tolerance. The question remains whether realistic damage can be safely sustained. The military requirements prescribe a set of rigid assumptions concerning initial, detectable, continuing and dependent cracks. As these are unrealistic, damage tolerance analysis provides 'numbers' only.

12.6. Other requirements

Other damage tolerance requirements exist for ships and for nuclear pressure vessels. Requirements for ships are issued by Shipping Bureaus, such as Lloyds of London, Veritas (Norway) and ABS (American Bureau of Shipping). Similar requirements exist for military ships. Essentially these are preventive requirements; no analysis is necessary. Ships of a certain size and over must be equipped with so called arrest strakes, which are located at the gunwale and at the bilge and sometimes mid decks. They are longitudinal strakes of a higher quality (higher toughness) material than the normal hull plating. The strakes are essentially of the same thickness as the plating. For a more detailed discussion of these see the example in Chapter 14, Section 3.

The damage tolerance requirements for nuclear pressure vessels are contained in the ASME boiler and pressure vessel code [5], Section XI and its Appendix

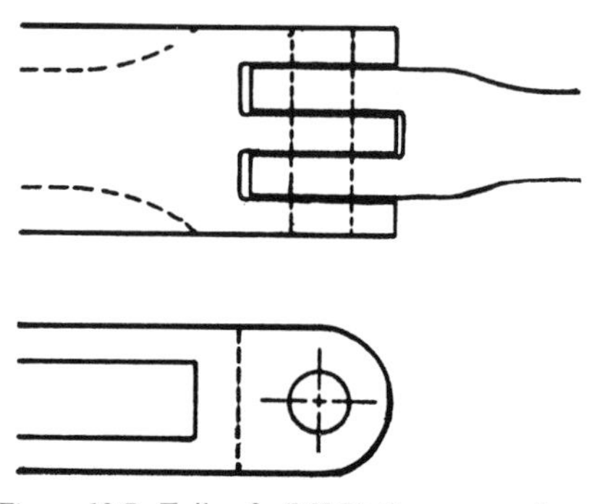

Figure 12.7. Fail-safe (MLP) three-prong lug.

A. Essentially the requirements provide acceptance limits for cracks detected in service. A great variety of possible crack configurations and locations are identified. The requirements then provide the crack sizes for each case that may be left unattended. Should a detected crack exceed the prescribed limits then one has two options

(a) Unconditional repair (repair weld).

(b) Perform analysis.

Since damage tolerance analysis is not mandatory; one has the option to conduct analysis if a detected crack exceeds a pre-set limit (the requirements give no justification for these pre-set limits).

If the option (b) above is selected, the following damage tolerance requirements apply:

$K <$ Arrest Toughness$/\sqrt{10}$ for upset conditions, and $K <$ Toughness$/\sqrt{2}$ for emergency and faulty conditions, where K is the stress intensity at these conditions. Strangely enough, these requirements are expressed in terms of the stress intensity and toughness. However, realizing that the fracture condition is $K =$ Toughness, in the upset condition, at the stress σ_{cu}, fracture would occur if $\beta_p \sigma_p \sqrt{\pi a_p} =$ Toughness. If the actual stress intensity must be smaller by $\sqrt{10}$, it follows that $\sigma_p/\sigma_{cu} \approx \sqrt{10} = 3.16$, assuming $\beta_p \approx \beta_{cu}$. This means that a safety factor of 3.16 must remain with regard to upset conditions. Thus the requirement can be stated in terms of the minimum permissible residual strength σ_p. It follows from the above that $\sigma_p = 3.16\,\sigma_{cu}$, in accordance with the previous discussions in this book. This case is displayed in Figure 12.8a using the same nomenclature as before.

At the same time the stress intensity must be less than the toughness divided by $\sqrt{2}$ for emergency conditions (stress σ_{ce}). Using the same arguments as above the minimum permissible residual strength, σ_p, must provide a safety factor of $\sqrt{2} = 1.41$ with regard to upset conditions. This is shown in Figure 12.8c.

A different toughness is used in the two cases, namely the arrest toughness (Chapter 9) and the regular toughness. Since the former is less than the latter, there are effectively two residual strength curves in play as shown in Figure 12.8a, c. Naturally, this does not change the principle of the analysis. In either case the permissible crack size a_p, follows from the residual strength diagram as shown. Obviously, it is impossible to satisfy both requirements exactly at the the other superfluous. If it can be foreseen which of the two generally is the severest, the requirement can be simplified.

The requirement presents an alternative. Instead of the above one may satisfy the following requirements:

$a_p \leqslant a_{cu}/10$ upset and operating conditions;

$a_p \leqslant a_{ce}/2$ emergency conditions.

Where a_{cu} is the critical crack (causing fracture) at upset conditions and a_{ce} the

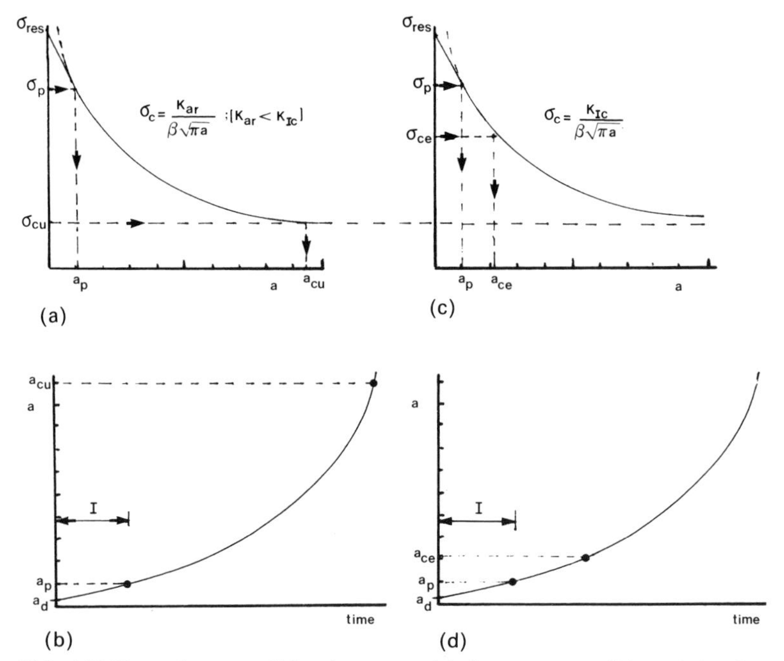

Figure 12.8. ASME requirements. Subscript *cu* = critical at upset conditions; Subscript *ce* = critical at emergency conditions; I = time till next inspection (shut down). (a–b) Upset conditions; (c–d) emergency conditions.

critical crack in emergency conditions. In the ASME code a_p is denoted as a_f. This provides with $a_{cu} = 10\,a_p$: $K = \beta_{cu}\sigma_{cu}\sqrt{\pi \times 10\,a_p} = \beta_p\sigma_p\sqrt{\pi a_p}$, so that $\beta_p\sigma_p/\beta_{cu}\sigma_{cu} = \sqrt{10} = 3.16$. This requirement will be identical to the one stated previously ($\sigma_p/\sigma_{cu} \approx 3.16$) only if $\beta_p = \beta_{cu}$. Since a_{cu} is a longer crack than a_p, in general $\beta_{cu} > \beta_p$ so that the requirement leads to a safety factor somewhat smaller than 3.16. The same arguments hold for a_p and a_{ce}. Both sets of requirements apparently attempt to set the same conditions and only one is necessary.

The requirement is specifically for a case where a crack is detected in service. One may then prove by analysis that this crack is not dangerous during further operation until the next shut-down. Essentially, this is contradictory to the general idea of damage tolerance presented in this book. The analysis is intended to provide the information to ensure timely crack detection (e.g. inspection interval) and repair. For this reason a crack then must be repaired when detected. Instead in the ASME requirements the damage tolerance analysis is used to decide whether a structure with a KNOWN crack can be left in service without repair. Besides, analysis is not used to determine the inspection interval. This is an important difference with the approach in aircraft where cracks must be repaired and where the analysis is used to ensure detection and repair, not to determine whether it is 'safe' to fly with a known crack. The above is not meant as a critique but to point out the difference in approach.

Once a crack is detected and analysis is preferred above immediate repair, crack growth must be analyzed as well. Fatigue crack growth must be calculated starting at a_d, which is the crack actually present and discovered, and continuing over the period until the next inspection (shut down), and using a load history as e.g. in Figure 6.2. Over this period the crack may not grow beyond a_p as determined by the criteria discussed above. This condition is shown in Figure 12.8b, d (Note again that a_p is denoted as a_f in the ASME code).

In most structures (bridges, ships, offshore, airplanes, cranes, etc.) inspections can be scheduled at almost any time and be dictated by the damage tolerance analysis. These inspection schedules may be bothersome, but in essence this bother is only a consideration of cost. A nuclear reactor can be inspected only during shut down periods which are dictated by many other considerations as well as cost. Thus the inspection interval necessarily is determined more by criteria other than damage tolerance.

The time to the next inspection thus being predetermined (Figure 12.8) the normal process is more or less reversed. Crack growth is calculated as it will occur during the service period until the next inspection. The growth may not exceed a_p as determined above. If the calculated growth does not take the crack beyond a_p, no repair is required and the crack may remain 'in service' until the next shut down. Should the calculated growth go beyond a_p then the crack may not be left in the structure and a repair be made immediately.

The requirements do not leave anything to chance; in Appendix A to the requirements [4], the analysis procedure is fully prescribed and even the toughness and rate data are prescribed. (Use of other analysis and data is subject to approval by authorities.) The prescribed analysis procedure is mostly in agreement with the discussions in this book. The stress distribution at the crack location must be obtained. If it has large gradients it may be approximated by procedures as shown in Figure 8.22: a uniform stress and (a number of) bending moment(s). The stress intensity is to be obtained by superposition as described here in Chapter 8. Surprisingly however, superposition of the stress intensity due to pressure inside the crack for flaws at the inner surface (Figure 8.7) is not required.

The mentioned appendix also prescribes that flaws be assumed elliptical (for effect see Chapter 9) and it provides the geometry factors for elliptical cracks in the same way as in Figure 8.3. However, it uses the obsolete procedure of modifying Q by accounting for a plastic zone corrrection to the crack size (Chapter 9). Plastic zone corrections were once thought [5] to be a way to account for e.g. the tangent to the residual strength diagram (Chapters 3 and

10). The procedure has long been abandoned as impractical and inadequate. If a plastic zone correction is applied to K, it should be done so generally, and therefore also in the determination of K_{Ic} in toughness tests. Although the procedure is obsolete it will provide somewhat conservative results, and as such is not objectionable in a requirement. However, the requirement ignores the fact that for a small a the elastic fracture mechanics approach is unconservative and that a tangent must be used (Chapters 3 and 10) or possibly a collapse approach.

As mentioned, the appendix to the requirements even prescribes the toughness and rate data to be used. The given data are very conservative, so that the actual safety factors are likely to be much higher than those discussed earlier in this section. Operators wanting to use more realisitic data must seek approval from the authorities.

12.7. Flaw assumptions

In most cases fracture control is to be planned for anticipated (i.e. postulated) cracks. It is not known in advance where cracks will occur, only where they might occur. Establishing the potential crack sites requires a diligent review of critical locations, stress concentrations, eccentricities and so on (Chapter 10). Also crack shapes or damage configurations must be postulated. As the shape and configuration are of great influence on crack growth it is this ASSUMPTION that may overshadow all efforts for preciseness.

If fracture control is to be based upon inspections, the problem may be somewhat simpler, especially if larger damage is covered. This is shown by the examples in Figure 12.9. Since only detectable cracks are of interest no assumptions would be necessary with regard to damage development. The configurations of detectable damage for visual inspection are rather obvious.

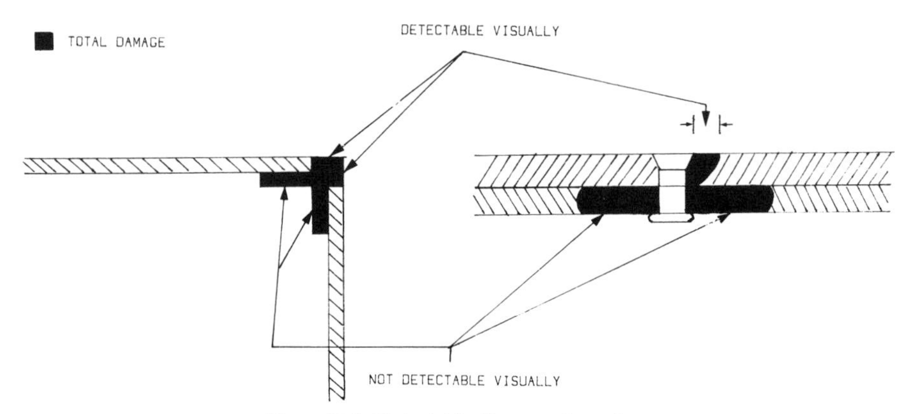

Figure 12.9. Detectable (from top) cracks.

However, if for these same cases inspections would be by, e.g., X-ray, some ASSUMPTIONS on flaw shape development would be necessary, since X-raying can potentially reveal cracks in hidden layers and components. In such a case several scenarios of damage development would have to be postulated, as it cannot be foreseen which of these would lead to the shortest life (inspection interval).

If initial flaws are postulated, the problem is of a different nature. Initial defects in welded structures can be defined. Weld defects such as porosity, undercut, lack of fusion, lack of penetration can be identified. On the basis of the weld quality control criteria it can be concluded which size defects might pass quality control inspection, and be present in the new structure. Such defects can be treated as initial cracks (Chapter 14).

In other cases initial flaw assumptions are a more delicate problem. Indeed sometimes they may be based on quality control experience (or criteria) if they are of sufficient size to represent a crack, as for example in the case of castings. But it is questionable what would be the size of an initial flaw in a crankshaft built under stringent quality control.

In the case of the military airplane requirements this problem was addressed as follows [6]. A total of 2000 holes in a wing that had been subjected to a fatigue test (known loads) were broken open to reveal any cracks. Of these, 119 holes were found cracked; the crack sizes were established. Crack growth analysis was then performed for each crack starting at a very small size, using the appropriate β and stress history for each location. The calculated crack growth curves were shifted so that the final crack size matched the one in the test (Figure 12.10). It followed which size of crack should have been assumed present at the start of

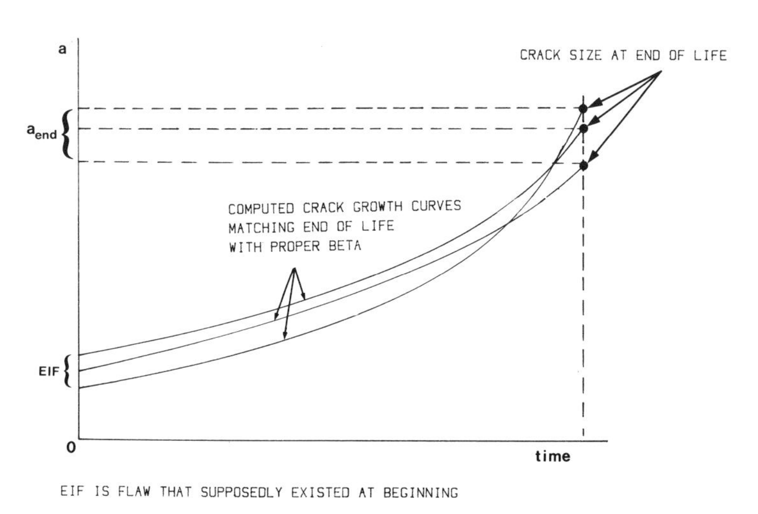

Figure 12.10. EIF (equivalent initial flaw) determination.

the test to produce the final crack size as observed. These initial cracks can be considered equivalent Initial Flaws (EIF). The statistical distribution of these 119 EIF's was obtained and extrapolated to extreme values, which led to the flaw sizes of 0.02 and 0.05 inch (Section 12.5). There were also 1881 uncracked holes. Clearly, the EIF for these holes was much smaller than for the other 119. Eliminating these from the statistical distribution has biased the results.

In subsequent efforts many specimens with holes were tested and analyzed [6] in the same manner as above, to determine the EIF. Attempts were made to correlate the EIF with hole quality (roundness, scratches, reaming, burrs, etc.). Most calculated EIF's were on the order of 0.001 to 0.002 inch; significant correlations with hole quality failed to emerge. Other attempts to correlate the EIF with e.g. inclusions and second phase particles [7] were inconclusive as well. In view of the above, the 0.02 and 0.05 inch initial flaws assumptions are arbitrary and have no bearing upon initial quality control.

From a practical point of view, there is no objection to assumed flaw sizes and shapes if they lead to safe structures. The only danger is that the numbers have tended to become a standard, which is extremely unfortunate and objectionable as they are ARBITRARY. Specification of initial cracks and detectable sizes in requirements and regulations are tied to the present state of technology.

12.8. Sources of error and safety factors

Contrary to common beliefs, the short-comings of fracture mechanics methods are NOT the important error drivers. Rather these are the input and the assumptions. In this section the various sources of error will be identified, and estimates made of their maximum possible effects; actual errors are often smaller.

Error sources can be classified in six main categories:

(a) Intrinsic shortcomings of fracture mechanics.
(b) Uncertainty and assumptions in data input.
(c) Uncertainty due to flaw assumptions.
(d) Interpretations of and assumptions in stress history.
(e) Inaccuracies in stress intensity.
(f) Intrinsic shortcomings of computer software.

In each of these categories, there are a number of factors contributing to inaccuracies in the analysis. They are listed in Table 12.2. Many of these already received attention previously. Therefore Table 12.2. and the following discussion make references to other chapters for details and illustrations; they are intended only to provide an overview of the error sources and their effects.

As the calculated life H is decisive for fracture control and safety, Table 12.2. provides rough estimates of how particular errors may affect the calculated life H. This estimate is given as a factor on life (not as a percentage error). The numbers are somewhat subjective but of the right order of magnitude (from analysis experience).

The error due to the use of LEFM for the residual strength analysis occurs mainly at small crack sizes (tangent approximation) and for small structures (collapse) as discussed in Chapter 3 and 10, but accounts for these make this error quite acceptable. For longer cracks and larger structures the error is very small (Figure 5.29). Note that this is the error due to the procedure alone, and does not include the one due to data scatter. The error will hardly be less if EPFM is used; collapse will still be a problem especially for small cracks and components. However, errors in a_p have only a small effect on the life H, as was shown already in Figure 5.29; most of the life is in the early stages of growth.

Retardation models are not ideal, but as shown in Figures 5.19 and 5.22 well calibrated models provide results in which the error in life is generally only around 10% (1.1) with few exceptions running as high as 30% (1.3). This is under conditions where β, da/dN, stress history and calibration factors are known accurately; i.e. it is the intrinsic error of the models. The fact that retarded crack growth analysis is generally less accurate is due to other factors which will be considered separately.

The errors due to data input are larger. Misinterpretation of scatter and force-fitting by unsuitable equations may introduce a factor of 2 to 3. But even careful assumptions may well cause a factor of 1.5–2. The situation is worse for mixed environments where the data for the separate environments must be used to obtain a weighted average. By itself this may be an acceptable engineering approach, but estimating the mixture and sequence of environment requires judgement and assumptions, which can easily introduce a factor of 2. These issues were discussed extensively in Chapter 7.

Flaw assumptions are other big drivers of errors. By assuming a 'conservative' circular instead of elliptical flaw, one may 'casually' introduce factors of 2 or 3. As many flaws are not elliptical, the assumption of ellipticity by itself causes errors (Chapter 9). Assumptions for initial flaw size may introduce equally large factors (Figure 12.6 and Exercises 1–5). Flaw development assumptions, continued cracking assumptions when cracks run into holes (Figure 12.5) and so on, are equally influential.

Every load history is an approximation (Chapter 6). Loads and number of occurrences must be approximated (number of levels). Decisions have to be made about clipping and truncation. Since the simple clipping of a few loads can have dramatic effects on H (Figure 6.23), the decision on clipping should not be made by load experts but by damage tolerance experts. Improper sequencing is another error driver. Randomizing the history while in reality it is semi-random (mild weather-storms), can cause great differences. All these are introduced by assumptions.

Errors in stress intensity are drivers of intermediate importance. Crack

Table 12.2. Error sources

Category	Cause of error	Comment	Possible factor on calculated H	
(a) Intrinsic shortcomings of fracture mechanics	1. LEFM approximation	Small error in a_p for small cracks or small parts	1.1–2	Figure 5.29
	2. Retardation model	Small error if well calibrated	1.1–1.3	Figure 5.19
(b) Data input	3. K_{Ic} or K_c, J_R	Small error in a_p (20%)	1.1–1.2	Figure 5.29
	4. da/dN-data	Normal scatter	1.1–1.5	Figure 7.16
	5. Assumption 90% band	Erroneous apparent scatter	1–2[a]	Figure 7.17 Figure 7.18
	6. Assumption variable environment;	E.g. weighted averages	1–2[a]	Table 7.3
	7. Equations for da/dN	Unnecessary force fits	1–2[a]	Section 7.7
(c) Assumptions	8. Direction (e.g. LT versus SL)	Wrong data applied	1–2[a]	Figure 7.8
	9. Size	Important for initial flaw only	1–3[a]	Figure 12.6
	10. Shape	Surface flaws	1–3[a]	Figure 9.3
	11. Development	E.g. multiple cracks, load transfer etc., continuing damage	1.1–2[a]	Figure 12.5
(d) Interpretation of stress history	12. Sequence	Semi-random vs random	1–2[a]	Figure 6.19
	13. Truncation	Improper truncation	1.1–1.3	Figure 6.25
	14. Clipping	Assumptions	1.5–3[a]	Figure 6.22 Figure 6.23

Table 12.2. (Continued)

Category	Cause of error	Comment	Possible factor on calculated H	
(e) Stress intensity	15. Actual load values	Measurement, analysis 15% $(1.15)^4$	1–1.75	Chapter 6
	16. Stresses	Assumptions, boundary conditions, load transfer 10% $(1.1)^4$	1–1.5[a]	
	17. β	10% $(1.1)^4$	1–1.5	Chapter 8
(f) Computer software	18. Integration scheme	Minor if small steps	1–1.1	Chapter 12
	19. Double precision	Usually minor; Large possible	1–(2)	Chapter 12
	20. State of stress for retardation	No large error if calibrated	1–1.5	Chapter 12
	Total possible:		2.7–1.37 000	

[a] Fully, or partly due to assumptions.

growth is roughly proportional to the 3rd or 4th power of K. Since $K = \beta\sigma\sqrt{\pi a}$, all errors in life are proportional to the errors in β and σ to approximately the 4th power. A 10% error in stress causes a factor of $(1.1)^4 = 1.46$ on life. Errors in stress stem from errors in loads and stress analysis.

The calculated loads contain an error, a 10% error being quite acceptable. Subsequently these loads are used for stress analysis. No matter how sophisticated the latter, the error is unlikely to be much less than 10%, especially in places of importance (stress concentrations, eccentricities, load transfer). In finite element analysis complex structures are often crudely modeled at such places, boundary conditions are assumed, fasteners represented by assumed springs, three-dimensional cases approximated in 2-dimensions, etc. Admittedly, this situation can be improved, but the cost may be prohibitive for analysis covering hundreds of potential crack locations, or even for one crack in a common hammer. FEM has a potential for good accuracy, but in general applications an accuracy of 10% is all that may be expected.

The stresses may be obtained within 10% for the given load, but also the load contains an error. Hence, the final stress may have an error larger than 10%, possibly 15%. This causes a factor between (0.85)–(1.15) or between 0.52 and 1.75 on life, the expected life being 1.

Also the error in β is included in the stress intensity. If this error can be reduced from e.g. five to three percent, the gain is only from a factor on life of $(1.05)^4 = 1.2$ to $(1.03)^4 = 1.13$, a small improvement indeed in comparison with other factors. If the inaccuracies in loads and stresses together account for a factor of 1.5 or more, while a simple assumption on flaw shape may cause a factor of 2, it is hardly a worthwhile effort to obtain β for this assumed flaw within three percent through a costly analysis if a simple procedure (Chapter 8) can yield a 5–10% accuracy. And if the assumption of a circular surface flaw (for conservatism or otherwise), introduces a factor of 2 to 3, it is not realistic to demand a high accuracy for β.

Finally there are errors due to specific computer modelling. These may be due to (1) the integration scheme, (2) rounding errors, and (3) equations for retardation and state of stress.

A crack growth calculation per se is but a simple numerical integration which does not give rise to large errors. Integration scheme errors can be introduced only in the case of constant amplitude where integration is performed in steps, because for variable amplitude loading integration is performed cycle-by-cycle anyway. If integration in constant amplitude is done in large steps, the accuracy is less. This was demonstrated by the hand-calculation and other examples in Chapter 5. But it was also shown there that integration is an intrinsically accurate process, as opposed to differentiation. Numerical integration procedures such as the Runge–Kutta and Simpson rules were devised in the pre-computer era when hand-calculations forced large steps. With the introduction of the computer, the step-size does approach zero (as it should), because the computer can perform many steps in a short time. Since integration is very forgiving in the first place (Chapter 5), these small step sizes are adequate and produce negligible errors, in particular when the result is seen in the context of the other errors discussed above.

The use of single instead of double precision can sometimes cause significant errors, especially in variable amplitude loading where da (one cycle) is very small, but also in constant amplitude loading with very small steps (in that case smaller steps give a LESS accurate answer than larger steps). This is an intrinsic problem of numerical computers. Personal computers provide eight significant figures in single precision and 16 in double precision, while mainframe computers generally work with 16 significant figures in single and 32 in double precision. The following examples are for personal computers; they apply equally to mainframes if one just changes the numbers.

An output given as e.g. 831 259 cycles is erroneous, if the accuracy is a factor of 2. The number should be 830 000, the error being much larger than 1 259/830 000 = 0.1%. Similarly an input for $da/dN = C_p K^{m_p}$ as 9.4327 E-10 $K^{3.7234}$ is unrealistic. Considering the accuracy of the data an input of 9.45 E-10 $K^{3.7}$ is more than adequate. Giving $m = 3.7234$ is implying that $3.7233 < m < 3.7235$. Clearly m is not known that accurately: at best $3.6 < m < 3.8$.

However, double precision has nothing to do with the accuracy of input and output; it defines the number of significant figures carried in the computations, not in input and output. If the computer must evaluate 1.79 E-10 $\times$ 2.73 E-11, the result is 4.8867 E-21. Note that these numbers provide 12, 13 and 25 decimals respectively, and the result is evaluated properly. The number of decimals is not important. In single precision the product 1879.43284 $\times$ 3.83 will be evaluated as 1879.4328 $\times$ 3.83 because the first number has nine significant figures of which only eight are carried. The difference is insignificant for engineering calculations. Therefore, throughout most of the crack growth analysis, single precision is MORE than adequate for multiplications, divisions, power, logs, etc.

However, double precision may become important in addition of large and small numbers and in subtractions of large numbers. This situation occurs when the small crack growth in one cycle is added to a large crack (a + da). For example, in a particular cycle da is evaluated as 7.45 E-8 = 0.000 000 074 500 000. This occurs PROPERLY in eight significant figures; leading zeros do not count. If the crack size is 12, the results will be a + da = 12.000 000 + 0.000 000 0745 = 12.000 000. As 12.000 000 has eight significant figures, da will be rounded off and not be counted. It will appear as if there is no growth. This might occur in a similar way in 10 000 000 successive cycles. The total growth would then have been 10 000 000 $\times$ 7.45 E-8 = 0.745, so that $a + da = 12.745$. However, in each cycle the growth was rounded off and after the 10 000 000 cycles a is still 12.

Double precision will mend this problem, but only to a degree: $a + da = 12.000\,000\,000\,000\,00 + 0.000\,000\,0745 = 12.000\,000\,07450000$, and indeed after a 10-million cycles the size will be 12.745. However if da appears to be 7.45 E-16, this crack growth will still be ignored. The problem occurs in mainframes also, but it is less important because 32 significant figures are carried in double precision. Hence, double precision is useful, but there is a limit to accuracy. Fortunately, the above problem seldom arises, but the use of double precision is recommendable at one place in the software, namely where $a + da$ is evaluated.

Other inaccuracies are introduced when the computer model cannot cope with the simultaneous growth of two axis of a surface flaw. Then the user is forced to make the circular flaw assumption, the inaccuracy of which was discussed above and at other places. (Chapter 9 shows how this problem can be circumvented, at least partially.) Serious errors may occur if the code can perform random loading only, thus ignoring semi-randomness as discussed in Chapter 6.

The way retardation is treated may affect the results considerably. All models use F_{ty}, but F_{ty} is but an arbitrarily defined number. One can argue whether the model requires the use of the 0.01% yield strength, the 0.02% yield strength, or the cyclic yield strength, where the latter can be defined qualitatively only. All models make use of a plastic zone equation which contains arbitrary numbers for plane stress and plane strain. Different computer codes use different numbers and equations. A proper code will check the state of stress in every cycle. The latter depends upon thickness, K_{max} and F_{ty}.

Even if using the most sophisticated retardation models, all computer codes contain assumptions, with regard to retardation. Consequently, retardation calibration parameters are not transferable between codes. Calibration must be performed using the same code as used for subsequent analysis. If the models are thus calibrated it does not matter what the code's assumptions are. The same assumptions used in calibration will be used in analysis, and inaccuracies due to assumptions are compensated for by the same assumptions in the analysis. But if these calibration factors are used with other computer codes the results will be different. As long as codes are used in a consistent manner, only small errors occur. Naturally, if the computer code provides more options, it is more versatile and can provide somewhat better results.

In the case that all errors discussed are active (which depends upon the complexity of the problem), the total factor on life (Table 12.2) would be between 2.7 and 137 000, with a logarithmic average of 600. This can hardly be called an error; it is a total misrepresentation. Naturally, errors generally will not operate in the same direction, and some will compensate others. However, it can be seen readily that the reliability of the result is affected much more by assumptions than by shortcomings of fracture mechanics or computer software. It is not worthwhile to improve the strong links in a chain; the weak link must be improved. The geometry factors, fracture mechanics concepts and calibrated retardation models are not the weak links. Improving these will hardly improve the result. The weak links are the assumptions involved in rate data, clipping, flaw size, flaw shape and so on.

There is only one way in which the magnitude of the inaccuracies due to assumptions can be assessed, namely, by repeating the analysis using different assumptions. It should be second nature to a damage tolerance analyst to perform calculations a number of times to evaluate the effects of assumptions with regard to stresses, loads, stress history, clipping levels and so on. Once the analysis is set up, such evaluations amount to no more than a number of similar computer runs.

'Garbage in, garbage out' is a worn phrase, but it needs more repetition. No answer to engineering problems is more suspect than the one generated by a computer. Although the computer is perfect, and good computer programs are nearly perfect, the result is still dependent upon input and assumptions. The effects of assumptions should be assessed. Only then can the problem be bounded and an impression of the 'true answer' obtained. A single analysis is never adequate.

The common practice of making 'conservative' assumptions everywhere is ASSUMING that all errors work in the same direction. Table 12.2. shows that the answer could be off by a mere factor of 137 000. Realism and sound judgement are necessary and it is better to use best estimates, than conservative estimates. Even with the best estimates the answer will be in error, but it will be closer to the truth. Analysis to assess the sensitivity to assumptions is required. In the end, one must admit ignorance; the result is dubious, but this holds for any other engineering analysis. The magnitude of the safety factor should depend upon the total 'uncertainty', as in conventional design. Regulating societies and/or authorities would have to establish rules and recommendations. Where such information is lacking, engineers will have to decide on a case-by-case basis.

12.9. Misconceptions

A number of misconceptions have crept into the engineering world about fracture mechanics and damage tolerance analysis. These have all been proven wrong in this book, but it seems worthwhile to briefly review a number of them.

The most persistent and most damaging misconception is that fracture mechanics is inaccurate, almost to the point of being useless. Some of the concepts are indeed less than ideal when considered from a fundamental point of view. However, fracture mechanics is an engineering tool for damage tolerance analysis. Almost no engineering method is ideal, but if it provides needed answers, the method is useful.

Fracture mechanics can provide useful anwers with reasonable accuracy when it is used judiciously in the manner described in this book. Anyone can use household tools such as hammer, saw and screw driver, but it requires expertise to produce a piece of furniture with these tools. Similarly, it requires expertise to obtain useful answers with fracture mechanics. This book shows that the acclaimed inaccuracy of fracture mechanics is mostly due to unknowns, assumptions, and inaccuracy in input: the tool (fracture mechanics) is not the cause. Hammer, saw and screw driver cannot be blamed for poor results if used on third grade knotted wood. Fracture mechanics is of little help if the user does not have (or refuses to obtain) basic information on loads, stresses and material data.

Another persistent misconception is that linear elastic fracture mechanics can be used only if there is plane strain. This is probably brought about by the facts that (1) only plane-strain toughness tests have been standardized, and (2) short popular summaries of LEFM emphasize plane strain. As shown in Chapters 3 and 10, the procedure for using LEFM is the same whether there is plane stress, plane strain or a transitional state of stress. If there is no plane strain the toughness is usually high, and the higher the toughness the sooner LEFM leads to errors. But, as shown, in such cases reliable approximations can be made. For very short cracks the fracture strength will tend to infinity, or at least will be close to the yield strength. In such cases the tangent from F_{col} or F_{ty} provides a good approximation as was clearly demonstrated. Also the possibility that failure occurs by net section yield or collapse must be considered. But for this case good engineering solutions are available as well.

Often forgotten is the fact that the same problem exists in plane strain. For small cracks the fracture strength will still tend to infinity no matter how low the toughness. In small structural components, failure will still be by collapse. Thus, the problems of small cracks and collapse must be faced in plane strain as well. As a matter of fact, they even must be faced in elastic-plastic fracture mechanics: the fracture stress still tends to infinity for $a \rightarrow$ zero, and collapse will still occur in small components. Whether there is plane stress or plane strain, whether LEFM or EPFM is used, approximations for small cracks and net section collapse are necessary. In all cases these conditions MUST be evaluated together with the fracture strength on the basis of K or J. The condition first satisfied (at the lowest stress) is the failure strength. If collapse or net section yield prevails, the fracture strength based on K or J (also in EPFM) is too high.

This leads to the third misconception, namely that the use of LEFM is always conservative. If for $a \rightarrow 0$ the calculated fracture strength is infinite; one can hardly maintain that this answer is conservative. If the calculated fracture strength is much higher than the stress for collapse, failure occurs by collapse and not at the calculated LEFM fracture strength (the latter's result is unconservative). Again, the same problem exist in EPFM.

In many practical cases, the cracks of interest are surface flaws or corner cracks, which indeed are in plane strain and should be treated as such (that thickness has no relevance here was discussed extensively in Chapters 3 and 7). Thus the plane strain toughness is needed in such cases. But if the toughness is high, the ASTM standard of $B > 2.5(K_{Ic}/F_{ty})^2$ may require such a large specimen that K_{Ic} cannot be measured. Ergo, fracture mechanics cannot be applied; another misconception.

In the first place, the number 2.5 used in the standard is rather arbitrary and not rigorous as it is often considered (Chapter 7). But apart from that, the value of K_{Ic} can be reasonably well estimated from a specimen that is too thin, as was discussed in Chapter 7 and in the solution to Exercise 6 of Chapter 3. Insight and ingenuity go a long way in obtaining engineering solutions. Certainly, in such cases the ASTM standard is not satisfied, but the standard is there for convenience, not to make engineering impossible. Of course, one MUST check whether collapse occurred in the test; but if it did the 'apparent' toughness following from the test is too low (conservative). Actually a check for collapse should always be made, whether mentioned in the standard or not.

In EPFM and J_R-tests the problem of constraint still exists. The LEFM test for K_{Ic} puts much emphasis on thickness and state of stress, but the EPFM test for J_R is unrealistic (Chapter 4). This raises the impression that constraint is of less importance in EPFM. But as shown above, whether there is plane stress or plane strain the 'procedures' still apply provided collapse is recognized as a failure criterion.

The present standard for the J_R-test analysis is based on a collapse condition. In that case the obtained J is too low; as a matter of fact it is only an 'apparent' J_R. Collapse and fracture are competing conditions, and the one satisfied first will prevail. The true J_R-test should be on larger specimens (no collapse) and be evaluated with $J = H\sigma^{n+1}a/F$. Indeed, a new standard for the EPFM test is badly needed (Chapters 4, 7).

In some cases, where approximations must be used in LEFM, it is better to use EPFM. This is not a misconception, but certainly a statement reflecting more academic than pragmatic wisdom. J_R-curves are difficult to measure so that the data are inaccurate. Thus, it is questionable whether the answers EPFM provides, are any more accurate than those obtained with LEFM using the appropriate approximations as discussed. If the fracture stress can be calculated within 10% using a simple procedure, one would not want to use a 'sophisticated' and complicated procedure, to produce results to the same or poorer accuracy. This may not always be the case, but from an engineering point of view, fundamental rigor does not count; only results do.

As far as crack growth analysis is concerned, a few persistent misconceptions exist as well. Should crack growth be based on K or J? In fatigue crack growth most of the life is at small ΔK (Chapter 5), i.e. from $K = 5$–$20\,\text{ksi}\sqrt{\text{in}}$, and most of that part at $K = 5$–$10\,\text{ksi}\sqrt{\text{in}}$. Even in a material with a yield strength as low as $F_{ty} = 50$ ksi, the plastic zone size at $\Delta K = 15\,\text{ksi}\sqrt{\text{in}}$ ($R = 0$), is only 0.004 inch; during most of the crack growth it is even smaller. This is a very small plastic zone indeed. The data clearly show that it is not necessary to question the use of K. The high growth rate regime affects only a small portion of the life; it does not change the life H to a significant degree. Considering the general accuracy, using J in this regime amounts to a third order correction and is of no practical interest.

Retardation models are inaccurate and hence crack growth analysis is useless; another misconception. Certainly, the first part of the statement is true. Retardation models are inaccurate, but they can be calibrated empirically, and then they work satisfactorily. This is based upon empiricism, but also the da/dN data are empirical, F_{ty} is empirical, and even E is empirical. A calibrated retardation model provides useful results. Any errors due to other unkowns in load and stress input, and clipping, overshadow those due to the models.

A final misconception is that test data can be obtained only from standard specimens. If that were true they could be applied only to standard specimens and not to structures. If the data can be applied to configurations other than test specimens, they can be obtained from any non-standard specimen, as long as the β for the specimen is known (Chapter 7). Some specimens have been standardized for which a very accurate β is available. This is a matter of convenience only, it is not a restriction.

12.10. Outlook

Speculations about the outlook for the future must take due account of the points discussed in the previous two sections. At present it is possible with judicious use of fracture mechanics, combined with small crack approximations and collapse analysis, to predict the failure strength of a structure in most cases within about 10%. This is as good as is desirable for engineering analysis. Buckling strength, or for that matter the strength of uncracked structures, cannot be predicted with greater accuracy. If a more rigorous fracture theory emerges in the future, the resulting engineering analysis will not be better than it is now. Scatter in material data will not be less. It will be equally difficult to predict the actual loads on a structure. Hence, the predicted fracture case will still be within about 10% only.

Most likely more refined crack growth and retardation models will be developed in the future. But these will not improve crack growth analysis much. The inherent large scatter in da/dN will remain; it is 'in the nature of the animal'

(Chapter 7). Loads and stresses in complex structures will still contain errors, and since cracks grow rates are proportional to some power of stress/load, small errors in the latter are magnifed by this power: $(1.1.)^4 = 1.46$. Thus, the accuracy of crack growth analysis as presently obtained in engineering application will not be much improved. It will remain equally difficult to make projections in the future of stress histories and changing environments. Predictions will remain predictions.

From a fundamental point of view it is desirable that research continue and more is learned, better hypotheses and procedures developed. From a practical point of view, the present situation cannot be much improved upon. The exponential growth of the number of researchers in the field has come at an inopportune time; their efforts could be better spent in other, new areas. After a period of slow and consistent growth, subsequent exponential growth in any bull market signifies that collapse is near. Many agencies have discovered this, and research funds are decreasing. In 1987 more papers were published on fracture mechanics than in the entire decade of 1960–1970. However, the results were less worthwhile. Research should certainly be continued, but should concentrate on problems of real engineering interest (dynamic fracture, composites, etc.).

Fracture mechanics has become an established tool. It is not perfect, but it provides engineering answers previously unobtainable. Engineers feel uncomfortable because of the possible errors (again, largely due to input), but this can be cured by experience. Compare the situation as it was 100 years ago. A bend member was designed by calculating the bending stress with $\sigma = Mh/I$ (elastic) and by sizing the member with a safety factor. Obviously, the elastic analysis was in error, and there was a great deal of uncertainty with respect to the (inaccurate) results. But structures were designed on this basis, because there was no alternative. With the years came experience. Certainly, there were misphaps, but in the end the procedures were made to work, and eventually the necessary safety factors were established. Today nobody questions this procedure; 100 years of engineering experience has shown that it works. Even today, virtually all structues are designed on this very basis: an 'inadequate' elastic analysis and a safety factor. Plastic analysis is certainly possible but it is also more complicated; it is more fundamental, but does not lead to better results in the general design of load-bearing structures. Compare LEFM and EPFM in this light.

Fracture mechanics should be considered against this background. Once a century of experience is obtained, it will be as common as present day design analysis. Mishaps will occur, but become fewer when experience accumulates. Experience will be obtained only through application. Naturally, safety factors are necessary. As before, experience must show how large these should be.

Fracture mechanics is useful now. Waiting for further technical improvement will be in vain. Better methods will evolve, but the engineering results will improve only marginally. (Elastic-plastic analysis has not displaced regular elastic design analysis). The time of application of fracture mechanics is now. Inexperience and unjustified fears are no excuse. If they had been 100 years ago with elastic design analysis, there would have been little progress in enginneer-ing. But a century ago, engineers were willing 'to stick their neck out'. This is true, now as much as a century ago. Fracture mechanics is a new tool to prevent fractures. As such it can only lead to improvements. Damage tolerance analysis may not prevent all fractures, but it can prevent many, which is sufficient justification for its use.

Fracture prevention is not glamorous. It can never be proven that a fracture was prevented. If the fracture does not happen, some may think that the effort was 'money down the drain', but in today's litigious society a fracture may cost much more than fracture prevention. If fracture mechanics is not used and a fracture occurs, lawyers will be quick in pointing out that 'the best available' techniques were not used. Even though fracture mechanics is not perfect, it is the best available. The time for its application is now.

12.11. Exercises

1. A damage tolerance analysis is performed by assuming an initial edge crack of 0.1 inch. $K_c = 60\,\text{ksi}\sqrt{\text{in}}$ and $\mathrm{d}a/\mathrm{d}N = 2\,E\text{-}9\,\Delta K^{3.2}$, $F_{ty} = 70\,\text{ksi}$, $B = 0.2$ inch. Assume constant amplitude loading to 15 ksi at $R = 0$. The minimum permissible residual strength must provide a safety factor of 2. The item will be replaced after$H$; inspections cannot be performed. If initial quality control can assure discovery of an 0.04 inch flaw, what is the extra factor on life obtained. Assume $\beta = 1.12$ throughout ($W = 100$ inch).

2. Repeat Exercise 1 for a through-crack at a hole of one inch diameter in a wide plate.

3. Compare the two factors on life obtained in Exercises 1 and 2. How would you change the damage tolerance specification (minimum permissible residual strength) or the assumptions to provide more consistent safety regardless of the type of crack?

4. The rate data in Exercise 1 are the average of a scatter band that covers a factor of 1.5 on $\mathrm{d}a/\mathrm{d}N$. The toughness is an average number from tests with a data scatter of 13%. The stresses as given are expected to have an accuracy of five percent. Estimate the upper and lower bound of the life in Exercises 1 and 2 (estimate error in β yourself). Where do the largest inaccuracies come from?

5. Using the information obtained in Exercises 1 through 4 do you have reason to change the damage tolerance specification or the assumptions? Given there are 10 000 cycles per year what would be the replacement times? Would you consider inspection?

References

[1] *Airworthiness requirements FAR 25b.* U.S. Federal Aviation Administration.
[2] *Damage tolerance requirements for military aircraft*, MIL-A-83444.
[3] G.L. van Oosten and D. Broek, *Fatigue cracks approaching circular holes*, Delft Un. rept (1973).
[4] Anon. ASME boiler and pressure vessel code; Section XI; *In service inspection of nuclear power plant components, plus Appendix A, Analysis of flaw indications*, ANSI/ASME, American Society of Mechanical Engineers, New York, Issued annually.
[5] D. Broek, *Elementary engineering fracture mechanics*, 4th ed. Nijhoff (1985).
[6] J.P. Gallagher et al., *USAF damage tolerant design handbook*, AFWAL-TR 82-3073.
[7] R.C. Rice and D. Broek, *Evaluation of equivalent initial flaws for damage tolerance analysis*, Naval Air Dev. Center NADC-77250-30 (1978).

CHAPTER 13

After the fact: fracture mechanics and failure analysis

13.1 Scope

Despite careful fracture control, service failures will continue to occur; but without it more fractures would be experienced. Engineering journals from the turn of the century discuss the large numbers of failures then occurring. Presently, with many more structures in service, the number of structural failures is relatively low (although these few get much more publicity). This is due to better design, but not in the least due to fracture control and quality control. Fracture mechanics and damage tolerance analysis can further improve the situation, but cannot eliminate all failures.

When a service fracture occurs, a failure analysis is usually performed. Every fracture, in principle, contains all the evidence about its cause, although this information is sometimes hard to extract. The broader the scope of the failure analysis, the greater the likelihood that the scenario and cause can be reconstructed. In this respect fracture mechanics can provide a great deal of information as discussed in this chapter.

Fractography is an indispensable part of the failure analysis as it is usually the only means by which the failure mechanism can be established, but the knowledge that failure was caused by fatigue or stress corrosion does not solve the problem. The purpose of the failure analysis is to arrive at 'solutions' that will prevent subsequent failures. In the author's experience the majority of failures is due to design and production deficiencies; few are due to material defects. The remedies often lie in design and structural changes. Quantitative fractographic analysis and fracture mechanics can be of help. This chapter does not review fractographic features as excellent texts on the subject are available [e.g. 1, 2, 3, 4]. Instead, its purpose is to review the structural and design aspects of failure analysis, and to show the role fracture mechanics can play. In this respect some quantitive measurements of the microscopic features are reviewed; also a few fractographic features observable with the naked eye will be discussed, as they may be of help to the damage tolerance analyst involved in failure analysis.

13.2 The cause of service fractures

A load-bearing structure is designed to sustain the maximum anticipated service loads with a safety factor between 1.5 and 3, depending upon the type of structure. There is usually uncertainty about the maximum anticipated service loads; the safety factor covers these, as well as inacurracies in stresses, possible below-average material strength, unknown residual stresses, dimensional tolerances, and – to some extent – small defects escaping quality control. Because of the safety factor, defect free structures should never fail below or at the maximum service load. Indeed they almost never do; a true 'overload failure' is rare.

The causes of fracture are: (a) The remote possibility of a true overload failure; (b) Development of cracks during service either due to a material defect not detected during quality control, or (more often) due to poor detail design (notches and eccentricities) so that conventional design analysis was inadequate; (c) Crack development due to extreme circumstances (e.g. temperature, and residual stresses) not accounted for in the design. As a rule, fractures are precipitated by cracks. A crack may be considered a partial failure. The final failure – the complete separation – is caused by fracture (Chapter 1). The propagating fracture is often referred to as a fast propagating CRACK, but the word crack is used here (and throughout this book) to pertain to partial failures developing slowly in time.

Crack growth can occur by a variety of mechanisms; the most prominent of which are, fatigue, stress corrosion and creep, or combinations of these (Chapter 1). By themselves these do not cause a fracture. The latter is a consequence of the crack, and occurs by rupture, cleavage, or intergranular separation. Cracks impair the strength, so that a fracture eventually may occur at the operating stress (service failure).

Fractographers sometimes use the term 'overload fracture', sometimes only to distinguish between 'crack' and 'fracture'. This is confusing because an overload can be interpreted as a load higher than the (maximum) service load. The vast majority of fractures occur at service loads: these may be the high loads in the spectrum, but in general, they are not overloads from the design point of view. When the verdict from a fractographic analysis is 'overload fracture' was the cause, it usually signifies that the crack or defect that precipitated the fracture could not be identified, for example because it was very small.

A true overload fracture can occur only due to (1) Extreme abuse by the user, causing stresses higher than $j\sigma_{max}$, where j is the safety factor and σ_{max} the maximum service stress anticipated during design (Figure 11.1); (2) Gross underestimate of the maximum service load so that the structure was under-designed; or (3) Poor design with sharp notches, misfits etc, so that conventional design analysis was inadequate.

With few exceptions service fractures are brittle from the engineering point of view (Chapter 2). Yet, the great majority occurs by ductile rupture. The fractographer's definition of a brittle fracture pertains to whether or not plastic deformation is required for fracture mechanism. Cleavage (Chapter 1) does not require plastic deformation, although some plasticity may occur: the fractographer calls this brittle fracture. Rupture is the result of plastic deformation: the fractographer calls it ductile, although virtually no plastic deformation may occur. This dichotomy was already discussed in Chapter 2. In an unnotched bar pulled to fracture ample plastic deformation will occur throughout; the fracture is ductile. In the case of a crack, plastic deformation is confined to the fracture path (Figure 2.12). The fracture is brittle, because there is little overall plasticity; yet the fracture mechanism may be ductile rupture. Since most service fractures are due to cracks almost all are brittle from the engineering point of view, regardless of the fracture mechanism.

Fracture is the direct cause of the failure, but the actual culprit is usually a crack or defect without which the fracture would not have occurred. Hence, a failure analysis should determine the cause of the CRACKING (fatigue, stress corrosion, etc). Each mechanism has certain characteristic features by which it can be recognized [1–5]. However, the mechanism by itself does not explain the failure; the real question is why this mechanism could become operative.

There are five fundamental causes for the start of cracking, namely

(a) Material defects.
(b) Manufacturing defects.
(c) Poor choice of material or heat treatment.
(d) Poor choice of production technique.
(e) Poor (detail) design.

One might add 'poor quality control', but this is not a fundamental cause: it is the secondary reason by which others become possible.

If the failure analysis can identify one of the above to have been operative, the remedy is at hand. Manufacturing defects can be introduced by blunt tools, overheating during machining, welding, etc. Material selection is an obvious culprit. The material might be perfect for fatigue, but propensity for stress corrosion cracking due to the particular heat treatment might have been overlooked. Local heat treatments such as carburizing, nitriding and surface hardening, almost always cause a volume change of the surface layer, so that they introduce residual stresses not accounted for in analysis. As for the production technique, the problem of grain orientation was discussed in Chapter 7, on the basis of Figure 7.8. It is often a cause of 'unexpected' failures.

Poor detail design is a major cause of service failures. That sharp notches should be avoided is commonly understood, but hidden stress concentrations are sometimes not recognized. A classical example was discussed in Chapter 10 on the basis of Figure 10.2. Statically the design is adequate: when loaded to failure, plastic deformation will ensure even distribution of the load over the

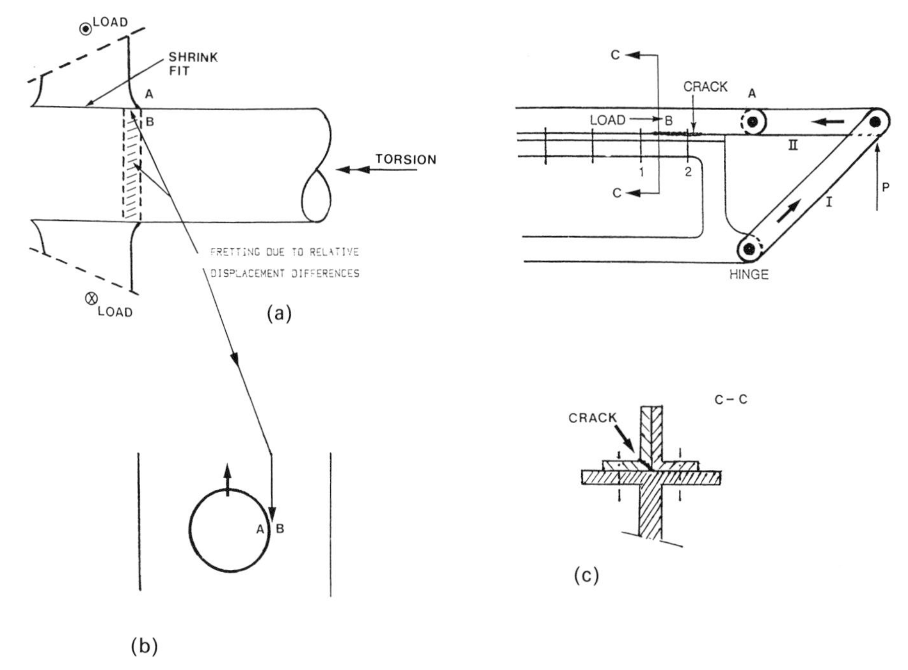

Figure 13.1. Cracks due to relative displacements (fretting; a, b) or consequent secondary stresses (c).

bolts. However, during elastic service loading, the center bolts transfer no load, causing cracking at the highly loaded outer bolts.

Other design details may cause secondary displacements, not accounted for in design or fatigue analysis. Consider the shrink fitting in Figure 13.1a. Eventually, the load must be transferred from the shaft to the shrunk-on part. However, at *A* the shrunk-on part is still relatively stress-free (no stress and no strain). The shaft is strained, so that there will be relative movement between *A* and *B*, which may cause fretting and subsequent fatigue cracks. The same happens at bolt shafts, under bolt heads, and so on.

Secondary stresses due to displacements may cause problems as well. The loading of parts I and II in Figure 13.1b causes (small) upward displacements of *A* and therefore of *B*. As *B* cannot undergo vertical displacements because of the bolts, secondary bending stresses in the bolted flanges may cause cracking. Elimination of some bolts may solve the problem, because there will be some freedom for vertical displacements if bolt 1 and 2 are omitted.

Determining the cracking mechanism is a significant part of the failure analysis, but the starting point of the crack must be found in order to establish the reason why it became operative. All fundamental causes listed above must then be considered. At this time there is an essential task for the damage tolerance analyst. Questions regarding the origin of stresses, loads, load path, primary and secondary displacements/stresses, must be answered. Answers to these are often found in general area deformations (occurring during the fracture process); deformation of adjacent parts will reveal directions of acting loads, and stresses. It should be checked whether these are compatible with the design assumptions and with the cracking mechanism. Analysis of stress fields for 'unanticipated' stress concentrations, secondary stresses and displacements (Figure 13.1) is necessary.

13.3. Fractography

The tools for fractography are loupe and stereo microscope, optical microscopes, electron microscopes, X-ray analysers, and image analysers. The non-fractographer should be aware that an electron fractograph (high magnification photograph) shows only a very small area. This may be adequate, but it is sometimes deceiving because it may not be representative of the whole.

The main cracking mechanisms were discussed in Chapter 1. Fatigue damage and fatigue cracking in service take place under nominal elastic stresses, but a fatigue crack cannot initate without plastic deformation, however local and minute (Chapter 1). Generally, such plastic deformation will occur at the tip of a notch or at a stress raiser (including particles in the material). Once a crack is initiated it grows by a mechanism similar to the example in Figure 1.4. A regular repetition of blunting and sharpening causes the formation of distinct lines on the fracture surface, the fatigue striations. One striation is formed during each cycle as shown in the electron fractograph in Figure 1.5. This opens the possibility to measure the rate of growth by measurement of the striations spacing.

Striations as regular as those in Figue 1.5 can be formed only when the material can accomodate the mechanism of Figure 1.4 by opening and closing in a uniform manner over some distance along the crack front, as shown in Figure 13.2. If the material's deformation possibilities are insufficient to open

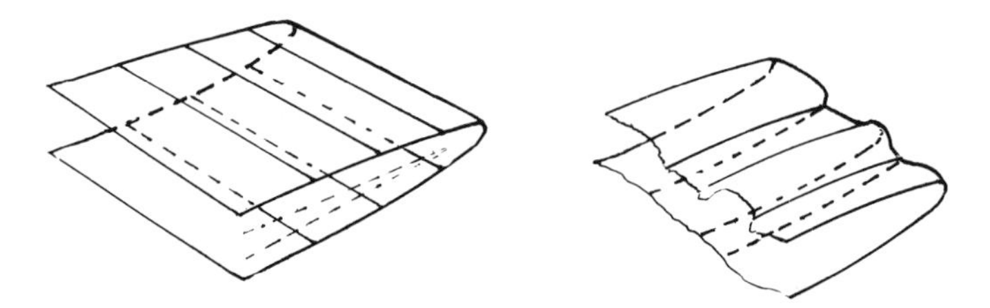

Figure 13.2. Formation of reguiar and ill-defined striations Left: uniform opening and closing over some distance creates lines on crack surface (regular striation as in figure 1.5) Right: Non-uniform opening and closing (ill-defined striations as in Figure 13.3).

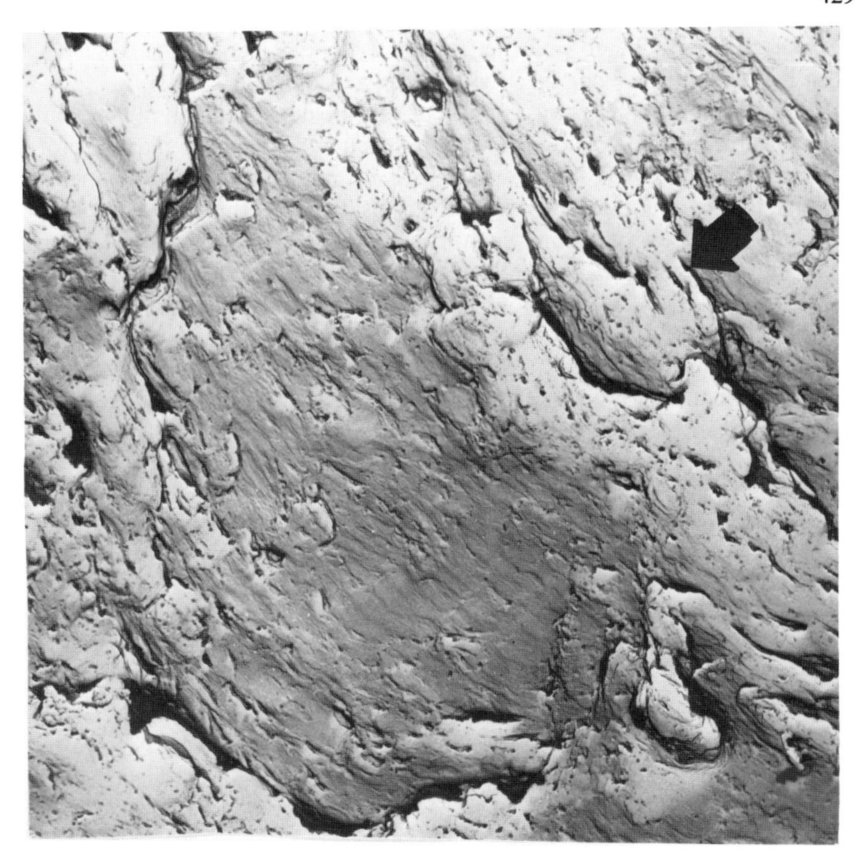

Figure 13.3. Patch of rather well-defined striations (center) and 'chopped' striations (top left and bottom right); 4340 steel;3500 ×.

and close the crack uniformly, the striations become ill-defined as illustrated in Figure 13.2 (right). This is the case in many steels (Figure 13.3). Nevertheless striations (even if ill-defined) can provide the rate of growth by their average spacing.

Fractographic features of other cracking mechanisms are more variable and differ from material to material. A stress corrosion crack often follows the grain boundaries because the chemical composition at the grain boundaries is different from that inside the grains. The fractographer will recognize the features by which to identify a stress corrosion crack.

Fracture mechanisms were discussed in Chapter 1 as well. To a degree the shape of dimples depends upon the (local) stress gradient. Thus, the dimple shape can be used occasionally for a qualitative assessment of the local stress field, but dimple shape can be deceiving as it depends upon the angle of view [4]. The dimple rupture surface is irregular and (in contrast to the glittering cleavage surface) diffuses light, so that the fracture looks dull grey to the naked eye; as such it often can be recognized without microscopic aid.

Whether fracture occurs by cleavage or rupture depends upon rate of loading, temperature, and state of stress. Roughly speaking, if sufficient plastic deformation can occur to relieve stresses, cleavage will not occur. At low temperatures and/or high-loading rates the yield strength is higher. If the state of stress is one with high hydrostatic tension, yielding is postponed to stresses higher than the uniaxial yield strength (Chapter 2). Hence, the above conditions tend to confine plastic deformation and promote cleavage if the stress peaks at or above the cleavage strength. In other cases local plasticity will ameliorate conditions through lower stresses and larger plastic strains to set the rupture process in motion. In many alloys it is virtually impossible to induce a cleavage fracture.

Of particular interest for the damage tolerance interpretation of the fracture surface are transverse cracks, such as shown in Figure 13.4. They are an indication of high hydrostatic tension (plane strain), or high σ_z as shown in Figure 13.4, possibily combined with a low fracture toughness in Z-direction (ST or SL).

13.4. Features of use in fracture mechanics analysis

Fracture surface topography can be measured [6, 7] from fractographs in much the same way as terrestial topography from aerial photography. The projectional displacement of identical features in two photographs taken at different angles, permit the determination of the height of the feature due to the viewing angle difference, as shown in Figure 13.5. The larger the viewing angle difference, the larger the relative displacement, and the more accurate the procedure.

The height is determined from (Figure 13.5):

$$\left.\begin{aligned} p &= CD - CE = \delta\cos\theta + h\sin\theta \\ &\quad - (\delta\cos\theta - h\sin\theta) = 2h\sin\theta \\ h &= \frac{p}{2\sin\theta} \end{aligned}\right\} \qquad (13.1)$$

For fracture mechanics analysis the most fruitful application of this technique is in the area of transition between crack and fracture, as in the example given in Figure 13.6. A topographic measurement [8] provides the size of the Crack Tip Opening Displacement, CTOD. Stereo pictures show the blunting in a different projection (different size; Figure 13.7), so that CTOD can be measured by Equation (13.1). Using fracture mechanics equations, the CTOD can be

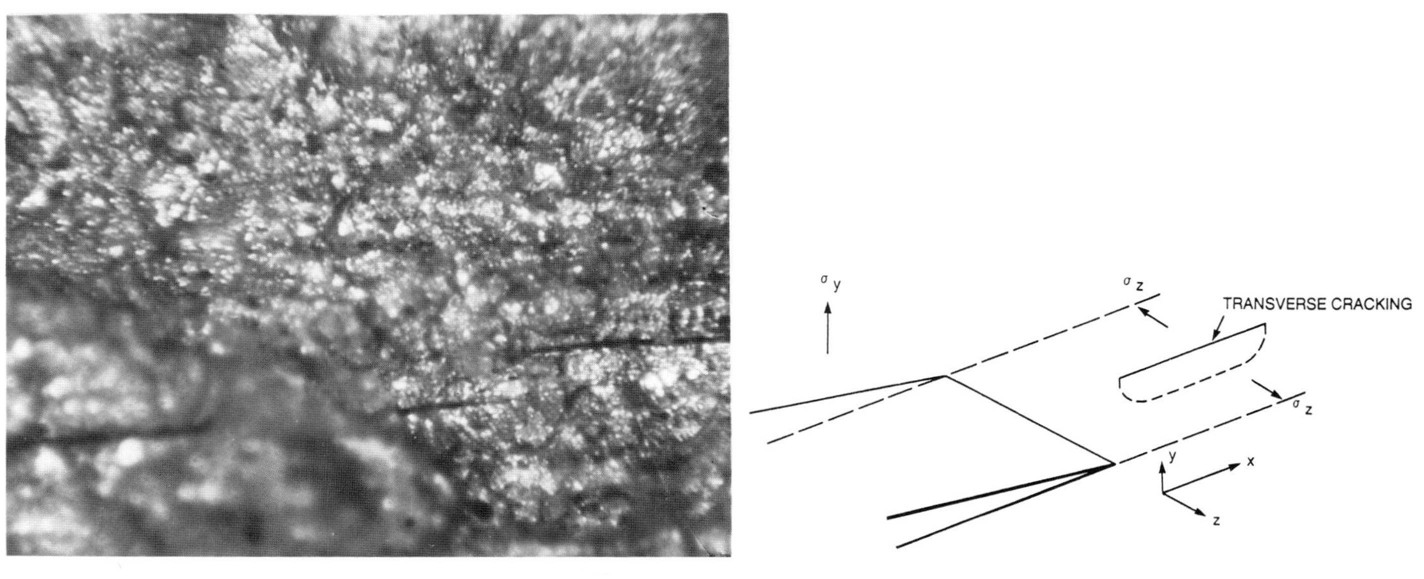

Figure 13.4. Transverse cracks due to high σ_2 Left: crack surface at 50 × (optical microscope).

related to the applied stress as discussed in next section; if the CTOD is measured from the topography, the toughness can be calculated.

In the case of a fatigue crack, striation counts are a source of valuable information. The technique is acclaimed inaccurate, but this is generally due to insufficient numbers of measurements, and impatience of the fractographer. Fractographs display an extremely small part of the crack, while crack growth rates depend strongly upon local circumstances. In a crack growth test, growth is measured as an average over hundreds of grains. The same averaging must be done in striation counts, so that the procedure literally requires hundreds of fractographs. The slow photographing process in many scanning microscopes is the reason why striation counts are often limited to a few dozen fractographs; this does not provide useful data. Therefore the use of replicas and transmission microscopes is preferable, because photographing is semi-automatic; a modern transmission microscope can easily produce a hundred fractographs per hour, provided film is used instead of plates (which is adequate for the purpose).

There is basically one striation per load cycle. Whether striations occur only here or there, are chopped up, or regular (Figures 1.5 and 13.3), their average spacing does provide the rate of crack propagation. A typical result is shown in Figure 13.8. (Each measurement of striation spacing is an average of many measurements.) Numerical integration of the rates provides the crack propagation curve as a function of the number of cycles, as shown in Figure 13.8 and Table 13.1. Such a curve can be compared with the one obtained by the fracture mechanics analysis.

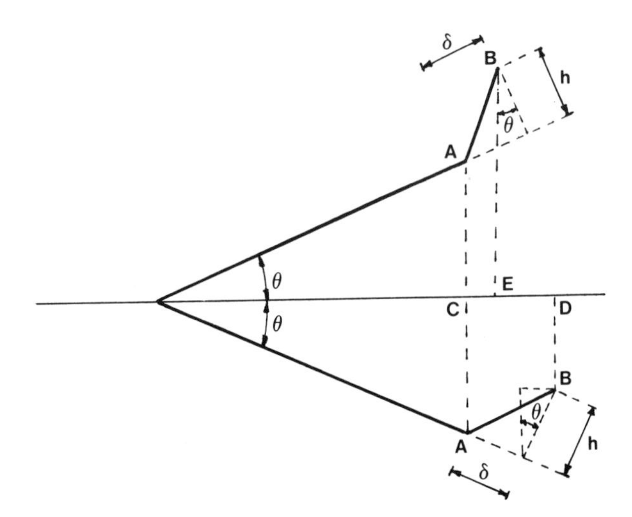

Figure 13.5. Height measurement from stereo micrographs. Features A and B project at different distances on photographic plane (horizontal).

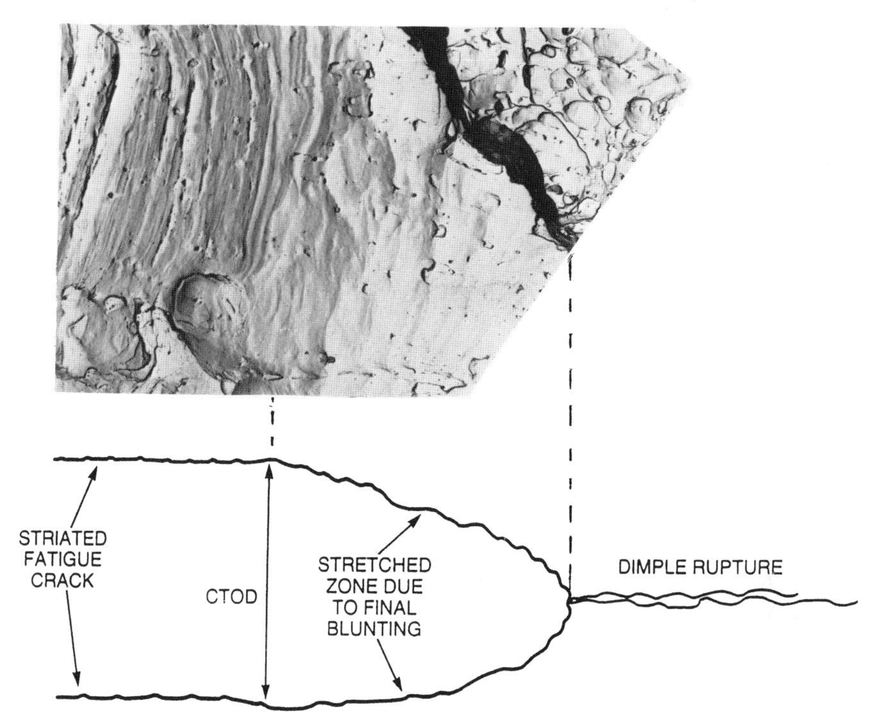

Figure 13.6. Blunting of crack tip before fracture (slip) forming stretched zone characterized by wavy slip lines.

Crack growth often is suspended for long periods because the structure is not being loaded (not in use). The already present part of the crack will be subject to environmental influences and may undergo slight discoloration. Then a ring shaped mark (beach mark) will be visible on the crack surface, delineating the shape of the crack (front) at a given point in time (see Figure 13.9). Changes in load will change the rate of crack growth. Higher growth rates (higher stress) result in a rougher crack surface, than low rates (low stress). A smoother crack surface has more luster because it is more reflective. Also this will result in a delineation on the crack surface, creating another beach mark. Beach marks may occur on any crack surface, whether fatigue or stress corrosion. They always signify a change in circumstances, either load or environment. An absence of beach marks is an indication of uniform circumstances throughout the cracking phase. As beach marks signify the positions of the crack front at certain times, they are useful in correlations with the stress-time-environment history.

Fracture is a fast process. Its macroscopic features are reflections of fracture speed and hence of stress level and fracture toughness. Many fracture surfaces (both cleavage and rupture) show chevron marks as depicted in Figure 13.10:

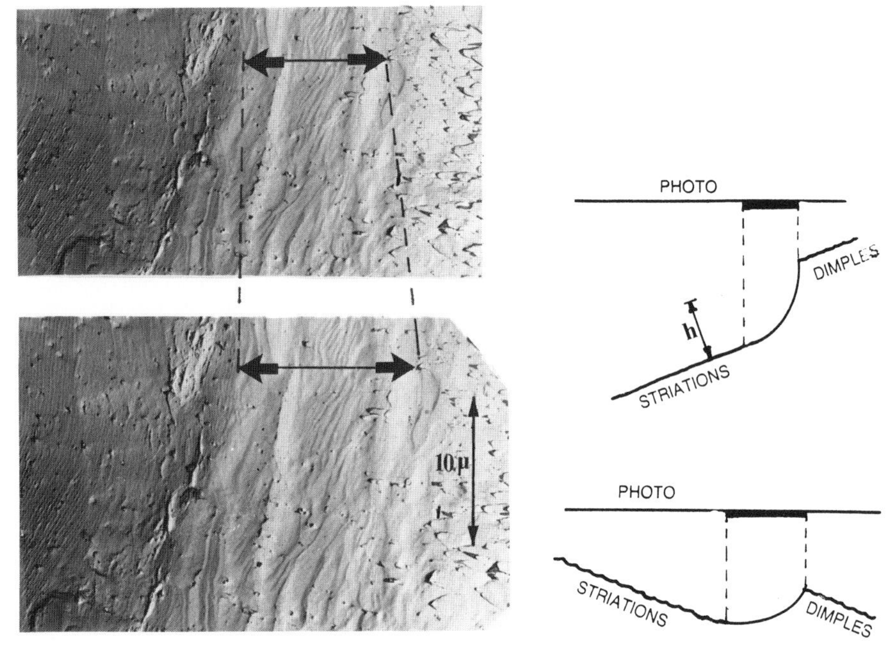

Figure 13.7. Measurement of h (1/2 CTOD) in accordance with Figure 13.5. Stereo photographs at + and − 6 degrees.

they point to the origin of the fracture. The sharper the points of the chevrons, the faster the fracture (low toughness; high stress). The chevrons are not indicative of the shape of the fracture front; instead they are more or less a mirror image of the fracture front. They show increasing height (roughness) towards the outer surface: in the interior the state of stress is often triaxial, but at the surface σ_z is always zero, so that there is a gradual decrease of the hydrostatic tension towards the surface. The higher the hydrostatic tension the lower the fracture resistance (plane strain versus plane stress). Consequently the fracture tries to run faster in the center giving a smoother fracture surface, but naturally the fracture at the free surface must move along with that in the interior.

If the stresses are high and/or the material's fracture resistance low, the energy released during fracture is high. When sufficient energy is available, two or more fractures can be maintained simultaneously (a branch is formed [5]). A higher energy surplus causes more branching. High energy surplus occurs in the case of low fracture resistance (windows shatter when fracturing). Branching provides another indication of the fracture origin. This is particularly helpful when shattering has occurred.

Much information can be gained from the general condition of the broken

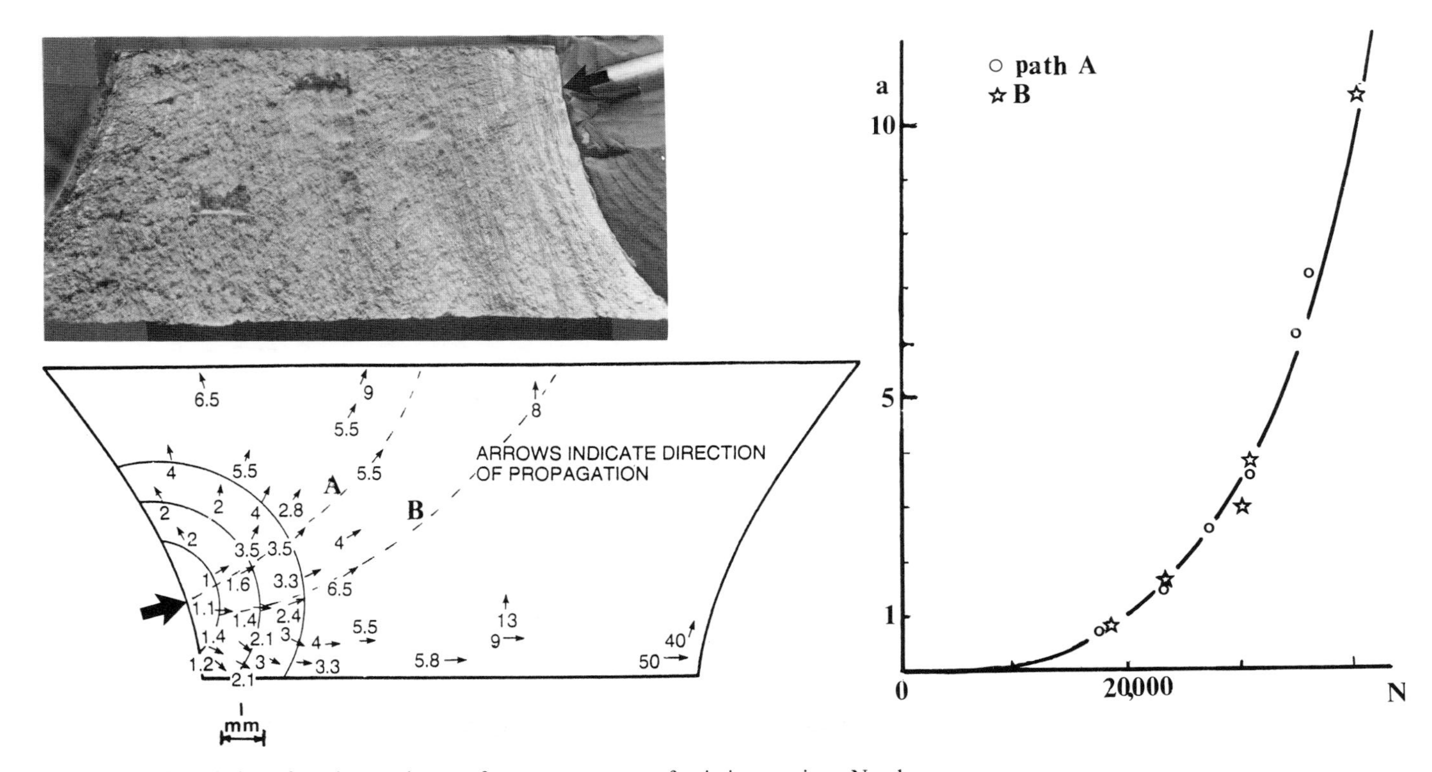

Figure 13.8. Calculation of crack growth curve from measurement of striation spacings. Numbers in bottom left drawing are measured striation spacings in units of 0.1 microns (see Table 13.1).

Table 13.1. Calculation of crack growth curve along path A in Figure 13.8

a (mm)	Striation spacing μ/c	Δa (mm)	$a_{average}$ over increment	da/dN average over increment μ/c	da/dN average mm/c	$\Delta N = \frac{\Delta a}{da/dN}$ average	$N = N + \Delta N$
0	0						0
		0.8	0.4	0.05	5.0×10^{-5}	16000	
0.8	0.1						16 000
		0.93	1.27	0.13	1.3×10^{-4}	7150	
1.73	0.16						23 150
		1.20	2.33	0.26	2.6×10^{-4}	5150	
2.93	0.35						28 300
		3.27	4.57	0.45	4.5×10^{-4}	7270	
6.20	0.55						35 570
		1.13	6.77	0.75	7.5×10^{-4}	1510	
7.33	0.94						37 080

component and surrounding structure. Deformations that occurred AFTER fracture, secondary cracks and fractures, branches, etc., can be used to reconstruct the fracture event and provide information on the direction or magnitude of the loads. In general, as much of the surrounding structure should be preserved for the analysis as practical. Considering the magnifications, fractographic features are very small. The slightest damage to the fracture surface will obliterate them: fitting the pieces together will accomplish this; acid deposits of fingerprints do the same.

One of the most important preliminaries to a failure analysis is the protection of the fracture surface by means of clear acrylic lacquer spray or acid-free oil; these can be removed later with organic solvents. If a reconstruction is to be made on the spot, the pieces should be laid out (without the fracture surfaces touching) in the position from whence they came. Photographs from many different angles should be made for documentation. If it is not strictly necessary to perform the reconstruction in the field, it should be done in the laboratory.

13.5. Use of fracture mechanics

Failure analysis as discussed, will show the mechanism of cracking and the mechanism of fracture, and whether or not material defects or manufacturing defects were involved. Fractography generally will show the crack size at which final fracture occurred because cracks have a different surface roughness than fracture. Examples are shown in Figure 13.9.

Fracture occurs in accordance with the concepts discussed in Chapter 3 and 4. If operational stresses are known, the equations in Chapters 3 and 4 can be used to calculate the toughness from the crack size at fracture, which, if compared with the anticipated toughness, will show whether or not the material

Figure 13.9. Beach marks on various crack surfaces.

had a toughness higher or lower than expected and whether the material was in accordance with the specification.

If the stresses are not known on the other hand, the toughness may be obtained from data handbooks. The operating stresses then can be calculated. Depending upon the material and toughness either LEFM (Chapter 3) or EPFM (Chapter 4) must be used. The best way to proceed is to calculate the entire residual strength diagram in the manner discussed in Chapter 10; collapse should naturally be accounted for. Given the crack size at fracture as determined from fractography, the fracture stess can be obtained from the residual strength diagram.

If neither fracture stress nor toughness is known, an estimated toughness value may be used, e.g. for comparative materials or estimated on the basis of a Charpy-value (Chapter 7). Instead stretched zone measurements may be used to obtain the toughness; at the least these may be of help to corroborate the toughness estimate. The size of the stretched zone (Figures 13.6 and 13.7) is related to the crack-tip opening displacement (CTOD) at the time of fracture; it can be measured using the techniques described in the previous section on the basis of Figures 13.5–13.7 and Equation (13.1). The CTOD at fracture is related to the toughness as (Chapter 4):

$$G = \frac{K_c^2}{E} = F_{ty}\,\mathrm{CTOD} \tag{13.2}$$

so that the toughness can be calculated from the measured stretched zone size. Careful measurement can give reliable toughness values [8]. For example, consider the case of Figure 13.7, which is for a material with $F_{ty} = 65$ ksi and $E = 10\,000$ ksi. The distance between the two features indicated by the large arrows can be measured from which the value of p in Equation 13.1 is obtained. Using the micron-scale shown in Figure 13.7, the value of p in Equation (13.1) is found as 2.7 microns = 0.0027 mm. Since $\theta = 6$ degrees = 0.10 radians, the height h is found from Equation (13.1) as $h = 0.0027/0.2 = 0.0135$ mm. As CTOD = 2 h (Figure 13.6), it follows that CTOD = 0.027 mm = 0.00106 inch. Then with the aid of Equation (13.2) the toughness is found as $\sqrt{10\,000 \times 65 \times 0.00106} = 26\,\text{ksi}\sqrt{\text{in}}$. It was pointed out already that micrographs exhibit only a small area, so that several measurements of different areas are required to obtain an average. If more measurements are made using the fractographs of Figure 13.7 an average toughness value of $28\,\text{ksi}\sqrt{\text{in}}$ emerges.

With the toughness known, fracture stresses can be calculated from $\sigma = K_c/\beta\sqrt{\pi a_c}$, where a_c is the crack length at fracture; β is determined by one of the procedures discussed in Chapter 8. It then can be determined whether the fracture took place due to an excessively high load or whether the local stresses were too high.

Fatigue-striation counts (Figure 13.8) generally provide a reasonable account

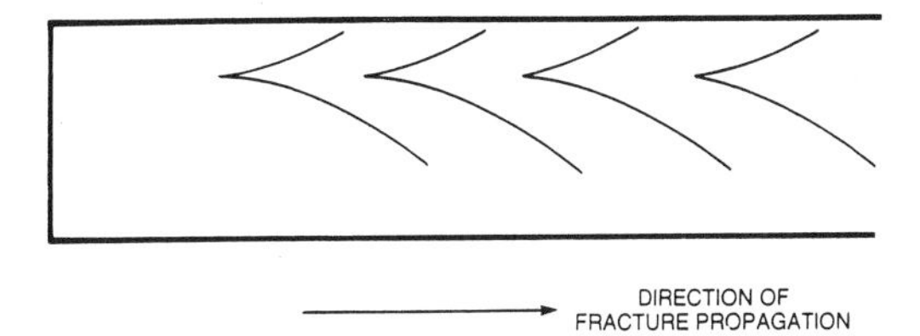

Figure 13.10. Chevrons on surface of high speed fracture (high stress/low toughness).

of the crack-growth rates and crack-growth curve. If the crack-growth rate data of the material and the stresses are known, the stress intensity can be inferred, and a comparison made between actual and anticipated properties for a conclusion about the adequacy of the material. Conversely, if the stresses are not known, the measured rates and the rate properties can be used to estimate the acting stresses. From the amount of crack growth (crack size at fracture), known stresses, and growth-rate properties, a reasonable insight can be obtained regarding the question of misuse (e.g. continual high loading). The time to failure and final crack size are determined using fracture mechanics, as discussed above. When the results are not in accordance with the observations, the analysis can be repeated to determine how much higher (or lower) the stresses would have had to be to produce the cracking time and crack size at fracture as observed.

The most direct way to use the information is as a basis for determining inspection intervals for similar components, using the procedures discussed in Chapter 11. Indirect use can be made by comparing the curve with a calculated one using the concepts described in Chapters 5, 6 and 10. This can provide information regarding the adequacy of the analysis (e.g. load history, σ, da/dN). If no corroboration is found the analysis can be performed for different load histories or da/dN data, to evaluate the cause of the problem. Other indirect use can be made by determining ΔK–da/dN at various crack sizes from the striation measurements. Comparison of these results with the anticipated rate properties can reveal whether the material might have been below standard. (For an example see Exercises 3, 4, 5.)

A change in loading or environment during the cracking process is likely to leave a beach mark. If information on the nature of these changes is available, the crack size at which they occurred can be used to obtain information on rate properties or stresses at that time. Conversely, without knowledge of the nature of the changes, information (a change in loading in particular) can be obtained from known growth-rate properties: the time for growth between beach marks can be calculated, and the stress required to produce the observed crack sizes estimated.

If fractography shows the presence of initial defects – whether mechanical or metallurgical – crack-growth analysis can be used, starting from the initial defect. Again, depending upon the objective, information can be obtained on stresses (known material properties) or material properties (known stresses).

The stress intensities due to different stress systems are additive as long as all stress systems are of the same mode (Chapter 8). Residual stresses due to welding, cold deformation, surface treatments, or heat treatments can thus be accounted for. Although this is useful in principle, the magnitude of the residual stresses is often unknown. It may be assumed that yielding at the crack tip is so extensive that residual stresses can be ignored for the fracture analysis. For fatigue-crack growth, the residual stress (static) causes only a change in R-ratio (Chapter 9). In the case of sustained-load stress corrosion, the stress intensities are additive. These guidelines can be used to evaluate the effect of residual stress. The procedures can at least provide an estimate of the effect, but engineering judgement is required. Residual stresses can be accounted for (Chapter 9), but the real problem is to know their magnitude.

13.6. Possible actions based on failure analysis

The purpose of a failure analysis is to determine how other failures can be prevented. If the fracture appears to be an incidental case only, no action may be required at all. However, if it appears symptomatic, one or more of the following measures may be taken:

(a) Relief of residual stress.
(b) Relief of secondary displacement.
(c) Selection of other material.
(d) Redesign.
(e) Change in manufacturing procedures.
(f) Replacement of similar components (or life limitation).
(g) Inspection on the basis of detectable crack size and maximum permissible crack size (based on crack growth analysis); repair when crack is detected.
(h) Retrofit of load bypass (second member; Chapter 9).

These actions are in accordance with those discussed in Chapters 11 and 12.

Some of these will put part of the burden on the operator. A life limitation may be preferable, if the costs of the part or component and of replacement are low. For costly parts or components a retrofit may be a good alternative. If the operator is the 'general public' only replacement may be possible, as inspection is only an alternative for professional equipment. The possiblility of taking no action at all should not be overlooked: the cost of fracture may be less than the cost of fracture control.

13.7. Exercises

1. Using the same procedure as in the text calculate the toughness by making

measurements at two or more locations in Figure 13.7 and estimate the average toughness.

2. If the crack at fracture in Exercise 1 was 20 mm long and $\beta = 0.7$, what was the nominal stress at fracture.

3. Measurements of striation spacings at 8000 × provide spacings of 0.03, 0.08 and 0.15 inch respectively at crack sizes 0.1, 0.2, 0.3, and striation spacings of 0.06 and 0.09 inch at 2000 × at crack sizes of 0.4 and 0.5 inch resp. Determine the growth rates and estimate the crack growth curve.

4. Assuming that $\beta = 1$ throughout, and assuming that a materials handbook provides $da/dN = 2E\text{–}8\,\Delta K^3$ for the material, estimate the stress range in Exercise 3 if the loading was of constant amplitude. $R = 0$.

5. Using the rate data quoted in Exercise 4, calculate the crack growth curve, and compare the result with that of Exercise 4.

6. Suppose that the stress range in the case of Exercise 4 is known to be 15 ksi, could there be a problem with the material?

7. It appears that in the component concerned in Exercise 4 a stress concentration with $k_t = 1.5$ exists. Re-evaluate the results of Exercises 5 and 6. Which remedial action would you take.

8. A fracture occurs in a single lap joint with three bolts due to a crack. The joint transfers a load of 100 000 lbs. The material has a yield strength of $F_{ty} = 50$ ksi, the bolts a shear strength of 100 ksi. The joint was designed with a safety factor of 3 against the above numbers. No action was taken in the original design for equal load distribution during elastic loading. The width of the joined plates is two inches. What is the configuration of the original design. Redesign the joint for optimal crack resistance still using three bolts. Would four bolts in the original design provide a good alternative?

9. A service fracture appears to have been caused by a stress corrosion crack. The crack size at fracture was 1 inch ($\beta = 0.7$), the material's toughness is 50 ksi $\sqrt{\text{in}}$, while $K_{Iscc} = 7$ ksi $\sqrt{\text{in}}$. What was the operating stress and which size of defect would have to be eliminated in quality control to prevent recurrence.

10. A failure analysis leads to the hypothesis that the crack was caused by a material deficiency. No crack is apparent. A crack larger than 0.05 inch would have been identified. The specified toughness of the material is 60 ksi $\sqrt{\text{in}}$, $F_{ty} = 80$ ksi. At the location of the crack there is a stress concentration with $k_t = 2$. The nominal stress was calculated to be 25 ksi. Is the hypothesis tenable?

11. A material's yield strength is 90 ksi and its crack propagation properties at $R = 0$ can be described by $\mathrm{d}a/\mathrm{d}N = 8 \times 10^{-9}(K)^{3.5}$ inch/cycle if K in ksi $\sqrt{\text{in}}$. Failure analysis of a very large failed part made of this material shows that fracture occurred due to a fatigue crack penetrating 0.6 inches from the surface ($\beta = 0.64$). At a magnification of 2000 × the striation spacing at the very front of the fatigue crack is 0.12 inches. The loading was of constant amplitude with occasional overloads of twice the maximum of the constant amplitude loading ($R = 0$ in both cases). Calculate the material's toughness and the maximum stress during the constant amplitude loading.

References

[1] D.A. Ryder, *Elements of fractography*, AGARDograph 155–71 (1971).

[2] A. Phillips et al., *Electron fractography handbook*, AFML-TDR-64-416 (1965).

[3] Various authors, Failure analysis and prevention. *Metals Handbook*, Vol. 11, 9th ed. ASM (1986).

[4] D. Broek, Some contributions of electron fractography to the theory of fracture, *Int. Met. Reviews, Review* **185**, Vol. 9 (1974) pp. 135–181.

[5] D. Broek, *Elementary engineering fracture mechanics*, Nyhoff, 1985.

[6] J.F. Nankivell, Minimum differences in height details in electron stereomicroscopy, *Brit. J. Appl. Phys.* **13** (1962) pp. 126–128.

[7] D.C. Wells, Correction of errors in electron stereomicrography, *Brit. J. Appl. Phys.* **11** (1960) pp. 199–200.

[8] D. Broek, *Corelation between stretched zone size and fracture toughness*, ICF conference, Munich (1973).

CHAPTER 14

Applications

14.1. Scope

Many examples of practical damage tolerance problems and analysis were given throughout this book, especially in the Exercises and their solutions. This chapter provides a number of examples in which application of damage tolerance analysis and fracture control can be seen in the larger context of structural design and operation. In some cases use is made of fictitious situations and numbers, but they are derived from actual cases. Sometimes insufficient input data are available, which is common in engineering. These conditions were not changed artificially. Instead, the problems are represented as they might appear in engineering, so that judgement and approximations can be illustrated.

Only scant details are provided on the actual analysis since the analysis procedures were amply discussed in the forgoing chapters. Instead, the problems are treated in their general context, so that the reader may appreciate how input, assumptions, and judgement affect results, accuracy and subsequent fracture control decisions. It would have been easy to fill this whole book with examples from the author's experience concerning aircraft, but in order to avoid the suggestion that damage tolerance analysis only works for aircraft, all of the examples are for other structures. It stands to reason that not all areas of technology can be covered. However, the examples are of such variety that similarities to other problems may be found and ideas be transferred.

14.2. Storage tank (fictitious example)

(a) *Problem definition*

A storage tank for non-flammable chemicals (Figure 14.1) developed a complete fracture, spilling its contents. Failure analysis revealed a pre-existing defect in a vertical weld with dimensions as shown in the figure. At the time of the incident the recorded temperature was -20 F. Many other tanks of the same

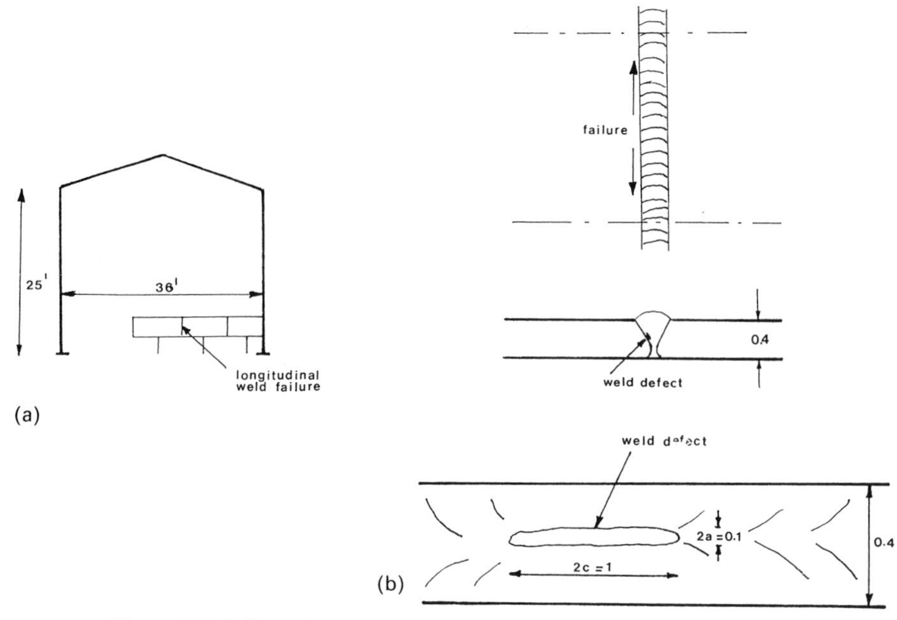

Figure 14.1. Failure of storage tank. (a) Tank; (b) Fracture and weld defect.

design operate in areas with different climates. The cost of the incident, including lost revenues, clean-up operation, and replacement is estimated at 30 man-year equivalents. A fracture control plan costing less than a fracture, would be implemented.

(b) *Loads, environment, stresses*
Loads due to wind, snow and sleet are negligible; the only loading is the hydrostatic load due to the contents. The chemical being non-corrosive, stress corrosion is unlikely. The tanks are emptied approximately 12 times per year; they are expected to have a remaining economical life of 50 years. The lowest anticipated service temperature is -40 F in the northern regions.

The chemical has a specific weight of 0.029 lb/cu in. Since the height of the tank is 300 in (25 ft) the hydrostatic pressure at the bottom is $p = 300 \times 0.029 = 8.7$ psi. With the dimensions given in Figure 14.1 the hoop stress is: $\sigma = pR/B = 8.7 \times 216/0.4 = 4698$ psi $= 4.7$ ksi. The chemical can expand freely in the empty top part so that no pressurization due to expansion occurs. Due to the fixity at the base, thermal stresses can occur due to shrinkage at low temperature. Erection took place at 70 F; hence there was a temperature differential of 90 degrees at the time of failure. Assuming full constraint, this would cause a thermal stress of $\sigma = E\alpha\Delta T = 30.000 \times 7 \times 10^{-6} \times 90 = 18.9$ ksi. Hence, the total stress could have been as high as $4.7 + 18.9 = 23.6$ ksi.

(c) *Material data*

The material's yield strength at -20 F is $F_{ty} = 55$ ksi. Toughness data not being available, Charpy specimens containing longitudinal welds were cut from the failed tank (weld crosswise in the specimen at the location of the notch). These tests showed the transition temperature at 10 F, and a Charpy energy of 2 ft/lbs at -20 F. Using the conversion equations discussed in Chapter 7, the toughness is estimated to be K_{Ic} $12\sqrt{2} = 17$ ksi$\sqrt{\text{in}}$. Naturally, the weld defect is in plane strain, but fracture would first cause a through-the-thickness crack. For through-cracks plane strain occurs at a thickness larger than $2.5 \times (17/55)^2 = 0.24$ inch, so that with the given wall thickness the toughness estimated above for fast loading, it is considered a useful number because at break-through due to fracture the high fracture speed must be accounted for in considering leak-before-break (Chapter 9).

(d) *Analysis*

The weld-defect here is a lack of fusion defect, with dimensions as in Figure 14.1. It can be considered as an elliptical flaw with $a/2c = 0.05$. The geometry factor is $\beta = \beta_{FFS}/\phi$ (Figure 8.3). Since this is a buried defect the back free surface correction of 1.12 is not required. In determining β_{FFS} it must be realized that the distance from crack center to the wall is equal to half the wall thickness, so that the value of a/B used in Figure 8.3c must be $0.05/0.2 = 0.25$. This leads to $\beta = 1.15/1.05 = 1.1$.

After breakthrough there is a through-the-thickness crack of $2a = 1$ in. For such a crack in a pressurized container $\beta = \sqrt{1 + 1.61a^2/RB} = \sqrt{1 + 1.61 \times 0.5^2/(18 \times 12 \times 0.4)} = 1.002$. Since this is approximately equal to 1 superposition to the thermal stress can be accomplished by directly adding the stresses (same geometry factor).

With a total stress of 23.6 ksi the stress intensity at fracture of the embedded flaw was $K = 1.1 \times 23.6 \times \sqrt{\pi 0.05} = 10.3$ ksi$\sqrt{\text{in}}$, which is considerably lower than the estimated toughness. The latter would require a stress of $17/(1.1\sqrt{\pi} \times 0.05) = 39$ ksi so that either an additional residual stress of about 15 ksi was present or the toughness was lower than estimated. The stress intensity of the one inch through crack was $23.6\sqrt{\pi 0.5} = 29.6$ ksi$\sqrt{\text{in}}$, which is larger than the toughness, so that fracture should indeed occur after break-through of the internal defect.

It may be assumed that the residual stresses were largely relieved after break-through. Using the stress intensity for the through crack, the stress intensity and residual strength can be calculated for various locations in the tank. For example at half of the height of the tank the hoop stress is reduced to $4.7/2 = 2.35$ ksi. Assuming (very conservatively) that the thermal stress decreases linearly to the top, because fixation is at the bottom, the corresponding thermal stresses is 9.3 ksi. If the toughness is 17 ksi $\sqrt{\text{in}}$ a through cracks of

$2a = 2(17/12.6)^2/\pi = 1.6$ inches can be sustained (leaks).

Fatigue crack growth may occur during emptying and refilling. The fatigue crack growth properties are estimated to be $da/dN = 10^{-9}\Delta K^4$. With refill cycles of 12 times per year, there will be 600 cycles in the remaining service life of 50 years. Assuming that these will coincide with the thermal cycles (extremely conservative) the stress intensity of the internal defect is $10.3\,\text{ksi}\sqrt{\text{in}}$ as calculated above. Using the above, crack growth in 50 years' $\Delta a = 10^{-9} \times (10.3)^4 \times 600 = 0.007\,\text{in}$, which is negligible. However, many more thermal cycles of smaller magnitude may occur, so that fatigue crack growth should be calculated more accurately. Assuming daily thermal cycles of 30 F, the thermal stress cycle is $18.9/3 = 6.3$ ksi. The crack growth curve for this case is shown in Figure 14.2.

(e) *Fracture control plan*

A toughness of $40\,\text{ksi}\sqrt{\text{in}}$ or more is estimated for temperatures above 10 F (transition temperature). With a toughness that high, through cracks of several inches can be sustained (leak). On this basis the following fracture control plan is established.

No action to be taken for tanks in those climates where the lowest anticipated temperature is above 10 F. In tanks in colder climates all longitudinal welds in the lower half are to be X-rayed. Any defects longer than 0.1 inch are to be repaired (ground out and repair welded).

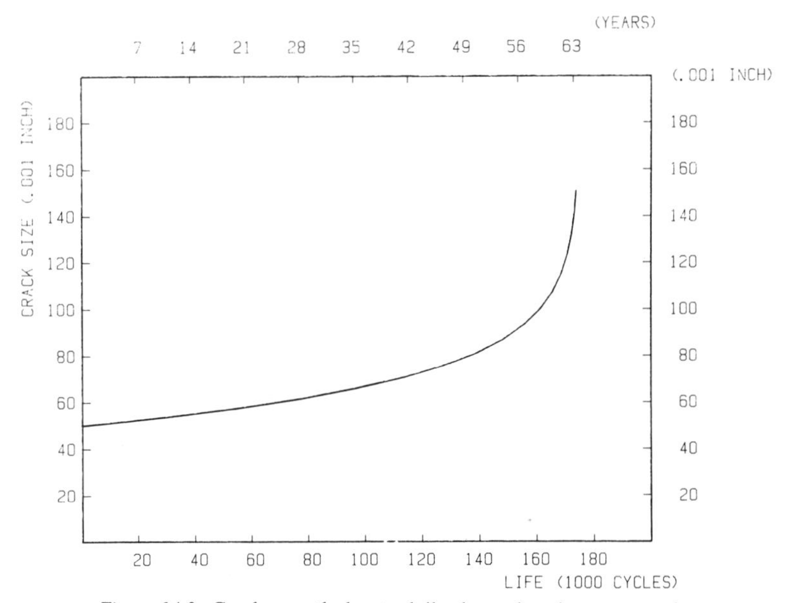

Figure 14.2. Crack growth due to daily thermal cycle as assumed.

The costs for inspection and repairs are estimated at an equivalent of three man years. The fracture control plan is implemented because the cost is negligible with respect to the anticipated cost of fracture.

NOTE: The above is a fictitious example. Numbers may bear no relation to reality. Certain aspects of the problem may have been overlooked.

14.3. Fracture arrest in ships

(a) *Problem definition*

Of some 2000 Liberty ships built during World War II, more than 100 broke in two; many others had serious cracks. Since then, ship design requirements have been changed to include mandatory fracture arrest strakes. These strakes are currently applied [1, 2] at the gunwale in deck and side plating (Figure 14.3) and at the turn of the bilge. In very large ships additional strakes are required. In general, the strake material is of a higher grade than the normal plating material. The effectiveness of the strakes is assessed in this section. Also the feasibility of hybrid arresters is assessed.

The purpose of an arrester is to stop a running fracture within the continuous structure, so that the structure with the arrested crack is still capable of sustaining appreciable loads until remedial action can be taken. This may be accomplished by means of heavy stingers, strakes or hybrid arresters.

The first type is a heavy stringer or girder. Load transfer from the cracked plate to the girder reduces the crack tip stresses and thus K (Chapter 9). An arrester of the second type is a welded-in strake of a material of higher toughness than the base plate. When the fracture enters the strake, it meets with a higher

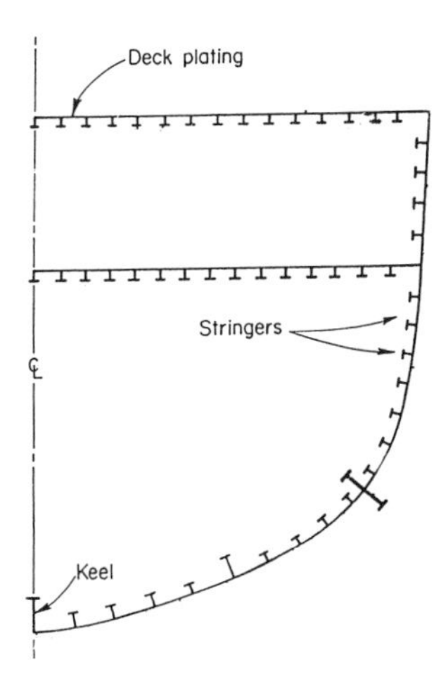

Figure 14.3. Cross section of ship hull.

toughness which may cause arrest. A hybrid arrester is a combination of the two other types, i.e. a heavy girder plus a high toughness strake.

Early research on arresters was conducted by the naval industry [3–7]. Then the aircraft industry began developing crack arrest technology [8–12], carrying it to a state of near perfection, and accounting for such things as stringer and fastener plasticity (Chapter 9). Modern airplane designs take full advantage of the arrest capability of skin-stringer structure. The present assessment of hybrid arresters draws upon aircraft technology.

(b) *Background*

A Navy ship is designed [13, 14] for a Standard Design Wave with a wave length equal to the Length Between Perpendiculars (LBP) and a wave height equal to 1.1 LBP, the LBP being practically equal to the ship length. Two cases are considered, one where the top of the wave is at midship (hogging) and one where the trough of the wave is at midship (sagging), as shown earlier in Figure 6.10. The Standard Design Wave provides the distribution of the upward forces on the ship; the downward forces can be obtained from the estimated weight distribution. These provide the bending moment.

The material allowables used are as follows (requirements [13, 14]):

Material	F_{tu} (ksi)	F_{ty} (ksi)	Max. allowable total stress (ksi)	Max. allowable bending stress (ksi)	Toughness at LAST (estimated) (ksi$\sqrt{\text{in}}$)
Mild steel	60	34	27	19	50
ABS-Grade E					90
HY-80	100	80	55	23.5	175
HY-130		130			220

If the ship is built of mild steel the maximum allowable live stress (due to applied loads) is 19 ksi. The maximum allowable for the 'total stress' is 27 ksi, which means that there is an allowance of 8 ksi for secondary stresses, due to dead weights, misfits, slamming, etc. These secondary stresses are not evaluated in the design. Their existence is acknowledged and a margin is provided by limiting the bending stress to 19ksi. Essentially, this provides a safety factor of $34/27 = 1.26$ against yield and of $60/27 = 2.22$ against F_{tu} (compare 1 and 1.5 respectively for airplane structures).

Shell plating, and the size and spacing of longitudinals (stringers) are selected to produce the section modulus required to sustain the bending at the Standard Design Wave with stresses less than or equal to the allowable; the hull is then roughly sized. The design, in particular the bottom, is checked for hydrostatic pressure and elastic stability to determine the final size and spacing of the longitudinals. The loop is closed by calculating the section modulus of the final design and by checking whether the bending stress is not in excess of the allowable. Naturally this summary is simplifying matters. Merchant ship hulls are designed in essentially the same manner, but the Standard Design Wave is different.

(c) *Arrest criterion*
Before the usefulness of arresters can be assessed, a criterion for arrest must be set. No arrester will be effective under all circumstances. Experience shows that ships may incur damage during severe weather [15] and that even vessels fitted with special steels may experience fractures. Yet, requiring arrest under the worst conceivable circumstances leads to inordinately heavy, closely spaced arresters, if arrest is at all feasible. The requirement would be unrealistic, because of the low probability that the worst circumstances concur with the presence of a critical defect. The toughness of ship steels depends strongly upon temperature. Hence, the fracture arrest capacity at the Lowest Anticipated Service Temperature (LAST) is the most relevant. Normally, a ship spends only a few percent of its life at the LAST, whereas the Design Wave is expected to occur only once in the ship life [13].

Measured and calculated stress spectra of commercial ships [16, 17] show that the once per lifetime (10^9 waves) peak-to-peak-stress range is on the order of 18–25 ksi. For one year of operation, the maximum range would be 14–19 ksi. If approximately two-thirds of the range is in tension, the secondary stress about 5 ksi, then the once-per-year tension stress would be 15–17 ksi. Hence, a good criterion would be that arrest must be possible at the LAST at stresses between 14 and 17 ksi.

(d) *Effectiveness of arrester strakes*
Linear elastic fracture mechanics apply at the LAST, especially because large cracks are concerned (Chapters 3 and 12). Dynamic effects are ignored initially, but their possible consequences will be discussed later. Four materials are considered, namely mild steel (MS), ABS Grade E steel, HY-80 and HY-130. The toughness of these materials at LAST were estimated as shown in the table above.

Using $\beta = 1$ for wide plates (Chapter 3), the residual strength curves can be calculated as in Figure 14.4. At the time a fracture in MS runs into an arrester strake, the problem becomes essentially identical to the case in which the whole plate is of the arrester material (both have the same modulus and thickness). Then consider a MS deck plate with arrester strakes. Both operate at the same stress. A crack of 4-inch length ($a = 2$ inches) in MS would be critical at 20 ksi (point A in Figure 14.4). It could be arrested by an ABS Grade E strake if the strake was no more than three inches away from the original crack tip, because at 20 ksi the strake is critical at $a =$ 5-inch (point D). Similarly, HY-80 and

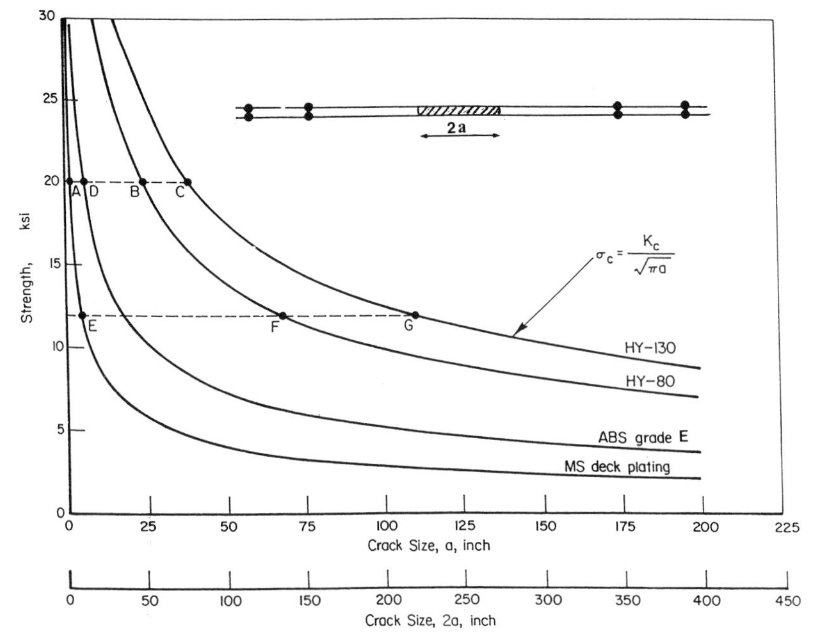

Figure 14.4. Arrest capability of various strake materials.

HY-130 strakes would provide arrest if the strakes were 23 and 28 inches away respectively (points B and C). If arrest were required at a stress of 12 ksi, a somewhat longer crack (point E) could be tolerated in the MS. Arrest would occur in HY-80 arrester strakes at a distance of approximately 70 inches or in HY-130 strakes at a distance of about 115 inches (points F and G).

On this basis residual strength diagrams for plates with arresters can be constructed. If arrest were required for example at a stress of 14 ksi, the diagram of Figure 14.5 would apply. For this situation, HY-80 arrester strakes at 100-inch spacing would be required. The width of the strakes was taken here as 25 inches, but the necessary strake width depends upon the dynamic effects (see later). Figure 14.5 shows that any crack smaller than four inches ($2a = 8$ inches) will not be arrested because it will be critical at a stress higher than 14 ksi (e.g. point C). When a fracture precipitated by a smaller crack (higher stress) has propagated across the MS bay and reaches the arrester, the arrester cannot sustain the damage at this higher stress (point D). Cracks longer than 4 inches will be critical at stresses below 14 ksi. For example, a fracture starting at point A runs to B, where it is arrested. In the above it was implicitly assumed that the crack in the MS would occur in the center of the bay, but in the absence of dynamic effects, the location of the crack (if it has two tips) is immaterial. A fracture starting at both tips would still propagate through the whole bay.

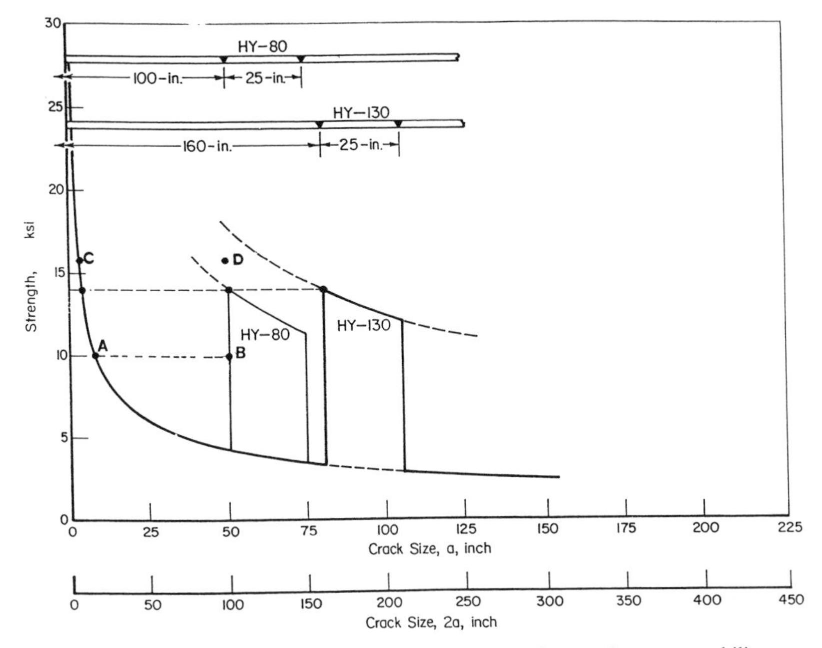

Figure 14.5. Required spacing of two types of strakes for equal arrest capability.

The examples apply to plates, but a hull consists of plate and stringers. (Figures 14.3). A running fracture in the plate can and will enter the welded stringers, severing them in the process. The load originally carried by these longitudinals also has to bypass the crack, which will increase the stress intensity factor, and therefore, reduce the arrest capability of the strakes. In other words Figures 14.4 and 14.5 may draw too optimistic a picture of the effect of strakes. The figures show that crack arrester strakes as used presently have little effect. If a crack would occur anywhere in the deck plating in Figure 14.3, it could essentially run through the entire deck (e.g. 60 ft) if arresters were present only at the gunwale.

For a 60 foot crack size, the residual strength of HY-80 is 5.2 ksi; arrest in general then would be possible only at stresses below 5.2 ksi.

(e) *Principle of hybrid arresters*

For the discussion of load transfer arresters the reader is referred to Chapter 9. A summary follows below. Consider a plate with arrester stringers and a central crack shown in Figure 14.6. Since there are only small stress gradients, the nominal stress in the arrester is equal to that in the plate. As such the arresters are effective in sharing the load on the uncraked structure and they can be accounted for in the sizing of the structure for the Standard Design Wave. In

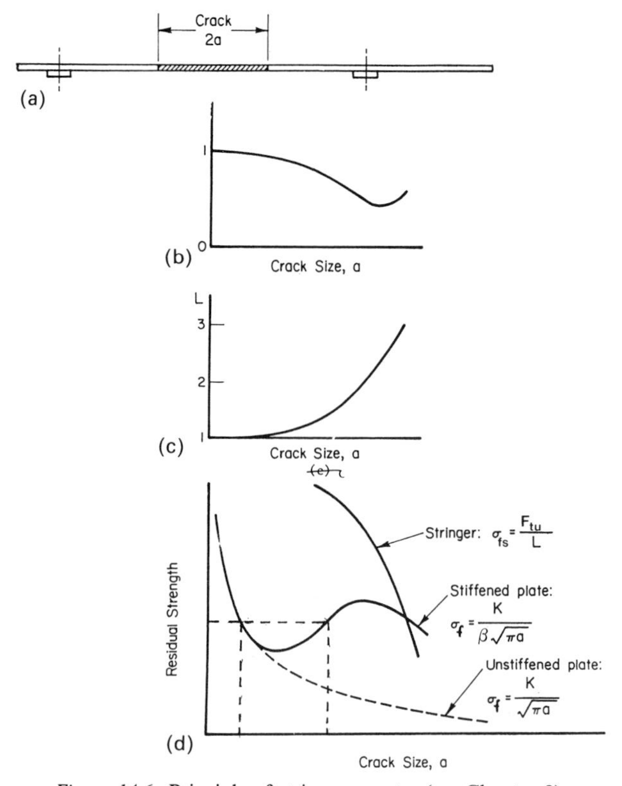

Figure 14.6. Principle of stringer arrester (see Chapter 9).

other words, they are not just an addition for the mere purpose of fracture arrest. In the absence of arresters, $\beta = 1$. The load originally carried by the cracked material has to bypass the crack. Some of this load will be transferred to the arresters which will cause a reduction of the stress intensity factor: $\beta < 1$. (Chapter 9). This causes the arrester stress to increase to $\sigma_l = L\sigma$, where L increases with crack size.

If the plate material has a toughness K_c, the residual strength of the unstiffened plate follows from $\sigma_{fr} = K_c/\beta\sqrt{\pi a}$ with $\beta = 1$ (dashed line in Figure 14.6). The residual strength of the plate with arresters follows from the same equation with $\beta(a) < 1$. Since β decreases with crack size, the residual strength curve shows the typical relative maximum. The nominal stress at which arrester fracture occurs is obtained as $\sigma_{fa} = F_{tu}/L$, where F_{tu} is tensile strength of the arrester material. See Chapter 9 for more details and a discussion of the arrester critical case.

The residual strength diagram is determined by β, L, K_c, and F_{tu} (of the stringer material), where β and L depend upon arrester spacing and stiffness, and upon the fastening system. They are also affected by plastic deformation in arrester and fastener holes. All these effects can be accounted for in the calculation. If the fastener spacing is smaller, load transfer is more effective, which results in larger L and lower β. Indeed, β is lower for welded arresters, but in that case the fracture can readily enter the arrester and a failed arrester is worse than none at all (Chapter 9).

A hybrid arrester is a combination of a strake and a load transfer arrester (Figure 14.7). The residual strength diagram of a strake arrester is shown schematically in Figure 14.7a, that of a stringer arrester in Figure 14.7b. Combining these leads to the residual strength diagram of the hybrid as in Figure 14.7c. The effect of the strake is a further increase of the relative maximum.

For a mere assessment of their effectiveness the use of handbook solutions [18, 8] for β is satisfactory (Figure 14.8). This case is for a fastener spacing equal to 1/12 of the bay width; smaller fastener spacings would give somewhat better results. The discussion will be for riveted hybrids; welding, if performed in a certain way, is feasible as will be discussed. The value of β depends upon the stiffening ratio.

Ship plating is already stiffened by stringers. These longitudinals are relatively light, but more important, will be penetrated by the fracture because they are welded to the plate. Unlike riveted stringers they are not effective as arresters. Special heavy stringers or girders are required. In order to avoid confusion about the use of the word stringer, they are called arrester in previous paragraph and in the following discussion. When the arrester must come into action, the 'hatched' plate and stringers in figure 14.9 will all be broken. The broken stringers cause an increase in β, depending upon their size and spacing. As a first approximation, an increase of β by a factor 1.2 was assumed on the basis of calculations for broken stringers [18]. This factor may be somewhat low, but the β from Figure 14.8 is too high, because the fastener spacing for the large bay widths considered here will certainly be less than 1/12th of the bay width. Thus,

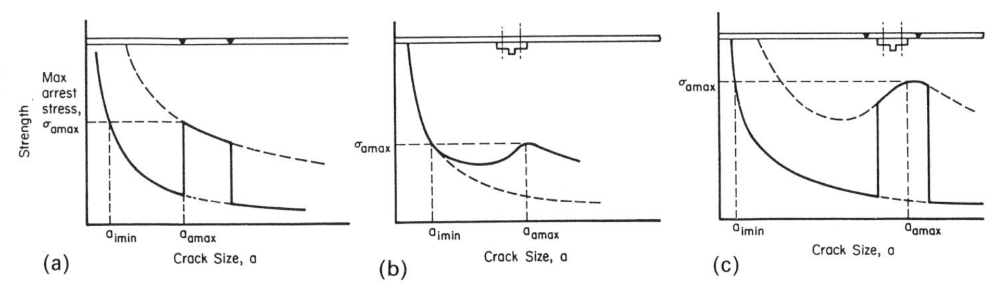

Figure 14.7. Principle of hybrid arrester. (a) Tough strake; (b) Stringer or Doubler; (c) Hybrid.

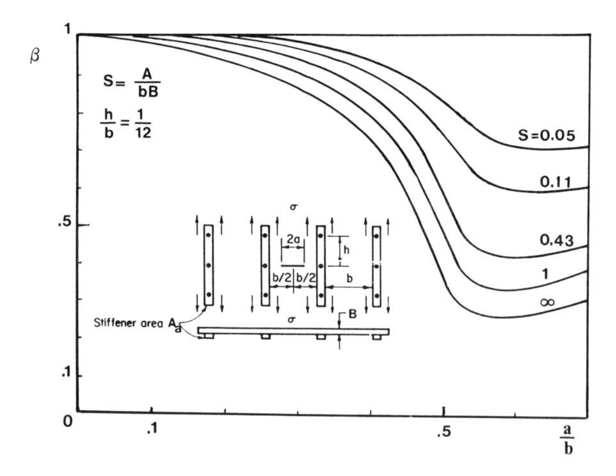

Figure 14.8. Geometry factors.

errors in the above assumptions will compensate one another. As the broken stringers in Figure 14.9 are part of the total load path interruption, they must be accounted for in arrester stiffening ratio. In order to avoid complications due to curvature and hydrostatic pressure flat deck plates were considered only. For an assessment as aimed at here, it is sufficient if β is within about 10% of the actual value; for actual design more rigorous procedures [9, 12] can be used. Plate critical conditions were assessed only; because of the high F_{tu} of HY-80 and HY-130 arrester critical cases will be rare.

(f) *Effectiveness of hybrid arresters*

The performance of two hybrid arrester designs is shown in Figure 14.10. Design B has a crack arrest capacity of 14 ksi. In order to obtain the same crack arrest stress by means of HY-80 strakes only, the strake spacing would have to be 50 inches (Figure 14.5), whereas the hybrid arresters could be spaced at 160 inches (Figure 14.10). If the same design (B) were built of HY-130, the arrest capability would go up to 18 ksi (Figure 14.11).

An example may show how the design of arresters would proceed. Let the problem be that of designing arresters with a spacing of 300 inches for a required arrest stress of 15 ksi. The two material candidates are HY-80 and Hy-130. Figure 14.8 is assumed applicable, but for an actual design more specific curves are needed [9, 12]. It is assumed that the general design specifies the deck plating to be 20.4 lbs/sq ft with deck stringers of 10.2 lbs/ft spaced at 30 inches (ship plating is specified in weight per foot). According to Figures 14.8 and 14.9, the minimum value of β occurs at $a/b = 0.6$. Therefore, the arrested crack size should be $a = 0.6 \times 300 = 180$ inches. It is assumed that the broken stringers

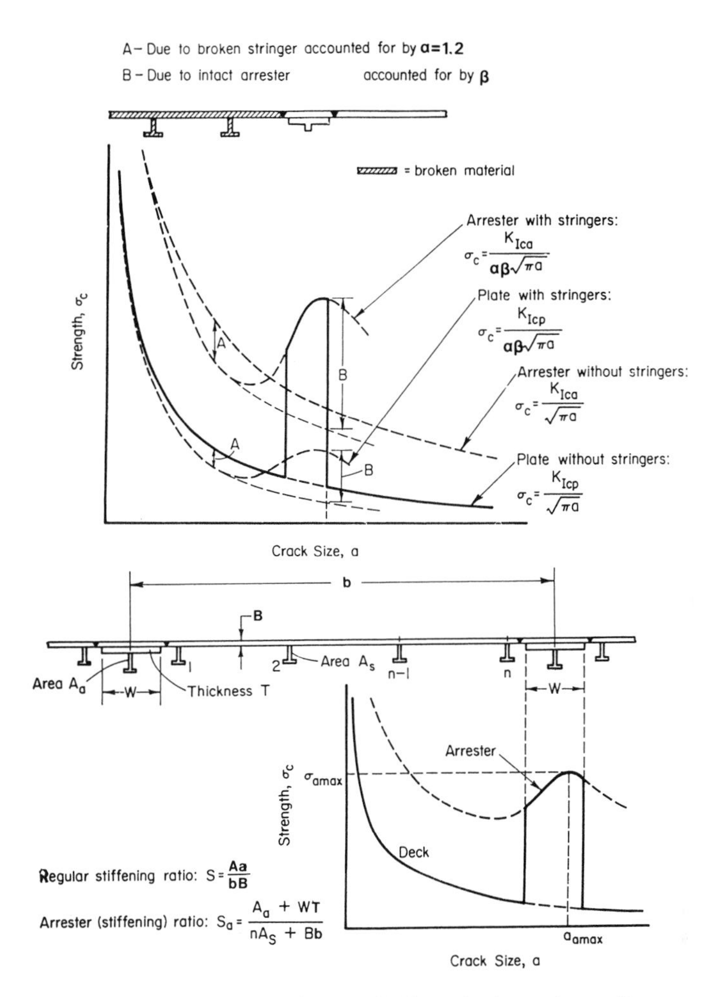

Figure 14.9. Effect of deck stringers. Top: increased K due to broken stringers. Bottom; redefinition of arrester ratio.

account for a factor of 1.2 as before. In order for this crack to be arrested, the stress intensity $K = 1.2\beta\sigma\sqrt{\pi a}$, must be less than the toughness ($K_c = 175\,\text{ksi}\sqrt{\text{in}}$ for HY-80). Since arrest must occur at $\sigma = 15\,\text{ksi}$, the required β follows from:

$$\beta = \frac{K_c}{1.2\sigma\sqrt{\pi 2}} = \frac{175}{1.2 \times 15 \times \sqrt{\pi \times 180}} = 0.41.$$

For $\beta = 0.41$ at $a/b = 0.6$, the required stiffening ratio is $S = 0.43$, according to Figure 14.8.

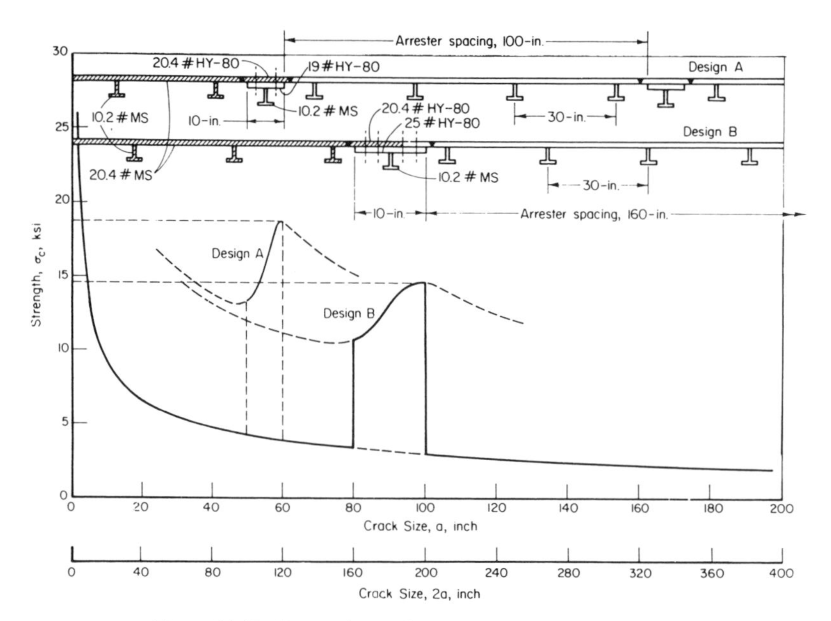

Figure 14.10. Comparison of two arrester designs in HY-80.

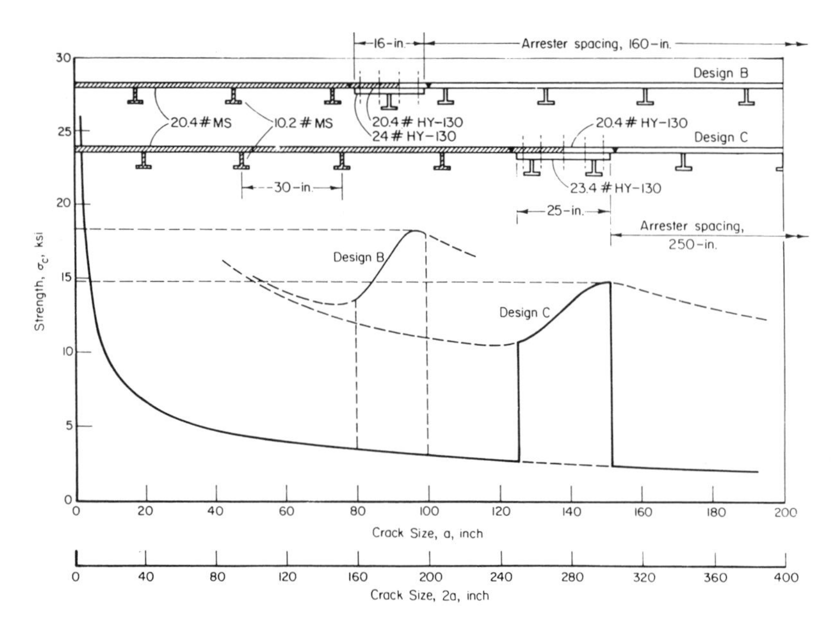

Figure 14.11. Comparison of two arrester designs in HY-130.

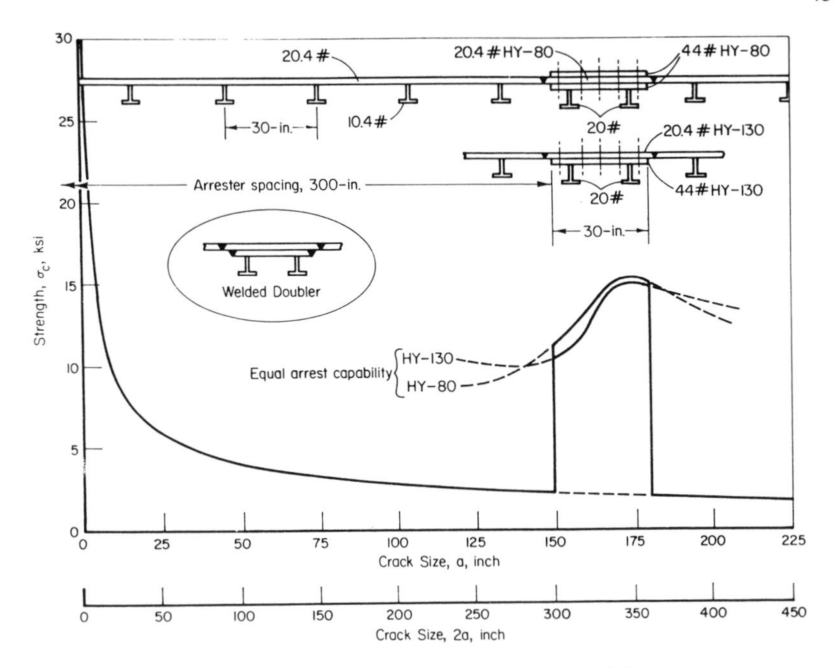

Figure 14.12. Two designs with equal arrest capabilities.

The deck plating being 20.4 lbs/sq ft and the spacing between arresters 300 inches, the total weight of the plating is $20.4 \times 300/12 = 510$ lbs/ft. Attached to this are nine stringers of 10.2 lbs/ft with a total weight of 92 lbs/ft. Hence, the weight of the cracked bay is $510 + 92 = 602$ lbs/ft. With a required arrester stiffening ratio of $S = 0.43$, the required arrester weight is $0.43 \times 602 = 259$ lbs/ft.

The arrester width is taken as e.g. 30 inches. The arrester strake is of equal weight (equal thickness) as the deck plating and the arrester doubler has the same width as the strake. If the arrester is provided with two stringers of 20 lbs/ft (40 lbs/ft total), the doublers have to provide $259 - 40 = 219$ lbs/ft. Then the doubler plating should be $219/30/12 = 88$ lbs/sq ft. Since such heavy plating might be impractical, two doublers each of 44 lbs/sq ft could be used. Given that $S = 0.43$, the β-values for other crack sizes can be derived from Figure 14.8 and the residual strength calculated from $\sigma_{fr} = K_c/(1.2\beta\sqrt{\pi a})$. The residual strength diagram is so obtained (Figure 14.12). Also the design for HY-130 is shown ($K_c = 220\,\text{ksi}\sqrt{\text{in}}$).

The example shows that the design procedure is straightforward if the required arrest stress is specified. Almost any arrest requirements can be met, but the resulting design may be expensive and impractical.

The analysis is somewhat crude, but is meant only as an assessment of feasibility. As such it shows that hybrid arresters are feasible. Required spacings would be on the order of 200–300 inches for an arrest capability of about 15 ksi, in accordance with the criterion discussed (once per year wave at LAST). At temperatures above LAST the arrest stress would be higher. In the examples, the load tranfer part of the arresters is riveted to the strake. This would prevent the fracture from entering into the arrester and from rendering it useless. However, welding the doubler plates to the strakes along the edges of the doubler may a viable solution, as the filled weld (Figure 14.12 insert) would provide relatively little opportunity for the fracture to enter the doubler. In addition, the weld would provide small (zero) 'fastener' spacing which would reduce β, which can be exploited to increase arrester spacing or to increase the arrest stress. However, the fillet weld must have sufficient shear strength to permit load tranfer from the cracked plate to the doubler (Chapter 9). A doubler plate is necessary. The required stiffening should not be effected by using just a thicker strake because the running fracture would then sever the reinforcement.

(g) *Dynamic effects*

Dynamic effects may be due to (1) the material properties, (2) the stress intensity, and (3) the kinetic energy. The material properties (including toughness) are affected by the rate of deformation. If the dynamic toughness (fast moving fracture) is known, its effect can be accounted for. Also the stress intensity is different than in the quasi-static case, but at fracture velocities observed in steels, the effect is small. However, the contribution of kinetic energy could change the conclusions reached.

The rate of displacement, v, of the material at each side of the fracture is very high and proportional to the fracture speed $\mathring{a}$: the material has a kinetic energy $1/2\,\mathrm{mv}^2 = 1/2Cm\mathring{a}^2$, where m is the specific mass, and C a proportionality factor depending upon location. The elastic energy release rate during fracture is G, the energy required for fracture R (Chapter 3). The surplus $(G - R)$ is converted into kinetic energy, so that

$$\int_{a_1}^{a_2} (G - R)\ \mathrm{d}a = \int_V \frac{1}{2} Cm\mathring{a}^2\ \mathrm{d}V,$$

where V is the volume of material involved. Since G and R are function of a, the equation can be solved [12] to obtain the fracture speed $\mathring{a}$.

When the fracture is to be arrested, this kinetic energy has to be dissipated (the displacements MUST come to a halt). Since the kinetic energy enters the energy conservation equation, fracture must be analyzed in terms of the energy balance. It can be shown (Chapter 4) that the elastic energy release per unit fracture extension, da, under constant stress is equal to $G = K^2/E = \pi\beta^2\sigma^2 a/E$. The fracture energy per unit fracture extension is $R = K_c^2/E$. It will be assumed here that R is approximately constant, and it will be shown later that this

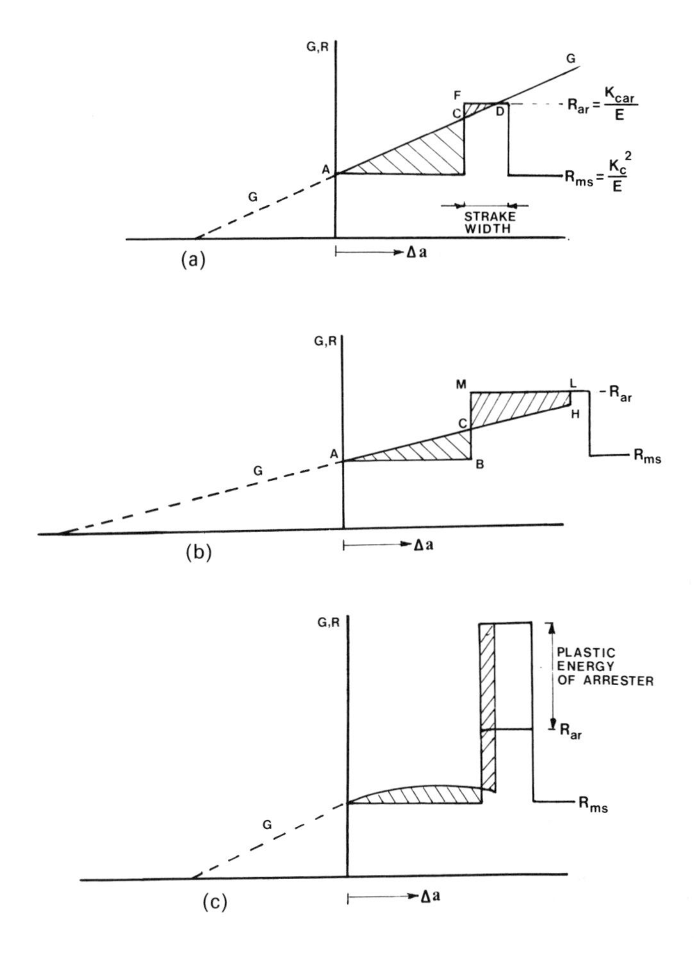

Figure 14.13. Dynamic effect. (a) Arrester strake of insufficient width; (b) Strake of sufficient width; (c) Hybrid.

assumption does not affect the conclusions for the case of hybrid arresters.

If the structure contains a crack arrester strake of a high toughness material, the fracture energy (K_{car}^2/E) suddenly increases as shown in Figure 14.13a. Assuming that $\beta = 1$, the energy release G is represented by a straight line. The kinetic energy, $(G - R)\,\mathrm{d}a$, is represented by the area between the G and R lines. When the fracture enters the arrester this kinetic energy is given by the area ABC (Figure 14.13a). At that time the energy release rate (driving energy) is less than the fracture energy. In other terms: $G < R_{ar}$, which means $K^2/E < K_{car}^2/E$, or $K < K_{car}$. The latter criterion was used in the foregoing analysis, but this condition may not be sufficient for arrest.

The material around the crack is still displacing at a high rate and 'contains' a kinetic energy represented by the area ABC. This energy may now drive the fracture. While fracture continues, G increases to point D where $G > R_{ar}$, so that fracture is driven again by G; there is no arrest. While the fracture propagates from C to D, the energy release is smaller than the fracture energy, ($G < R_{ar}$). The difference, $R_{ar} - G$, has to come from the kinetic energy. Therefore, the area CDF represents the part of the kinetic energy used for fracture from C to D. Since CDF < ABC, arrest does not occur; the fracture slows down only, because of the loss of kinetic energy (CDF), but it continues because $G > R_{ar}$ at point D. For the fracture to be arrested, all kinetic energy must be used (displacements must stop). This means that the arrester strake must have sufficient width, as illustrated in Figure 14.13b. Because area CHLM = area ABC, all kinetic energy would be used (all displacement motion stopped) and arrest would occur at H.

The case of a hybrid arrester is depicted in Figure 14.13c. Due to the load transfer to the arrester β decreases with increasing a. Therefore the G-line ($G = \pi\beta^2\sigma^2 a/E$) slopes down as shown. Without dynamic effects arrest would occur at C. This was the arrest condition used in the foregoing analysis ($G < R_{ar}$ or $K < K_{car}$), but if kinetic energy can be used the fracture may run through the strake as before. However, in the case of a hybrid arrester there is a much more effective energy dissipator than the high toughness trake, namely the doubler plate with attached stiffeners (both uncracked). If fracture were to continue (large local displacements), the doubler and stiffeners would have to undergo large strains in the region crossing the fracture path. The plastic deformation energy of this uncracked material constitutes an effective increase in the fracture resistance R, as shown in Figure 14.13c.

As a result, the hybrid arrester can easily absorb all kinetic energy. Consider for example, arrester design B of Figure 14.11. The energy diagram for this arrester is shown in Figure 14.14. Using $K_{car} = 174\,\text{ksi}\sqrt{\text{in}}$ for HY-80 as in the foregoing, provides $R_{ar} = K_{car}^2/E = 175^2/30000 = 1020\,\text{lbs/in}$, as shown in Figure 14.14. According to Figure 14.11 the maximum static arrest stress for this design was 15 ksi; the G-line for this stress is shown in Figure 14.14, based upon the appropriate values of β and with the factor 1.2. The static solution predicts arrest at A as discussed. For the estimate of the plastic deformation energy that can be absorbed by the doubler and stringers, it is conservatively assumed that only one inch (length) of these takes part in the deformation. Also assuming, again conservatively, that the fracture strain is 30% (uncracked) and that there is no strain hardening, the 0.5 inch (thick) doubler can absorb 11000 lbs/(inch width), which is the effective R of the doubler. The effective R of the stringer can be estimated in the same manner, bringing the total effective R of the hybrid arrester to at least 15000 lbs/(inch width). Thus, the kinetic energy can be

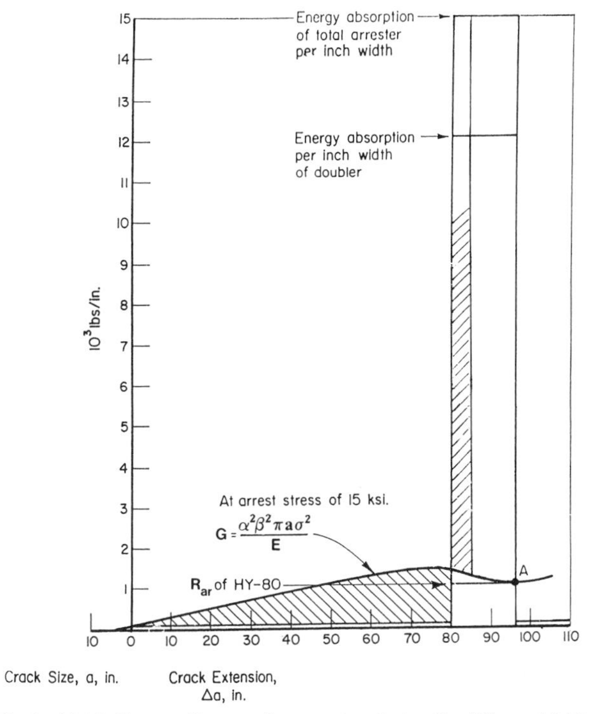

Figure 14.14. Energy diagram for arrester design B of figure 14.10.

absorbed; only about 40% of the energy absorption capacity is used arrest occurs. Since the estimate of the energy absorption capacity was low, the above numbers are conservative.

The example shows that the dynamic effect in the case of hybrid arresters is negligeable. In fact, the statically calculated arrest stress of 15 ksi still stands. It also is obvious now that the assumption of constant R has little effect upon the conclusion. Whether R is increasing or decreasing, it remains small in comparison to the effective R of the load tranfer arrester. The example also illustrates why quasi-static crack arrester analysis is in excellent agreement with experimental data [9–12], and why it is used so successfully in aircraft design.

(h) *Fracture control considerations*

Strake arresters as presently used have a limited arrest capability, but hybrid arresters can be effective. The cost of the latter may be a drawback. Considering costs of material and additional welding, it is estimated that hybrid arresters might increase cost of a ship by 1 to 1.5%. If fewer than 1 out of 100 ships break in two or have severe fractures (cost of fracture less than one percent of that of a ship), it may not be economical to spend 1–1.5% on fracture control.

Design of arresters and arrest capability must be based upon an arrest criterion to be specified in the design rules. As discussed it might not be rational to require arrest at the standard design wave at LAST, because of the extremely low probability of the simultaneous occurrence. A more realistic criterion may be similar to the one used here.

14.4. Piping in chemical plant (fictitious example)

(a) *Problem definition and fracture control objective*

Two vessels operating at low pressure are connected by an annealed austenitic stainless steel pipe with an elbow (Figure 14.15). The plant consists of a series of such arrangements. The temperature varies from 350F to 200F in the last one. Stress corrosion cracking may occur at the welds in the piping.

A leaking pipe can be tolerated. A pipe break would cause interruption of the flow and could result in dangerous overheating further down. The fracture control objective is to ensure that pipes will never be completely severed.

(b) *Loads, stresses and data*

The pressure is low; there is no significant stress due to pressurization. Essentially, the loading is due to thermal expansion. As the process is continuous, the temperature of the pipe is constant and hence the thermal stress is constant. Thermal expansion of one of the legs of the elbow has to be accomodated by bending of the other leg (Figure 14.16). This introduces a load P and a bending moment M. Because the legs are symmetric the angle of deflection is zero at the elbow. The loads, moments and stresses are as shown in Figure 14.16. The highest moment occurs at the welds closest to the vessels. If it is assumed that the pipes were installed at room temperature (70 F) then the maximum $\Delta T = 350 - 70 = 280$ F. Substiting this and other relevant numbers in Figure

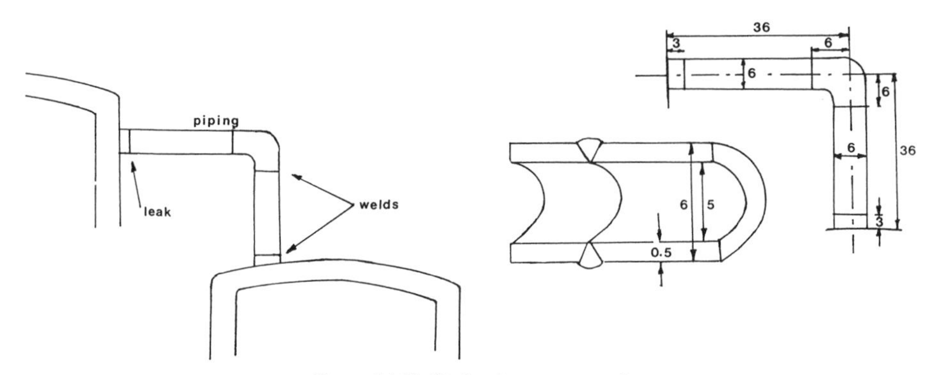

Figure 14.15. Piping between vessels.

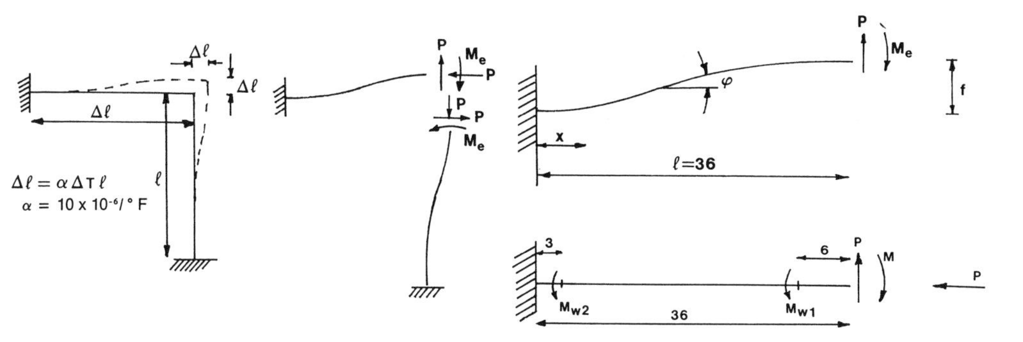

$$\varphi = \frac{Pl^2}{2EI} - \frac{M_e l}{EI} = 0 \text{ or } M_e = \frac{1}{2}Pl \qquad (\varphi = 0 \text{ at } x = l)$$

The deflection f at $x = l$ is equal to the thermal expansion of the other leg, which is $\Delta l = a\Delta Tl$.

$$f = \frac{Pl^3}{3EI} - \frac{M_e l^2}{2EI} = \alpha \Delta Tl$$

at Weld 1:

$$M_{w1} = M_e = P\lambda = 6\,\alpha\Delta T\, EI/l - 12\alpha\Delta T\, EI\frac{l}{6l^2} = 4\alpha\Delta T\, EI/l$$

at Weld 2:

$$M_{w2} = Pl - M_e = 12\,\alpha\Delta T\, EI\frac{11l}{12l^2} - 6\,\alpha\Delta T\, EI/l = 5\,\alpha\Delta T\, EI/l$$

elastic bending stress

$$\sigma_b = \frac{M_w}{I}\frac{D}{2} = \frac{5}{2}\alpha\Delta T\, ED/l = \frac{5}{2} \times 10 \times 10^{-6} \times 280 \times 30 \times 10^3 \times 6/36 = 35\,\text{ksi}$$

longitudinal compressive stress

$$\sigma_l = -\frac{P}{\frac{\pi}{4}(D^2 - d^2)} = \frac{-12\,\alpha\Delta T\, EI}{\frac{\pi}{4}(D^2 - d^2)\, l^2} = -3.0\,\text{ksi}$$

Figure 14.16. Stresses in piping.

14.16 provides $M_w = 5 \times 10 \times 10^{-6} \times 280 \times 30 \times 10^6 \times \pi/64(6^4 - 5^4)/36 = 384000$ in-lb, so that the acting stress is

$$\sigma = M \times \frac{D_0}{2} \Big/ \left[\frac{\pi}{64}(D_0^4 - D_i^4)\right] = 64 \times 384 \times 3/[\pi(6^4 - 5^4)] = 35\,\text{ksi}$$

Austenitic stainless steel operating at 200 F to 350 F is very ductile so that collapse may be the govering failure mode. Collapse conditions are evaluated in Figure 14.17 (see also Chapters 2, 3 and 10).

Collapse strength data for the steel are provided in Figure 14.18. The data are for 5/8-in plate, which is close enough to the 0.5-in wall thickness of the pipe. Obviously, the plate fractures occurred at constant net section stress, with a collapse strength of $F_{col} = 50$ ksi at 200 F and $F_{col} = 30$ ksi at 500 F. Therefore it is estimated that $F_{col} = 40$ ksi at the operating temperature. Also, the J_R curve and stress strain data are given in Figure 14.18.

(c) *Analysis*

The question is whether leaking pipes could cause complete fracture. Therefore, the crack of concern is a circumferential through-the-thickness crack (Figure 14.17). The stresses at the crack location are due to the bending moment M and axial compressive load P. The longitudinal compressive stresses due to P are small as compared to the bending stresses. Neglecting this compressive stress

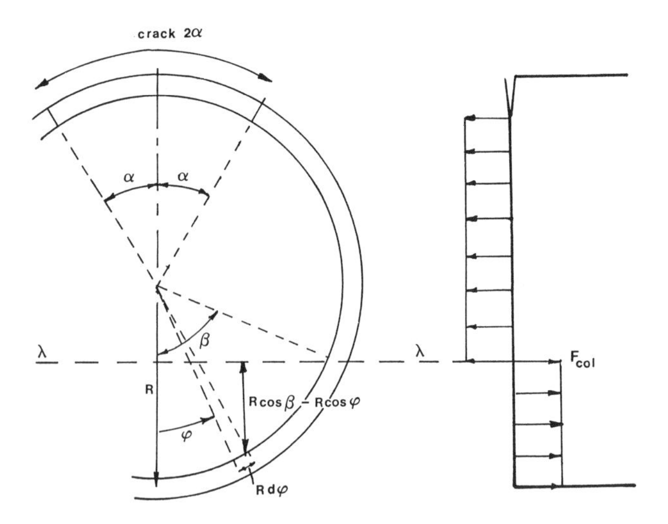

horizontal equilibrium

$$2\beta\ BF_{col} = (2\pi - 2\alpha - 2\beta)\ BF_{col} \quad \text{or} \quad \beta = (\pi - \alpha)/2$$

(collapse) moment M_{col} around the line $(\lambda - \lambda)$

$$M_{col} = 2\int_0^{\beta} BF_{col}\,(R\cos\varphi - R\cos\beta)R\,d\varphi + 2\int_{\beta}^{\pi-\alpha} BF_{col}\,(R\cos\beta - R\cos\varphi)\,R\,d\varphi$$

$$M_{col} = 2\,BR^2F_{col}\left(2\cos\frac{\alpha}{2} - \sin\alpha\right)$$

$$R = (D + d)/4 = (6 + 5)/4 = 2.75\,\text{in} \qquad M_{col} = 303\left(2\cos\frac{\alpha}{2}\sin\alpha\right)$$

Figure 14.17. Collapse conditions.

will be conservative and simplify the problem. The collapse analysis is shown in Figure 14.17. The residual strength can now be calculated for EPFM fracture and collapse. Results are shown in the residual strength diagram of Figure 14.19. Collapse is prevalent (lowest). Through cracks up to $2a = 4$ inches can be sustained, without fracture, at the acting stress of 35 ksi.

(d) *Fracture control plan*
A through crack of a certain angle will form by penetration of a stress-corrosion crack of equal angle. Stress-corrosion cracks will grow in the area of highest tension stress during the normal (elastic) loadings; hence it is unlikely that a stress-corrosion crack grows over an angle larger than 70 degrees before break through. Thus it can be concluded that leaks and no breaks will occur. (Note, however, that the bending moments should be determined at which surface flaws would break through. These moments should be lower than the moments for propagation of the through crack for leak-before-break under all circumstances, see Chapter 9.)

Since leaks are expected to be readily apparent, no fracture control plan would be necessary. Nevertheless, the following measures are implemented. (1) Installation of leak detectors at all pipes. (2) Daily walk-around visual inspection for leaks. Note: The above serves as an example. All numbers are fictitious. Some aspects of the problem may have been overlooked.

14.5. Fatigue cracks in railroad rails

(a) *Definition of the problem*
Fracture due to fatigue cracks in railroad rails is a common cause of derailment accidents in the USA. Reduction of the number of such fractures may be

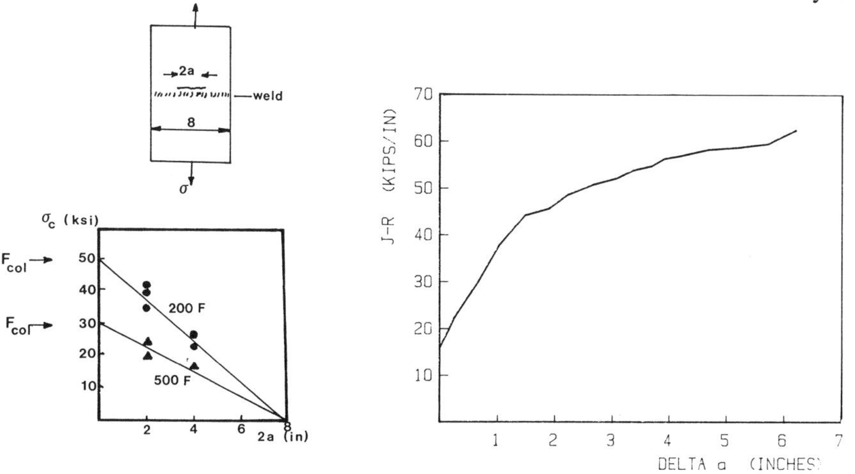

Figure 14.18. Material data (fictitious).

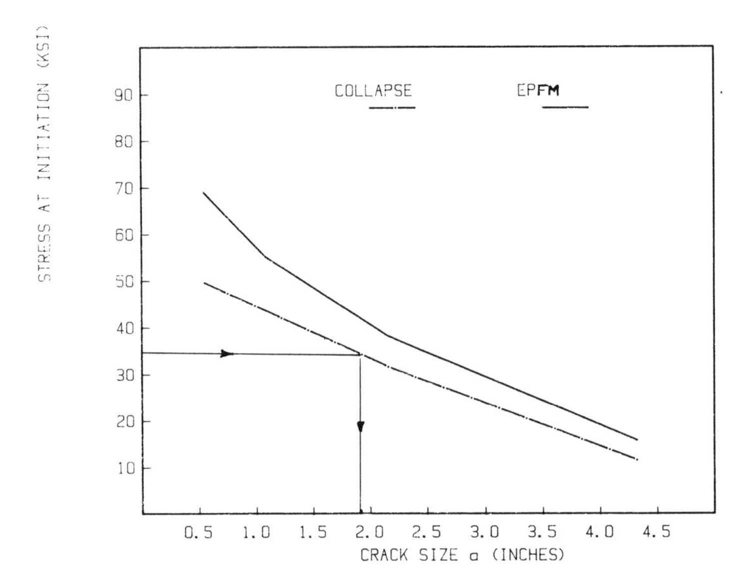

Figure 14.19. Residual strength diagram.

achieved by track maintenance, reduction of traffic (loads), or replacement of rails after timely detection of fatigue cracks through periodic inspection. Such fracture control measures can be effective only if adequate methods exist to predict the rate of crack growth, which requires knowledge of service loads, rail stresses, and of crack growth properties of rail material.

A comprehensive program to address this problem funded by the US Federal Railroad Administration [19–26], is outlined in Figure 14.20. Samples of rail obtained from track of various railroads were used to measure material data, including toughness and $da/dN - \Delta K, R$, as well as residual stresses. Tracks were straingaged to obtain loads and spectra. Tests, using simulated service loading (train-by-train loading) were performed to verify crack growth integration procedures. The objective was to arrive at a suitable damage tolerance requirement for railroad rails.

Rails are inspected for cracks by means of a car equipped with ultrasonic and magnetic inspection devices running along the rail. Cracks in different directions may initiate at various locations in the rail head; these are internal flaws of quasi-elliptical shape. Some are in a longitudinal plane, either vertical or horizontal, some in a transverse plane (commonly called detail cracks). An example of the latter is shown in Figure 14.21 [23, 25].

(b) *Loads and stresses*

Track was straingaged (Figure 14.20) at several locations to measure load-spectra for various kinds of usuage [19]. Finite element analysis was performed [20]

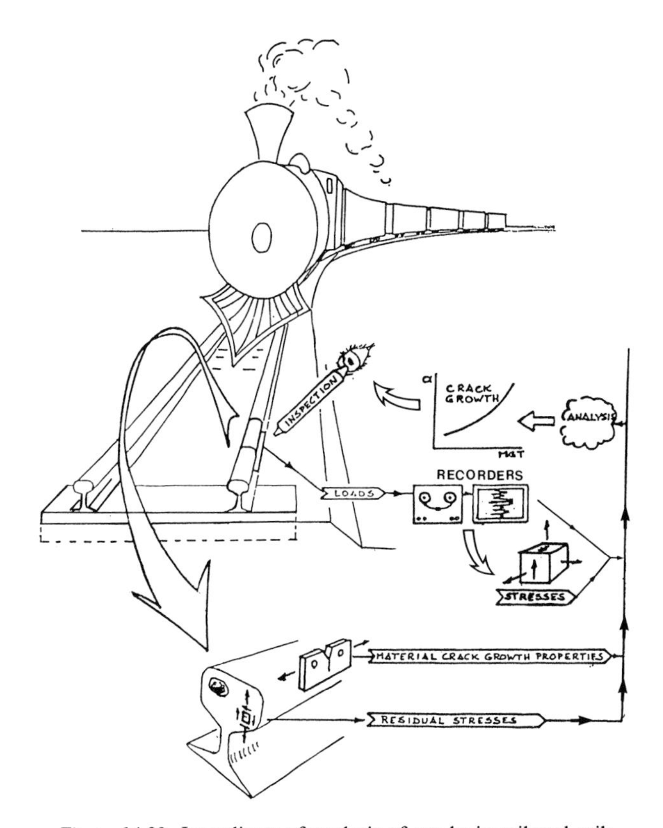

Figure 14.20. Ingredients of analysis of cracks in railroad rails.

to obtain the stresses in the rail from the loads, and subsequently the stress exceedance diagrams for different usage and locations in the rail head. A typical exceedance diagram [21], shown in Figure 14.22, provides stress exceedances during passage of 1 MGT (1 Million Gross Tons) of traffic. The diagram is non-linear and asymmetric, which is typical for man-induced exceedance diagrams (Chapter 6).

The stresses are given as ranges, $\Delta\sigma$, but since there is no live stress without a load, these are excursions from zero. When a wheel passes the rail, there will be downward bending, so that the stresses shown in the diagram are essentially compressive stresses (some tension occurs when the wheel is further away due to a tendency of the rail to be 'lifted' from the roadbed).

The question arises why fatigue cracks occur when the stresses in the rail head are compressive. The answer lies in residual stresses (Figure 14.23). Continuous 'pounding' by the wheels causes plastic deformation of the upper layers (3 mm) of the railhead, so that this layer is stretched (rolled out). However this layer,

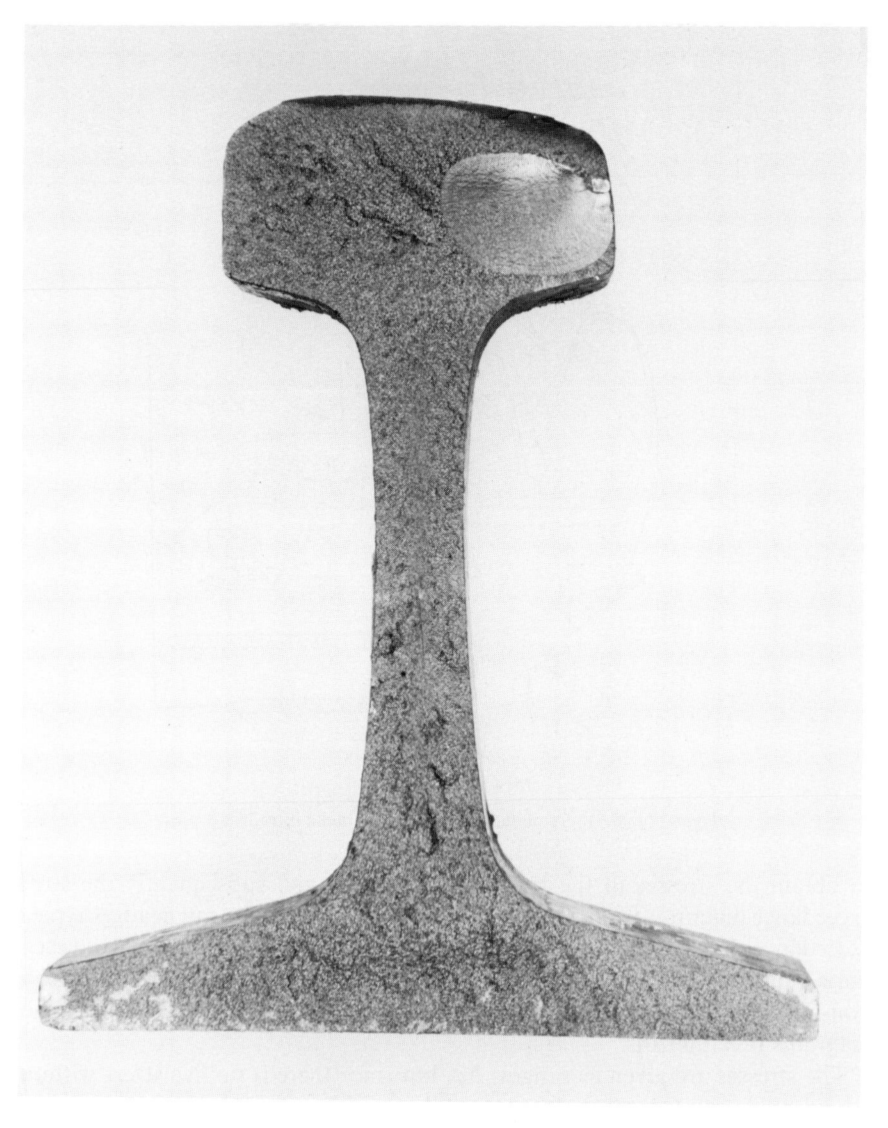

Figure 14.21. Fracture of railroad rail as a result of detail crack. Courtesy TSC and Battelle.

being attached to the head, must remain of the same length, which requires a (residual) compressive stress. Residual stresses must be in internal equilibrium, so that, there will be residual tensile stresses in the rest of the rail head as shown in Figure 14.23. The residual tension stresses are very high. Compressive stresses

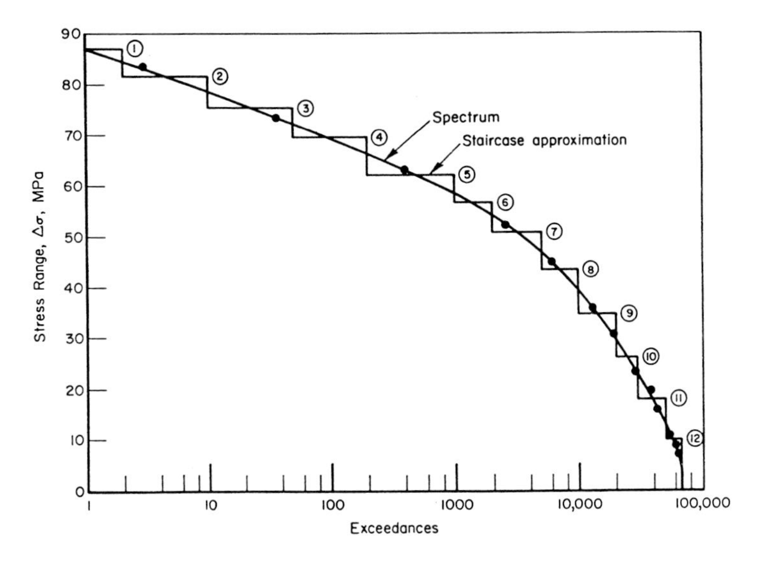

Figure 14.22. Stress exceedance diagram for 1 million gross tons (1MGTM) of traffic.

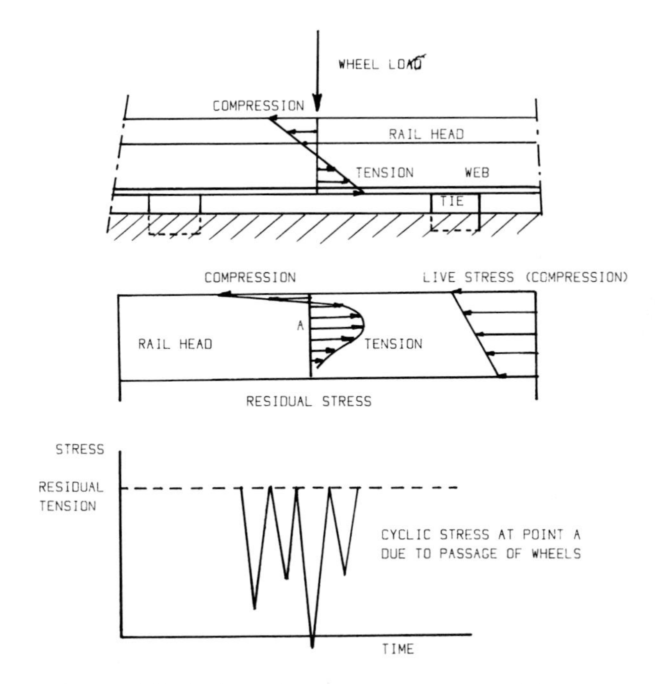

Figure 14.23. Residual stress and cyclic stress in rail head.

due to passing wheels will be superposed on the residual tension. Consequently, the cyclic stresses are actually in tension as shown.

Analysis requires knowledge of the residual stresses. These were measured [22] on samples cut from tracks with different traffic experience, which were straingaged (Figure 14.24) and subsequently sectioned. Residual stress fields can be reconstituted from the output of the gages after sectioning [22]. Various other techniques were used for the same purpose. The residual stress field is a very complicated tri-axial field with large gradients. Resulting tensile stresses are shown schematically in Figure 14.25.

The detail crack (transverse) usually starts in a horizontal plane, with high residual tension (Figure 14.26). This phase of cracking is called a 'shell'. If the shell crack perturbates into the vertical plane it experiences increasing longitudinal tension, so that the perturbation persists and the shell turns into a 'detail crack' [23]. The stages of development are shown schematically in Figure 14.26. (Other hypotheses have been advanced to explain the turning of the shell.)

The analysis problem concerns cracks of complex shape (β) in a stress field with large gradients. By using a reference stress, the stress gradients can be accounted for in β (Chapter 8), but the variation of β along the crack front must be accounted for by dependent crack growth analysis for more than one direction, because growth depends upon the changing crack shape (see surface flaws; Chapter 8). The case is more complicated than a normal surface flaw as there is no symmetry, and the analysis should be performed for several directions simultaneously (instead of two for common surface flaws).

(c) *Material data*

Almost 70 rail samples were collected from different locations, and characterized according to weight, year of production, chemical composition, and mechanical properties. Crack growth data were obtained from all samples, for various R-ratios, temperatures, and orientations.

Data for some individual rails and the scatter band of all samples are shown in Figure 14.27. Data trends for room temperature crack-growth in LT orientation are shown in Figure 14.28. The variability is substantial due to differences among rails. This is true variability as opposed to apparent scatter as discussed in Chapter 7. It will affect general purpose predictions; either statistical methods must be used to deal with this problem, or predictions must be made for specific (individual) rails.

(d) *Model calibration and analysis*

The objective called for a general procedure to predict crack growth in rails under service loading. This is a complex problem of a quasi-elliptical embedded flaw in a nonuniform stress field and a variable-amplitude stress history. The effects of variable-amplitude loading (retardation) were evaluated for through-

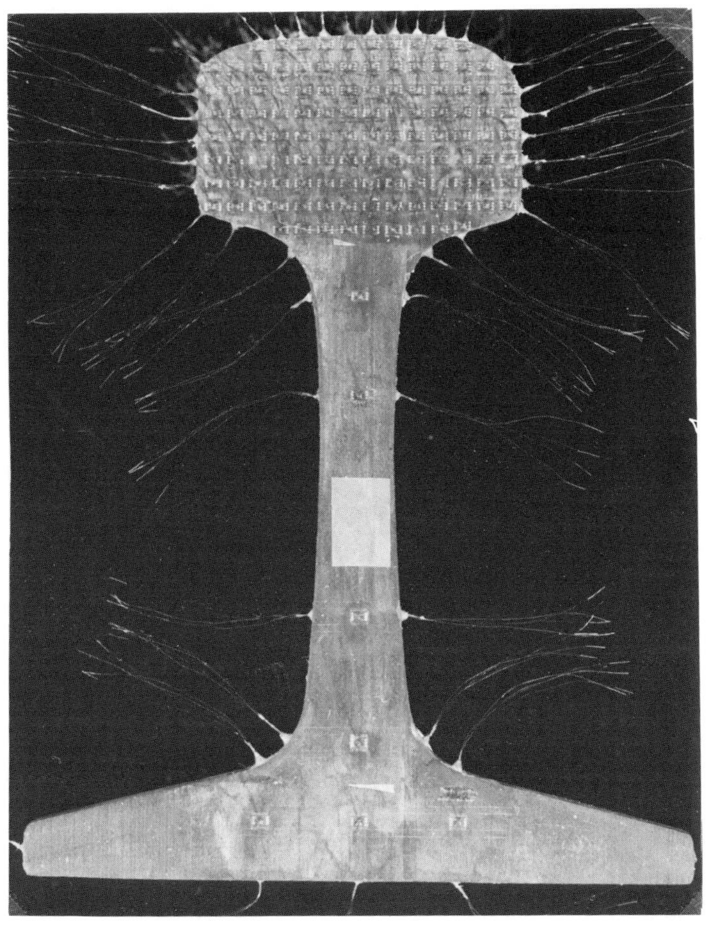

Figure 14.24. Strain gages on rail for measurement of residual stress. Courtesy TSC and Battelle.

the-thickness cracks subjected to simulated service stress histories. If crack growth under these circumstances can be predicted, the model can be generalized to include other complexities.

Stress exceedance diagrams (Figure 14.22) were used to generate semi-random service stress histories, employing the procedures discussed in Chapter 6. A stress history with six different trains (Figure 14.29) was established. The cycle content simulated actual trains according to length and weight. A mixture of 2-A1, 6-A2, 12-A3, 120-B, 20-C, and 10-D trains was used. Cyclic stresses due to passage of a car are shown in Figure 14.30, but simpler stress histories (elimination of small cycle) are desirable for analysis, although not necessary (for truncation see Chapter 6).

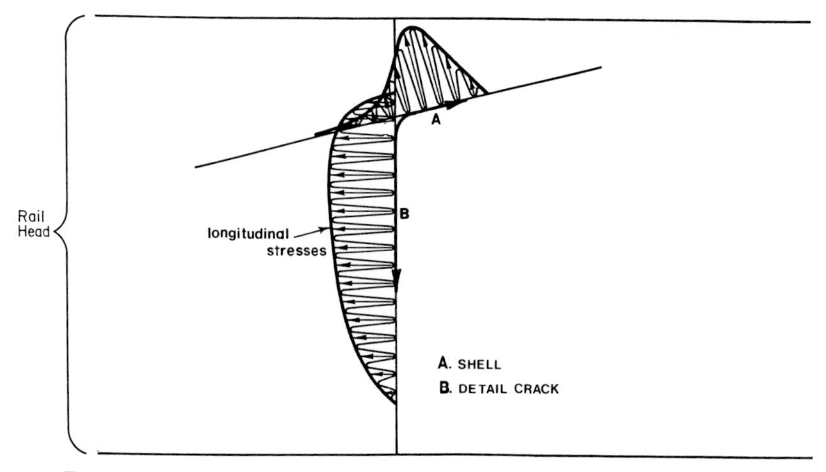

Figure 14.25. Cyclic stresses on shell crack, A, and detail crack (transverse), B.

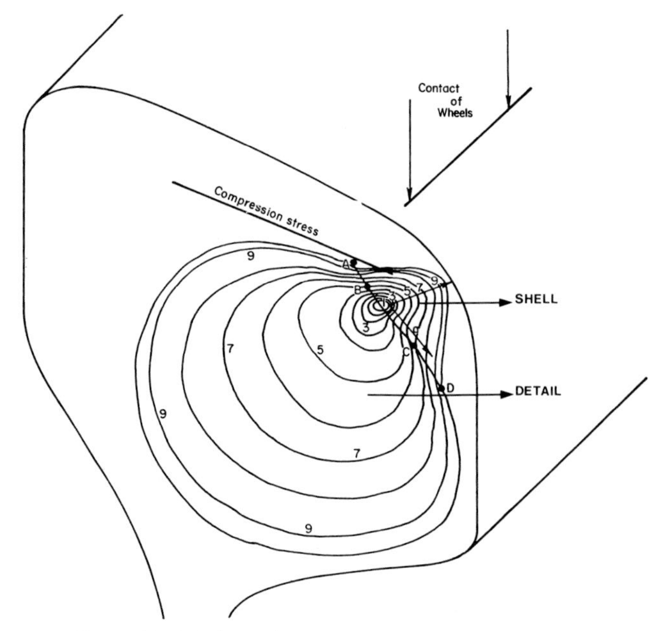

Figure 14.26. Schematic growth of shell and detail crack.

Load interaction (retardation) was negligible, and in view of the large variability in data, retardation was not included in the analysis. Predictions were made using average rate data, and the actual rate data of the particular rail used in a service simulation test. Some results and data are shown in Figure 14.31.

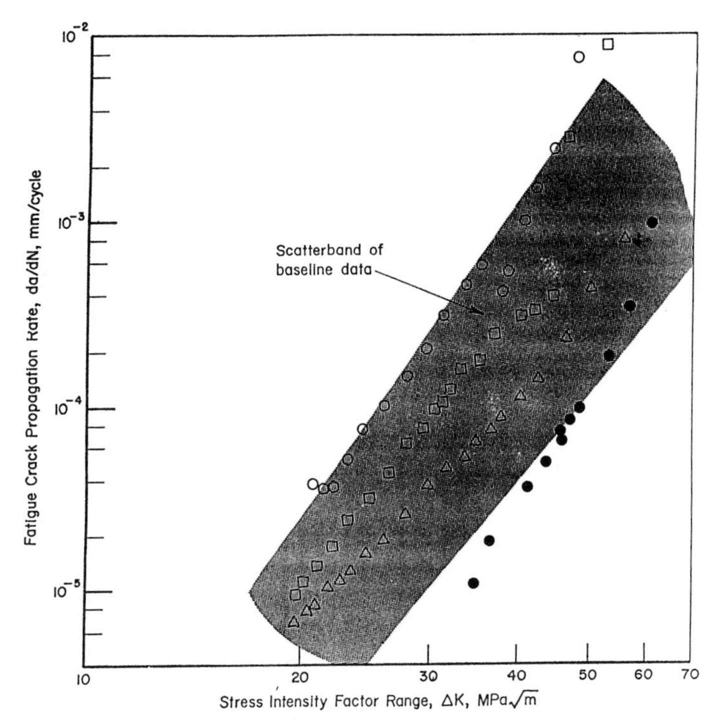

Figure 14.27. Variability of rate data among rail steels.

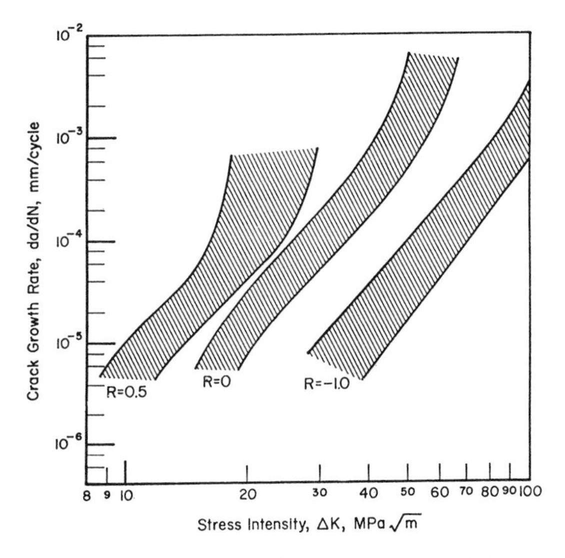

Figure 14.28. True scatter of rate data among rail steels.

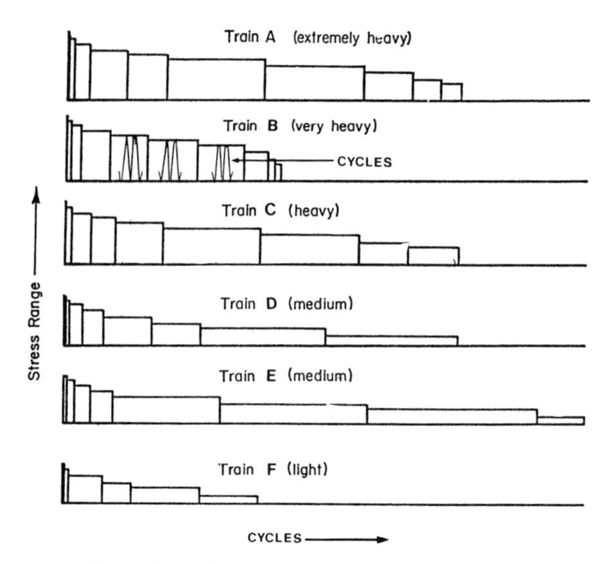

Figure 14.29. Simulated train loading histories.

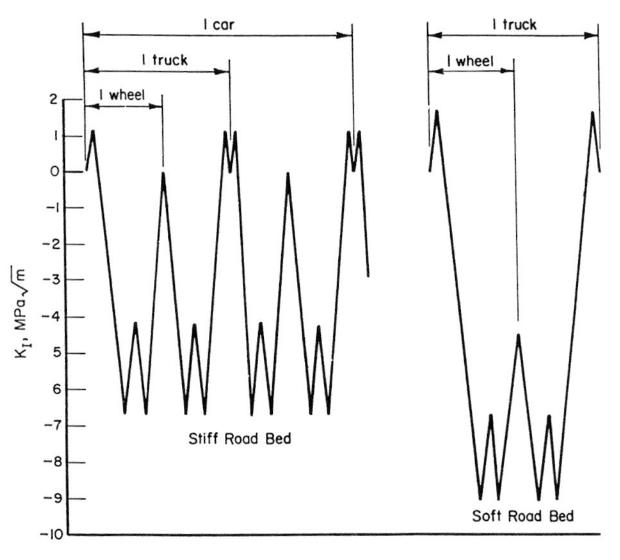

Figure 14.30. Cycles due to passage of wheels, trucks, and cars.

Predictions were generally within a factor of 2. This may not be very accurate, but the test data showed a similar variability. Obviously, if discrepancies are caused by material variability, they cannot be blamed on the prediction method.

Examples of analysis results for actual rail cracks are shown in Figure 14.32.

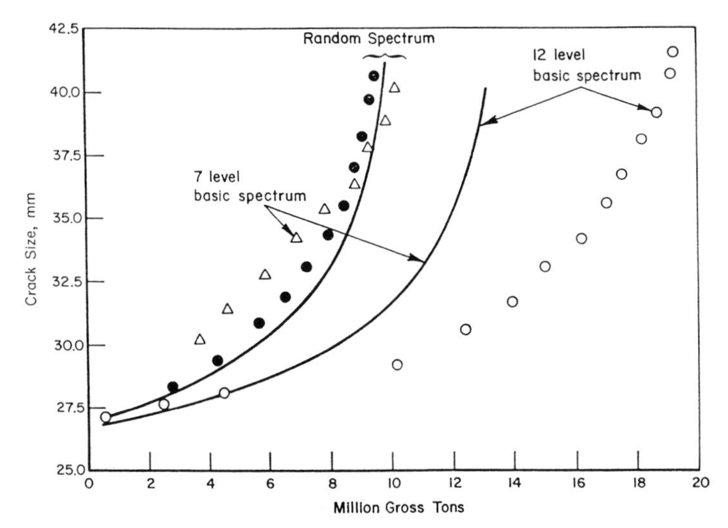

Figure 14.31. Predictions of crack growth (solid lines) and test data under simulated train-by-train loading.

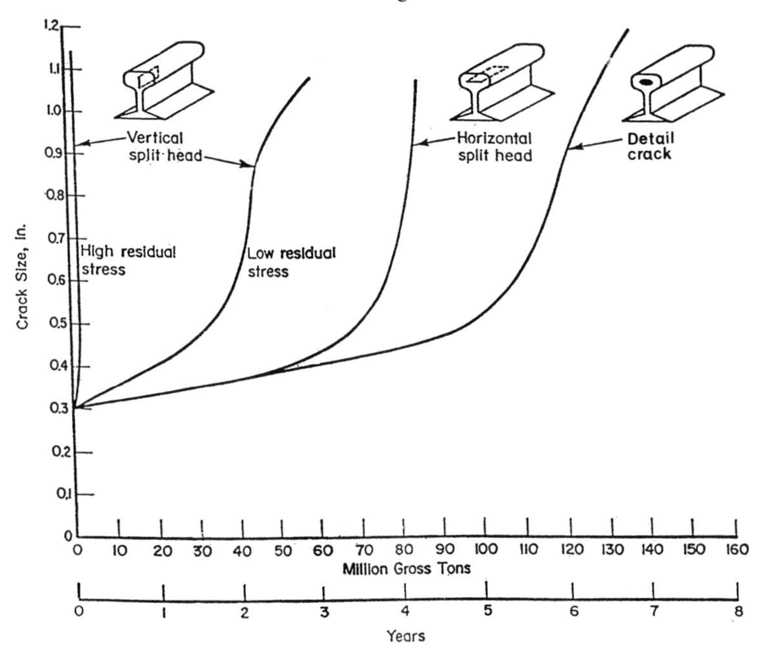

Figure 14.32. Predicted crack growth in rails.

The significant effect of residual stresses, is understandable because all cyclic stresses would be in compression in the absence of residual stresses (Figure 14.23).

(e) *Fracture control*
The use of the model (even with limited accuracy) does permit rational fracture control. As an example, consider a railroad in which the number of failures is considered unacceptable. The model can be exercised for the particular circumstances of that railroad, with a load spectrum for the prevailing traffic and track conditions. The analysis can be run for different conditions, such as reduced speed (loads), reduced traffic (load spectrum), and/or upgraded track and road red (stresses). Each of these measures will increase (reduce) the life by a certain factor. It then can be decided which measure is the most cost effective to obtain a certain factor of life improvement. Since the actual failure rate is known, the reduction factor can be applied to estimate the actual expected failure rate if one of the above measures is implemented.

The absolute accuracy of the model is not important if the ratios are predicted properly; the relative accuracy is significant; variability of parameters influencing crack growth is of no issue in relative assessments. Naturally, the absolute life calculations can be used to determine inspection intervals. With procedures as described in Chapter 11 to determine the interval, the variability can be accounted for by using a safety factor, which will reduce the inspection interval. Even though predicted crack growth has limited accuracy, the permissible increase (or reduction) of inspection intervals upon certain maintenance efforts could still be reliably analyzed, as they would be based on relative instead of absolute changes in predicted life, academic objections notwithstanding.

14.6. Underwater pipeline (Courtesy SNAM) [27, 28]

(a) *Definitions of the problem*
A submarine pipeline is laid from a barge as shown schematically in Figure 14.33. At *A* a new section of pipe is welded on, previous welds are X-rayed at *B*, protective coating applied at *C* and, to reduce buoyancy, a 1–2 inch concrete coating is applied as well. Then the barge is moved so that the next section can be built, the previous section being expelled. The line hangs down from the barge (prevention of buckling in the sharp bend being an important issue) until it hits bottom.

Free hanging spans may occur where the line encounters uneven bottom (Figure 14.33). Current flow (streams or tides), around the pipe causes turbulence (vorteces) as shown in the insert in Figure 14.33. Should the frequency of the vortex shedding be about the same as the natural bending frequency of a free span, bending excitation will occur, giving rise to cyclic stresses. The latter may cause fatigue; cracks growing from defects in the circumferential welds will eventually penetrate the wall. Repairs being impossible, any leak constitutes a failure. Consequently, the life of the line is governed by the time required for the growth of the largest possible weld defects

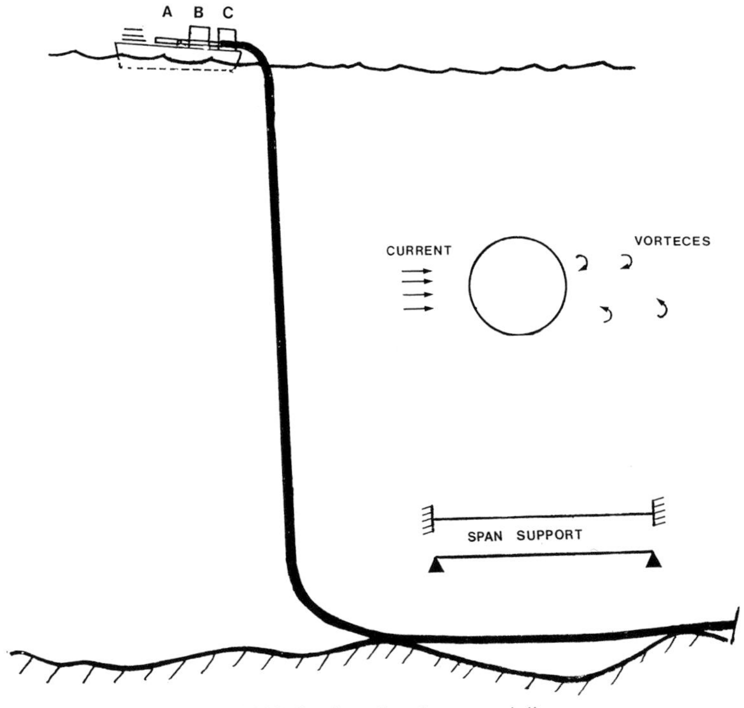

Figure 14.33. Laying of underwater pipline.

to wall penetration. Analysis of the problem requires detailed information on the amplitudes of vibration and consequent cyclic stresses, weld defect sizes, crack growth rates in a sea water environment, and feasible fracture control measures.

(b) *Amplitudes and excitation times*

First sonar measurements of bottom profiles were made to determine which span lengths might occur. Next came the calculation of the natural bending frequencies of all possible span sizes. Although this is trivial in principle, in practice it is not. The bending frequency depends strongly upon the end-fixity (Figure 14.33 insert). Depending upon the end support, and upon the (random) length of adjacent spans, the effective fixity can vary dramatically. As the future configuration is unknown, assumptions must be made.

VERITAS, the Norwegian shipping and offshore bureau, provides a procedure to determine vortex shedding frequencies as a functon of current velocity. From these, and from the natural frequencies of the spans, the current velocities causing excitation can be calculated. A typical result is shown in Figure 14.34. This information is necessary but insufficient. For the calculation

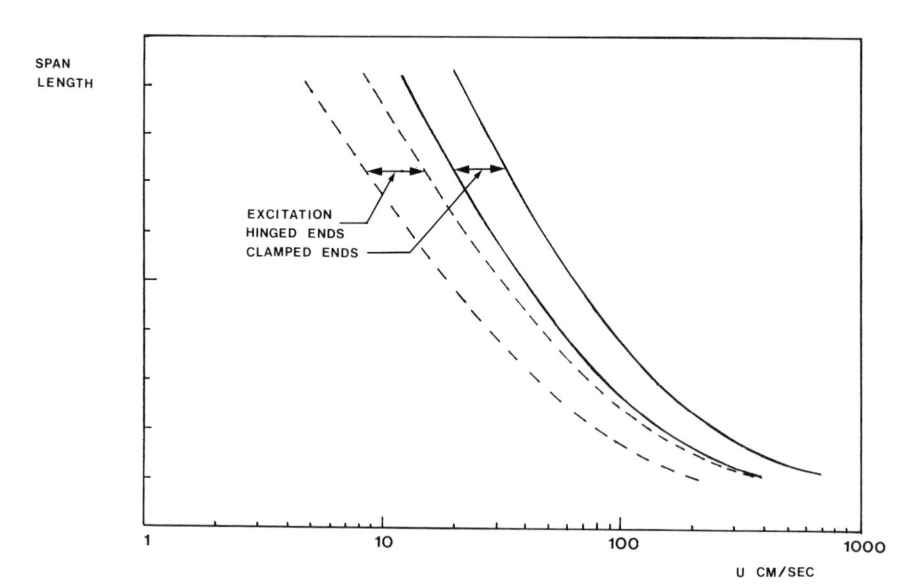

Figure 14.34. Excitation of span as a function of current velocity.

of stresses, the amplitude of the vibrations must be known. The response curve can be calculated, provided the damping is known as well. Hydrodynamic and material damping must be accounted for; many assumptions must be made. As this is not a 'fracture mechanics' problem, further details are beyond the scope of this book.

The above is a fine illustration of the arguments in Chapter 6 and 12 with regard to accuracy of the loads analysis. So many assumptions are involved in the solutions that the anticipated accuracy of the loads (amplitudes in this case) is certainly not better than 10 to 15%. It also illustrates how engineering depends upon assumptions. Vibration analysis is well founded, its rigor hardly questionable. Yet, practical results and accuracy are governed by assumptions. The same holds for fracture mechanics analysis. Deficiencies in the underlying theory do not determine its accuracy, but the engineering assumptions and judgements do.

(c) *Stresses*

The amplitude now being 'known', based upon reasonable assumptions, the stresses can be calculated; but the stress history has yet to be determined from the current spectrum. As spans of any size may occur anywhere, the current spectrum had to be measured at many points along the projected line. Only then was it be possible to determine the vibration response. With the information of Figure 14.34, the current spectra can be used to establish the excitation spectrum for any span. Using estimated fixity and damping, the critical current velocity and the maximum amplitude can be calculated. Vibration response curves show that excitation will still occur at frequencies (current velocities) close to the natural frequency, but the amplitudes are smaller (Figure 14.35). When this is accounted for, the stress spectrum can be obtained as in Figure 14.36. As it is not known which combinations will occur, this stress spectrum must be obtained for a range of span sizes and locations, so that a total of at least 100–400 cases must be analyzed.

A simple formula for bending stress permits calculation of the cyclic stresses, but crack growth is also affected by the mean stress (R-ratio; Chapter 5). The mean stress can be calculated from the pre-tension in the line, which depends upon the depth of laying (vertical suspension from barge), bottom friction, and so on. Again the problem is simple in principle, but assumptions must be made to obtain a solution. The results are strongly dependent upon this assumption.

(d) *Defect sizes*

Weld defect sizes could be estimated on the basis of weld quality control specifications (X-ray). But since the project was costly and the cost of a fracture beyond imagination, a more rigorous assessment of defect sizes was indicated. A rational procedure was to cut many specimens from a large number of welds produced on the barge under the most adverse (waves) conditions, but these should have passed X-ray quality control. The specimens were fatigue tested to failure. Examination of the fracture surfaces after failure revealed the defect that initated the cracking.

A large number of such tests established the defect size distribution as in Figure 14.37. In view of the severe consequences of a service failure, it was

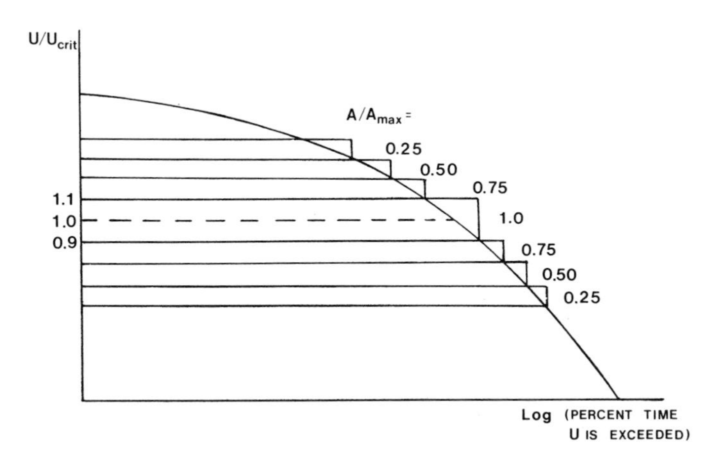

Figure 14.35. Amplitudes of vibration as a function of current. velocity.

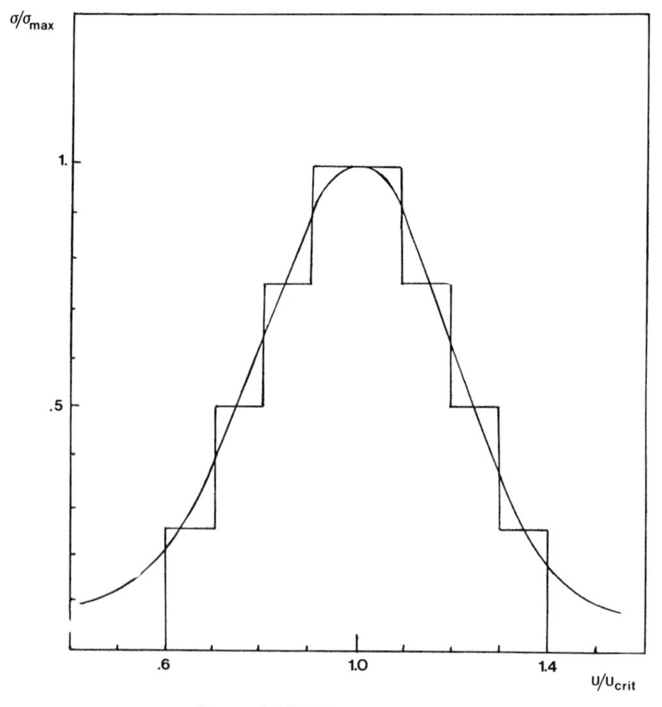

Figure 14.36. Stress spectrum.

prudent to base the analysis on the largest possible defect size, but a sensitivity analysis to assess the conservatism introduced by this assumption, is advisable.

(c) *Material data*

Defects may occur at various locations in the welds: at the outer surface (exposure to seawater), close to the inner surface (exposure to gas), or in the interior (vacuum). Crack growth rate data ($\mathrm{d}a/\mathrm{d}N - \Delta K$, R) were obtained for all these cases. As the sea-water environment is likely to produce the fastest growth, emphasis was on tests in sea-water, but the others were not discarded *a priori*, because life also depends upon crack location (β).

Fatigue crack growth tests to obtain da/dN data were performed on welded specimens (the cracks growing in the welds or HAZ), as cracks in practice will start in the welds; the specimens were machined from pipe welded on the barge. Typical data for crack growth in a seawater environment were shown in Figure 6.10. Schematically, the data can be represented as in Figure 14.38. A complication is that the rates depend upon cyclic frequency. Because the cyclic frequency varies with spans length, the data used in the analysis must be appropriate for the frequency of the relevant span size.

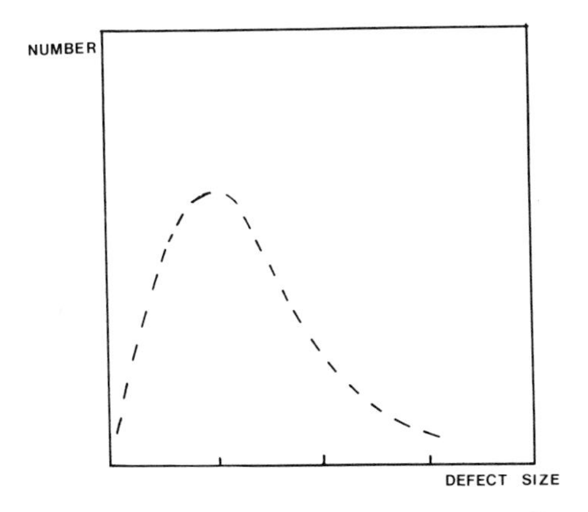

Figure 14.37. Weld defect sizes appearing from fatigue tests; Courtesy SNAM.

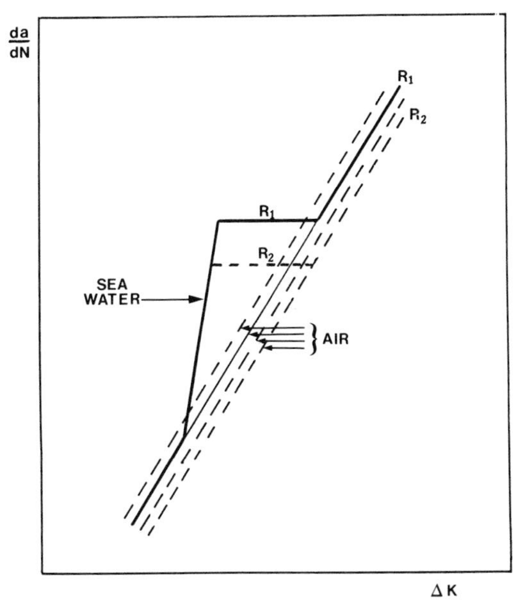

Figure 14.38. Schematic representation of rate data (fixed frequency).

(f) *Analysis and fracture control*

Crack growth was calculated for a range of span sizes, span locations, and crack locations, using a reliable computer program. In this case the crack growth curves were not of great interest; only the growth life until penetration was important. Examples of results are shown in Figure 4.39. Short spans are not excited and have infinite life. Very long spans are excited but only part of the

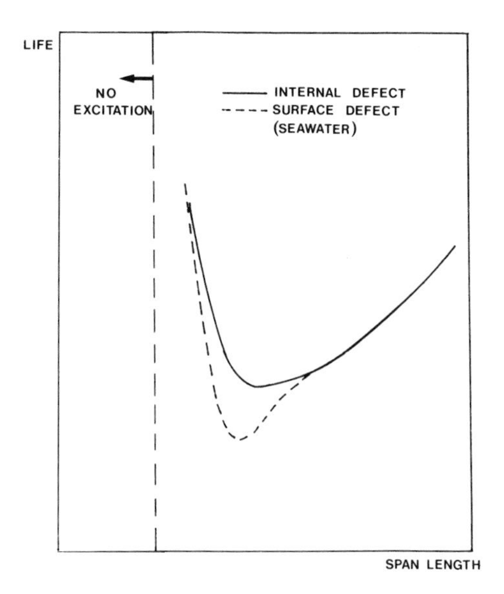

Figure 14.39. Life of various spans. Courtesy SNAM.

time, and their frequency is so low that few cycles are accumulated. Thus, their life (years; not cycles) is long. Note that the results may be quite different for other current spectra, line diameters, coatings etc.

In this case intermediate span sizes show the shortest lives. Should this shortest life (with adequate safety factors) be enough to cover the economic service life of the line, no specific fracture control measures would be necessary. Failing that, measures might be taken to reduce stresses, improve weldments, and so on. A sensitivity study might indicate that not much can be gained by such measures, so that other fracture control measures are indicated. Because inspection for cracks is impossible, one alternative is to inspect the line (by submarine) to find span sizes in the critical range. If any are discovered, they can be eliminated by insterting an artificial support. New span sizes in the critical range may develop (Figure 14.33) if the current washes away sand; should this occur then the life would still be limited. Hence, future inspections (submarine) would have to be scheduled at intervals equal to the life of the most critical span (with safety factor).

It needs no emphasis that, owing to the assumptions, the results of this damage tolerance analysis are not very accurate. Without damage tolerance analysis no information at all would be available. If responsible for the line, in which situation would be reader like to be: without any information, or with that provided by an imperfect damage tolerance analysis with appropriate safety factors? Obviously, the damage tolerance assessment is costly, because of the necessary data generation, but is far outreached by the cost of fracture. It needs no further arguments that extensive finite element analysis would be out of place in view of the many assumptions involved. Stresses in a pipe in bending and tension can be determined accurately enough with conventional procedures; errors due to the simple assumptions with regard to fixity will be much larger than those introduced by stress analysis.

14.7. Closure

The foregoing examples demonstrate the power of fracture mechanics and damage tolerance analysis, despite the fact that not all problems could be solved rigorously. Further development of fracture mechanics within the realm of engineering is worthwhile. So many assumptions are necessary, that concern with the shortcomings of fracture mechanics *per se* seem naive.

The last example in particular shows, that all engineering analysis needs assumptions; even 'rigorous' vibration analysis yields approximate answers only. The situation is no different for fracture mechanics and damage tolerance analysis.

Finally, claims that applications of fracture mechanics are more difficult for certain structures are negated by Table 10.1.

References

[1] I. Fioriti, *Fracture control of a naval destroyer*, Private Communication.

[2] Anon., *Rules for building and classing steel vessels*, American Bureau of Shipping (1978).

[3] R.J. Mosberg, *Behaviour of riveted and welded crack arresters*, Final Report on SR-134 to Ship Structures Committee, SSC-122 (1960).

[4] P. Romualdi et al., *Crack extension force near a riveted stringer*, Naval Research Lab. Rept No. 4956 (1957).

[5] T. Kanazawa et al., Study on Brittle Crack Arrester, Selected Papers from G of Soc., *Naval Architects of Japan* **11** (1973) pp. 135–147.

[6] T. Kanazawa et al., Fracture mechanics analysis and design of stiffener-type crack arrester, *Japan Shipbuilding and Marine Eng. (JSME)* **3, 6** (1968) pp. 10–19.

[7] M. Yoshiki et al., Fracture mechanics analysis of stiffener-type crack arrester, *JSNA* **118** (1965); *JSNA* **122** (1967) and *JSNA* **124** (1968).

[8] C.C. Poe, *The effect of riveted and uniformly spaced stringers on the stress intensity factor of a cracked sheet* AFFDL-TR-79-144, (1970) pp. 207–216.

[9] H. Vlieger, Residual strength of cracked stiffened panels, *Engineer Fracture Mechanics*, **5** (1973) pp. 447–478.

[10] T. Swift, Development of the fail-safe design features of the DC-10, *ASTM STP* **486** (1974) pp. 164–214.

[11] T. Swift, Design of redundant structures, *AGARD LSP* **97**, (1978) pp. 9/1–23.

[12] D. Broek, *Elementary engineering fracture mechanics*, Nijhoff, Fourth Edition (1986).

[13] I. Fioriti, Abbreviated hull structural design, Paper for Committee on Toughness Requirements for Materials in Weapon Systems (1978).

[14] T.C. Gillmer, *Modern ship design*, Second Edition, Naval Institute Press. Annapolis, Maryland (1975).

[15] G. Buchanan et al., Lloyd's register of shipping's approach to the control of the incidence of brittle fracture in ship structures, *Lloyd's Paper*, No. 56 (1969).
[16] E.V. Lewis, *Dynamic loadings due to waves and ship motions*, Paper presented at Ship Structure Symposium, Washington D.C. (1975).
[17] B.H. Barber et al., *Structural considerations in the design of the Solar class of coast guard icebreakers*, Paper presented at the Ship Structure Symposium, Washington, D.C. (1975).
[18] D.P. Rooke and D.J. Cartwright, *Compendium of stress intensity factors*, Her Majesty's Stationery Office (1976).
[19] D.R. Ahlbeck et al., *Evaluation of analytical and experimental methodologies for the characterization of a wheel/rail loads*, Fed. Railroad Adm. Rept. 76–276 (1976).
[20] T.G. Johns, and K.B. Davis, *A preliminary description of stress in railroad rails*, Fed. Railroad Adm. FRA-ORD-76-294 (1976).
[21] J.J. Groom, *Residual stress determination in railroad rails*. Dept of Transport. DOT-TSC-1426 (1979).
[22] D. Broek and R.C. Rice, Prediction of fatigue crack growth in railroad rails, *SAMPE Nat. Techn. Conf.* **9** (1977) pp. 392–408.
[23] R.C. Rice et al., *Post Service rail defect analysis*, Battelle Rept. to Dept of Transp. (1983).
[24] O. Orringer et al., Applied research on rail fatigue and fracture in the U.S., Theor. Appl. Fract. Mech. 1 (1984), pp 23–49.
[25] O. Orringer et al., Detail fracture growth in rails: test results Theor. Appl. Fract. Mech. 5 (1986) pp 63–95
[26] O. Orringer, Rail testing: strategies for safe and economical rail quality assurance, TRB symposium (1987).
[27] M. Celant et al. Fatigue characterization for probabilistic design of submarine pipelines, Corr. Science 23, 6 (1983) pp 621–636.
[28] M. Celant et al. Fatigue analysis for submarine pipelines, OTC paper 4233 (1982).

CHAPTER 15

Solutions to exercises

15.1. Solutions to exercises of Chapter 1

1. Material A: Safety factor of 3 against F_{tu}; allowable stress is $60/3 = 20$ ksi. Safety factor of 2 against F_{ty}; allowable stress is $50/2 = 25$ ksi. Lowest is applicable. Required cross section $A = 1000/20 = 50\,in^2$. $A = \pi D^2/4$. Required diameter $D = 7.98$ inch (8 inch). Ultimate design load: $P_u = AF_{tu} = 50 \times 60 = 3000$ kips.

Material B: Allowable stress lower of 50/2 and 80/3, i.e. allowable stress 25 ksi. Required area: $A = 1000/25 = 40\,in^2$; $D = 7.14$ inch; $P = 40 \times 80 = 3200$ kips.

2. Material A: Maximum service load is 1000 kip. Design load, $P_u = 2500$ kip. Required area $A = P_u/F_{tu} = 2500/60 = 41.7\,in^2$. At $1.3 \times 1000 = 1300$ kip no yielding is allowed: Check $\sigma = P/A = 1300/41.7 = 31$ ksi, is indeed less than F_{ty}. $D = 7.29$ inch.

Material B: Area $A = 2500/80 = 31.25\,in^2$. $\sigma = 1300/31.25 = 41.6$ is less than F_{ty} (OK); $D = 6.31$ inch.

3. Material A, problem 1: Minimum permissible residual strength $P_p = 1.2 \times 1000 = 1200$ kip. In terms of stress $\sigma_p = P_p/A = 1200/50 = 24$ ksi. Remaining safety factor $j = 1200/1000 = 1.2$.

Material B, problem 1: $P_p = 1200$ kip; $\sigma_p = 1200/40 = 30$ ksi, $j = 1.2$.
Material A, problem 2: $P_p = 1200$ kip, $\sigma_p = 1200/41.7 = 28.8$ ksi $j = 1.2$.
Material B, problem 2: $P_p = 1200$ kip; $\sigma_p = 1200/31.25 = 38.4$ ksi $j = 1.2$.

4. Fracture is final event. Cracking by fatigue, stress corrosion, creep, etc.

5. Known points: $a = 0$, $\sigma_{res} \approx F_{tu} = 60$ ksi;
$a = 2$, $\sigma_{res} = \sigma_p = 24$ ksi (see Exercise 3);
$a = D = 8$, $\sigma_{res} = 0$ (D = 8, Exercise 1).

Fair approximate curve through three points.

6. At 1500 kips stress is $P/A = 1500/50 = 30$ ksi. Fracture at about $a \approx 1.4$ inch (estimated from curve in exercise 5). At 1000 kips stress is $P/A = 1000/50 = 20$ ksi. Fracture at $a \approx 3$ inch (estimated from curve).

15.2. Solutions to exercises Chapter 2

1. $k_t = 1 + 2\sqrt{2/0.4} = 1 + 2\sqrt{5} = 5.47$;
   $k_\sigma = 5.47$, $k_\varepsilon = 5.47$ (elastic).
Full yield no strain hardening; $k_\sigma = 1$; $k_\sigma k_\varepsilon = k_t^2$; $k_\varepsilon = 5.47^2/1 = 29.94$.

2. Since $k_\sigma = 1$ net stress has become approximately uniform and equal to F_{ty}, i.e. 50 ksi; fracture load: $(W - 2a)BF_{ty} = (12 - 4)0.5 \times 50 = 200$ kips. In second case at collapse uniform net stress approximately 75 ksi; fracture load 300 kips. Nominal stress in the two cases: (1) $200/(12 \times 0.5) = 33.3$ ksi (2) $300/(12 \times 0.5) = 50$ ksi.

3. Circular hole $k_t = 3$.
Elliptical notch $k_t = 1 + 2\sqrt{0.8/0.1} = 6.66$.

4. Before hole drilling $k_t = 1 + 2\sqrt{a/0} = \infty$. After hole drilling $k_t = 1 + 2\sqrt{2a/d}$ ($\rho = 1/2d$).

5. Plane strain: yielding starts at about $3F_{ty}$, i.e. stress at yielding is $3 \times 30 = 90$ ksi, a little away from notch. At free surface of notch root there is plane stress ($\sigma_x = 0$); yielding starts at $\sigma = F_{ty} = 30$ ksi.

6. With $k_t = 4$, elastic stress (just before yielding) at crack tip is 30 ksi. Nominal stress $= 30/4 = 7.5$ ksi, i.e. yielding at notch tip begins at nominal stress 7.5 ksi.

7. No numerical answer.

8. No numerical answer.

9. (a) Center crack.
Linear curve from ($a = 0$, $\sigma_{fc} = F_{col} = 350$ MPa) to ($a = W/2 = 300$ mm, $\sigma_{fc} = 0$). With crack of 150 mm; $\sigma_{fc} = (W - 2a)F_{col}/W = (600 - 300) \times 350/600 = 175$ MPa.
   b. Edge crack.
Linear curve from ($a = 0$, $\sigma_{fc} = 350$ MPa) to ($a = W = 600$, $\sigma_{fc} = 0$) (constrained edges). With crack of 150 mm: $\sigma_{fc} = (W - a)F_{col}/W = (600 - 150)350/600 = 262.5$ MPa.

15.3. Solutions to exercises of Chapter 3

1. Stress at fracture $300/(20 \times 0.75) = 20$ ksi; $\beta \approx 1$ (Figure 3.3). Toughness $= K_{\text{facture}} = 1 \times 20 \times \sqrt{\pi \times 1} = 35.4\,\text{ksi}\sqrt{\text{in}}$. With $F_{ty} = 70$ ksi for collapse, the nominal stress at collapse would be $(20 - 2)70/20 = 63$ ksi; since fracture at 20 ksi collapse did not occur. Hence, calculated toughness is useful. For plane strain $B > 2.5\,(35.4/70)^2 = 0.64$ inch. Since $B > 0.64$ in, the above toughness is plane strain by ASTM criterion. Plastic zone $r_p \approx (35.4/70)^2/6\pi \approx 0.014$ inch.

2. $W = 5$ inch; $a = 2$ inch; $\beta = 2.1$. $\sigma_{fr} = 35.4/2.1\sqrt{\pi \times 2} = 6.73$ ksi. $\sigma_{fc} = (5 - 2)\,70/5 = 42\,\text{ksi} > \sigma_{fr}$; $\sigma_{\text{res}} = 6.73$ ksi. $W = 6$ inch, $a = 0.5$ inch; $\beta = 1.16$; $\sigma_{fr} = 24.4$ ksi; $\sigma_{fc} = 64$ ksi; $\sigma_{\text{res}} = 24.4$ ksi.

3. $W = 20$; $2a = 3$; $a = 1.5$; $\beta = 1.01$; $\sigma_{fr} = 31.9$ ksi; $\sigma_{fc} = 63.7\,\text{ksi} > \sigma_{fr}$; $\sigma_{\text{res}} = 31.9$ ksi. $W = 2$; $2a = 1$; $a = 0.5$; $\beta = 1.19$; $\sigma_{fr} = 46.9$ ksi; $\sigma_{fc} = 37.5\,\text{ksi} < \sigma_{fr}$; $\sigma_{\text{res}} = 37.5$ ksi (collapse).

4. $\sigma_{\text{res}} = 34.2$ ksi; $\beta = 1.07$; calculated K at fracture: $1.07 \times 34.2\sqrt{\pi \times 1} = 64.9\,\text{ksi}\sqrt{\text{in}}$. Collapse would occur at: $(W - 2a)BF_{ty} = (6 - 2)\,0.2 \times 50 = 40$ kip. Hence, fracture occurred by collapse (net section yield). Above K value is not the toughness. Equation (3.25) cannot be used because whole section is yielding; plastic zone is equal to remaining ligament.

5. With Equation (3.28): $K_Q = 33.97\,\text{ksi}\sqrt{\text{in}}$.
With Equation (3.29): $K_Q = 34.18\,\text{ksi}\sqrt{\text{in}}$. Same answer as it should be; $K_Q = 34\,\text{ksi}\sqrt{\text{in}}$. Valid because $B > 2.5\,(34/80)^2 = 0.45$ inch (actual 1 inch); $K_{Ic} = 34\,\text{ksi}\sqrt{\text{in}}$.

6. For plane strain $B > 2.5\,(50/100)^2 = 0.625$ inch. Thickness is less; therefore no plane strain. Toughness for 0.5 inch thickness is $K_c = 50\,\text{ksi}\sqrt{\text{in}}$. Plane strain toughness is less. Specimen measures plane strain if $0.5 > 2.5\,(K_{Ic}/100)^2$, i.e. if $K_{Ic} \leqslant 45\,\text{ksi}\sqrt{\text{in}}$. Hence, actual toughness is larger than $45\,\text{ksi}\sqrt{\text{in}}$ otherwise test would have been valid. Hence, $45 < K_{Ic} < 50$. Estimate $K_{Ic} = 47.5\,\text{ksi}\sqrt{\text{in}}$. (accurate within about five to six percent even if toughness is as low as 45, which is lowest possible). Highest possible $50\,\text{ksi}\sqrt{\text{in}}$; required specimen thickness 0.625 in or larger.

7. Center crack $\beta = 1.1$; $\sigma_{fr} = K_c/1.1\sqrt{\pi \times 0.75} = K_c/1.69$. Edge cracks $\beta = 1.13$ (Figure 3.3); $\sigma_{\text{fr}} = K_c/1.13\sqrt{\pi \times 0.6} = K_c/1.55$. Whatever K_c (same

for both cracks) fracture stress at center crack is lower; failure should occur at center crack.

8. Case a: $\sigma_p = 58$ ksi; $a_p = 0.6$ inch (on tangent);
$\sigma_p = 20$ ksi; $a_p = 5.1$ inch (σ_{fr} from K_c);
$\sigma_p = 10$ ksi; $a_p = 8.8$ inch (σ_{fr} from K_c);
Case b: $\sigma_p = 58$ ksi; $a_p = 0.5$ inch (collapse);
$\sigma_p = 20$ ksi; $a_p = 2.14$ inch (collapse);
$\sigma_r = 10$ ksi; $a_p = 2.57$ inch (collapse);
Case c: $\sigma_p = 58$ ksi; $a_p = 0.48$ inch (σ_{fr});
$\sigma_p = 20$ ksi; $a_p = 2.7$ inch (σ_{fr});
$\sigma_p = 10$ ksi; $a_p = 4.4$ inch (σ_{fr}).

15.4. Solutions to exercises in Chapter 4

1a. $A = \pi d^2/4 = \pi(0.4)^2/4 = 0.126\,\text{inch}^2$. $\sigma = P/A$. Draw straight line S_1 to determine E. For $P = 9.5$ kip, $\delta = 0.01$ inch. Hence, $\sigma = 9.5/0.126 = 75.4$ ksi. $\delta = \varepsilon L$, so that $\varepsilon = \delta/L = 0.01/4 = 0.0025$, hence, $E = 75.4/0.0025 = 30000$ ksi.

F_{ty} is defined as stress for plastic strain of 0.2%, i.e. $\varepsilon_p = 0.002$. This means $\delta_p = 0.008$. Draw line S_2 parallel to S_1 starting at 0.008 and intersect with curve; read load $P = 5.67$ kip; $F_{ty} = 5.67/0.126 = 45$ ksi. $P_{max} = 7.56$ kip; $F_{tu} = 7.56/0.126 = 60$ ksi. From points A through H read from curve and work out ($\varepsilon_{tot} = \delta/L$):

Point	P	Pd	σ	ε_{tot}	$\varepsilon_{el} = \sigma/E$	$\varepsilon_{pl} = \varepsilon_{tot} - \varepsilon_{el}$
A	5.67	0.014	45.0	0.0035	0.00150	0.00200
B	6.60	0.05	52.4	0.0125	0.00175	0.01075
C	6.8	0.1	53.9	0.0250	0.00180	0.02320
D	7.0	0.2	55.6	0.0500	0.00185	0.04815
E	7.3	0.3	57.9	0.0750	0.00193	0.07307
F	7.4	0.4	58.7	0.1000	0.00196	0.09804
G	7.5	0.5	59.5	0.1250	0.00198	0.12302
H	7.56	0.6	60.0	0.1500	0.00200	0.14800

1b. Plot $\sigma - \varepsilon_{tot}$ to obtain stress-strain curve. Plot $\log\sigma - \log\varepsilon_p$ to obtain F and n by determining slope and intercept from straight line through data points. $E = 30{,}000$ ksi, $n = 16.2$, $F = 3.9 \times 10^{29}$ ksi $\wedge$ 16.2, depending upon how you draw line.

1c. $\sigma_0 = 100$ ksi, $\varepsilon_0 = 100/30000 = 3.33E{-}3$; $\alpha = \sigma_0^n/\varepsilon_0 F = 100^{16.2}/(3.33E{-}3 \times 3.9 \times 10^{29}) = 193222$

1d. $\sigma_0 = 50\,\text{ksi}$, $\varepsilon_0 = 1.67E-3$, $\alpha = 5.12$.

1e. $\sigma = 50\,\text{ksi}$ for $\sigma_0 = 100\,\text{ksi}$:
$\varepsilon = 50/30000 + 50^{16.2}/3.9 \times 10^{29} = 0.0102$
$\varepsilon/\varepsilon_0 = 50/100 + 193222(50/100)^{16.2} = 3.067$
$\varepsilon = \varepsilon_0 \times 3.067 = 3.33\text{E}-3 \times 3.067 = 0.0102$
$\sigma = 55\,\text{ksi}$ for $\sigma_0 = 100\,\text{ksi}$:
$\varepsilon = 55/30000 + 55^{16.2}/3.9 \times 10^{29} = 0.042$
$\varepsilon/\varepsilon_0 = 55/100 + 193222(55/100)^{16.2} = 12.57$
$\varepsilon = 12.57 \times 3.33E-3 = 0.042$
$\sigma = 50\,\text{ksi}$ for $\sigma_0 = 50\,\text{ksi}$:
$\varepsilon/\varepsilon_0 = 6.12$
$\varepsilon = 6.12 \times 1.67E-3 = 0.0102$
$\sigma = 55\,\text{ksi}$ for $\sigma_0 = 50\,\text{ksi}$:
$\varepsilon/\varepsilon_0 = 25.08$
$\varepsilon = 25.08 \times 1.67E-3 = 0.042$
Note that results for $\sigma = 50$ and 55 ksi are identical (σ_0 can be selected arbitrarily). Check results in your $\sigma - \varepsilon$ diagram.

2. Reminder: $\sigma_{\text{true}} = \sigma_{\text{eng}}(1 + \varepsilon_{\text{eng}})$ and $\varepsilon_{\text{true}} = \ln(1 + \varepsilon_{\text{eng}})$; Same procedure as exercise 1 with different results (except for modulus E).

3. At $P = 7.4$ kips the plastic strain is 0.009804. Hence, the bar has become thinner. Constant volume requires that $(L + \Delta L)A = LA_0$ or $A_0 = A(1 + \varepsilon)$ so that $A = A_0/(1 + \varepsilon_p) = 0.126/1.09804 = 0.115\,\text{inch}^2$.

This new bar will yield at $P = 7.4$ kips, the cold worked material's $F_{ty} = 7.4/0.115 = 64.3$ ksi. The fracture load is 7.56 kips so that $F_{tu} = 7.56/0.115 = 65.7$ ksi.

This is the common way to strengthen materials by cold work (e.g. cold drawing).

4. $J_R = 8.72 \times 50^{17.2} \times 2/3.9\text{E}29 = 7.46$ kips/inch.
$\sigma_{fc} = 28 \times 55/32 = 48.1$ ksi (< 50 ksi; apparently collapse. J_R value not reliable).

5. $\sigma_{fr} = [(3.9\text{E}29 \times 7.46)/(13.5 \times 3)]^{1/17.2} = 46.6$ ksi.
$\sigma_{fc} = 18 \times 55/24 = 41.3$ ksi (< 46.6 ksi; collapse at 41.3 ksi).

6. $a/W = 0.125$; $a = 20 \times 0.125 = 2.5$ inch.

Case 1: $J = 193222 \times 100 \times 3.33\text{E}-3 \times 0.75 \times 2.5 \times 4.13[\sigma/0.75 \times 100]^{17.2} = 2.77E-27\sigma^{17.2} = 2$; hence $\sigma = 36.4$ ksi.

Case 2: $J = 5.12 \times 50 \times 1.67\text{E}-3 \times 0.75 \times 2.5 \times 4.13[\sigma/0.75 \times$

$50]^{17.2} = 2.80E-27\sigma^{17.2} = 2$; hence $\sigma = 36.4$ ksi.

Same result in both cases, independent of choice of σ_0, because $\varepsilon_0 = \sigma_0/E$ and α is adjusted accordingly.

7. Note that $J = 2A/bB$, $b = W - a$, P-δ-curve horizontal after rising part.

a	Δa	P	δ	$\Delta\delta$	ΔA	A_{tot}	b	J_R
1	0	5	0.15	0.15	$\frac{1}{2} \times 5 \times 0.15 = 0.38$	0.38	1.00	1.52
1.05	0.05	5	0.20	0.05	$5 \times 0.05 = 0.25$	0.63	0.95	2.65
1.15	0.15	5	0.27	0.07	$5 \times 0.07 = 0.35$	0.98	0.85	4.61

Plot J_R versus Δa for J_R curve.

15.5. Solutions to exercises of Chapter 5

1. $da/dN \approx \Delta a/\Delta N = (1.05 - 1)/7000 = 7.14 \times 10^{-6}$ inch/cycle.

2. $K_{max} = 12\sqrt{\pi \times 0.5} = 15$ ksi$\sqrt{\text{in}}$; $\Delta K = (1-R)K_{max} = 0.8 \times 15 = 12$ ksi$\sqrt{\text{in}}$. $da/dN = 1E-8(12)^2(15)^{1.5} = 8.4E-5$ inch/cycle.

Assuming K does not change over small $\Delta a = 0.51 - 0.5 = 0.01$ inch then da/dN remains constant. Then $\Delta N = 0.01/8.4E-5 = 120$ cycles.

3. $C = 7.83E-13$; $m = 4.54$ MPa$\sqrt{\text{m}}$, m/cycle;
$C = 4.28E-11$; $m = 4.54$ ksi$\sqrt{\text{in}}$, in/cycle.

4. For $\Delta\sigma = 16$ ksi; R $= 0$ ($\beta = \sqrt{\sec \pi a/w}$):

a	N	Δa	ΔN	$(da/dN) = (\Delta a/\Delta N)$	a_{ave}	$\Delta K = \beta\Delta\sigma\sqrt{\pi a_{ave}}$	K_{max}
0.1	0						
		0.005	1100	4.55E − 6	0.1025	9.08	9.08
0.105	1100						
1.5	i						
		0.05	100	5.00E − 4	1.525	35.53	35.53
1.55	i + 100						

Make a similar table for the case of $R = 0.5$, and plot da/dN versus ΔK on log-log.

5. From table in solution to exercise 4:
$\log 4.55E-6 = m_p \log 9.08 + \log C_p$;
$\log 5.00E-4 = m_p \log 35.53 + \log C_p$.
Solution as in Exercise 3: $C_p = 2.21E-9$; $m_p = 3.45$. Same for other case with $R = 0.5$; $\Delta\sigma = 10$ ksi; $R = 0.5$: $C_p = 6.49E-9$; $m_p = 3.45$.

6. Lines in Exercise 5 are parallel (Same m), so that Walker equation is certainly applicable $C_R = C_{R=0}/(1 - R)^{n_w}$ where $C_{R=0} = C_w$ (Equation 5.14), so that $(1 - R)^{n_w} = C_R/C_{R=0}$. Using the C values from Exercise 5 for $R = 0.5$ and $R = 0$ one obtains $(1 - 0.5)^{n_w} = 2.21/6.49$, so that $0.5^{n_w} = 0.34$ and $n_w = \log 0.34/\log 0.5 = 1.56$. As $\Delta K/(1 - R) = K_{max}$: $m_w = 3.45 - 1.56 = 1.89$. Hence, the Walker equation reads:

$$\frac{da}{dN} = \frac{2.21E-9}{(1 + R)^{1.56}} \Delta K^{3.45} \quad \text{or} \quad \frac{da}{dN} = 2.21E - 9\Delta K^{1.89} K_{max}^{1.56}.$$

7. Forman Equation:

1	2	3	4	5
$\Delta\sigma$	R	ΔK	da/dN	$\{(1 - R)K_C - \Delta K\}da/dN$
16	0	9.08	4.55E−6	3.23E−4
		35.53	5.00E−4	2.22E−2
10	0.5	5.68	2.50E−6	8.58E−5
		22.21	2.94E−4	5.23E−3

Columns 1 through 4 follow from results of Exercise 4. Plot columns 3 versus 5 on double log scale. ALL data must fall on a SINGLE straight line as can be deduced from Equaton 5.16. If they do not Forman equation does not apply. Indeed they do approximately. Determine C_F and m_F from line.

$$\log 8.58E-5 = m_F \log 5.68 + \log C_F;$$

$$\log 2.22E-2 = m_F \log 35.53 + \log C_F.$$

From which as in problem 3: $C_F = 4.58E-7$; $m_F = 3.03$. Equation:

$$\frac{da}{dN} = 4.58E-7 \frac{\Delta K^{3.03}}{(1 - R)80 - \Delta K}.$$

8. Naturally, all rate diagrams should be the same, otherwise equations would not fit the data. Example: for $\Delta K = 8\,\text{ksi}\sqrt{\text{in}}$ and $R = 0.5$ the Walker equation of Exercise 6 provides

$$\frac{da}{dN} = 2.21E-9 \times 8^{1.89} \times 16^{1.56} = 8.5E-6\,\text{inch/cycle}.$$

The Forman equation of Exercise 7 provides:

$$\frac{da}{dN} = 4.58E-7 \frac{8^{3.03}}{(1 - 0.5)80 - 8} = 7.8E-6\,\text{inch/cycle}.$$

9. Walker equation; $\beta = \sqrt{\sec \pi a/W}$; $\sigma_{max} = 15\,\text{ksi}$; $\Delta\sigma = 15 \times 0.7 = 10.5\,\text{ksi}$; da/dN from Exercise 6:

a	Δa	a_{ave}	$\Delta K = \beta\Delta\sigma\sqrt{\pi a_{ave}}$	K_{max}	da/dN	$\Delta N = \Delta a/da/dN$	N
0.5							0
	0.2	0.60	14.4	20.6	3.83$E-5$	5220	
0.7							5220
	0.3	0.85	17.2	24.6	7.07$E-5$	4243	
1.0							9463
	0.5	1.25	20.8	29.7	1.36$E-4$	3676	
1.5							13139
	0.5	1.75	24.6	35.1	2.42$E-4$	2066	
2.0							15205
	1.0	2.50	29.4	42.0	4.49$E-5$	2227	
3.0							17432

Computer result: $N_{tot} = 17717$ cycles.

10. Forman Equation; same β as in Exercise 9. $\Delta\sigma = 15 \times 0.7 = 10.5\,\text{ksi}$; da/dN from Exercise 7.

a	Δa	a_{ave}	ΔK	da/dN	ΔN	N
0.5						0
	0.2	0.60	14.4	3.56$E-5$	5618	
0.7						5618
	0.3	0.85	17.2	6.84$E-5$	4587	
1.0						10205
	0.5	1.25	20.8	1.28$E-4$	3966	
1.5						14111
	0.5	1.75	24.6	2.39$E-4$	2092	
2.0						16203
	1.0	2.50	29.4	4.84$E-5$	2066	
3.0						18269

Computer result: $N_{tot} = 18538$ cycles.

11. $\sigma_{max} = 11\,\text{ksi}$; $R = 0.2$; $\Delta\sigma = 0.8 \times 11 = 8.8\,\text{ksi}$ $\Delta K = 3 \times 8.8 \times \sqrt{\pi \times 0.1} = 14.8\,\text{ksi}\sqrt{\text{in}}$; $K_{max} = 18.5\,\text{ksi}\sqrt{\text{in}}$.

With Walker equation: $da/dN = 3.41E-5$ inch/cycle.

With Forman equation: $da/dN = 3.27E-5$ inch/cycle.

12. Overload $K_{max} = 3 \times 17\sqrt{\pi \times 0.1} = 28.6\,\text{ksi}\sqrt{\text{in}}$.
Current $K_{max} = 3 \times 11\sqrt{\pi \times 0.1} = 18.5\,\text{ksi}\sqrt{\text{in}}$; $\Delta K = 14.8\,\text{ksi}\sqrt{\text{in}}$.
$r_{p\,overload} = 0.00886$ inch ($\rho = r_{p\,overload}$).
$r_{p\,current} = 0.00371$ inch.

$da/dN = 2.21E{-}9\Delta K^{1.89} K_{max}^{1.56} = 3.41E{-}5$ inch/cycle (unretarded).
$da/dN = (371/886)^{1.2} \times 3.41E{-}5$ inch/cycle $= 1.20E{-}5$ inch/cycle (retarded).
After 0.005/inch growth:
Current $K_{max} = 3 \times 11 \times \sqrt{11 \times 0.103} = 18.8\,\mathrm{ksi}\sqrt{\mathrm{in}}$; $\Delta K = 15.0\,\mathrm{ksi}\sqrt{\mathrm{in}}$.
$\rho = 0.00886 - 0.003 = 0.00586$ inch; $r_{p\,current} = 0.00383$ inch.
$da/dN = 2.21E{-}9\Delta K^{1.89} K_{max}^{1.56} = 3.59E{-}5$ inch/cycle (unretarded).
$da/dN = (383/586)^{1.2} \times 3.59E{-}5$ inch/cycle $= 2.16E{-}5$ inch/cycle (retarded)
For $\gamma = 0$ unretarded rates.

13. For prevention $K \leqslant K_{Iscc}$

$$\sigma_p < 15/(3 \times \sqrt{\pi \times 0.1}) = 8.92\,\mathrm{ksi}.$$

14. Results ($a_i = 0.5$ inch in all cases; a_c is crack size at fracture); $\sigma = P/BW$.

P	σ	$K_i = \beta\sigma\sqrt{\pi a_i}$	$a_c = \frac{1}{\pi}\left(\frac{K_{IC}}{1.12\sigma}\right)^2$	Growth $a_c - a_i$	Time
18	12	16.84	1.58	1.08	5000
24	16	22.46	0.89	0.39	100
30	20	28.10	0.57	0.07	3
32	21	30.00	0.50	0.00	0

Plot K_i(linear) versus time (log). Draw asymptote for K_{Iscc}. Graph shows $K_{Iscc} \approx 16\,\mathrm{ksi}\sqrt{\mathrm{in}}$. Amount of growth shown in fifth column.

15.6. Solutions to exercises of Chapter 6

1. For results see Table 15.1 and Figure 15.1.

2. For results see Table 15.2 and Figure 15.2. The exceedance diagram is for 300 periods (days). In Exercise 1 it was represented by a total of 200 000 cycles, i.e. 200 000/300 = 667 cycles per period (day). In Exercise 2 this was represented by 100,000 cycles, i.e. 333 cycles per day (but with cycles of different magnitude). Computer crack growth analysis, for each case are shown in Figures 15.3 and 15.4. The data used were the same in each case. The results are different in numbers of cycles, (as they should) but not much in number of periods (days). Hence, both stress history representations of the same spectrum give about the same answer. Better agreement would be obtained if more levels were chosen (10–12).

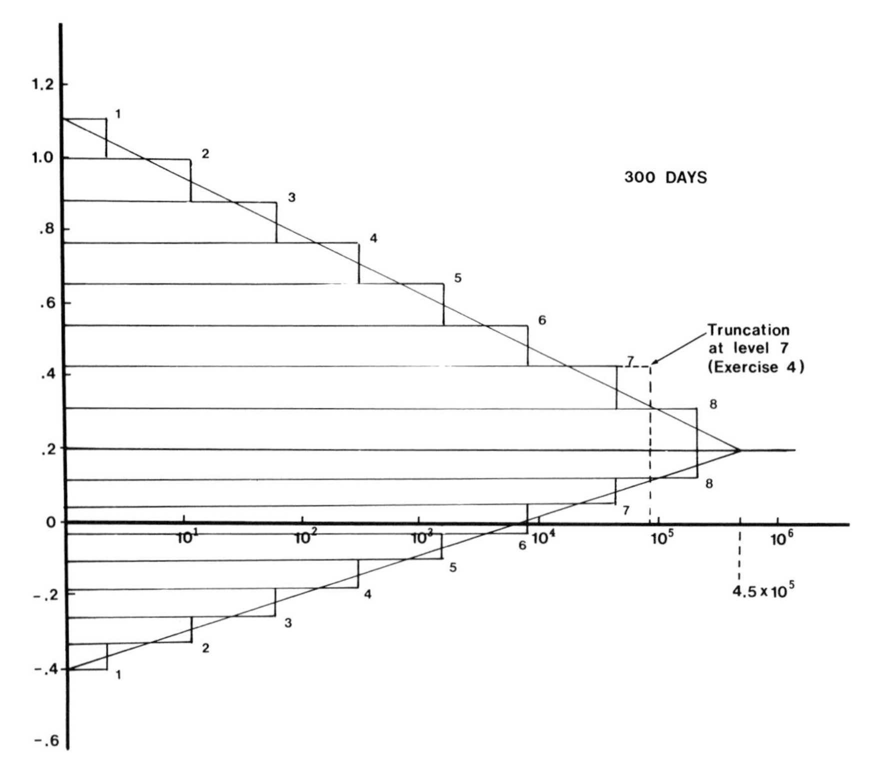

Figure 15.1. Results of Exercise 1 of Chapter 6.

3. Solutions in Figure 15.5 and Table 15.3 first 8 lines: 2 periods (days) type A, 5 types B, 34 types C, 130 types D, 129 E types E. Average total exceedances/day $= 4.5 \times 10^3/300 = 1500$, which provides 'pivot' point, P, for exceedance diagram of periods in Figure 15.5.

4. Solution Table 15.3, bottom lines.

Table 15.1. Results of Exercise 1 of Chapter 6 (see Figure 15.1).

Level	Relative σ_{min}	Relative σ_{max}	σ_{min}	σ_{max}	$\Delta\sigma$	Exceedances	Occurences (cycles)
1	−0.4	1.1	−6.0	16.5	22.5	2	2
2	−0.325	0.988	−4.9	14.8	19.7	11	9
3	−0.25	0.875	−3.8	13.1	16.9	63	52
4	−0.175	0.763	−2.6	11.4	14.0	316	253
5	−0.1	0.65	−1.5	9.8	11.3	1679	1363
6	−0.025	0.538	−0.38	8.1	8.48	7830	6151
7	0.05	0.425	0.75	6.4	5.65	43 300	35 470
8	0.125	0.313	1.9	4.7	2.80	200 000	156 700

Table 15.2. Results of exercise 2 of Chapter 6 (see Figure 15.2).

Level	Relative σ_{min}	Relative σ_{max}	σ_{min}	σ_{max}	Exceedances	Occurrences (cycles)
1	−0.4	1.1	−6.00	16.50	2	2
2	−0.37	1.05	−5.55	15.75	10	8
3	−0.24	0.83	−3.60	12.45	100	90
4	−0.17	0.74	−2.55	11.10	1000	900
5	−0.01	0.51	−0.15	7.65	10 000	9000
6	0.05	0.40	0.75	6.00	100 000	90 000

5. Instead of one cycle of level 1 and two of level 2, there will be three cycles of level 2 in period A.

6. Similar as Tables 15.1 and 15.2, but all stress levels, exceedances and cycle numbers in accordance, see Figure 15.6. Pivot point, P, at 1500 Exceedances.

7. All stresses will be multiplied by 21.5/15 = 1.43. Everything else remains the same as in Tables 15.1 and 15.2.

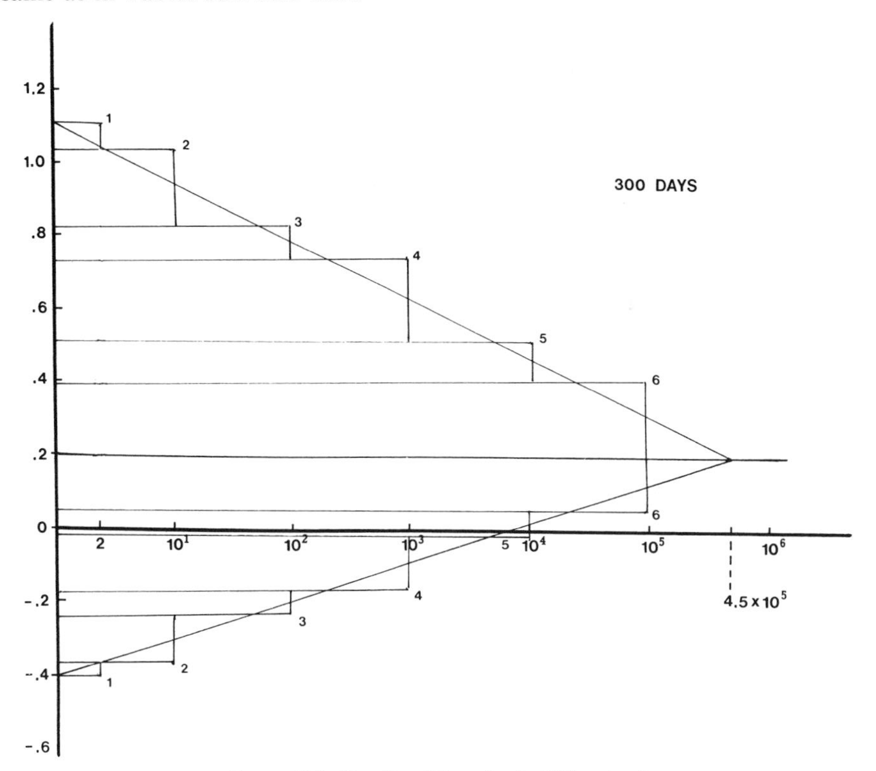

Figure 15.2. Results of Exercise 2 of Chapter 6

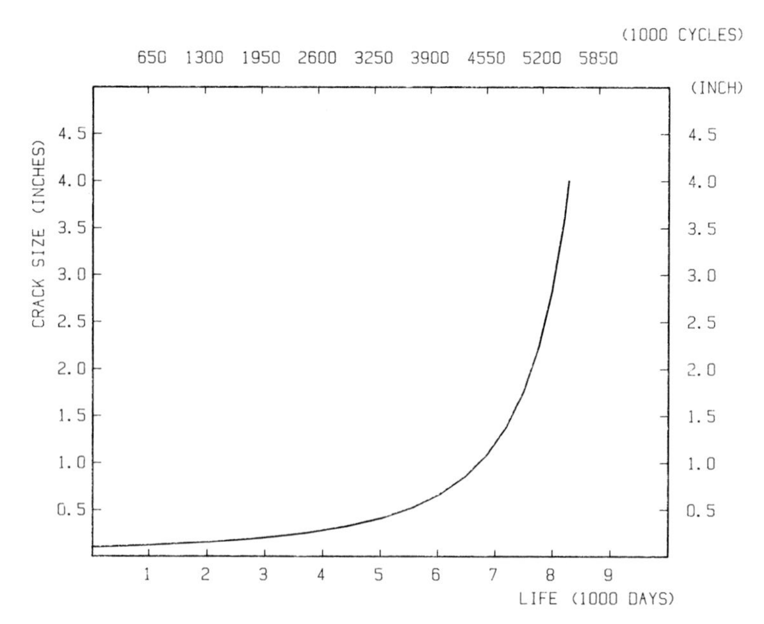

Figure 15.3. Same configuration and data as in next figure. Crack growth as in Exercise 1 of Chapter 6.

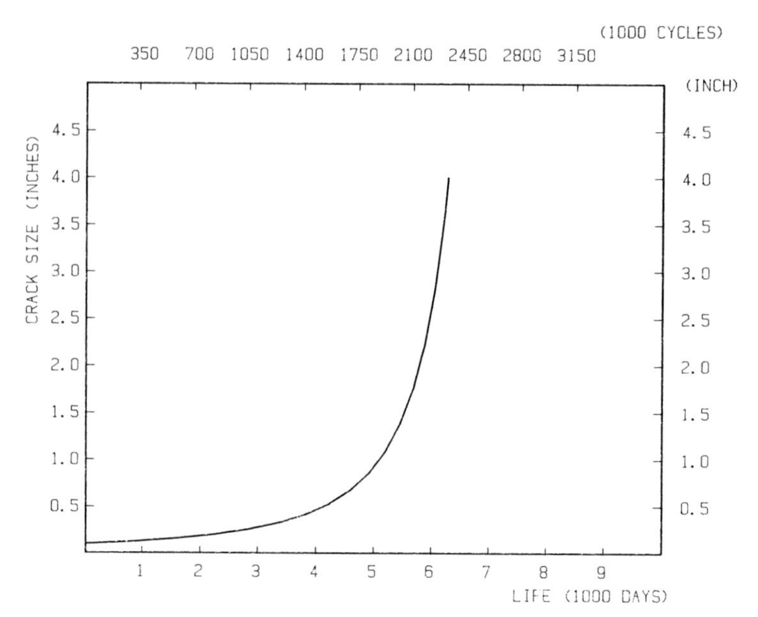

Figure 15.4. Same configuration and data as in previous figure. Crack growth in Exercise 2 of Chapter 6.

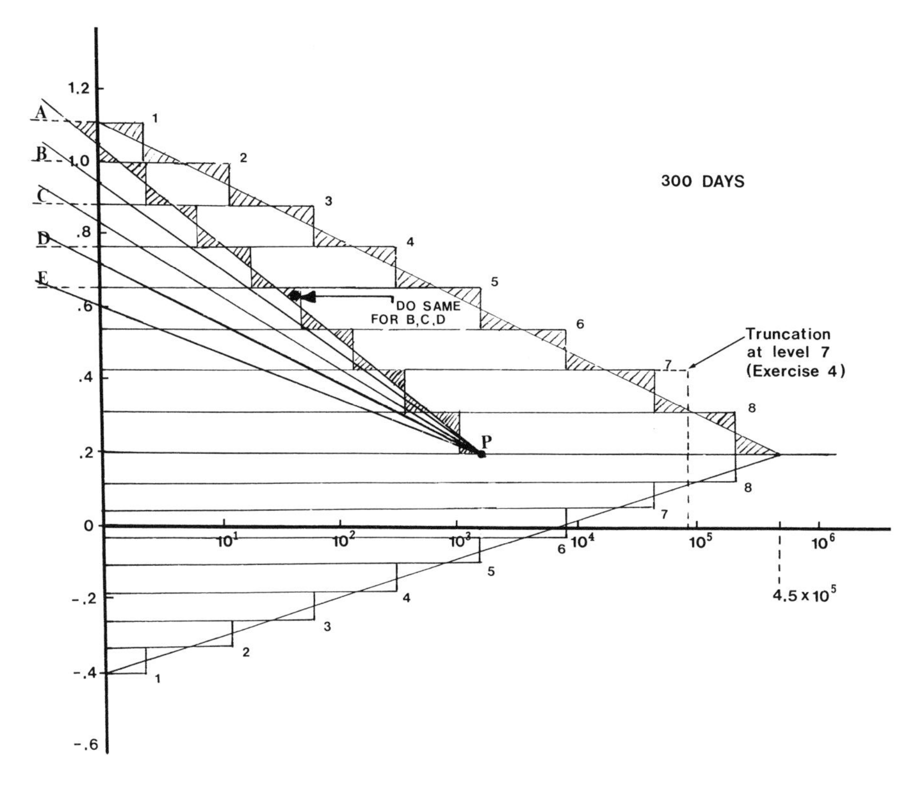

Figure 15.5. Solution to Exercise 3 in Chapter 6.

8. If there is no retardation the highest load only contributes a small amount of crack growth. Since there are only a few of these, their total contribution is extremely small. Without retardation they have no effect on the growth at lower loads.

15.7. Solutions to exercises of Chapter 7

1. Plane strain at $B = 2.5\,(40/80)^2 = 0.625$ inch.
Plane stress at $B \approx (40/80)^2/2\pi = 0.04$ inch. Assuming straight line between (0.04, 90) and (0.65, 40) as in Figure 7.4, toughness for 0.3 inch thickness is estimated as 68 ksi$\sqrt{\text{in}}$.

2. $K_c = \beta\sigma\sqrt{\pi \times 4.3}$; $K_{\text{eff}} = \beta\sigma\sqrt{\pi \times 4}$. Assuming β is about the same for $a = 4.3$ and 4 inches, $K_{\text{eff}} = \sqrt{4/4.3} \times 75 = 72.3$ ksi$\sqrt{\text{in}}$; $\sigma_{fc} = (16-8) \times 80/16 = 40$ ksi; $\sigma_{fr} = 72.3/1.19\sqrt{\pi \times 4} = 17.1$ ksi. Fracture was not by collapse, i.e. $K_{\text{eff}} = 72.3$ ksi$\sqrt{\text{in}}$.

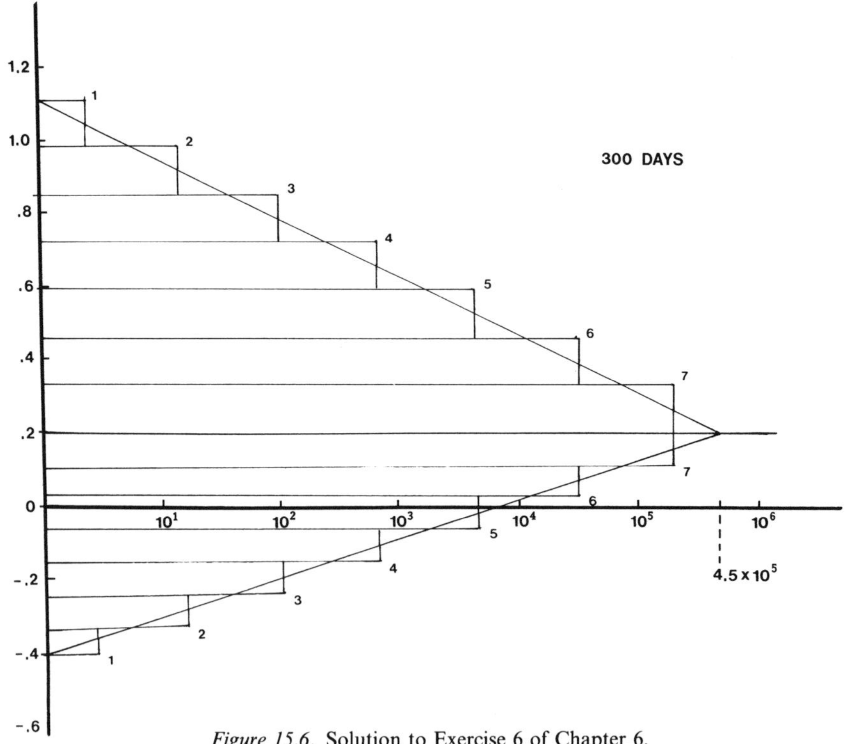

Figure 15.6. Solution to Exercise 6 of Chapter 6.

3. $\sigma_{fc} = (16-8)\ 35/16 = 17.5$ ksi; $\sigma_{fr} = 17.1$ ksi (see Exercise 2). Hence, fracture was probably by collapse; therefore $K_c \geqslant 75\,\text{ksi}\sqrt{\text{in}}$, and $K_{\text{eff}} \geqslant 72.3\,\text{ksi}\sqrt{\text{in}}$.

4. $K_{Ic} = 12\sqrt{16} = 48\,\text{ksi}\sqrt{\text{in}}$.
$\Delta T = 215 - 1.5 \times 70 = 110$ F, so that toughness might be useful for slow loading at $70 - 110 = -40$ F.

5. Estimate $K_{Ic} = 50 \times 60/75 = 40\,\text{ksi}\sqrt{\text{in}}$ for B. If yield strength values are reversed, estimate toughness B = toughness A.

6. Plane strain at $B = 2.5\,(40/70)^2 = 0.82$ inch.
Draw a straight line through (0.82, 40) and (0.5, 60) as in Figure 7.4 and extrapolate to 0.3 inch to obtain: $K_c = 72\,\text{ksi}\sqrt{\text{in}}$.

7. Should be about the same as in Table 7.1.

8. Should be about the same as in Table 7.2.

Table 15.3. Results of Exercises 3, 4, 5; See Figure 15.5

1 Level	2 Exceed-ances Table 15.1	3 Occur-rences Table 15.1	4 A-Exceed-ances Fig. 15.5	5 A-Occur-rences	6 Occur-rences in 2A $2 \times (5)$	7 Remain-der (3)–(6)	8 B-Exceed-ances Fig. 15.5	9 B-Occur-rences	10 Occur-rences in 5B $5 \times (9)$	11 Remain-der (7)–(10)	12 C-Exceed-ances Fig. 15.5	13 Occur-rences in C	14 Occur-rences in 34C $34 \times (13)$
1	2	2	1	1	2	0							
2	11	9	3	2	4	5	1	1	5	0			
3	63	52	7	4	8	44	3	2	10	34	1	1	34
4	316	253	17	10	20	233	10	7	35	198	3	2	68
5	1680	1364	46	29	58	1306	29	19	95	1211	14	11	374
6	7830	6150	124	78	156	5994	80	51	255	5739	56	42	1428
7	43 300	35 470	365	241	482	34 988	254	174	870	34 118	190	134	4556
8	200 000	156 700	1000	635	1270	155 430	866	612	3060	152 370	750	560	19 040
7 improper truncation	43 300	35 470	365	241	482	34 988	254	174	870	34 118	190	134	45 556
7 correct truncation	81 000	73 170	487	363	726	72 444	365	285	1425	71 019	273	217	7378

Table 15.3. (Continued)

15 Remainder (11)–(14)	16 D-Exceedances Fig. 15.5	17 D-Occurrences	18 Occurrences in 130D 130 × (17)	19 Remainder (15)–(18)	20 [a]For 129E Remains (19)	21 Occurrences in E (20)/129	22 Exceedances in E	23 Actual Occurrences in E (21) × 129	24 Difference Remainder (19)–(23)	25 Add to B (24)/5[b]	26 New B occurr. (9) + (25)	27 New B Exceed. Fig. 15.5	28 Extra in 5B 6(26)–(10)	29 [c]Remainder (24)–(28)
											1	1	0	
0											2	3	0	
130	1	1	130	0							7	10	0	
837	5	4	520	317	317	2	2	258	59	12	31	41	60	−1
4311	22	17	2210	2101	2101	16	18	2064	37	7	58	99	35	2
29 562	124	102	13 260	16 302	16 302	126	144	16 254	48	10	184	283	50	−2
133 330	650	526	68 380	64 950	64 950	503	647	64 887	63	13	625	908	65	−2
29 562	124	102	13260	16 302	16 302	126	144	16 254	48	10	184	283	50	−2
63 641	177	155	20 150	43 491	43 491	337	355	43 473	18	4	289	388	20	−2

[a] Total of 2A + 5B + 34C + 130D = 171 periods used; Remain: 300 − 171 = 129 periods E
[b] Or to other 'period'
[c] Differences negligeable (lower stresses; and consider log readings)

9. Walker: m_w, m_R, n_w remain the same (dimensionless).
Conversion of $C_w (m_R = m_w + n_w)$:

$$1 \frac{\text{in}}{(\text{ksi}\sqrt{\text{in}})^{m_R}} = (1000)^{-m_R} \frac{\text{in}}{(\text{psi}\sqrt{\text{in}})^{m_R}}.$$

C_w must be multiplied by $(1000)^{-3.07}$ if K in psi$\sqrt{\text{in}}$, i.e. $C = 3.88E-19$.

10. See solution to Exercise 9.

11. $C = 0.05 \times 1E-8 + 0.25 \times 3E-9 + 0.7 \times 5E-10 = 1.6\ E-9$.

15.8. Solutions to exercises of Chapter 8

1. $K = 1.48\sigma\sqrt{\pi a}$; $\beta = 1.48$; $\sigma_{fr} = 11.9$ ksi (max bending stress at fracture); Collapse $\sigma_{fc} = 1.5F_{ty}(W - a)^2/W^2 = 1.5 \times 60 \times (5/10)^2 = 22.5$ ksi (Figure 2.15); $\sigma_{res} = 11.9$ ksi.

2. $\sigma_{ten} = P/BW$; $\sigma_{ben}(\max) = MW/2I = 2PW/(BW^3/6) = 12P/BW^2$. $\sigma_{ben}/\sigma_{ten} = 18/W = 1.2$; $\sigma_{tot} = 2.2\sigma_{ten}$. For β see Table 15.4, columns 1 through 4.

3. See Table 15.4 columns 1 through 5.

4. See Table 15.4 columns 6 through 9 (note same results in columns 7 and 9).

5. See Figure 8.17; Compound with $\beta_w = \sqrt{\sec \pi a_{eff}/W'} = \sqrt{\sec \pi(a/2 + d/2)/W}$ (example see problem 6).

6. For $a/D = 0.3$ we have $\beta_{hole} = 1.47$.
$\beta_W = \sqrt{\sec \pi \times 0.65/6} = 1.03$. Total $\beta = 1.47 \times 1.03 = 1.52$; $\Delta K = 1.52 \times 10 \times \sqrt{\pi \times 0.3} = 14.76$ ksi$\sqrt{\text{in}}$; $K_{max} = 18.44$ ksi$\sqrt{\text{in}}$.
$da/dN = 2.4\ E-5$ inch/cycle.

7. Case 1: For $a \approx 0$ stress at hole is $\sigma_h = 3\sigma_L - \sigma_T = (3 - 0.33)\sigma_L = 2.67\sigma_L$.
Hence $\beta = 1.12 \times 2.67 = 2.99$.
Case 2: For $a \approx 0$ stress at hole is $3\sigma_L - \sigma_T = (3 + 0.5)\sigma_L = 3.5\sigma_L$.
Hence $\beta = 1.12 \times 3.5 = 3.92$.
In both cases $\beta = \sqrt{0.5 + D/2a}$ for large a. Draw curve for latter and fair into values for $a \approx 0$. Case 1: $\beta_h = 2.26$ (on faired) line; $\beta_w = 1.02$, total $\beta = 2.31$; $\Delta K = 2.31 \times 10\sqrt{\pi \times 0.1} = 12.9$ ksi$\sqrt{\text{in}}$; $K_{max} = 16.2$ ksi$\sqrt{\text{in}}$; $da/dN = 1.53\ E-5$ inch/cycle. Case 2: $\beta = 2.35 \times 1.02 = 2.39$; $da/dN = 1.72E-5$ inch/cycle.

Table 15.4. Solution to Exercise 2 (columns 1 through 4), Exercise 3 (columns 1 through 5), and Exercise 4 (columns 6 through 9) of Chapter 8.

$\sigma_{ten}/\sigma_{tot} = 1/2.2 = 0.455$; $\sigma_{ten}/\sigma_{tot} = 1.2/2.2 = 0.545$; $\sigma_{ben}/\sigma_{ten} = 1.2$.

1	2	3	4	5	6	7	8	9
a/w	β_{ten} Fig. 8.4a	β_{ben} Fig. 8.4b	σ_{tot} reference $\beta_1 = \left(\frac{\sigma_{ten}}{\sigma_{tot}}\beta_{ten} + \frac{\sigma_{ben}}{\sigma_{tot}}\beta_{ben}\right)$	σ_{ten} reference $\beta_2 = \left(\beta_{ten} + \frac{\sigma_{ben}}{\sigma_{ten}}\beta_{ben}\right)$	a	σ_{tot} $\sigma_{fr} = \frac{K_c}{\beta_1\sqrt{\pi a}}$	σ_{ten} $\sigma_{fr} = \frac{K_c}{\beta_2\sqrt{\pi a}}$	$\sigma_{tot} = 2.2\,\sigma_{ten}$
0.1	1.15	1.06	1.10	2.42	1	25.6	11.6	25.5
0.2	1.20	1.03	1.11	2.44	2	18.0	8.2	18.0
0.3	1.26	1.12	1.18	2.60	3	13.8	6.3	13.9
0.4	1.33	1.25	1.29	2.83	4	10.9	5.0	11.0
0.5	1.44	1.48	1.46	3.22	5	8.6	3.9	8.6
0.6	1.60	1.86	1.74	3.83	6	6.6	3.0	6.6

8. For large cracks and tension: $\beta_{hole} = \sqrt{1 + D/2a}$. Due to fastener force: $K_p = P/\sqrt{\pi(2a + D)}$, where P is per unit thickness; so that we have $K = \sigma W/\sqrt{\pi(2a + D)}$; $K_\sigma = \sqrt{1 + D/2a}\;\; \sigma\sqrt{\pi a}$. $K_{tot} = 0.8K_\sigma + 0.5 \times 0.2(K_p + K_\sigma) = \beta_t \sigma\sqrt{\pi a}$. Substitute expressions and work out β_t. Compound β with $\sqrt{\sec \pi(2a + D)/2W}$. Result in Table 15.5 first 2 columns.

9. See Table 15.5.

10. For $a \approx 0$: $\beta = 1.12(1 + 2\sqrt{d/r}) = 6.13$.

Table 15.5. Beta(a) for exercises 8 and 9/Chapter 8.

Loaded hole crack at both sides.
Through-the-thickness crack.

Loading: Bypass ratio = 0.8.

Stress intensity defined as K = beta × sigma × sqr(pi × a)
Sigma defined as: Nominal tension stress.

All dimensions in inches.

Thickness = 0.5
Width = 8
Hole diameter = 1

Crack.L inches	Beta	$\frac{K_c}{\beta\sqrt{\pi a}}$ ksi	After arrest ksi
0.100	3.022	29.517	29.517
0.200	2.455	25.694	26.768
0.300	1.924	26.768[a]	26.768
0.400	1.793	24.874	24.874
0.500	1.662	24.001	24.001
0.600	1.590	22.903	22.903
0.700	1.538	21.926	21.926
0.800	1.485	21.234	21.234
0.900	1.441	20.639	20.639
1.000	1.419	19.875	19.875
1.100	1.399	19.227	19.227
1.200	1.391	18.512	18.512
1.300	1.383	17.887	17.887
1.400	1.384	17.221	17.221
1.500	1.387	16.606	16.606
1.600	1.396	15.975	15.975
1.700	1.408	15.366	15.366
1.800	1.425	14.757	14.757

[a] Increasing residual strength (arrest conditions).

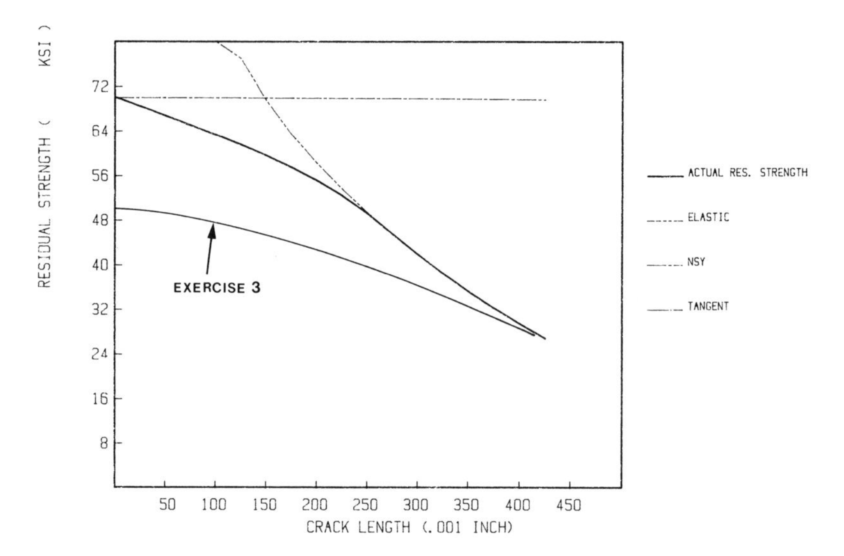

Figure 15.7. Residual strength diagram; solution to Exercise 1 of Chapter 9; $a/c = 1$.

For large a: β is β_e for edge crack (Figure 8.5) for $(a + d)/W$.
For example for $a = d = 0.5$ and $W = 5$: $a_{\text{eff}}/W = 0.1$; $\beta = 1.15$ from Figure 8.5a.

11. Same solution as in Exercise 10 with $r = D/2$ and $d = e + D/2$.

12. For small cracks $\beta \approx 1.5 \times 1.12 = 1.68$.
(For surface flaws multiply by β_{FFS}/ϕ; Figure 8.3).

13. Apparent stress intensity in elements is:
first element $K = 48.9\sqrt{2\pi \times 0.05} = 27.4\,\text{ksi}\sqrt{\text{in}}$;
second element $29.1\,\text{ksi}\sqrt{\text{in}}$, third element $31.6\,\text{ksi}\sqrt{\text{in}}$.

Plot versus r and extrapolate to $r = 0$ to final $K \approx 27\,\text{ksi}\sqrt{\text{in}}$. Taking highest stress as the reference stress: $\beta = 27/5\sqrt{\pi \times 2} = 2.15$.

14. $U_a = 1/2P\delta = 1/2 \times 10000 \times 0.0020 = 100$ in lbs/inch thickness.

$$U_{a+\Delta a} = 1/2 \times 10000 \times 0.021 = 105 \text{ in lbs/inch thickness.}$$

$$\mathrm{d}U/\mathrm{d}a = \Delta U/\Delta a = (105 - 100)/(1.01 - 1) = 500\,\text{lbs/in} = 0.5\,\text{kip/inch.}$$

$$K = \sqrt{E\,\mathrm{d}U/\mathrm{d}a} = \sqrt{5000} = 71\,\text{ksi}\sqrt{\text{in}}.$$

15. Since it is assumed that the case of Figure 8.25 is for $\mu = 0.4$ it follows that $A_s = 0.4 \times 8 \times 0.2 = 0.64\,\text{inch}^2$.
Example for $a/b = 0.5$: $\beta = 0.5$ from Figure 8.25d. $a = 4$ inch.

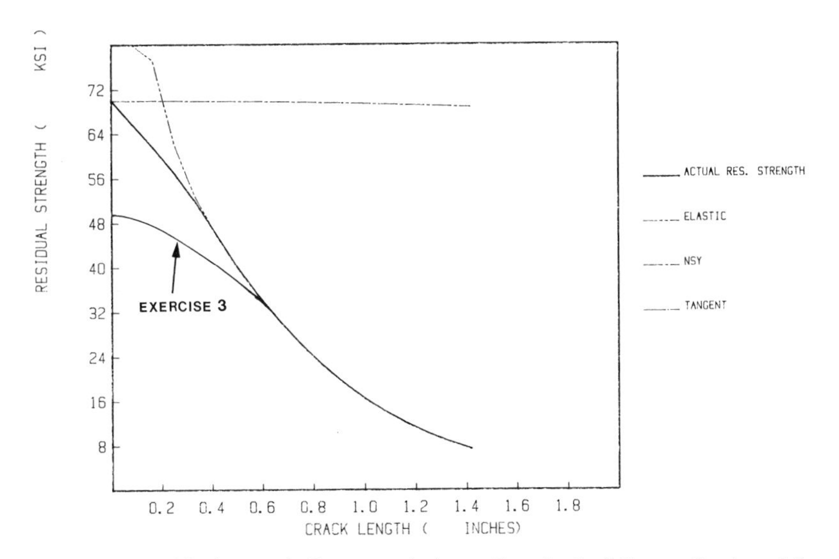

Figure 15.8. Residual strength diagram; solution to Exercise 1 of Chapter 9; $a/c = 0.3$.

From Equation (8.42): $L = 1 + 0.8\sqrt{2}(1 - 0.4)/0.64 = 2.06$ (compare with Figure 8.25e). Fastener load is according to Equation (8.43): $0.6(2.06 - 1)0.64\sigma = 0.41\sigma$ (in kips if σ is in ksi).

15.9. Solutions to exercises of Chapter 9

1. For external flaws: $K = \beta\sigma_{\text{hoop}}\sqrt{\pi a}$ (β from Figure 8.3).
For internal flaws: $K = K_{\text{press}} + \beta_p \, p\sqrt{\pi a} + \beta_h\sigma_{\text{hoop}}\sqrt{\pi a} = \beta(\sigma_{\text{hoop}} + p)\sqrt{\pi a}$ (since $\beta_p \approx \beta_h$; β form Figure 8.3). Results shown in Figures 15.7 and 15.8. (Subtract p from the given residual strength to get hoop stress at fracture for the internal flaws).

2. Hoop stress at the given pressure is $3 \times 5/0.5 = 30$ ksi. For through-the-thickness cracks there is still plane strain, because $B > 2.5\ (35/70)^2 = 0.625$ inch ($B = 0.5$ inch), so that K_{Ic} can be used.
(a) External cracks.
The circular crack will cause fracture at the hoop stress of 30 ksi when $a = c = 0.4$ inch (Figure 15.7). A circular crack will cause a through-the-thickness crack of length $2a = 2B$, i.e. $a = 0.5$ inch. Its stress intensity is $K = 30\sqrt{\pi \times 0.5} = 37.6\,\text{ksi}\sqrt{\text{in}}$. This is larger than K_{Ic}, so that fracture continues (break). The flaw with $a/c = 0.3$ causes fracture at $c = 0.7$ inch. The through-crack will then have a length of $2a = 1.4$ inch. Then $K = 30\sqrt{\pi \times 0.7} = 44.5\,\text{ksi}\sqrt{\text{in}}$, which is larger than K_{Ic}(break).

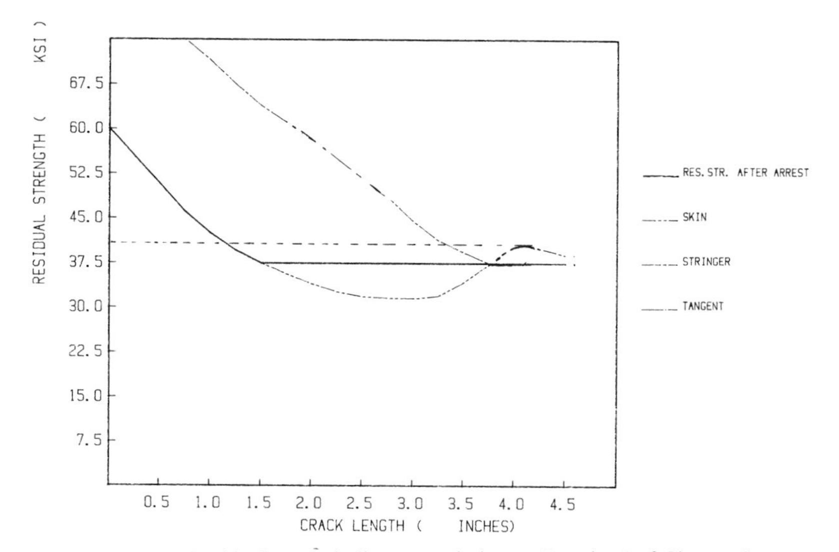

Figure 15.9. Residual strength diagram; solution to Exercise 5 of Chapter 9.

(b) *Internal flaws*
The same figures can be used, but as the figures provide $\sigma_h + p$ at fracture for internal flaws (see stress intensity superposition in Exercise 1), the critical crack size must be found for a stress of $\sigma_{\text{hoop}} + p = 35$ ksi. The circular flaw breaks through at $a = 0.36$ inch. It still gives a through crack $2a = 2B$. The stress intensity after break-through is $K = \sigma_{\text{hoop}}\sqrt{\pi a} = 30\sqrt{\pi \times 0.5} = 37.6\,\text{ksi}\sqrt{\text{in}}$ (break). The flaw with $a/c = 0.3$ is critical at $c = 0.59$ inch. Break through length is c, i.e. $K = 30\sqrt{\pi \times 0.59} = 40.8\,\text{ksi}\sqrt{\text{in}}$ (break).

3. For plane strain $B > 2.5\,(35/50)^2 = 1.225$ inch, which is not the case ($B = 0.5$). Hence, the given $K_c = 80\,\text{ksi}\sqrt{\text{in}}$ applies. Note that a new tangent from 50 ksi must be drawn in Figure 15.7 and 15.8 because of the lower yield.

The part through cracks are still in plane strain, so that otherwise Figure 15.7 and 15.8 apply. Following Exercise 3, the break-through crack sizes are:

External: $a/c = 1 a = 0.40$ inch. Through crack $a = B$; $K = 30\sqrt{\pi \times 0.5} = 37\,\text{ksi}\sqrt{\text{in}}$ ($< K_c$), fracture arrests (leak). For $a/c = 0.3$: $c = 0.7$ inch. Through crack $a = 0.7$ inch; $K = 30\sqrt{\pi \times 0.7} = 44.5\,\text{ksi}\sqrt{\text{in}} < K_c$ (leak).

Internal: Through crack $a = B$ for $a/c = 1$: $K = 37.6\,\text{ksi}\sqrt{\text{in}}$ (leak); $a/c = 0.3$: $K = 30\sqrt{\pi \times 0.59} = 40.8\,\text{ksi}\sqrt{\text{in}}$ (leak).

4. For $a/c = 0.3$ about same result as in Table 9.1 for $a/c = 0.33$. For $a/c = 1$ flaw remains circular.

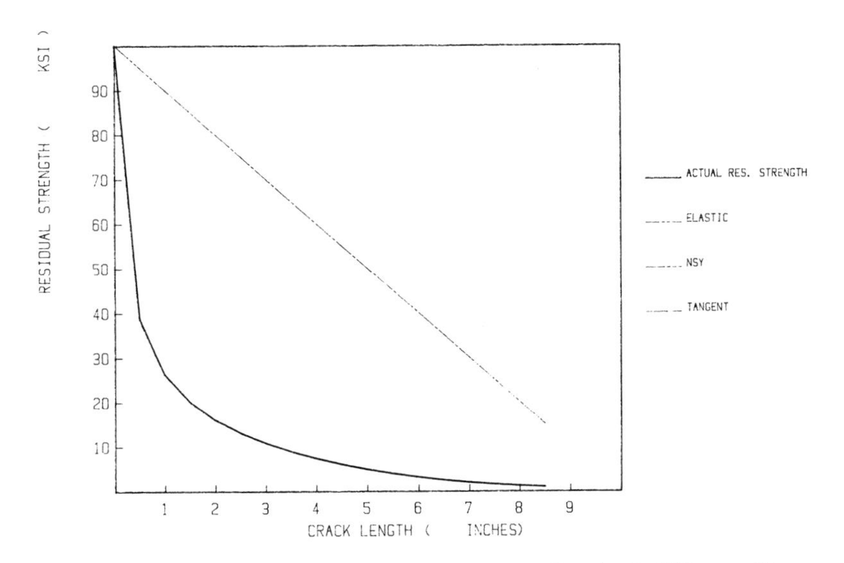

Figure 15.10. Residual strength diagram; solution to Exercise 2 of Chapter 10.

5. See Figure 15.9. Stringer critical. To make it stringer critical use stringers of another material; $F_{tu} = 75 \times 40/37.5 = 80\,\text{ksi}$. Fastener load at arrest (if skin critical): $P = 0.41 \times 40 = 16.4\,\text{kip}$ (Exercise 15 of Chapter 8). Fastener diameter is $D = \sqrt{4 \times 16.4/(100 \times \pi)} = 0.45\,\text{inch}$. Bearing stress: $\sigma_b = P/DB = 16.4/(0.45 \times 0.2) = 182\,\text{ksi}$; elastic; redistribution will occur.

6. From Figure 15.9:
(a) Stress for fracture at $a = 2\,\text{inch}$ is $\sigma_{fr} = 34\,\text{ksi}$.
(b) Arrest at $a = 3.5\,\text{inch}$.
(c) In Exercise 5 you calculated $\beta = 0.82$ for $a = 2\,\text{inch}$; $\Delta K = 0.82 \times 15 \times \sqrt{\pi \times 2} = 30.8\,\text{ksi}\sqrt{\text{in}}$; $K_{max} = 30.8/0.8 = 38.5\,\text{ksi}\sqrt{\text{in}}$; $\mathrm{d}a/\mathrm{d}N = 3E{-}9 \times 30.8^{2.1} \times 38.5^{0.9} = 1.07E{-}4\,\text{in/cycle}$.
(d, e) Same rate as in C, because residual strength curve is line of constant K ($K = K_c$), the stress intensities at $a = 2$ (fracture) and $a = 3.5$ (arrest) are equal (larger a at arrest but smaller β ($\beta = 0.62$ for $a = 3.5\,\text{inch}$).

7. Small a: $\beta \approx 1.12k_t = 1.12(1 + 2\sqrt{0.875/0.125} = 7.05$. Larger a: $\beta = 1.12\sqrt{1 + 0.875/a}$. Fair line from 7.05 at $a = 0$ to curve.

8. Before stophole $\sigma_{fr} = 50/1.12\sqrt{\pi \times 0.75} = 29.1\,\text{ksi}$. If e.g. crack of 0.02 inch has developed from hole:
$\sigma_{fr} = 50/7.48\sqrt{\pi \times 0.02} = 26.7\,\text{ksi}$.

9. Same answer as for Exercise 7.

10. Max stress is 30 ksi; minimum stress = 20 ksi.
$\Delta\sigma = 10$ ksi; $R = 0.67$; $\Delta K = 10\sqrt{\pi \times 1} = 17.7$ ksi$\sqrt{\text{in}}$
$\mathrm{d}a/\mathrm{d}N = 2 \times 10^{-7} \times 17.7^2/[(1 - 0.6)80 - 17.7] = 7.2E{-}6$ inch/cycle.

11. Stress intensity due to residual stress:
$K = 20\sqrt{\pi \times 1} = 35.4$ ksi$\sqrt{\text{in}}$, which is larger than K_{Iscc}. Crack will grow by stress corrosion.

12. Assuming $K_{2c} = K_{1c} = 80$ ksi$\sqrt{\text{in}}$ (circle):
$K_2 = 20\sqrt{\pi \times 2} = 50$ ksi$\sqrt{\text{in}}$. Allowable K_1 is then $K_1 = \sqrt{80^2 - 50^2} = 62.5$ ksi$\sqrt{\text{in}}$, hence $\sigma = 62.5/\sqrt{\pi \times 2} = 24.9$ ksi.
Assuming $K_{2c} = 0.8K_{1c} = 64$ ksi$\sqrt{\text{in}}$ (ellipse):
$K_1 = 80\sqrt{1 - 50^2/64^2} = 49.9$, hence $\sigma = 49.9/\sqrt{\pi \times 2} = 19.9$ ksi.

15.10. Solutions to exercises of Chapter 10

1. $a_p = 2$ inch.

2. See Figure 15.10.

3. $W = 30$ inch; $a_p = 0.38$ inch for $\sigma_p = 51$ ksi (tangent);
$a_p = 1.1$ inch for $\sigma_p = 35$ ksi (LEFM);
$a_p = 3.2$ inch for $\sigma_p = 20$ ksi (LEFM).
$W = 10$ inch $a_p = 0.38$ inch for $\sigma_p = 51$ ksi (tangent);
$a_p = 1.04$ inch for $\sigma_p = 35$ ksi (LEFM);
$a_p = 2.4$ inch for $\sigma_p = 20$ ksi (LEFM).
$W = 4$ inch $a_p = 0.30$ inch for $\sigma_p = 51$ ksi (collapse);
$a_p = 0.83$ inch for $\sigma_p = 35$ ksi (collapse);
$a_p = 1.33$ inch for $\sigma_p = 20$ ksi (collapse).

4. See Table 10.5 for example. In present case the numbers will be different. Plot σ_{fr} and σ_{fc} and draw tangent. Lowest of three prevails. Read a_p from curves for three values of σ_p.

5. $\sigma_c = 32$ ksi; stable fracture $\Delta a = 0.4$ inch.

6. Same answer as in Exercise 5.

7. $\sigma_c \approx 36$ ksi; $\Delta a \approx 0.3$ inch; J at fracture ≈ 0.68 kips/in.

8. $K = \sqrt{6800} = 82$ ksi$\sqrt{\text{in}}$.

15.11. Solutions to exercises of Chapter 11

1. Case 1: $H = 4800$ hrs; $I = 4800/6 = 800$ hrs.
   Case 2: $H = 4800$ hrs; $I = 4800/6 = 800$ hrs.

2. Solution in Table 15.6.

3. Solution in Table 15.7.

4. Computer results (may differ slightly from hand-calculations, because computer varies time of first inspection):
Case 1: $I = 500$ hrs, $P_{det} = 1$; $I = 1000$ hrs, $P_{det} = 1$; $I = 1500$ hrs, $P_{det} = 0.99$. For $P_{det} = 95\%$: $I = 1600$ hrs.
Case 2. $I = 500$ hrs, $P_{det} = 1$; $I = 1000$ hrs, $P_{det} = 0.98$; $I = 1500$ hrs, $P_{det} = 0.91$. For $P_{det} = 95\%$: $I = 1360$ hrs.

5. Critical stress at $a_p = 33$ mm is $\sigma_c = 50/\sqrt{\pi \times 0.033} = 155$ MPa. For $a_{proof} = 15$ mm: $\sigma_{proof} = 50/\sqrt{\pi \times 0.015} = 230$ MPa.
Interval $N_{ac} - N_{proof} = 4800–3400 = 1400$ hrs (Case 1).
Interval $= 4800 - 4700 = 100$ hrs (Case 2).
Both without safety factor

Table 15.6. Solution to Exercise 2 of Chapter 11.

Insp. #	Hours	Case 1					Case 2				
		a Fig. 11.7	P_{det} Fig. 11.8	P_{miss}	Cum. P_{miss}	Cum. P_{det}	a Fig. 11.7	P_{det} Fig. 11.8	P_{miss}	Cum. P_{miss}	Cum. P_{det}
1	800	6	0.6	0.4	0.4	0.6	5.2	0.5	0.5	0.5	0.5
2	1600	7.5	0.69	0.31	0.12	0.88	5.5	0.52	0.48	0.24	0.76
3	2400	10.0	0.74	0.26	0.03	0.97	6.0	0.60	0.40	0.096	0.904
4	3200	13.0	0.80	0.20	0.006	0.994	6.5	0.63	0.37	0.036	0.974
5	4000	20.0	0.87	0.13	0.0008	0.9992	7.5	0.69	0.31	0.011	0.989

Table 15.7. Solution to Exercise 3 of Chapter 11

Equation: $p = 1 - e^{-\{(a-5)/(8-5)\}0.5}$

Insp. #	Hours	Case 1					Case 2				
		a Fig. 11.7	P_{det} Eq.	P_{miss}	Cum. P_{miss}	Cum. P_{det}	a Fig. 11.7	P_{det} Eq.	P_{miss}	Cum. P_{miss}	Cum. P_{det}
1	800	6	0.44	0.56	0.56	0.44	5.2	0.23	0.77	0.77	0.23
2	1600	7.5	0.60	0.40	0.22	0.78	5.5	0.33	0.67	0.52	0.48
3	2400	10.0	0.72	0.28	0.06	0.94	6.0	0.43	0.57	0.29	0.71
4	3200	13.0	0.80	0.20	0.01	0.99	6.5	0.51	0.49	0.14	0.86
5	4000	20.0	0.89	0.11	0.001	0.999	7.5	0.60	0.40	0.06	0.94

6. At operating stress $a_c = (30/100)^2/\pi = 0.029$ inch.
For $\sigma_p = 150$ ksi: $a_p = (30/150)^2/\pi = 0.013$ inch.
Stripping depth: $\delta = 0.029 - 0.013 = 0.016$ inch every two years.

7. Cost of fracture is $C_{fr} = P \times T \times S_h$. Cost of fracture control: $C_{con} = S_a h\ TS$, assuming one replacement of critical part during economic life (durability). Do fracture control $C_{con} < C_{fr}$, otherwise fracture costs might be acceptable.

15.12. Solutions to exercises of Chapter 12

1. With $K_c = 70\,\text{ksi}\sqrt{\text{in}}$ and $\sigma_p = 2 \times 15 = 30$ ksi, we have $a_p = 1.01$ inch.
Life from $a = 0.04$ to 1.01 inch is 94,000 cycles.
Life from $a = 0.04$ to 0.1 inch is 47,000 cycles.
Hence life from 0.1 to 1.01 inch is 94,000 − 47,000 = 47,000 cycles. Safety factor on life by assuming $a = 0.01$ inch is 94,000/47,000 = 2.

2. Permissible crack size $a = 1.30$ inch.
Life from $a = 0.04$ to 1.30 inch is 22,000 cycles.
Life from $a = 0.04$ to 0.1 inch is 3000 cycles.
hence, life from 0.1 to 1.30 inch is 19,000 cycles. Safety factor on life by assuming $a = 0.1$ inch is 22000/19000 = 1.16.

3. By assuming a larger initial crack size, the extra safety for the crack at the hole (1.16) is much less than for edge crack (2), showing that assuming larger initial cracks for 'safety' is dubious; sometimes the extra safety is very small. For a consistent safety factor, it is better to assume the correct crack size (0.04 inch) in both cases, and then apply the same factor to both, e.g. factor of 2. In Exercise 1 life from 0.04 to 1.30 is 94,000 cycles. Factor of 2 gives $H = 47{,}000$ cycles. In Exercise 2, the same factor on 22,000 cycles gives $H = 11{,}000$ cycles. Now both cases are equally 'safe'.

4. Toughness is $60 \pm 0.065 \times 60 = 60 \pm 4\,\text{ksi}\sqrt{\text{in}}$. The actual toughness may be $56\,\text{ksi}\sqrt{\text{in}}$. Estimate error in β is two percent. If the stress is five percent too high, we must take $\sigma_p = 33$ ksi, so that with $K_c = 56\,\text{ksi}\sqrt{\text{in}}$, and $\beta = 1.14$ we have $a_p = 0.70$ inch.
Life from 0.04 to 1.01 inch is 94,000 cycles.
Life from 0.04 to 0.70 inch is 90,000 cycles.
Factor $L_a = 90/94 = 0.96$.
Factor on rate data is 1.5 total; from average to upper bound is 1.25. This could give a shorter life by a factor $L_r = 1/1.25 = 0.8$. If error in stress is 5 per cent, actual stress could be $1.05 \times 15 = 15.75$ ksi. Hence, possible factor on life due

to σ and β is $L_l = [15 \times 1.12/(15.75 \times 1.14)]^{3.2} = 0.81$. Hence, life may be shorter by $L_a \times L_p \times L_l = 0.96 \times 0.8 \times 0.81 = 0.62$, (i.e. $0.62 \times 94{,}000 = 58{,}280$ cycles). Factor of 2 will cover this (see Exercise 3), and in this case the factor of 2 obtained by assuming initial $a = 0.1$ instead of 0.04 inch will cover uncertainty as well.

For crack at hole toughness could be $56\,\text{ksi}\sqrt{\text{in}}$. Estimate error in β is 5 percent. For $\sigma_p = 33$ ksi this leads to $a_p = 0.76$ inch.
Life from 0.04 to 1.30 inch = 22,000 cycles.
Life from 0.04 to 0.76 inch = 16,000 cycles.
$L_a = 16/22 = 0.73$; L_r is same as in previous case;
$L_l = [15\beta/(15.75 \times 1.05\beta)]^{3.2} = 0.73$.
Total uncertainty factor: $0.73 \times 0.8 \times 0.73 = 0.43$.
A factor of 2 will not even cover this. In order to get the same 'safety' as in previous case a factor of $2 \times 0.62/0.43 = 2.9$ would be needed; i.e. $H = 22{,}000/2.9 = 7600$ cycles. Note however that the factor of 1.16 obtained by assuming initial $a = 0.1$ instead of 0.04 inch will certainly not cover uncertainty.

5. Replacement times: 47,000/10,000 = 4.7 years for case 1 and 7600/10,000 = 0.76 years = 6 months for case 2.

15.13. Solutions for exercises of Chapter 13

1. Average toughness: $K_{Ic} \approx 28\,\text{ksi}\sqrt{\text{in}}$

2. 20 mm = 0.8 inch; Nominal fracture stress $\sigma_{fr} = 30/0.7\sqrt{\pi \times 0.8} = 17$ ksi.

3. See columns 1 through 8 of Table 15.8.

4. See columns 9 and 10 of Table 15.8.

5. See columns 11 through 15 of Table 15.8.

6. If the actual stress range was 15 ksi, the growth rates were too high by a factor of $(15/9.97)^3 = 3.40$ (third power of ΔK). If the handbook data were correct then the material behaved as if $C = 6.8E{-}8$ instead of $C = 2E{-}8$ from the handbook. The material could be a 'bad lot', but may be the stress analysis was incorrect and the stresses were indeed 9.97 ksi. Possibly there was a hidden stress concentration, so that $\beta = 1.50$ instead of $\beta = 1$.

7. See solution to 6.

Table 15.8. Solutions to Exercises 3, 4 and 5 of Chapter 13.

1	2	3	4	5	6	7	8	9	10	11	12	13	14	15
a	Spacing	Magni-fication	da/dN 2/3	da/dN average	Δa	ΔN 6/5	N	ΔK[a]	$\Delta\sigma$ $\frac{\Delta K}{\sqrt{\pi a}}$	a average	ΔK[b]	da/dN Equation	ΔN 6/13	N
0.1	0.03	8000	$3.75E-6$				0	5.62	10.0					0
				$6.83E-6$	0.1	15 000				0.15	6.84	$6.4E-6$	15625	15 625
0.2	0.08	8000	$1.00E-5$				15 000	7.77	9.80					
				$1.44E-5$	0.1	7000				0.25	8.84	$1.4E-5$	7143	22 768
0.3	0.15	8000	$1.88E-5$				22 000	9.57	9.86					
				$2.44E-5$	0.1	4100				0.35	10.45	$2.3E-5$	4348	27 116
0.4	0.09	2000	$3.00E-5$				26 100	11.2	9.99					
				$3.75E-5$	0.1	2700				0.45	11.85	$3.3E-5$	3030	30 146
0.5	0.09	2000	$4.50E-5$				28 800	12.8	10.21					
							Average $\Delta\sigma$ =		9.97					

[a] $da/dN = 3E-8\Delta K^3$, so that $\Delta K = (da/dN/2E-8)^{1/3}$; use da/dN from column 4.
[b] $\Delta K = \Delta\sigma_{average}\sqrt{\pi a_{average}}$.

8. See Figure 12.8.

9. Operating stress: $50/0.7\sqrt{\pi} = 40$ ksi.
A defect of $a = (7/0.7 \times 40)^2/\pi = 0.02$ inch would not grow by stress corrosion at the given stress and the given K_{Iscc}. Hence, quality control should prevent parts with larger defects from entering into service.

10. The stress intensity at a crack of 0.05 inch is $K = 2 \times 25\sqrt{\pi \times 0.05} = 20$ ksi$\sqrt{\text{in}}$. This is much lower than the toughness; the material could not be that much worse (specified toughness 60 ksi$\sqrt{\text{in}}$. It might be somewhat below standard, but more than likely the stress concentration was higher than expected, and probably the stresses s well. May be the loads were higher than anticipated, but also stress analysis and stress concentration factor are suspect.

11. If fracture would have been approached gradually at lower stress, the striation spacing would have increased toward fracture. If striations suddenly end (Figure 13.7), the fracture was due to one of the higher loads. The crack growth rate just before fracture was: $\mathrm{d}a/\mathrm{d}N = 0.12/2000 = 6E-5$. It follows from the crack growth equation that $\Delta K = (6E-5/8E-9)^{1/3.5} = 12.8$ ksi$\sqrt{\text{in}}$. Hence, $\Delta\sigma = 12.8/0.64\sqrt{\pi \times 0.6} = 14.6$ ksi. The occasional overloads ($R = 0$) are $\sigma_{\max} = 29.2$ ksi, and $K_{\max} = 25.6$ ksi$\sqrt{\text{in}}$. If fracture occurred at the peak of the overload $K_{Ic} = 25.6$ ksi$\sqrt{\text{in}}$. It could have occurred before the load reached its maximum, so that $K_{Ic} \leqslant 25.6$ ksi$\sqrt{\text{in}}$.

Subject Index

p 247 – eq 8.5 – elliptic integral has square root $\to \{ \quad \}^{1/2}$